TRANSACTIONS

OF THE

AMERICAN SOCIETY

OF

CIVIL ENGINEERS

(INSTITUTED 1852)

VOLUME 150

1985

NEW YORK
PUBLISHED BY THE SOCIETY
1986

NOTE—The Society is not responsible for any statement made or opinion expressed
in its publications.

ISBN:0-87262-514-1
Manufactured in the United States of America

CONTENTS

Foreword

Transactions of the American Society of Civil Engineers, Vol. 150, 1985 contains abstracts for all ASCE journal papers and technical notes, *Civil Engineering—ASCE* feature articles, special publications, and Manuals and Reports on Engineering Practice published by the Society during 1985. In addition, the president's annual address to the Society, abstracts for all papers published by ASCE that won 1985 awards or prizes, and memoirs of distinguished deceased members are presented. A keyword index and an author index are provided to help the user locate an abstract by subject or author.

The user will notice that abstracts for all 1985 Society journal papers are grouped together according to journal. For example, abstracts for all papers and technical notes published in the *Journal of Construction Engineering and Management* are together, beginning with the first paper published in that journal. The journals are in alphabetical order. Abstracts for all *Civil Engineering—ASCE* feature articles follow the journals. Abstracts for all special publications and Manuals and Reports appear next in alphabetical order according to title.

Each abstract entry includes an entry number, title of paper, article, or book, author's name and affiliation, and complete bibliographic information. Following each abstract is a list of discussions regarding that document which have appeared to date. Sample entries are shown below.

Sample entry for journal paper

[1]830   [2]**Comments on Cross-Flow Principle and Morison's Equation**

[3]**C. J. Garrison**, Member, ASCE, [4](Consultant in Marine Hydrodynamics, Pebble Beach, Calif. 93953)

[5]*Journal of Waterway, Port, Coastal and Ocean Engineering*, Vol. 111, No. 6, November, 1985, pp. [6]1075-1079

[8]Some of the recent test results on steady and oscillatory flow past inclined circular cylinders are discussed. More specifically, the experimental error in results presented by Sarpkaya, et al, for oscillatory flow is described and the formula for the necessary correction factor is given. Contrary to earlier conclusions based on such erroneous results, the corrected results support the cross-flow principle very well, in general, in the case of the inertia coeeficient and at the larger Reynolds number and Keulegan-Carpenter number in the case of the drag coefficient.

[9]Discussion: **T. Sarpkaya**, (Distinguished Prof. of Mech. Engrg., Naval Postgraduate School, Monterey, Calif. 93943) WW Nov. '85, pp. 1087.

1. Entry Number. 2. Title. 3. Primary Author (and membership grade, if applicable). 4. Author's affiliation. 5. Journal title, volume number, issue number, month, year. 6. Pagination. 7. Reprint Number. 8. Abstract 9. Discussion, Errata, and/or Closure.

Sample entry for special publication

[1]**1005**   [2]**Rehabilitation, Renovation, and Reconstruction of Buildings**
[3]Proceedings of a workshop sponsored by the National
Science Foundation and the American Society of Civil
Engineers, New York, Feb. 14-15, 1985

[4]**Lynn S. Beedle**, (editor), Honorary Member, ASCE, (Dir. of Lab., Lehigh Univ., Fritz Engrg. Lab., Bethlehem, PA)

[5]*New York: ASCE*, 1985, [6]112pp.

[7]This document constitutes the proceedings for a National Science Foundation funded workshop to discuss research needs pertaining to the buildings side of infrastructure. It includes discussion of research needs associated with assessing building condition, repair and rehabilitation techniques, implementation and project monitoring. Papers cover the owners' perspective, international experiences and specific discussions relating to the technological areas of steel and cast iron, concrete, masonry and cladding. In addition, consensus lists of research needs are discussed in the individual reports (steel/cast iron, concrete, masonry and cladding) which resulted from the working group discussions at the workshop. Two clear conclusions can be reached. They are, that much research is needed and that a great deal of information is already available but not being used.

1. Entry number. 2. Title. 3. Subtitle. 4. Editor. (If this book was the work of a committee, the committee name would appear here.) 5. Publisher Information. 6. Number of pages. 7. Abstract.

In the subject index the user will find under each keyword a list of titles pertaining to that subject. Each title is followed by the entry number for that abstract. In the author index, the document title and entry number are given if the author listed is the primary author of the document. For secondary authors, the primary author and entry number are given. Sample entries are shown below.

Sample subject index entry

[1]**Crossflow**
[2]Comments on Cross-Flow Principle and
Morison's Equation, [3]830
1. Keyword (subject heading). 2. Title. 3. Entry number.

Sample author index entry for primary author

[1]**Garrison, C. J.**
[2]Comments on Cross-Flow Principle and
Morison's Equation, [3]830
1. Author's name. 2. Title. 3. Entry number.

AMERICAN SOCIETY OF CIVIL ENGINEERS

Founded November 5, 1852

TRANSACTIONS

President's Year End Report

Richard W. Karn

It's been a fast year since I stood before you in San Francisco at the beginning of my presidential term and outlined some of the areas that I thought were important to the Society. I am not going to dwell on the accomplishments during the past year. These are summarized in the "President's Message" in the convention program which each of you has (Appendix A). I am very pleased with our accomplishments, achieved through exceptional efforts by ASCE members and staff.

I would like to commence my remarks this morning by thanking the many people who made this a significant year for ASCE, but the list is unending. It includes staff, Board members, committee people, Section members and so many that I cannot possibly name them all so I will simply say thanks to all of you. Your efforts have made it a great year.

Last year in my inaugural address I emphasized two specific subjects. They were Strategic Planning and Quality in Engineering.

As all of you know, the Strategic Plan has been adopted by the Board of Direction and is in the process of being implemented. As I reviewed the reports submitted by committees in the agenda for the Board meeting last weekend, and as I have met with Section and committee leaders throughout the nation, I have been impressed by the references to the Strategic Plan initiatives in their programs. All of you are to be commended for taking a serious interest in the strategic planning process and for moving to implement various elements of the plan.

With respect to the current plan, our emphasis is now changed to one of implementation. The responsibility of monitoring implementation has been assigned to the Committee on Society Objectives, Planning and Organization (COSOPO). They have a difficult task ahead of them. It will take the same effort to implement the initiatives as it did to develop the plan. The planning effort has helped us to focus our ideas and our energies on common objectives. Implementing programs within that framework will make those plans become realities.

In the meantime, we have embarked on the second cycle of strategic planning by the appointment of a new committee. It will be holding its first meeting early next month. It will build upon the first plan and create even stronger programs for our Society.

The second subject that I stressed last year was a program for Quality in Engineering, which cuts completely across all segments and functions of ASCE. It touches on education, practice, professionalism and all that is civil engineering. We have made some

Presented at the Annual Convention of the American Society of Civil Engineers, Detroit, Michigan, October 23, 1985.

notable progress in these areas, the most significant of which is in developing a program to prepare a Manual of Professional Practice for Quality in the Constructed Project. Following the strong statements of support that we have received from other design disciplines, the construction industry and owners of large civil engineering projects, this will be an interdisciplinary undertaking. It is a major effort and will consume several years of concentrated work by ASCE members and representatives of these other groups.

The mechanism has been put in place to continue the momentum of our fast start on this program, and to carry it to completion. Bob Bay has taken a strong leadership role in this program and will carry it forward through the year of his presidency.

The quality problem is not new. ASCE President Lee Walker said in November, 1976 "We have a fascination with 'loss prevention' as a mechanical process, rather than an approach to ensure mutual understanding of who has the responsibility for the risk."[1] Our Quality Program addresses this issue. We are making progress and I believe that implementation of our program will have a profound effect on all of us involved in the construction industry.

These two programs, Strategic Planning and Quality, still require our concerted efforts, if we are to bring them to fruition for the benefit of our profession and society as a whole.

There is no shortage of challenges for the years ahead. Some of them are not new challenges, but ones that have been discussed in the past. Richard Tatlow, President of ASCE in 1968, commented "Engineering has splintered into many fragments—is it too late to bring them together?"[2] He went on to discuss the underlying reasons behind the formation of specialty civil engineering organizations and how we might be able to attract more organized groups back to ASCE. The Board of Direction, meeting this weekend, spent a considerable amount of time discussing how affiliations can be developed with other civil engineering groups that can preserve the individuality of those groups, while building and strengthening the areas of commonality. The Board thinks that such an objective can be attained—and has approved an implementation program. Although the challenge is not new, we now have a mechanism to attack it and have stated our resolve to do so.

Many ASCE presidents have spoken about the need for unity in engineering. Arthur Fox in 1976 said "We need to help the engineering profession get its umbrella organization sorted out."[3] John Rinne in 1973 indicated that unity in the engineering profession has been a long-sought goal, "yet we manage to create new unity organizations to add more confusion than unity."[4] Richard Tatlow in 1968[2] and Enoch Needles in 1956[5] commented that our intersociety relations are important to our future and the future of our country.

We are at a critical point in establishing effective intersociety relationships. The American Association of Engineering Societies (AAES) was formed nearly six years ago in an aura of optimism about its role in providing the unity umbrella for engineering. Unfortunately the optimism faded and the reality of differing objectives caused disaffection by many of its members and a serious retrenchment.

AAES has now been reorganized, its headquarters moved to Washington, and it is embarking on a scaled-down and more realistic series of programs. I believe that it has turned the corner, and that if carefully nurtured and wisely managed, AAES can grow into the role of providing the unity in engineering that we aspired to when it was formed six years ago. ASCE has not wavered in its support of the need for unity in engineering. It has supported and offered leadership to AAES during this rebuilding period. Today, we have the opportunity of creating the unity organization we have long desired. We need even stronger support for AAES to implement our goal of unity.

The recent tragic earthquakes in Mexico City revealed to us another weakness in ASCE's programs. We found that we had no organized way to respond to the requests and needs of our colleagues in Mexico or to other requests for aid and assistance. We were aware of this shortcoming because of similar requests with regard to the famine in parts of Africa.

I had commenced some discussions with our fellow civil engineers in the United Kingdom and learned that they have developed an organization named REDR—Engineers for Disaster Relief, which has a register of approximately 450 engineers who are available on a loan basis to disaster relief agencies for temporary work. They have been quite successful, as over 70 engineers have undertaken assignments in disaster areas since REDR was organized in 1980.

In July I asked the attendees at the Irrigation and Drainage Division Specialty Conference in San Antonio if they were interested in such a program in our country and if ASCE or some other organization should get involved. I have received numerous responses from members who were at that meeting—all positive.

As a result of the Mexican earthquake and the response of our members to the REDR possibility, the Board of Direction has authorized President Bay to appoint a task committee to study what ASCE's role should be in disaster relief. So here is another challenge that our Society is rising to meet.

There are many other challenges for ASCE and I know that Bob Bay and Dan Barge will be leading us in programs to meet those challenges. Enoch Needles in his Presidential address entitled "Our Exciting Profession"[5] delivered in Knoxville, Tennessee on June 7, 1956, indicated that he could find no brief definition of a profession. However he said that all definitions covered four basic truths: (1) Proper education and training; (2) ethics in practice; (3) service in behalf of our fellow men; and (4) an avowal to serve as a professional person.

Nearly thirty years later, the same four basic truths are required of a professional. Membership and participation in ASCE can provide avenues of attaining the first three. Through ASCE, we stress the need for proper education and training and put emphasis upon competence through enforcement of registration laws. Through ASCE, we stress the importance of ethics and enforce our beliefs through a program of self-discipline. As civil engineers, we serve the public more directly than other disciplines because our projects have more direct daily impact on people's lives.

An individual decision and dedication is necessary to accomplish the fourth basic truth. President Needles said "If we as civil engineers have these qualities in our individual activities, we are professional men, we know it, and no one can take that knowledge away from us. To that degree, our professional status is for our own determination."[5]

The opportunity to be a professional arises daily. It is how we react that determines whether we are or not.

I am proud to be a member of the civil engineering profession. I am proud of our Society. I think there can be no better way to serve mankind than through our profession and I deeply appreciate the opportunities that you have given me to provide such service. Thank you.

Appendix A
President's Message

Exceptional efforts by many members and staff made 1984–85 an outstanding year for ASCE. Some of our achievements this year can be summarized as follows:

Strategic Plan—Adopted by Board after widespread member input, stimulated creative thinking, established long range guidelines, incorporates new programs, provides for biennial re-examination;

Education—C.E. Education Conference at Ohio State (April 1985), completed comprehensive CE education and academic salary surveys, increased number and popularity of continuing education courses;

Professional Affairs—Workshop on "Quality in the Constructed Project" (November 1984), started a manual of professional practice on quality, cooperated in study of all-risk insurance concept, increased membership from 94,000 to over 100,000, handled over 9,000 calls on 800-line;

Technical Affairs—Formed Materials Engineering Division and Council of Forensic Engineering, improved interface between Technical Divisions and Sections and Branches, secured $1 million grant from EPA, held 24 conferences and conventions including ISSMFE and Waterpower '85;

Publications—Produced first Buyers' Guide in *Civil Engineering Magazine* (December 1984), started new *Journal of Engineering Management* (January 1985), published 67 books compared to 31 in FY '84;

Washington Affairs—Doubled size of Key Contact Program, contacted employers of government engineers, responded rapidly and effectively to attempts to weaken Brooks Act and to National Science Foundation reorganization;

Public Communications—Improved public image by ad campaign "Civil Engineers Make the Difference" with nearly 10 million exposures, proposed new direction for conventions, initiated Statue of Liberty Records Project;

Finance & Administration—Purchased and installed new Burroughs A9F Computer System, meeting or ahead of schedule to bring computer on-line and provide better member support, ended FY '85 with a budget surplus even though programs were increased.

With these significant accomplishments, we are making our Society the most effective professional engineering organization. The path has been blazed, the tools are in place and the future is limited only by the extent to which we expend our energies.

It has been a privilege and an honor to serve as President. I anticipate even greater achievements by ASCE in the years ahead.

Richard W. Karn
President

Appendix B
References

1. "When the Going Gets Tough, the Tough Get Going," Leland J. Walker, President, ASCE 1976–77, *Civil Engineering*, ASCE, Vol. 46, No. 11, Nov., 1976, pp. 65-66.

2. "ASCE's Future Is Our Business," Richard W. Tatlow III, President, ASCE 1967–68, *Civil Engineering*, ASCE, Vol. 38, No. 6, June, 1968, pp. 62-63.

3. "Presidential Address," Arthur J. Fox, Jr., President, ASCE, 1975–76, *Engineering Issues—Journal of Professional Activities*, Vol. 102, No. EI3, Proc. Paper 12263, July, 1976, pp. 315-319.

4. "ASCE at Midterm or Challenges of 1972," John E. Rinne, President, ASCE 1972–73, *Civil Engineering*, ASCE, Vol. 43, No. 7, July, 1973, p. 87.

5. "Our Exciting Profession," Enoch R. Needles, President, ASCE 1955–56, Address at the Summer Convention, Knoxville, TN, June 7, 1956, *Transactions*, ASCE, Vol. 121, 1956, pp. 1398-1410.

1985 ASCE AWARDS AND PRIZES

1

Arthur M. Wellington Prize

Thaw Settlement Analysis for Buried Pipelines in Permafrost

Kenneth J. Nyman, Assoc. Member, ASCE (Sr. Geotech. Engr., ARCO Pipeline Co., Alaskan Div., Los Angeles, Calif.)

Proceedings of the Conference on Pipelines in Adverse Environments II, edited by Mark B. Pickell. New York: ASCE, 1983, pp. 300-325

This paper focuses on determination of allowable thaw settlement for the design of cross country pipelines in permafrost areas. The purpose of the thaw settlement evaluation is to aid in the selection of the proper pipeline construction mode which will ensure pipeline integrity and minimize impact on the environment and adjacent structures. The potential thaw settlement of each pipeline segment is calculated in accordance with the project design criteria and compared to allowable design values. The appropriate construction mode to mitigate adverse settlement is selected for segments where the potential thaw settlement exceeds the allowable design values. Geothermal-geotechnical analyses, which are based on soils information obtained from investigations along the pipeline route, are performed to predict thaw depth over the design life of the system and resulting cumulative thaw settlement potential. The effectiveness of this method is dependent on a well defined soil properties database along the alignment. Design values for allowable thaw settlement are determined from stress analysis for appropriate combinations of pipeline settlement configurations and soil conditions.

2

ASCE State-of-the-Art of Civil Engineering Award

Methods Used to Evaluate Highway Improvements

James M. Witkowski, Assoc. Member, ASCE, (Asst. Prof., Dept. of Civ. Engrg. and Engrg. Mech., The Univ. of Arizone, Tucson, Ariz. 85721)

Journal of Transportation Engineering, Vol. 109, No. 6, November, 1983, pp. 769-784

This study provides information to state-level planners for rational decision–making regarding the enhancement of the analytical process for setting highway improvement priorities. Arizona's analytical procedure, a sufficiency rating scheme typically used by many states, is reviewed in light of the current problems facing decision-makers. Four computerized alternative analytical techniques representing the state-of-the-art are also reviewed. Sufficiency rating techniques are found to supply inadequate information regarding project benefits and the attainment of the improvement program objectives. Also sufficiency ratings and utility functions are considered technically unsupportable because of their subjective base. Option-evaluation techniques are found to supply valuable information regarding the user benefits generated by highway improvements. However, these techniques are limited in their scope of application and cannot incorporate all of the various project types or program goals into the improvement priority process. What is needed is a method which can evaluate the trade-offs between various types of improvement projects on a comprehensive objective basis and evaluate improvement priorities against the multiple constraints and objectives which exist.

3   Collingwood Prize

Reinforced Embankments: Analysis and Design

R. Kerry Rowe, Member, ASCE, (Assoc. Prof., Faculty of Engrg. Sci., Univ. of Western Ontario, London, Ontario, Canada N6A 5B9)

Journal of Geotechnical Engineering, Vol. 110, No. 2, February, 1984, pp. 231-246

A numerical technique for the analysis of geotextile reinforced embankments is outlined. This technique permits consideration of soil-reinforcement interaction, slip at the soil-fabric interface, plastic failure within the soil and large deformations. The applicability of the approach has been assessed by examining the observed and predicted performance of a number of reinforced embankments constructed on soft foundations. In this paper, the application of the approach is illustrated by reference to an embankment constructed on a soft peat deposit. Finally, the paper presents a practical design procedure which involves the use of simple design charts. The use of this design procedure is illustrated by means of a worked example using a typical set of design charts.

4   J. C. Stevens Award

Discussion of: Unsteady Seepage Analysis of Wallace Dam

Jacob Bear, Member, ASCE, (Prof., Dept. of Civil Engrg., Technion-Israel Institute of Technology, Haifa 32000, Israel) and **P. C. Menier**, Assoc. Member, ASCE, (Grad. Student, Dept of Civ. Engrg., Technion-Israel Institute of Technology, Haifa 32000, Israel)

Journal of Hydraulic Engineering, Vol. 110, No. 5, May, 1984, pp. 668-669

In the original paper ("Unsteady Seepage Analysis of Wallace Dam," by Mustafa M. Aral and Morris L. Maslia, *Journal of Hydraulic Engineering*, Vol. 109, No. 6, June, 1983, pp 17-35) the authors made use of the definition and concept of relative hydraulic conductivity, K_r, in dealing with unsaturated flow. The writers of this discussion point out that for an isotropic porous medium, the definition of relative hydraulic conductivity as the ratio of effective hydraulic conductivity, $K(\theta)$, to the hydraulic conductivity at saturation, K_s, raises no problem. However, certain problems arise when this concept is extended to unsaturated flow in anisotropic porous media, and the writers recommend that the concept of relative hydraulic conductivity (or relative permeability) should not be used when dealing with anisotropic media. This recommendation is supported with examples.

5   J. James R. Croes Medal

Horizontal Response of Piles in Layered Soils

George Gazetas, Member, ASCE, (Assoc. Prof. of Civ. Engrg., Rensselaer Polytechnic Inst.,Troy, N.Y. 12181) and **Ricardo Dobry**, Member, ASCE, (Prof. of Civ. Engrg., Rensselaer Polytechnic Inst., Troy, N.Y.12181)

Journal of Geotechnical Engineering, Vol. 110, No. 1, January, 1984, pp. 20-40

An inexpensive and realistic procedure is developed for estimating the lateral dynamic stiffness and damping of flexible piles embedded in arbitrarily layered soil deposits. Starting point is the determination of the pile deflection profile for a static force at the top using any reasonable method -- beam-on-Winkler foundation, finite elements, well-instrumented pile load tests in the field, etc. Material as well as radiation damping due to waves emanating at different depths from the pile-soil interface are rationally taken into account; the overall equivalent damping at the top of the pile is then obtained as a function of frequency by means of a suitable energy relationship. The method is applied to study the dynamic behavior of three different piles embedded in two idealized and one actual layered soil deposit; the results of the method, obtained by hand computations, compare favorably with the results of three dimensional dynamic finite element analyses.

6   Karl Emil Hilgard Hydraulics Prize

Two-Equation Turbulence Model for Flow in Trenches

Ben J. Alfrink, (Project Engineer, Delft Hydraulics Laboratory, Laboratory De Voorst, Postbox 152, 8300 AD Emmeloord, The Netherlands) and **Leo C. van Rijn**, (Project Engineer, Delft Hydraulics Laboratory, Laboratory De Voorst, Postbox 152, 8300 AD Emmeloord, The Netherlands)

Journal of Hydraulic Engineering, Vol. 109, No. 7, July, 1983, pp. 941-958

Steady recirculating flow is described by means of a mathematical model based on the full unsteady Reynolds equations and the k-ε model for turbulence closure. The model is applied to the flow in a steep-sided trench perpendicular to the main flow direction. Inlet profiles are taken with reference to developed channel flow. For the wall boundary a local equilibrium is assumed, yielding among others a logarithmic behaviour for the mean flow velocity. The constants of the k-ε model are related to the roughness conditions. A sensitivity study is reported to identify the relative importance of the various constants and inlet conditions. Numerical results are compared with laboratory experiments.

7   Moisseiff Award

Plate Instability for W Shapes

John L. Dawe, Member, ASCE, (Assoc. Prof. of Civ. Engrg., Univ. of New Brunswick, Fredericton, N.B., Canada E3B 5A3) and **Geoffrey L. Kulak**, Member, ASCE, (Prof. of Civ. Engrg., Univ. of Alberta, Edmonton, Alberta, Canada T6G 2G7)

Journal of Structural Engineering, Vol. 110, No. 6, June, 1984, pp. 1278-1291

The analytical technique utilizes computer computational methods and it is verified by comparing predicted buckling behavior with the behavior observed in 53 laboratory tests conducted by independent researchers. The analytical technique results in good agreement of predicted behavior with that observed during testing procedures. The technique is thus validated and is considered to be a reliable method for studying a broad range of parameters which affect the local buckling behavior of W shapes used as columns, beams, and beam-columns.

8   Local Buckling of W Shape Columns and Beams

John L. Dawe, Member, ASCE, (Assoc. Prof., Civ. Engrg., Univ. of New Brunswick, Fredericton, N.B., Canada E3B 5A3) and **Geoffrey L. Kulak**, Member, ASCE, (Prof., Civ. Engrg., Univ. of Alberta, Edmonton, Alberta, Cnada T6G 2G7)

Journal of Structural Engineering, Vol. 110, No. 6, June, 1984, pp. 1292-1304

An analytical method, previously developed and programmed for computer use, is used to carry out an extensive study of the various parameters affecting the local buckling behavior of W shape columns and beams. The effects of flange and web interaction, residual stresses, plate geometry, and inelastic action are included in the study. Specific parameters investigated include web slenderness, flange slenderness, yield stress, residual stress, strain-hardening modulus, and member length. Local buckling curves of critical loads versus web slenderness are presented for various values of flange slenderness for beams and columns. The results of the study indicate that the present values of flange slenderness = 100 (260 SI) and web slenderness = 225 (670 SI), as presently specified for columns in the Canadian Standard, CAN3-S16.1M78, are inappropriate. Values of flange slenderness = 72 (189 SI) and web slenderness = 300 (788 SI) are indicated. The results also show that the presently specified values of flange slenderness = 100 (260 SI) and web slenderness = 690 (1,810 SI) for Class 3 (non-compact) sections may be too liberal with respect to the flanges and conservative with respect to the web. Values of flange slenderness = 72 (189 SI) and web slenderness = 800 (2,100 SI) are indicated by the study. The effects of all the parameters investigated during the study are shown in the form of local buckling curves as described here.

9   Norman Medal

Filters for Silts and Clays

James L. Sherard, Fellow, ASCE, (Consulting Engr., San Diego, Calif.), **Lorn P. Dunnigan**, Member, ASCE, (Head, Soil Mechanics Lab., National Technical Center, Soil Conservation Service, U.S.D.A., Lincoln, Neb.) and **James R. Talbot**, Member, ASCE, (National Soil Engr., Soil Conservation Service, U.S.D.A., Washington, D.C.)

Journal of Geotechnical Engineering, Vol. 110, No. 6, June, 1984, pp. 701-718

An investigation was made of the filters needed in dams for fine-grained clays and silts. The "critical" downstream filter in a central core dam should be capable of controlling and sealing a concentrated leak through the core, and should also be stable in conventional laboratory filter tests under a relatively high gradient, such as 1,000. Two different types of laboratory tests were developed to simulate the action of a critical filter (slot and slurry tests). Both gave identical and reproducible results. For fine-grained clays a sand filter with D_{15} of 0.5 mm is conservative. For sandy clays and silts the filter criterion $D_{15}/d_{85} \leq 5$ is conservative and reasonable. The Atterberg limits of a clay have no significant influence on the needed critical filter. For nondispersive and dispersive clays having similar particle size distribution the needed critical filters are the same. For "noncritical" filters, such as filters upstream of a clay core, quantitative filter criteria are not necessary.

10   Raymond C. Reese Research Prize

Reinforced Concrete Pipe Columns: Behavior and Design

X. L. Liu, (Grad. Asst., School of Civ. Engrg., Purdue Univ., West Lafayette, Ind. 47907) and **W. F. Chen**, (Prof. and Head of Struct. Engrg., School of Civ. Engrg., Purdue Univ., West Lafayette, Ind. 47907)

Journal of Structural Engineering, Vol. 110, No. 6, June, 1984, pp. 1356-1373

The results of a series of 14 full-scale tests on reinforced concrete pipe columns are reported. For each of these tests, a complete set of synchronously measurements, covering the entire range of loading, are acquired. The tests were controlled by deflection increments when the loads were near and beyond the maximum, or peak value. These data are utilized in developing a design method for reinforced concrete pipe columns. Based on the 14 full-scale tests, the mechanics of progressive failure of reinforced concrete pipe columns is demonstrated through the explanation of the maximum load point, or the P_{max} point, which represents the nonlinear stability load of a column, and the maximum strength point, or the M_{max} point, which corresponds to the ultimate strength of the critical cross section. Based on the difference of the P_{max} point and the M_{max} point, a deflection formula to estimate the maximum load carrying capacity of a reinforced concrete pipe column is presented. It is recommended, as an alternative method, for the design of reinforced concrete pipe columns.

11   **Rudolph Hering Medal**

Fundamentals and Theory of Air Scour

Appiah Amirtharajah, Member, ASCE, (Prof., Dept. of Civ. Engrg. and Engrg. Mechanics, Montana State Univ., Bozeman, Mont. 59717)

Journal of Environmental Engineering, Vol. 110, No. 3, June, 1984, pp. 573-590

An original theoretical analysis of the dynamics of air scour from a fundamental and microscopic viewpoint is presented. Concepts in soil mechanics and porous media hydraulics are combined to explain the complex flow patterns which emerge when air and water flow concurrently through porous media. An equation which predicts the formation and collapse of air pockets within the bed is developed by equating the air pressure within a bubble to the soil stresses in an active Rankine state plus the pore-water pressures. This condition of collapse-pulsing which occurs at particular combinations of simultaneous air and subfluidization water flows has been associated with the probable optimum condition for air scour. The theory compares well with four sets of experimental data collected on two different graded sands. The theoretical expression has been condensed into a simple design equation for practical application.

12   **Samuel Arnold Greeley Award**

Reservoir Modeling to Develop AWT Nitrate Standard

John P. Hartigan, Member, ASCE, (Sr. Engr., Camp Dresser & McKee, 7630 Little River Turnpike, Annandale, Va. 22003.)

Journal of Environmental Engineering, Vol. 109, No. 6, December, 1983, pp. 1243-1258

A continuous simulation reservoir model was used to develop streamflow-related effluent standards for nitrate-N discharges into a water supply impoundment. The nitrogen removal standard for advanced wastewater treatement (AWT) plant represents relaxation of year-round 1.0 mg/L total N effluent standard intended for lake eutrophication management. During years with average or above-average streamflow conditions, AWT discharges must reach 40 mgd (151.6×10^6 L/day) before even limited AWT nitrogen removal operations may be necessary to meet the reservoir's 5 mg/L nitrate-N standard. During years with relatively low streamflows, the 5 mg/L nitrate-N standard can be met with limited or no AWT nitrogen removal operations by a 10 mgd (37.9×10^6 L/day) discharge while extensive nitrogen removal operations during most of all of the low flow period are typically required for discharges in excess of 20 mgd (75.8×10^6 L/day). Streamflow-related effluent standard and AWT operating rule curves permit annual operation and maintenance cost-savings of approximately one to two million dollars, depending upon AWT discharge and streamflows, and deferral of nitrogen removal unit expansion for 10-15 years.

13   Simon W. Freese Environmental Engineering Award
and Lecture

Particles, Pretreatment, and Performance in Water Filtration

Charles R. O'Melia, Member ASCE, (Prof., Dept. of Geography and Environmental Engrg., Johns Hopkins Univ., Baltimore, MD 21218)

Environmental Engineering, edited by James C. O'Shaughnessy. New York: ASCE, 1985, pp. 1125-1130

Relationships among raw water quality, pretreatment facilities, and the design of packed bed filters are presented and applied. The particle size, particle concentration, particle surface characteristics, and solution chemistry in the raw water supply have important and predictable effects on filter design. An integrative approach to water treatment plant design, from raw water quality to filter bed performance, will facilitate process evaluation and has the potential for providing a basis for optimal plant design.

14   Thomas A. Middlebrooks Award

State of the Art: Rock Tunneling

Edward J. Cording, Member, ASCE, Prof. of Civ. Engrg., Univ. of Illinois, Urbana, IL

Tunnelling in Soil and Rock, edited by K. Y. Lo. New York: ASCE, 1984, pp. 77-106

In the past twenty years, improvements have been made in the quality of geotechnical information on tunnel projects. Even with good exploratory information, it remains difficult to predict the effect of rock conditions on support and lining behavior, particularly for deep rock tunnels. With the advent of tunnel boring machines, rates of advance have increased dramatically. They have been increasingly used in difficult ground conditions where rock stability around the machine must be controlled to maintain advance rates. Large underground chambers, both deep and shallow, are being excavated in difficult ground conditions, taking advantage of the ability to stabilize the openings with tied-back and continuous supports. Many more construction tools and procedures are available, and there is a growing understanding of the conditions in which their use is appropriate. The evaluation of performance is important for understanding the often subtle rock behavior effecting tunnel excavation and support. Index properties can be correlated with performance if the mechanisms controlling behavior are understood. Two contrasting groups of case histories in which field measurements were carried out are summarized: small, deep tunnels in highly stressed ground, and large chambers at shallow depth.

15   Thomas Fitch Rowland Prize

Segmental Post-Tensioned Dolphin at Tagrin Point

Samir G. Mattar, Member, ASCE, (Project Mgr., Alberta Public Works, Supply and Services, Calgary, Alberta, Canada)

Journal of Construction Engineering and Management, Vol. 109, No. 3, September, 1983, pp. 276-285

The nature and extent of resource constraints on projects in developing countries are described in the context of the construction of the dolphin at Tagrin Point in Sierra Leone, West Africa. The interplay between what is possible and what is feasible given the constrained resources is considered in terms of various design alternatives. A new type of ship breasting structure evolving from the construction constraints proves to be functional and promising. The segmental, post-tensioned dolphin not only meets the structural and functional requirements but, in making use of appropriate modern technologies, resolved the constructional problems on this project and in so doing, opens up a new field in the design and construction of berthing facilities, with the possibility of changing many concepts and previous constraints.

16   Wesley W. Horner Award

Modeling Ground-Water Flow at Love Canal, New York

James W. Mercer, Member, ASCE, (President, Geotrans, Inc., P.O. Box 2550, Reston, Va. 22090), **Lyle R. Silka**, (Sr. Hydrologist, GeoTrans, Inc., P.O. Box 2550, Reston, Va. 22090) and **Charles R. Faust**, (Vice Pres., GeoTrans, Inc., P.O. Box 2550, Reston, Va. 22090)

Journal of Environmental Engineering, Vol. 109, No. 4, August, 1983, pp. 924-942

A ground-water modeling study was conducted at the Love Canal area, Niagara Falls, New York, as part of a comprehensive evaluation of contamination at the area. The hydrogeology underlying Love Canal consists of a shallow system of silts and fine sands, underlain by confining layers of lacustrine clays and glacial till, which are underlain by the Lockport Dolomite. A two-dimensional computer model of the Lockport Dolomite was constructed to help define the ground-water flow system. Both history matching and a sensitivity analysis were conducted. Assuming the dolomite was contaminated, Monte Carlo simulation and uncertainty analysis were used to estimate contaminant travel times to the upper Niagara River. Results indicate that travel times would average 1000 days with less than 5% probability of less than 100 days or greater than 10,000 days. Analysis of remedial action for the Lockport Dolomite using best estimates of hydrologic parameters indicates that three interceptor wells at the south end of the canal, pumped at only 1155 ft^3/d (32.3 m^3/d), should reverse the flow of groundwater to the river and provide an adequate barrier to migration of potential contaminants to the river.

1985 ASCE PUBLICATIONS

17   **Construction Productivity Improvement**

David Arditi, Assoc. Member, ASCE, (Assoc. Prof. of Civ. Engrg., Illinois Inst. of Tech., Chicago, Ill. 60616)

Journal of Construction Engineering and Management, Vol. 111, No. 1, March, 1985, pp. 1-14

Construction productivity has been on the decline in the last decade. The results are presented on a survey of the Engineering News-Record 400 largest contractors to obtain their views on where productivity improvements would most help and to compare the trends with a similar survey carried out in 1979. Data were collected on the general company characteristics of the responding contractors, and on the contractors' opinions on potential areas for productivity improvement in the office and in the field. Findings indicate that immediate research should concentrate on improving marketing practices, planning and scheduling, labor-management relations, site supervision, industrialized building systems, equipment policy and engineering design; and that governmental regulations have lost the immediate urgency attached to them in 1979. It is also recommended that similar surveys be conducted every 3 to 4 years to identify new trends and to steer research in the appropriate direction.

18   **Raise Boring in Civil and Mining Applications**

William R. Nash, Member, ASCE, (Project Dir., McCarthy Brothers Construction Co., P.O. Box 20036, Brentwood Station, St. Louis, Mo. 63144)

Journal of Construction Engineering and Management, Vol. 111, No. 1, March, 1985, pp. 15-30

Drilling and blasting is the traditional method of shaft excavation. This is labor intensive and time consuming. With the advent of machine boring, the RBM, or raise-boring machine, was developed to increase safety and decrease the cost of shaft construction. Compared to conventional drill-and-blast shaft excavation methods, raise boring offers many advantages: (1) Safety; (2) increased productivity; (3) disturbance of the rock unit from its equilibrium state is not as great as when using drilling and blasting methods; (4) excavation crew size is smaller; and (5) resultant bottom-line reduction in cost of shaft construction.

19   **Productivity Analysis of Construction Operations**

Amir Tavakoli, Member, ASCE, (Asst. Prof., School of Engrg., Southern Illinois Univ. at Edwardsville, Edwardsville, Ill. 62026)

Journal of Construction Engineering and Management, Vol. 111, No. 1, March, 1985, pp. 31-39

This paper proposes an interactive system for analysis of construction operations. The analysis is carried out in the context of various work modules which address: 1. Quantity development; 2. resource definition; and 3. production and cost analysis. The quantity work module generates quantities based on information available in the design documents. The resource definition module receives and stores data regarding the labor-equipment combination to be used to execute work tasks. This module provides the user with a set of standard useful construction process models. For each construction operation to be analyzed, the terminal describes the standard models. The user makes input of a set of parameters for process keyname, quantity, work task durations, number of resources, production capacity of each unit, and cost per hour of each unit to the standard model to be used. Using input from the resource definition module, the productivity and cost analysis module generates production rates and unit costs based on process simulation using CYCLONE methodology.

20   Pioneering Construction Engineering Education

Bonnie S. Ledbetter, (Asst. Prof., History Dept., Texas A&M Univ., College Station, Tex. 77843)

Journal of Construction Engineering and Management, Vol. 111, No. 1, March, 1985, pp. 41-51

In the 1930's, a small number of educators and contractors began to advocate that colleges and universities develop programs in construction engineering. Through the Great Depression and World War II, they debated what the nature of construction engineering education ought to be. Following World War II, the climate among educators and contractors was receptive to the ideas which the pioneers in construction engineering education had been advocating for 15 yrs. Colleges and universities began to offer construction engineering options in their civil engineering curricula.

21   Valence of and Satisfaction with Job Outcomes

William F. Maloney, Member, ASCE, (Asst. Prof., Civ. Engrg. Dept., Univ. of Michigan, Ann Arbor, Mich. 48109) and **James M. McFillen**, (Assoc. Prof., Management Dept., Bowling Green State Univ., Bowling Green, Ohio 43403)

Journal of Construction Engineering and Management, Vol. 111, No. 1, March, 1985, pp. 53-73

Approximately 2,800 unionized construction workers were surveyed and responses were received from 703, representing all crafts except Boilermakers. The workers were asked about the importance they attach to various job related factors and their satisfaction with each factor. The 28 individual factors were reduced to seven using a factor analysis technique. The most important set of factors were those relating to the intrinsic nature of the work: working like a craftsman, performing challenging work, etc. The set of factors with which the workers was most satisfied was that of Performance Level: high productivity; quantity; and doing your work in a craftsmanlike manner. Individual factors that require attention on the part of contractors to improve worker motivation and satisfaction were identified using a 2×2 matrix that allowed the combination of importance and satisfaction scores. Overall satisfaction with the job was measured and a multiple regression analysis revealed that satisfaction with intrinsic factors makes the greatest contribution to general job satisfaction.

22   Construction Claims: Frequency and Severity

James E. Diekmann, Member, ASCE, (Assoc. Prof. and Civ. Engr., Univ. of Colorado, Boulder, Colo.) and **Mark C. Nelson**, (Grad. Stud, Construction Engrg. and Management Program, Univ. of Colorado, Boulder, Colo.)

Journal of Construction Engineering and Management, Vol. 111, No. 1, March, 1985, pp. 74-81

There has been increasing emphasis in recent years on construction contract claims and disputes. This paper examines the frequency of occurrence of 427 separate construction claims which were experienced on 2 federally funded and administered construction projects. The data examined include various claim types, the frequency of their occurrence and the average cost of these claims. Also, various factors thought to influence the frequency of claims occurrence were investigated. This study indicates that the largest proportion of change orders and modifications originate with the owner of the project or with those responsible to the owner.

23   On Site Performance Improvement Programs

Alexander Laufer, Member, ASCE, (Sr. Research Engr., Bldg. Res. Station, Technien—Israel Inst. of Tech., Haifa, Israel)

Journal of Construction Engineering and Management, Vol. 111, No. 1, March, 1985, pp. 82-97

The paper presents the various phases in a comprehensive On Site Performance Improvement Program (OSPIP) of a medium size construction project which includes: problem identification; data collection; data analysis and planning the change content; planning the change process; and measuring and evaluating the results. Methods, techniques, and means to put the various program phases into practice are examined, and the problems likely to be encountered are discussed. An OSPIP case study of a medium size construction project is described in detail. It deals with production rate, safety, and manpower turnover problems, improvement plans to overcome them, and how implementation was realized in each of the problem areas. The paper suggests that OSPIP in medium size projects is not only economically feasible and worthwhile, but highly recommendable.

24   Production Estimating for Draglines

Rita F. Stewart, Student Member, ASCE, (Student, Louisiana Tech. Univ., Civ. Engrg. Dept., Ruston, La. 71272) and **Cliff J. Schexnayder**, Member, ASCE, (Adjunct Prof., Louisiana Tech. Univ., Civ. Engrg. Dept., Ruston, La. 71272)

Journal of Construction Engineering and Management, Vol. 111, No. 1, March, 1985, pp. 101-104

Dragline productin estimates, as derived from published tables, are compared to actual field production. Several equipment texts present dragline production estimating tables. The base data source for most of these tables are studies performed by the Power Crane and Shovel Association, PCSA. Only a small variance existed between the actual and estimated production.

25   Network Scheduling Variations for Repetitive Work

James J. O'Brien, Fellow, ASCE, (Chf. Executive Officer, O'Brien-Kreitzberg & Assoc., Inc., Merchantville, N.J.), **Fred C. Kreitzberg**, Fellow, ASCE, (Pres., O'Brien-Kreitzberg & Assoc., Inc., San Francisco, Calif.) and **Wesley F. Mikes**, Fellow, ASCE, (Sr. Vice Pres., O'Brien-Kreitzberg & Assoc., Inc., Merchantville, N.J.)

Journal of Construction Engineering and Management, Vol. 111, No. 2, June, 1985, pp. 105-116

Observant CPM schedulers in the early years of CPM application, the 1960s, noted that there were often familiar steps inherent in scheduling similar projects. This similarity occurred in the development of housing neighborhoods, such as New Town of Columbia, Maryland; schools, as in the Philadelphia School District Program involving 200 schools; and in the review process for projects during the preconstruction phase in large cities, namely Philadelphia and New York. In these situations, the development of a prototypical network provided a cost-effective method of applying network scheduling to major programs. These applications are described in the form of case studies, and a recent application to the King Khalid Military City Housing Program is noted. An alternate manual approach to repetitive scheduling in high-rise construction, the vertical production method developed in the 1970s, is also described.

26   Total Quality Management for Construction

Jerald L. Rounds, Member, ASCE, (Assoc. Prof., Construction Engrg., Dept. of Civ. Engrg., Iowa State Univ., Ames, Iowa) and **Nai-Yuan Chi**, (Grad. Student, Furukawa Lab., Dept. of Architecture, Kyoto Univ., Japan)

Journal of Construction Engineering and Management, Vol. 111, No. 2, June, 1985, pp. 117-128

Traditional approaches to quality control in the construction industry are inadequate and should be replaced with the Total Quality Control concept implemented through the Quality Control (Q.C.) Circle as developed in Japan and currently in wide use throughout the manufacturing industry. The term "total quality control" is defined, and four total quality control principles are set forth on the basis of this definition. The evolution of quality control is traced from the nineteenth century to today to explain the decline in quality standards and to illustrate the need for a new approach. Unique characteristics of the construction industry are described as they relate to the Q.C. circle concept. Implementation of this concept will result in higher quality, lower costs, and increased productivity in the construction industry.

27   Climatic Effects on Construction

Enno Koehn, Member, ASCE, (Assoc. Prof., School of Civ. Engrg., Civ. Engrg. Building, Purdue Univ., West Lafayette, Ind. 47907) and **Gerald Brown**, (U.S. Army Construction Engrg. Lab. (CERL), Champaign, Ill.)

Journal of Construction Engineering and Management, Vol. 111, No. 2, June, 1985, pp. 129-137

The results of an investigation designed to determine the relationship between overall construction productivity, and temperature and humidity are presented. Data from the following activities or crafts, or both, were employed: excavation (manual), erection, masonry, electrical, carpentry, laborers, and excavation (equipment). Two nonlinear equations were determined, one for cold or cool weather ($R^2 = .62$) and another for hot or warm weather ($R^2 = .64$). The overall findings indicate that below -10°F and above 110°F it is difficult to achieve efficient construction operations. In addition to productivity data, the health hazards, possible preventive measures, and acclimatization of workers to severe environments are examined.

28   Pipe Jacking Method for Long Curve Construction

Yoshihiko Nomura, (Engr., Ibaraki Electrical Communication Lab., Nippon Telegraph and Telephone Public Corp., Tokai, Ibaraki-ken, 319-11, Japan), **Hiroshi Hoshina**, (Staff Engr., Ibaraki Electrical Communication Lab., Nippon Telegraph and Telephone Public Corp., Tokai, Ibaraki-ken, 319-11, Japan), **Hiroshi Shiomi**, (Engr., Ibaraki Electrical Communication Lab., Nippon Telegraph and Telephone Public Corp., Tokai, Ibaraki-ken, 319-11, Japan) and **Takao Umezu**, (Engr., Ibaraki Electrical Communication Lab., Nippon Telegraph and Telephone Public Corp., Tokai, Ibaraki-ken, 319-11, Japan)

Journal of Construction Engineering and Management, Vol. 111, No. 2, June, 1985, pp. 138-148

The D301 pipe jacking method is the first to accomplish long spans, curves and high speed construction in the field of small diameter (300 mm) tunneling. The design of the system components are described: tunneling machine with directional control ability, jacking machine, power unit, and control unit which contains a microprocessor. As a result of field tests in about 100 m in length with a 200 m curvature radius, the following capabilities are confirmed: the span length is up to 100 m, the curvature radius is down to 200 m, it is useable in either cohesive or sandy soils having N-values up to 10, and the construction speed is up to 3 m/hr.

29   Steel Construction Evaluation by MLR-Strategies

Bruce Forde, (Research Asst., Dept. of Civ. Engrg., Univ. of British Columbia, Vancouver, British Columbia, Canada) and **Siegfried F. Stiemer**, (Assoc. Prof., Dept. of Civ. Engrg., Univ. of British Columbia, Vancouver, British Columbia, Canada)

Journal of Construction Engineering and Management, Vol. 111, No. 2, June, 1985, pp. 149-156

Improvements in design, achieved through a better understanding of fabrication costs, may cause an overall cost reduction for steel structures. Multiple Linear Regression (MLR) provides the mechanism for transformation of data collected by an information system to data which is used in the evaluation of structural steel fabrication costs. The integration of fabrication control and analysis provided by this system permits its implementation in existing fabrication environments, and presents a desperately needed technological gain for the fabrication process. Traditional design approaches which have relied heavily on experience can now be evaluated and improved in terms of cost competitiveness, prior to or during fabrication by the proposed MLR strategy.

30   Product Specification Practices and Problems

C. William Ibbs, Jr., Assoc. Member, ASCE, (Asst. Prof., Dept. of Civ. Engrg., Univ. of Illinois, Urbana, Ill.)

Journal of Construction Engineering and Management, Vol. 111, No. 2, June, 1985, pp. 157-172

The technical product specifications of construction contracts and the associated submittal review processes are shown to be involved in a major portion of all serious project disputes. A large number of actual publicly-funded water and wastewater treatment facility building projects were examined with subsequent analysis of those reviews indicating that these disputes are detrimental in terms of additional project cost, schedule delays, and overall project disruption and loss of goodwill. In particular, the proprietary "brand name or equal" product specification method is seen to be commonly at the heart of these product disputes. Several different solution strategies were expressed as statistical hypotheses and tested for effectiveness. Among other propositions, the managerial strategies of required bid listing of proposed products, timely dispute resolution, and clarifying the submittal review process proved to be realistic and effective tools for reducing both the incidence and severity of product-related contract disputes. Exact details are put forward along with sensitivity ranges.

31   Application of Small Computers in Construction

Task Committee on the Application of Small Computers in Construction

Journal of Construction Engineering and Management, Vol. 111, No. 3, September, 1985, pp. 173-189

The deliberations of the ASCE Task Committee on the Application of Small Computers in Construction are presented. The paper first identifies needs that are stimulating the application of small computers in construction, and then examines their present and potential utilization in several representative application areas. Applications include accounting and payroll, estimating, field office administration, contract language retrieval, electronic communications, scheduling, process simulation, graphical reporting, computer-aided design, and process control. The role that ASCE might take in supporting the effective application of small computers in construction practice is considered. Possibilities include participation in the development of standards, education of practitioners, stimulation of innovative techniques, liaison with other organizations, input to computer manufacturers and software developers about specific needs and requirements in construction, studying the organizational and behavioral aspects of computers in construction, and input to ASCE awards committees about meritorious work in this area. It is concluded that an ongoing ASCE Construction Division committee is needed to monitor and influence the effective application of this rapidly evolving small computer technology in construction.

32 Automation and Robotics for Construction

Boyd C. Paulson, Jr., Member, ASCE, (Prof. and Assoc. Chmn., Dept. of Civ. Engrg., Stanford Univ., Stanford, Calif. 94305)

Journal of Construction Engineering and Management, Vol. 111, No. 3, September, 1985, pp. 190-207

The potential for automated real-time data acquisition, process control and robotics for remote, large-scale field operations, such as those on construction engineering projects, is addressed. Classifications of technologies for automation and robotics in such operations include hard-wired instrumentation, remote sensing, analog and digital telecommunications, optical (laser, infrared and fiber-optic) data transmission, monitoring via microcomputer-based instrument control and data recording, on-site process control for fixed plants, partial or fully automatic control of mobile equipment, fixed-based manipulators, mobile robots, communications between on-site computers and automated machinery, electronic ranging and detection, and video-image pattern recognition. Combining selected technologies with microcomputer-based software could facilitate analysis, design and control decision-making, and could provide a means of coordinating various discrete automated components or machines that must work together to perform field tasks. This paper also mentions categories of needs for such technologies on field operations, and potential barriers to implementation. Progress will depend on the interest and support of researchers qualified to advance this field.

33 Vibration Criteria for Historic Buildings

Walter Konon, Member, ASCE, (Assoc. Prof. of Civ. and Environmental Engrg., New Jersey Inst. of Tech., Newark College of Engrg., Newark, N.J. 07102) and **John R. Schuring**, Member, ASCE, (Asst. Prof. of Civ. and Environmental Engrg., New Jersey Inst. of Tech., Newark College of Engrg., Newark, N.J. 07102)

Journal of Construction Engineering and Management, Vol. 111, No. 3, September, 1985, pp. 208-215

The 2.0 in./sec (50 mm/s) peak particle velocity criterion traditionally used to protect structures from construction-induced vibration damage is nonconservative for historic and sensitive older buildings. The relevant parameters which must be considered in establishing vibration criteria for historic and sensitive older buildings are examined. Existing criteria by past investigators are reviewed and a new criterion for this class of structure is recommended.

34 Optimization Model for Aggregate Blending

Said M. Easa, Member, ASCE, (Assoc. Prof., Dept. of Civ. Engrg., Lakehead Univ., Thunder Bay, Ontario, Canada) and **Emre K. Can**, Assoc. Member, ASCE, (Asst. Prof., Dept. of Civ. Engrg., Lakehead Univ., Thunder Bay, Ontario, Canada)

Journal of Construction Engineering and Management, Vol. 111, No. 3, September, 1985, pp. 216-230

An optimization model of aggregate blending is presented that considers gradation, cost, and design requirements and is applicable to the blending of any number of aggregates. The model is formulated as a quadratic programming problem that minimizes the mean deviation from midpoint specification limits, subject to constraints on the preceding requirements. The model is applied to a numerical aggregate blending problem. Sensitivity analysis is performed to show how the model can also be used to minimize cost or to provide a trade-off between mean deviation and cost. Extensions of the model to accommodate special practical cases are examined.

35   Tender Evaluation by Fuzzy Sets

Van Uu Nguyen, Member, ASCE, (Lect., Dept. of Civ. and Mining Engrg., Univ. of Wollongong, NSW 2500, Australia)

Journal of Construction Engineering and Management, Vol. 111, No. 3, September, 1985, pp. 231-243

The process of evaluating tenders is considered to be largely dependent on subjective judgment when cost is not the only criterion used. A systematic procedure based on fuzzy set theory and multicriteria modeling is proposed for the selection of bid contracts. The proposed procedure is suitable for a general tender evaluation process that may involve many decision-making parties and noninteractive multiple criteria. Illustrative examples are given for cases involving three major criteria: cost, present bid information, and past experience of tenderers.

36   General Contractors' Management: How Subs Evaluate It

George S. Birrell, Member, ASCE, (Assoc. Prof., Dept. of Civ. Engrg., Case Western Reserve Univ., Cleveland, Ohio 44106)

Journal of Construction Engineering and Management, Vol. 111, No. 3, September, 1985, pp. 244-259

A large set of criteria by which top quality subcontractors evaluate the managerial performance of general contractors under whose management they have worked during the construction process and by which they may differentiate their bids to different generals for the same future project are listed and described. These criteria can also be seen as the intrinsic managerial, cost- and time-sensitive factors by which general contractors or any manager of construction could improve performance, competitiveness and profitability. It also describes separate, "most important" and "super-important" subsets of the foregoing criteria/factors for office and site staff, etc., and the range of effects of the generals' good and bad managerial performance against these criteria/factors on the costs and duration of the subcontractors' work. The cost and duration effects, etc., that each lump sum bidding general contractor and appointed construction management agent has on the subcontractor's work, are compared, and the carefully formatted research process which produced these results is outlined.

37   Robotics in Building Construction

Abraham Warszawski, (Head, Building Research Station,, Technion I.I.T., Haifa, Israel) and Dwight A. Sangrey, (Prof. and Head, Dept. of Civ. Engrg., Carnegia Mellon Univ., Pittsburgh, Pa. 15213)

Journal of Construction Engineering and Management, Vol. 111, No. 3, September, 1985, pp. 260-280

The paper examines possible applications of robotics to building construction. First, the main features of industrial robots and their applications are described. Then, building activities are separated into basic components, and the performance requirements from a robot, necessary for their execution, are specified. A conceptual description of four types of construction robots is derived from these performance requirements. An adaptation of the construction process and of building components for efficient application of these robots is then analyzed. Some special problems associated with robotization of construction process are also explored.

38   Portland Cement Concrete Thin-Bonded Overlay

Richard J. Wibby, (Student, Dept. of Civ. Engrg., Louisiana Tech. Univ., Ruston, La.)

Journal of Construction Engineering and Management, Vol. 111, No. 3, September, 1985, pp. 281-292

An innovative approach to Portland Cement Concrete (P.C.C.) thin-bond overlay is presented. It includes a review of both equipment and methods. A complete presentation of this type of construction is offered, including step-by-step detailing of the construction process. A cost comparison between a P.C.C. thin-bond overlay and an asphalt concrete overlay is also examined.

39   Expert Systems for Construction Project Monitoring

M. McGartland, (Engr., Bechtel Power Div., Gaithersburg, Md. 20877) and **C. Hendrickson,** Assoc. Member, ASCE, (Assoc. Prof., Dept. of Civ. Engrg., Carnegie-Mellon Univ., Pittsburgh, Pa. 15213)

Journal of Construction Engineering and Management, Vol. 111, No. 3, September, 1985, pp. 293-307

Potential applications of knowledge based expert systems in the area of construction project monitoring and control are described. Originally developed from research in artificial intelligence, these systems are computer programs that can undertake intelligent tasks currently performed by highly skilled people. While some project monitoring can be accomplished by algorithmic procedures, the capability of knowledge based expert systems to deal with ill-structured problems and to be extensively modified over time make them desirable for application in this area. Sample applications and heuristic rules in scheduling and inventory control are provided.

40   SIREN: A Repetitive Construction Simulation Model

Donncha P. Kavanagh, Member, ASCE, (Grad. Student, Dept. of Civ. Engrg., Univ. of Missouri-Rolla, Rolla, Mo. 65401)

Journal of Construction Engineering and Management, Vol. 111, No. 3, September, 1985, pp. 308-323

SIREN (SImulation of REpetitive Networks) is a computer model of repetitive construction such as the construction of multi-story buildings, housing estates, linear projects, etc. The user interactively inputs a precedence diagram for the repetitive unit (e.g., one floor of a skyscraper) and additional "sub-networks" that are not part of the repetitive sequence (e.g., first floor of skyscraper). From this information, the computer generates the whole network. Data is input via an IBM-PC at which point extensive error checking is carried out. The model itself is coded in the GPSS language and runs on a remote mainframe computer. It simulates the various crews as they queue to carry out activities. A working schedule and cumulative cost curve are produced and statistics are gathered on crew and equipment utilization, all being output graphically. A Monte-Carlo simulation is also included as probability distributions may be associated with the duration of each activity. This yields confidence intervals on cumulative costs throughout the project and on milestone attainment.

41   **A Simulation Model to Forecast Project Completion Time**

Hira N. Ahuja, (Prof., Faculty of Engrg., Tech. Univ. of Nova Scotia, Halifax, Nova Scotia, Canada) and **V. Nandakumar**, (Sr. Proj. Engr., Metallurgical and Engrg. Consultants (India) Ltd. (MECON), India)

Journal of Construction Engineering and Management, Vol. 111, No. 4, December, 1985, pp. 325-342

Project cost is most sensitive to its schedule. The construction project environment comprising dynamic, uncertain, but predictable, variables such as weather, space congestion, workmen absenteeism, etc., is changing continuously, affecting activity durations. The reliability of project duration forecast can be enhanced by an explicit analysis to determine the variation in activity durations caused by the dynamic variables. A computer model is used to simulate the expected occurrence of the uncertainty variables. From the information that is collected normally for a progress update of the tactical plan and by simulating the project environment, the combined impact of the uncertainty variables is predicted for each progress period. By incorporating the combined impact in the duration estimates of each activity, the new activity duration distribution is generated. From these activity duration distributions, the probability of achieving the original project completion time and of completing the project at any other time is computed.

42   **Decisions in Construction Operations**

Bilal M. Ayyub, Assoc. Member, ASCE, (Asst. Prof., Dept. of Civ. Engrg., Univ. of Maryland, College Park, MD 20742) and **Achintya Haldar**, Member, ASCE, (Asst. Prof., School of Civ. Engrg., Georgia Inst. of Tech., Atlanta, GA 30332)

Journal of Construction Engineering and Management, Vol. 111, No. 4, December, 1985, pp. 343-357

A method for selecting the most desirable construction strategy is proposed in this paper. A decision analysis framework is developed considering the information on relative risk, along with the information on cost, benefit, and consequences of each construction strategy. Many factors affect the safety of construction operations. Labor skill, supervisors' experience and attendance, condition of falsework, weather conditions, type of equipment, level of operator's experience, etc., may be considered as the main factors that affect the safety of construction operations. From an engineering point of view, these factors are always desirable to be in their best states. However, it may not be practical in many cases due to time and money constraints. Consequently, the risk of failure and cost of construction operations, and the consequences of failure need to be estimated in order to decide about the optimum construction strategy. The decision problem is complicated since the state of the factors are generally expressed in linguistic terms. The theory of fuzzy sets is used to translate these terms into mathematical measures and to estimate the risk of failure. The proposed method is illustrated with the help of an example.

43   **Stochastic Priority Model for Aggregate Blending**

Said M. Easa, Member, ASCE, (Assoc. Prof., Dept. of Civ. Engrg., Lakehead Univ., Thunder Bay, Ontario, Canada) and **Emre K. Can**, Assoc. Member, ASCE, (Asst. Prof., Dept. of Civ. Engrg., Lakehead Univ., Thunder Bay, Ontario, Canada)

Journal of Construction Engineering and Management, Vol. 111, No. 4, December, 1985, pp. 358-373

Aggregate blending models that incorporate the optimization of two objectives with their priority levels are presented. The two objectives include the minimization of the mean deviation (or mean absolute deviation) from the midpoint of specification limits and the minimization of the unit cost of the blend. The models are applicable to any number of aggregates and can be used to provide the optimum proportions corresponding to a given priority level or to establish trade-off curves between mean deviation and cost. The stochastic elements of aggregate gradations are formulated and incorporated into the models. Both the deterministic and stochastic models are applied to a numerical aggregate blending problem, and extensions of the models to accommodate some practical cases are presented.

44   Fair and Reasonable Markup (FaRM) Pricing Model

Foad Farid, Assoc. Member, ASCE, (Asst. Prof. of Civ., Environmental, & Architectural Engrg., Univ. of Colorado, Campus Box 428, Boulder, CO 80309) and **L. T. Boyer**, Member, ASCE, (Prof. of Civ. Engrg., Univ. of Illinois, Urbana, IL 61801)

Journal of Construction Engineering and Management, Vol. 111, No. 4, December, 1985, pp. 374-390

The Fair and Reasonable Markup (FaRM) is the smallest markup that satisfies the Required Rate of Return (RRR) of the contractor for the particular (or at least the general risk-class of) project at hand. The model is based on reasonable and easily-accessible information, and will result in a Minimum Acceptable Price (MAP). The firm cannot accept the project at a price below this MAP without diminishing the "equityholders' wealth." A modified version of the FaRM Pricing Model for certain contracts under which home-office overhead expenses must be recovered through FaRM is also presented. Once the FaRM Pricing Model has been implemented, contractors can make more intelligent pricing decisions. Instead of using a subjective markup, which may ignore the cash-flow differences of various jobs, contractors using FaRM Pricing Model can bid lower on projects which are more attractive and become more competitive while satisfying their RRR. This should result in lower costs to owners. Conversely, by bidding higher on the less-attractive jobs, contractors will still maintain their RRR should they obtain the contract.

45   Formwork Design

Richard C. Ringwald, (Asst. Prof., Construction Engrg., Iowa State Univ., Ames, IA 50011)

Journal of Construction Engineering and Management, Vol. 111, No. 4, December, 1985, pp. 391-403

The American Concrete Institute's methodology for designing formwork adequate for its loading is condensed, organized, and simplified into a two-page, step-by-step format of equations and instructions that cover most formwork design situations. Plotted design curves are shown and their limitations discussed.

46   Impacts of Constructability Improvement

James T. O'Connor, Assoc. Member, ASCE, (Asst. Prof., Dept. of Civ. Engrg., Univ. of Texas, Austin, TX 78712)

Journal of Construction Engineering and Management, Vol. 111, No. 4, December, 1985, pp. 404-410

An analysis of the construction resource utilization trade-offs, which occur from constructability improvements, provides insight into the constructability improvement process. Matrices of construction and engineering impacts likely to result from constructability improvements are presented. Constructability improvements collected on a large industrial construction project are analyzed for their impact to the job. Frequencies of occurrence of both desirable and undesirable impacts are noted, as are the cost-significances of the various impact types. Constructability strategies and methods for achieving the most cost-beneficial impacts are presented. The likelihood of delays may be decreased most effectively by increasing engineering information availability and understandability. The amount of required construction manpower may be most effectively decreased by simplifying the design, combining design elements, and seeking optimal design-originated construction techniques such as optimal construction systems, modularization, and improved design details. Of course, additional engineering effort may be required. Construction activity durations may also be most effectively decreased by seeking optimal design-originated construction techniques.

47 An Information System for Building Products

G. Bülent Coker, (Research Sci. and Assoc. Prof., Building Research Inst., Bilir sokak no. 17, Kavaklidere, Ankara/Turkey)

Journal of Construction Engineering and Management, Vol. 111, No. 4, December, 1985, pp. 411-425

In a building process that enables a proper flow of product information among the "Users of Information" (Agents), a "Product Information System" is proposed. The system established in connection with a set of criteria involves five subsystems (Thesaurus, Summarized Data, Manufacturers, Compatibility Tables, and Detailed Data) linked with each other. Agents looking for product information can enter the system through the subsystem Thesaurus or Manufacturers, and they can decide on the choice of appropriate products at the first stage (Summarized Data), whereby they can find access to the subsystem Detailed Data afterwards for the detailed information needed. The complete code presents a medium for a common language among the users of product information.

48 Evaluation of Craftsman Questionnaire

Luh-Maan Chang, Member, ASCE, (Asst. Prof., School of Building Constr., Univ. of Florida, Gainesville, FL) and John D. Borcherding, Member, ASCE, (Asst. Prof., Dept. of Civ. Engrg., Univ. of Texas at Austin, Austin, TX)

Journal of Construction Engineering and Management, Vol. 111, No. 4, December, 1985, pp. 426-437

This paper is concerned with evaluating the validity of the Craftsman Questionnaire for determining lost manhour estimates. The validity of the Craftsman Questionnaire has been questioned since it was introduced in the construction industry. To examine its validity, the background of the Craftsman Questionnaire is revised. Secondly, an analysis of the data collected from past studies is conducted. Herein, sources of inaccurate estimates are revealed. Thirdly, a research method called criteria validity is employed to investigate the validity of the Craftsman Questionnire by comparing its measures with those of Work Sampling. Although the comparison of results do not reach an absolute conclusion at this stage, the experiment still gives evidence of the validity of employing Craftsman Questionnaires to determine lost time estimates.

49 Leaching of Tetrachloroethylene from Vinyl-Lined Pipe

Avery H. Demond, (Grad. Research Asst., Dept. of Civ. Engrg., Stanford Univ., Stanford, Calif. 94305)

Journal of Environmental Engineering, Vol. 111, No. 1, February, 1985, pp. 1-9

Tetrachloroethylene has been used in the manufacture of vinyl-lined asbestos cement pipe installed in water distribution systems in certain New England communities. Tetrachloroethylene was used as a solvent for Piccotex, a vinyl-toluene α-methyl styrene co-polymer, which was then sprayed on the interior of the finished asbestos cement pipe. The migration of tetrachloroethylene through the vinyl lining was modeled as a diffusive process in which the diffusion coefficient was an exponential function of concentration. The parameters of the model were found by fitting a numerical solution of the diffusion equation to data from laboratory experiments involving the evaporation of tetrachloroethylene from a Piccotex layer. The conclusion was reached that significant quantities of tetrachloroethylene may be leached into the water supply for a time period on the order of years.

50   Unified Analysis of Thickening

Don E. Baskin, (Grad. Research Asst., Dept. of Civ. Engrg., Univ. of Illinois, Urbana, Ill. 61801) and **Makram T. Suidan**, (Assoc. Prof., Dept. of Civ. Engrg., Univ. of Illinois, Urbana, Ill. 61801)

Journal of Environmental Engineering, Vol. 111, No. 1, February, 1985, pp. 10-26

Zone settling data can generally be linearized with solids concentration when the logarithm of the settling velocity is plotted as a function of either the concentration or the logarithm of the concentration. The resulting mathematical expressions were used to develop two generalized dimensionless solutions that describe the steady-state performance of thickeners. Graphical solutions of these models were also developed. These solutions could be used for controlling the performance of a thickener operating under transient conditions.

51   Time Series Models for Treatment of Surface Waters

Paul J. Ossenbruggen, Member, ASCE, (Assoc. Prof. of Civil Engrg., Univ. of New Hampshire, Durham, N.H.)

Journal of Environmental Engineering, Vol. 111, No. 1, February, 1985, pp. 27-44

The use of probability models derived from a long-term (30 yr) color record forms an approach of forecasting raw water color and assigning an alum dose for treatment of surface waters. Three probabilistic models are developed which incorporate the uncertainties associated with forecasting raw water color and alum dose. These forecasts are based upon the risk or probability associated with underdosing. The effect of seasonal and daily color variation is incorporated into one time series model. In the other time series model, daily color variation is considered exclusively. All models are compared for treatment performance. The methods of model identification, parameter estimation, and forecasting are presented for each model. The results indicate that the two time series models are feasible methods because they lead to the judicious use of alum without excessive overdosing.

52   Dispersion Model for Waste Stabilization Ponds

Chongrak Polprasert, (Assoc. Prof. of Environmental Engrg., Asian Inst. of Tech., Bangkok, Thailand) and **Kiran K. Bhattarai**, (Doctoral Student, Environmental Engrg., Asian Inst. of Tech., Bangkok, Thailand)

Journal of Environmental Engineering, Vol. 111, No. 1, February, 1985, pp. 45-59

The partially-mixed or dispersed-flow equation is proposed for use in pond design for organic and bacterial reductions. This equation considers the effects of hydraulic detention time (θ), reaction rate coefficients, and the dispersion factor including short-circuiting, and other hydraulic transport processes which influence the pond's performance. Some hydraulic models of dispersion are reviewed. A dispersion prediction formula for ponds is proposed in this paper which relates the value of the dispersion number to θ, kinematic viscosity, and the pond's geometric shape, i.e., length, width, and depth.

53   Nitrogenous Wastewater Treatment by Activated Algae

S. K. Gupta, (Lect., Environ. Sci. and Engrg. Group, Dept. of Chemistry, Indian Inst. of Tech., Powai, Bombay 400076, India)

Journal of Environmental Engineering, Vol. 111, No. 1, February, 1985, pp. 61-77

A biological treatability study by activated algae process was performed with synthetic wastewater containing a high concentration of nitrogen. It was found that the wastewater could be processed at all nitrogen removal rates. The yield coefficient and decay coefficient for heterotrophic bacteria were 0.06 (COD basis) and 0.019^{-1} (COD bases) respectively. The yield coefficient and decay coefficient for nitrifiers were 0.06 and 0.02 day^{-1} respectively. NH^+_4-N seemed to inhibit bacteriological growth as the yield coefficients values were significantly lower. Nitrification was observed at all the nitrogen loadings. Diffusion of NH_3 into the atmosphere was the dominant mechanism of nitrogen removal. The results demonstrated a symbiotic relationship between algae and bacteria.

54   Performance of Surface Rotors in an Oxidation Ditch

Roger V. Stephenson, Assoc. Member, ASCE, (Predoctoral Research Assoc., Dept. of Civ. Engrg., Iowa State Univ., Ames, Iowa), **E. Robert Baumann**, Fellow, ASCE, (Anston Marston Prof. of Engrg., Iowa State Univ., Ames, Iowa) and **William L. Berk**, (Vice Pres., Lakeside Equipment Corp., Bartlett, Ill.)

Journal of Environmental Engineering, Vol. 111, No. 1, February, 1985, pp. 79-91

Oxygen transfer tests were conducted on surface rotor aerators installed in a plant scale oxidation ditch. The effects of rotor rotational speed on the standard oxygen transfer rate, the net power input, and the standard aeration efficiency are presented. Initial nonuniform distribution of deoxygenating chemical in a circulating tank can result in poorly behaved data. The circulation pattern in an oxidation ditch is accurately modeled by the first order oxygen transfer rate equations applied to a completely mixed batch reactor. Oxygen transfer rates estimated by the nonlinear regression and best fit log deficit methods of data analysis vary less than $\pm$ 10%, and the results of the two analytical methods are identical for practical purposes. A rotor cover which restricts the spray pattern of the rotor can result in increased oxygen transfer rates at the expense of an increase in power and a decrease in ditch flow velocity.

55   Residence Time in Stormwater Detention Basins

Stephen J. Nix, Assoc. Member, ASCE, (Asst. Prof., Dept. Civ. Engrg., Syracuse Univ., Syracuse, N.Y. 13210)

Journal of Environmental Engineering, Vol. 111, No. 1, February, 1985, pp. 95-100

Residence time has been the traditional indicator of the ability of primary and secondary clarifiers in wastewater treatment plants to remove pollutants from the waste stream. The steady-state definition of residence time used in these applications has also been applied to stormwater detention basins in which the flows are unsteady. A computer simulator is used to show that theoretical residence in basins receiving unsteady flows can deviate significantly from the steady state values.

56   **Deterministic Linear Programming Model for Acid Rain Abatement**

J. H. Ellis, (Asst. Prof., Dept. of Geography and Environmental Engrg., Johns Hopkins Univ., Baltimore, Md. 21218), **E. A. McBean**, (Prof., Dept. of Civ. Engrg., Univ. of Waterloo, Waterloo, Ontario, Canada N2L 3G1) and **G. J. Farquhar**, (Prof., Dept. of Civ. Engrg., Univ. of Waterloo, Waterloo, Ontario, Canada N2L 3G1)

Journal of Environmental Engineering, Vol. 111, No. 2, April, 1985, pp. 119-139

A deterministic Linear Programming Model is presented for development of acid rain abatement strategies in eastern North America. Pollutant (SO_2) sources are categorized as either controllable or noncontrolled, (i.e., area sources). The model determines the least-cost set of SO_2 removal levels at each of the 235 largest point sources in eastern North America, such that stipulated maximum wet sulfate deposition rates are not exceeded at 20 pre-determined sensitive receptor locations. Emphasis is placed on the model's flexibility with respect to input data requirements and its ability to conveniently and efficiently accommodate varying constraint formulations. Model applications are presented which involve five different constraint formulations.

57   **Land Treatment of Wastes: Concepts and General Design**

Raymond C. Loehr, Member, ASCE, (H.M. Alharty Centennial Prof. of Civ. Engrg., Univ of Texas, Austin, Tex. 78712) and **Michael R. Overcash**, (Prof. Chemical Engrg., North Carolina State Univ., Raleigh, N.C. 27650)

Journal of Environmental Engineering, Vol. 111, No. 2, April, 1985, pp. 141-160

Land treatment is a managed treatment and ultimate disposal process that involves the controlled application of a municipal or industrial waste to a soil or soil-vegetation system. This technology is based upon scientific and engineering principles and field experience. This paper describes the limiting constituent concept that is the basis of successful design and operation of land treatment systems, identifies how the concept can help determine pretreatment, monitoring, and feasibility decisions, and uses four case studies to illustrate the applicability of the technology to industrial wastes and to sludges.

58   **Metal Distribution and Contamination in Sediments**

Kevin D. White, (Research Asst., Dept. of Civ. Engrg., Virginia Polytechnic Inst. and State Univ., Blacksburg, Va.) and **Marty E. Tittlebaum**, (Assoc. Prof., Dept. of Civ. Engrg., Louisiana State Univ., Baton Rouge, La.)

Journal of Environmental Engineering, Vol. 111, No. 2, April, 1985, pp. 161-175

Heavy metals analysis was performed on the sediments from various south Louisiana waterways. Objectives of the study were to determine vertical heavy metal distributions, to evaluate a proposed statistical method used to make qualitative assessments of cultural contamination based on trace metal-conservative metal relationships, and to attempt to determine heavy metal contamination in sampled sediments. Cores were obtained from three distinct areas in south Louisiana. Area 1 consisted of waterways located in heavily industrialized areas. Area 2 sample consisted of a 110 m deep core obtained just off the present Mississippi River delta. Area 3 samples were obtained from a shallow urban lake. The vertical metal concentration distribution of the deep core revealed very consistent concentrations for metals throughout the entire depth. The relative atomic variation (RAV) method, used to reduce the grain size effects on heavy metal accumulations in sediments by correlating element pairs, was evaluated. The evaluation of the method in this study uncovered serious limitations. Metal contamination was determined by comparing both absolute concentrations and trace metal-conservative metal concentration ratio values to their respective deep core background values.

59 **Optimal Expansion of Water Distribution Systems**

Pramod R. Bhave, (Prof. and Head, Dept. of Civ. Engrg., Visvesvaraya Regional Coll. of Engrg., Nagpur 440 011, India)

Journal of Environmental Engineering, Vol. 111, No. 2, April, 1985, pp. 177-197

A method for the optimal expansion of water distribution systems subjected to a single loading pattern is developed. The method optimally decides the pumping heads at new source nodes, if any, and also optimally selects the links which need strengthening to improve the delivery capability of the existing system. The method is iterative and converges rapidly to a local but fairly good optimal solution. Optimization of an entirely new distribution system is a particular characteristic of this general approach. Examples of the method are given, including its application to the expansion of the New York City water supply tunnel system. The possible extension of the method to time-varying demand patterns is also described.

60 **Horizontal Elutriator Preliminary Test Results under Nonstill Air Conditions**

Dennis D. Lane, Assoc. Member, ASCE, (Assoc. Prof. of Civ. Engrg., Univ. of Kansas, Lawrence, Kans.), **Terry E. Baxter**, Assoc. Member, ASCE, (Research Asst., Univ. of Kansas, Lawrence, Kans.), **Thomas Cuscino**, (Sr. Environmental Engr., Midwest Research Inst., Kansas City, Mo.) and **Chatten Cowherd, Jr.**, (Section Head, Midwest Research Inst., Kansas City, Mo.)

Journal of Environmental Engineering, Vol. 111, No. 2, April, 1985, pp. 198-208

Recently, a high-volume horizontal elutriator with a flared inlet was developed to monitor inhalable particulates in the ambient air. Preliminary results of a study to evaluate the performance of the horizontal elutriator under nonstill air conditions are presented. Collection efficiency data were obtained for monodisperse aerosols with particle sizes of 4.0, 9.25, 20.0, and 30.7 um aerodynamic diam. Experimental results indicate that the device does not meet the established performance criteria. The main cause of nonperformance is prrobably a combination of exterior turbulent propagation into the initial part of the elutriator, nonisokinetic sampling conditions and particle diffusion.

61 **Calculating Effectiveness of Water Conservation Measures**

Thomas M. Walski, Member, ASCE, (Research Civ. Engr., U.S. Army Corps of Engineers Waterways Experiment Station, Vicksburg, Miss.), **William G. Richards**, (Project Manager, Roy F. Weston, Inc., West Chester, Pa.), **Deborah J. McCall**, (Asst. Project Engr., Roy F. Weston, Inc., West Chester, Pa.), **Arun K. Deb**, Member, ASCE, (Vice Pres., Roy F. Weston, Inc., West Chester, Pa.) and **Joe Miller Morgan**, Member, ASCE, (Prof., Civ. Engrg. Dept., Auburn Univ., Auburn, Ala.)

Journal of Environmental Engineering, Vol. 111, No. 2, April, 1985, pp. 209-221

Implementation of water conservation programs can result in significant savings to water utilities. Engineers and planners working with utilities need to be able to predict the reduction in water requirements (i.e., effectiveness of water conservation) that will result from implementation of water conservation measures. A procedure that was developed to help predict water conservation effectiveness for typical conditions is presented.

62   Nomogram for Velocities of Partial Flows in Sewers

B. R. N. Gupta, (Prof., Faculty of Civ. Engrg., Coll. of Military Engrg., Pune-411 031, Maharashtra State, India)

Journal of Environmental Engineering, Vol. 111, No. 2, April, 1985, pp. 225-230

For the design of sewers nomograms are available in the literature for various flow equations. These nomograms can be used only for the condition when the depth of flow is full diameter of the sewer. The sewers are invariably designed to flow only for partial depths. For design of sewers involving determination of velocities and discharges under partial flow conditions another graph, depicting the relationship between ratios of depths of flow and ratios of discharge and velocity when flowing partially and those when flowing full, is required in conjunction with the above nomograms. The conventional method for solving a problem under partial flow conditions of a sewer involves six steps consisting of arithmetical calculations and reference to more than one chart or graph. The time and effort required for a designer are thus considerable. A nomogram is developed to eliminate all the tedious procedures involved in the conventional method. The required parameters can be directly read on the nomogram with reasonable accuracies for flows between 0.25 and 0.75 diameters.

63   Modified Approach to Evaluate Column Test Data

Sunil Agrawal, (Research Asst., Dept. of Civ. Engrg., Univ. of Windsor, Windsor, Ontario, Canada) and **Jatinder K. Bewtra**, Member, ASCE, (Prof., Dept. of Civ. Engrg., Univ. of Windsor, Windsor, Ontario, Canada)

Journal of Environmental Engineering, Vol. 111, No. 2, April, 1985, pp. 231-234

In the conventional method of calculating the fractional removal of flocculent particles, it is assumed that the average settling velocity for different particle sizes in the lower depths in settling basins is the same as in the upper depths. This assumption is not quite correct, because the particles continuously flocculate or shear during the entire settling depth, and hence the average settling velocity changes both with depth and time. In the modified method, a different approach for calculating the fractional removal of particles is proposed to overcome the deficiency in the conventional method. Instead of considering a vertical constant time line, a horizontal constant depth line for that particular time is used. This approach is more reasonable because the settling velocities of different flocculating particles are all averaged over the same depth.

64   Specific Resistance Measurements: Nonparabolic Data

G. Lee Christensen, Member, ASCE, (Prof., Civ. Engrg. Dept., Villanova Univ., Villanova, Pa. 19085) and **Richard I. Dick**, Member, ASCE, (Joseph P. Ripley Prof. of Engrg., School of Civ. and Environmental Engrg., Cornell Univ., Ithaca, N.Y. 14853)

Journal of Environmental Engineering, Vol. 111, No. 3, June, 1985, pp. 243-257

A computer-aied data acquisition system was developed for the generation of slurry filtration data. The system has some advantages over the traditional Buchner funnel apparatus, including: (1) Operation at pressure differences up to 200 psi (1.4 MPa); (2) a more desirable aspect ratio for the filtration vessel; (3) the ability to record filtrate volumes at intervals of 1 s or less; (4) improved accuracy and precision; (5) avoidance of the need for constant attention to data collection during filtration experiments; and (6) convenient computerized manipulation of data. The data acquisition system was used to investigate nonparabolic filtration behavior. The results show that nonparabolic filtration data are associated with filtration runs accompanied by sedimentation prior to filtration, sedimentation concomitant with filtration, or slurry-media interactions. In many instances, nonparabolic filtrations can be redone to produce parabolic filtration data by applying pressure immediately following the introduction of the slurry into the filtration vessel, thickening the slurry prior to filtration to reduce the significance of sedimentation with respect to filtration, and choosing a different support medium.

65   Specific Resistance Measurements: Methods and Procedures

G. Lee Christensen, Member, ASCE, (Prof., Civ. Engrg. Dept., Villanova Univ., Villanova, Pa. 19085) and Richard I. Dick, Member, ASCE, (Joseph P. Ripley Prof. of Engrg., School of Civ. and Environmental Engrg., Cornell Univ., Ithaca, N.Y. 14853)

Journal of Environmental Engineering, Vol. 111, No. 3, June, 1985, pp. 258-271

A computer-aided data acquisition system was used with three sludges, two chemical-conditioning systems and nine support media to investigate the effect of slurry solids concentration on specific resistance, the effect of initial time and filtrate volume readings on filtration data, and factors affecting the precision and accuracy of specific resistance measurements. The results of the study show that: (1) The specific resistance of a flocculent slurry is a very strong function of suspended solids concentration at low concentrations and relatively independent of concentration at high values; (2) improper determinations of the onset of filtration or errors in filtrate volume measurements produce slurry filtration data that do not fit the standard model (nonsynchronous data); and (3) specific resistance values are not as accurate and translatable as they might be because of differences in methods used in different laboratories. Several suggestions for improving the translatability of specific resistance measurements are presented.

66   Flux Use for Calibrating and Validating Models

Joseph H. Wlosinski, (Aquatic Ecologist, Environmental Lab., U.S. Army Engr. Waterways Experiment Station, Vicksburg, Miss. 39180)

Journal of Environmental Engineering, Vol. 111, No. 3, June, 1985, pp. 272-284

Predicted flux, or process rates, must be considered along with predicted state variables when calibrating or validating water quality models. Concentrating only on changes of state variables and ignoring predicted fluxes could result in the implementation of inappropriate management strategies. To illustrate this, the Corps of Engineers' one-dimensional reservoir water quality model, CE-QUAL-R1, was calibrated using data collected at DeGray Lake, Arkansas. Initial predictions of dissolved oxygen, using coefficient values obtained from the literature, were not satisfactory. Evaluation of the flux predictions for oxygen showed which processes were most important in regulating dissolved oxygen, and showed those that were incorrect when compared to measured values. During the ensuing calibration simulations, as the flux predictions were brought more in line with measured values, the predictions of dissolved oxygen improved markedly.

67   Flow under Tilt Surface for High-Rate Settling

Sadataka Shiba, Member, ASCE, (Research Assoc., Faculty of Engrg. Sci., Dept. of Chem. Engrg., Osaka Univ., Toyonaka, Osaka, Japan 560)

Journal of Environmental Engineering, Vol. 111, No. 3, June, 1985, pp. 285-303

A clear layer flow rising under inclined surfaces (particle-free flow stratified on the suspension), which can be utilized for particle settling enhancement, has been theoretically analyzed in order to apply it to design and operation of high-rate settlers in water and wastewater treatments. The flow proves to be driven by settling convection due to the density difference between the clear liquid and the suspension. A mathematical model for the flow simulation has been obtained by making use of the boundary layer approximation and the clear-layer-averaged inertia terms in the momentum equations. A dimensionless number, Γ, which controls the model and is given by the ratio of Grashof number, Gr, to Reynolds number, R, plays an important role in the flow. The flow velocity and the layer thickness predicted by the analytical solutions agree fairly well with previous experimental results observed in a batch settler to verify the analysis.

68   Expression for Drinking Water Supply Standards

Devendra Swaroop Bhargava, Fellow, ASCE, (Reader in Pollution Control, Div. of Environmental Engrg., Dept. of Civ. Engrg., Univ. of Roorkee, Roorkeee, U.P., India 247667)

Journal of Environmental Engineering, Vol. 111, No. 3, June, 1985, pp. 304-316

Various authorities and regulating agencies have set standards for deciding the suitability of a water for drinking purposes. These standards prescribe the permissible concentrations of quality variables. When some variables exceed the permissible levels, a decision for permitting further use of the water supply has to be based on the importance of those variables with exceeded concentrations. It is proposed that standards for a drinking water supply should be set through a single number representing the integrated effect of all the variables, keeping due regard to the importance of each variable. Such an integrated water quality index (WQI) would help in decision making. Models and curves have been presented to evolve a WQI for drinking water supplies. It is suggested that water with a WQI lower than 90 should not be permitted. The acceptable quality, therefore, should be in the 90-100 range of the WQI.

69   Calcium Sulfate Solubility in Organic-Laden Wastewater

Iris Banz, (Engr., Risk Assessment Tech., Westinghouse Electric Corp., Pittsburgh, Pa. 15230) and **Richard G. Luthy**, Member, ASCE, (Prof., Dept. of Civ. Engrg., Carnegie-Mellon Univ., Pittsburgh, Pa. 15213)

Journal of Environmental Engineering, Vol. 111, No. 3, June, 1985, pp. 317-335

Calcium sulfate solubility product and ion pair dissociation constant were measured in clean water, and these results were employed in tests with a pretreated coal conversion process wastewater to assess the tendency for organic matter in the wastewater to function as a complexing agent for calcium. It was demonstrated that wastewater organic matter interacted with calcium to form a calcium-organic complex. The extent of this interaction in wastewater was as significant as that for formation of the $CaSO_4°$ ion pair in assessing solubility of $CaSO_4$. It was shown that the organic matter complexed with calcium to an extent comparable to humic acid, and that the complexing strength was similar to that predicted for citrate when compared on an equivalent COD or TOC basis. The results of this study are important for evaluating $CaSO_4$ scale-forming reactions if wastewater is to be reused as makeup water to an evaporative cooling tower.

70   Stream Dissolved Oxygen Analysis and Control

David A. Todd, (Research Sci., Environmental Sci. and Engrg., Rice Univ., Houston, Tex. 77251) and **Philip B. Bedient**, Assoc. Member, ASCE, (Assoc. Prof., Environmental Sci. and Engrg., Rice Univ., Houston, Tex. 77251)

Journal of Environmental Engineering, Vol. 111, No. 3, June, 1985, pp. 336-352

Buffalo Bayou, a major stream in Houston, Texas, was the subject of a detailed water quality study. The stream is heavily stressed by domestic waste loads and has shown significant degradation in water quality since 1970. Currently, the stream reaches critically low DO values through much of its length during low flow, high temperature periods. The QUAL-II(TX) model was calibrated, verified, and applied to simulate worst case conditions and to test a series of stream protection alternatives. A combined program of sewage plant regionalization, stream aeration, and flow augmentation was found to be tthe most cost-effective solution.

71   Biofilm Growths with Sucrose as Substrate

Ju-Chang Huang, Member, ASCE, (Prof. of Civ. Engrg. and Dir., Environmental Research Center, Univ. of Missouri, Rolla, Mo.), **Shoou-Yuh Chang**, Assoc. Member, ASCE, (Asst. Prof. of Civ. Engrg., Univ. of Missouri, Rolla, Mo.), **Yow-Chyun Liu**, (Grad. Research Asst., Dept. of Civ. Engrg., Univ. of Missouri, Rolla, Mo.) and **Zhang-Peng Jiang**, (Visiting Scholar in Environmental Research Center, Univ. of Missouri, Rolla, Mo.)

Journal of Environmental Engineering, Vol. 111, No. 3, June, 1985, pp. 353-363

This study was conducted to: 1) evaluate the effect of DO on cell yield in a fixed film reactor using 1,000 mg/l sucrose as a substrate, 2) evaluate the correlations of the biofilm thickness and density with DO and their resultant substrate stabilization rates, and 3) examine the response of biofilm communities as a result of DO and biofilm thickness changes. Data obtained from this study indicate that DO has only a minor effect on the cell yield. However, the thickness of aerobic biofilm is definitely related to DO, or thickness (mm) = $(2.08 \times DO)/(.2 + DO)$. The biofilm density is also related to its thickness. At a DO of 5 mg/l or lower, the biofilm texture is firm and has a wet density of 27 – 48 mg/cu cm. At a higher DO (5 – 16 mg/l), the biofilm becomes porous and filled with air pockets, with its density being reduced to 25 mg/cu cm. The biological community in biofilm at a high DO environment (16 mg/l) is predominantly short rods grouped in a chain structure. At a low DO environment (0.5 mg/l), however, the prevalent forms are large rods, none of which are in chain grouping.

72   Effect of Ambient Air Quality on Throughfall Acidity

Carl W. Chen, Member, ASCE, (Prin., Systech Engrg., Inc., 3744 Mt. Diablo Blvd., Suite 101, Lafayette, Calif. 94549), **Steven A. Gherini**, (Tetra Tech, Inc., 3746 Mt. Diablo Blvd., Suite 300, Lafayette, Calif. 94549), **Robert A. Goldstein**, (Electric Power Research Inst., Inc., Palo Alto, Calif.) and **Nicholas L. Clesceri**, Member, ASCE, (Rensselear Polytechnic Inst., Troy, N.Y.)

Journal of Environmental Engineering, Vol. 111, No. 3, June, 1985, pp. 361-372

Observations at Woods Lake-watershed in the Adirondacks (New York) indicate that precipitation is further acidified by passage through coniferous canpoies; conversely, passage through deciduous canopies has a net alkalizing effect. Both effects are dependent upon the levels of air quality. If the aerosol concentrations were to decrease to 35% of the current levels, both types of canopy would have a net alkalizing effect on incident precipitation. If the aerosol concentrations doubled, both canopy types would be net acidifiers of throughfall. The experimental procedure of varying aerosol concentration in an enclosed chamber may provide a means for measuring foliar exudation.

73   **Multi-Objective Decision-Making in Waste Disposal Planning**

Robert D. Perlack, (Research Staff., Energy Div., Oak Ridge National Lab., P.O. Box X, Oak Ridge, Tenn. 37831) and **Cleve E. Willis**, (Prof., Dept. of Agricultural and Research Economics, Univ. of Massachusetts, Amherst, Mass. 01003)

Journal of Environmental Engineering, Vol. 111, No. 3, June, 1985, pp. 373-385

A multiple objective programming model of the Boston sludge disposal problem was formulated in which the objectives of net economic benefits, environmental impact, and variability of impacts were included. A generating technique was used to solve for the noninferior solutions. To reduce information overloading on the part of decision-makers, and redundancy among solutions, cluster analysis was used to prune the noninferior set of solutions.

74   **Simultaneous In-Stream Nitrogen and D.O. Balancing**

John J. Warwick, Member, ASCE, (Asst. Prof. of Environmental Sci., Univ. of Texas at Dallas, Grad. Program in Environmental Sci., Richardson, Tex. 75083-0688) and **Archie J. McDonnell**, Member, ASCE, (Dir. of Land and Water Research Inst., Pennsylvania State Univ., University Park, Pa.)

Journal of Environmental Engineering, Vol. 111, No. 4, August, 1985, pp. 401-416

A one-dimensional, pseudo unsteady state, water quality model (SNOAP) was developed to simultaneously solve both nitrogen and dissolved oxygen mass balance equations. Expressions are derived to estimate ammonia-N exsolution, bacterial, ammonia-N uptake, and aquatic plant ammonia-N assimilation in terms of standard water quality indices. The SNOAP model calculates hourly values for all nitrogen species reaction rates (including nitrification and denitrification) based on observed in-stream variation of organic-N, ammonia-N, nitrite-N, nitrate-N, dissolved oxygen, temperature, and pH. An iterative solution algorithm is required to simultaneously solve the nitrogen and dissolved oxygen mass balance equations. This iterative approach is stable and results in relatively quick convergence. Erroneous values of observed nitrogen species or dissolved oxygen variation, or both, may result in an impossible situation for simultaneous nitrogen and dissolved oxygen mass balance solution. This non-convergence feature is a significant attribute, in that it allows for a systematic screening of measured water quality data.

75   **Nitrogen Accountability for Fertile Streams**

John J. Warwick, Member, ASCE, (Asst. Prof. of Environmental Sci., Univ. of Texas at Dallas, Grad. Program in Environmental Sci., Richardson, Tex. 75083-0688) and **Archie J. McDonnell**, Member, ASCE, (Dir. of Land and Water Research Inst., Pennsylvania State Univ., University Park, Pa.)

Journal of Environmental Engineering, Vol. 111, No. 4, August, 1985, pp. 417-430

The Stream Nitrogen and Oxygen Analysis Program (SNOAP) is a one-dimensional, pseudo unsteady state, water quality model, which simultaneously solves both nitrogen and dissolved oxygen mass balances equations. The SNOAP model was successfully calibrated and verified for three, independent, data intensive water quality surveys. The modeled stream system was unsteady due to significant upstream impact from wastewater treatment plant effluent discharges. Denitrification was found to be the dominant total nitrogen sink. Average denitrification rates calculated by the SNOAP model range from 22.0–43.6 mg-N/m^2/hr. The computed values compare quite favorably with measured laboratory sediment denitrification rates which ranged from 28.2–39.2 mg-N/m^2/hr. In addition, the rate of ammonia-N decay was found to be the most accurate nitrogen species estimator of in-stream nitrification rates.

76   **Ground-Water Protection in San Francisco Bay Area**

Don M. Eisenberg, (Sr. Engr., California Regional Water Quality Control Board, S. F. Bay Region, 1111 Jackson St., Room 6040, Oakland, Calif., 94607), **Adam W. Olivieri**, (Sr. Engr., California Regional Water Quality Control Board, S. F. Bay Region, 1111 Jackson St., Room 6040, Oakland, Calif., 94607), **Martin R. Kurtovich**, (Assoc. Engr., California Regional Water Quality Control Board, S. F. Bay Region, 1111 Jackson St., Room 6040, Oakland, Calif. 94607), **Peter Johnson**, (Assoc. Engr., California Regional Water Quality Control Board, S. F. Bay Region, 1111 Jackson St., Room 6040, Oakland, Calif. 94607) and **Lori Pettegrew**, (Environmental Specialist, Aqua Terra Resources, Oakland Calif.)

Journal of Environmental Engineering, Vol. 111, No. 4, August, 1985, pp. 431-440

In the San Francisco Bay Region, the Regional Water Quality Control Board is now regulating approximately 140 sites where toxic chemicals have contaminated ground water. Many of these were located through the Board's Underground Tank Leak Detection Program, which required subsurface investigations at sites with underground solvent tanks. New ordinances and state laws are expected to result in the discovery of hundreds more contamination sites in the next few years. The Regional Board is developing a ranking methodology to assign priorities among cases and a set of guidelines for establishment of clean-up objectives on a site-by-site basis. Appropriate regulatory mechanisms have been selected to enforce the necessary investigation and clean-up activities.

77   **Hazardous Waste Disposal as Concrete Admixture**

Ronald E. Benson, Jr., Assoc. Member, ASCE, (Asst. Prof. of Civ. Engrg., The Citadel, Military College of South Carolina, Charleston, S. C.), **Henry W. Chandler**, Assoc. Member, ASCE, (Materials Engr., Law Engrg. Testing Co., Greenville, S. C.) and **Kenneth A. Chacey**, (Environmental Engr., Southern Div. Naval Facilities Engrg. Command, Charleston, S. C.)

Journal of Environmental Engineering, Vol. 111, No. 4, August, 1985, pp. 441-447

Sand blasting residue containing cadmium and lead was used as an admixture for concrete. Concrete mixes with up to 15% waste added were found to retain the heavy metals such that none was detected following the EP Toxicity extraction procedure. Degradation of concrete strength as a function of amount of waste material added was observed. Results were correlated to determine the amount of additional cement needed to add along with the waste to maintain design strength.

78   **Attenuation Versus Transparency**

Steve W. Effler, (Engr., Upstate Freshwter Inst., Inc., Box 506, Syracuse, N. Y. 13214)

Journal of Environmental Engineering, Vol. 111, No. 4, August, 1985, pp. 448-459

The differential response of the two commonly used measures of light penetration in water, Secchi disc transparency (SD, m) and the attenuation coefficient for downwelling irradiance (K_d, m^{-1}), to the light attenuating processes of absorption and scattering is described. It is demonstrated that the commonly invoked constancy in the product $K_d \cdot SD$ should only be expected when the relative contributions of scattering and absorption to K_d remains uniform. A number of scenarios and documented phenomena are presented for which preferential effects on scattering or absorption, and therefore variable values of $K_d \cdot SD$, are expected. Measurement errors also contribute to variability in $K\partial d \cdot SD$, particularly at low (<1.5m) SD values. Paired observations of $K\natural_d$ and SD presented from four different systems, representing a wide range of K_d and SD values, demonstrate the variability of the magnitude of $K_d \cdot SD$ within individual systems. Secchi disc transparency is probably not a reliable indicator of K_d in many systems. It is recommended that the uncertainty associated with invoking the constancy of $K_d \cdot SD$ in various ecological modeling efforts, in which K_d measurements are missing or are temporally limited, be accommodated.

79   Performance of Expanded-Bed Mathenogenic Reactor

Yi-Tin Wang, Assoc. Member, ASCE, (Research Assoc., Dept. of Civ. Engrg., Univ. of Illinois at Urbana-Champaign, 208 N. Romine St. Urbana, Ill.) and **Makram T. Suidan**, Assoc. Member, ASCE, (Assoc. Prof., Dept. of Civ. Engrg., Univ. of Illinois at Urbana-Champaign, 208 N. Romine St., Urbana, Ill.)

Journal of Environmental Engineering, Vol. 111, No. 4, August, 1985, pp. 460-471

A completely mixed, expanded-bed, anaerobic granular activated carbon reactor was operated continuously on synthetic wastewaters in which acetate was the only organic carbon source. Steady-state performance was achieved for four influent acetate concentrations namely: 800, 1,600, 3,200, and 6,400 mg. Steady-state removal efficiencies of acetate, COD, and DOC exceeded 98, 97 and 98%, respectively. Biological utilization was the major removal mechanism for acetate in the anaerobic filter. This process was demonstrated to be capable of purifying low strength wastewaters down to levels that are acceptable for final discharge.

80   Outfall Dilution: The Role of a Far-Field Model

David A. Chin, Assoc. Member, ASCE, (Asst. Prof., Dept. of Civ. and Architectural Engrg., Univ. of Miami, Coral Gables, Fla. 33124)

Journal of Environmental Engineering, Vol. 111, No. 4, August, 1985, pp. 473-486

Effluent dilutions predicted by near-field models do not realistically reflect near-field conditions in tidally influenced coastal waters since they neglect the advection of the pollutant cloud over the source. A theoretical solution has been developed for the case in which a conservative pollutant is released from a line source into a uniform unsteady flowing ambient where diffusion is only significant in the cross-stream direction. Considering the case of a sinusoidal variation of velocity along the steamwise principal axis, it has been shown that the expected concentration at a source located in a typical flow field may increase by as much as 15% per cycle over that predicted by the near-field model. A hypothetical outfall was analyzed using real data and a dispersion model which accounts for the near-, as well as far-field mixing. The near-field model was found to underestimate the average dilution at the source by about 21%. Extreme values were also underestimated. The results of this study indicate that a complete dispersion model should be used in predicting dilutions in outfall mixing zones of coastal regions.

81   Adsorption of Organic Vapors on Carbon and Resin

K. E. Noll, Member, ASCE, (Prof., Dept. of Environmental Engrg., Illinois Inst. of Technology, Chicago, Ill. 60616) and **A. A. Aguwa**, (Environmental Engr., PRC Engineering, Chicago, Ill. 60601)

Journal of Environmental Engineering, Vol. 111, No. 4, August, 1985, pp. 487-500

A comparative study was conducted on the adsorption/desorption of organic vapors (toluene and para-xylene) on a synthetic resin (XAD4) and activated carbon using a gravimetric method involving the use of a quartz spring expansion. While the two sorbents can effectively remove the organic vapors, it was observed that activated carbon adsorbs more organic vapor than synthetic resin at low concentrations; but at higher industrial level concentrations, the resin adsorbs more organic vapor. The rate of adsorption is higher than that of desorption for the two organic vapors regardless of the sorbent. However, the resin showed higher reversible adsorption. The effective intraparticle diffusion coefficients (D_e) were observed to be strongly dependent on the solute concentration. As such, rate studies ought to be conducted around the expected concentration range that is expected during application. Pore diffusion dominated the adsorption/desorption of both organic vapors on the XAD4 resin. For the carbon system, pore diffusion dominated the adsorption but surface diffusion contributed to the desorption process. This is believed to be higher interaction of the solutes with activated carbon.

82   Conditioning of Anaerobically Digested Sludge

Anne I. Cole, (Environmental Engr., Camp Dresser & McKee Inc., Raleigh, N. C.) and **Philip C. Singer**, (Prof., Dept. of Environmental Sci. and Engrg. School of Public Health, Univ. of North Carolina, Chapel Hill, N. C.)

Journal of Environmental Engineering, Vol. 111, No. 4, August, 1985, pp. 501-510

Synthetic organic polyelectrolytes can be used to condition sludges to enhance their dewaterability. When conditioning biological sludges, the charge on the polymer has a significant impact on the effectiveness of the polymer as a conditioner. The objectives of this investigation were to determine the most effective type of polymer product for conditioning anaerobically digested sludge prior to dewatering, and to investigate how the chemical characteristics of the polymer influence the way it interacts with the sludge particles. Capillary suction time and particle electrophoretic mobility measurements were employed to achieve this objective. The results indicate that cationic polymers with molecular weights in excess of 10^6 appear to be required for effective conditioning. However, effective sludge conditioning can be achieved with either high or low change density cationic polymers and that dosing to achieve charge neutralization of the sludge particles is not a prerequisite for effective sludge conditioning.

83   Removal of Radon from Water Supplies

Jerry D. Lowry, Member, ASCE, (Assoc. Prof., Dept. of Civ. Engrg., Univ. of Maine, Orono, Me. 04469) and **Jeffrey E. Brandow**, (Environmental Engr., New York Dept. of Environmental Conservation, Albany, N. Y.)

Journal of Environmental Engineering, Vol. 111, No. 4, August, 1985, pp. 511-527

Granular activated carbon (GAC) adsorption is extremely effective for the removal of radon from water supplies. A high removal efficiency is achieved through an adsorption/decay steady state that can result in near-background radon levels with currently available commercial GAC units. An adsorption/decay steady state analysis based upon secular equilibrium shows that the life of the GAC will be in terms of decades with respect to radon. A variety of GAC products were evaluated for radon removal and laboratory and field data are presented to document the steady state performance for water supplies ranging in concentration from 1,500–750,000 pCi/L. Because radon and its daughters build to steady state levels corresponding to the influent mass loading, the GAC bed becomes a source of low level gamma radiation. An empirical relationship is presented to estimate the maximum activity of the bed as a function of the influent radon concentration. This relation and other field measurements show that the low level activity of a GAC bed does not present a problem for the vast majority of high radon wells.

84   Design Considerations for a Novel Landfill Liner

Alexander C. Demetracopoulos, Assoc. Member, ASCE, (Asst. Prof., Dept. of Civ. and Environmental Engrg., Rutgers Univ., Piscataway, N. J. 08854) and **Lily Sehayek**, (Instr., Dept. of Civ. and Environmental Engrg., Rutgers Univ., Piscataway, N. J. 08854)

Journal of Environmental Engineering, Vol. 111, No. 4, August, 1985, pp. 528-539

A novel landfill leachate liner design consisting of a series of V-shaped contiguous elements constructed from an impervious membrane, a boxed soil layer saturated with clean water, and a gravel layer is proposed. Certain idealized boundary conditions are stipulated and an analytical solution is given for the diffusional mass transport of leachate through the saturated layer. This is used to assess the maximum value and time-of-occurrence of leachate concentration at the impervious membrane. It is shown that these two important quantities are sensitive to the diffusion coefficient and the thickness of the saturated layer. A method is formulated for the estimation of the diffusion coefficient, accounting for the type of contaminant, the effect of the porous medium, and the possibility of contaminant adsorption on the soil matrix. The overall procedure would be valuable in estimating the potential of destruction of the impervious membrane from the chemicals contained in the leachate.

85   Revegetation Using Coal Ash Mixtures

Charles N. Haas, (Assoc. Prof., Pritzker Dept. of Environmental Engrg., Illinois Inst. of Tech., Chicago, Ill.) and Joseph J. Macak, III., (Prin. Environmental Engr., Commonwealth Edison Co., Chicago, Ill. 60603)

Journal of Environmental Engineering, Vol. 111, No. 5, October, 1985, pp. 559-573

To ascertain whether bottom ash, or mixtures of fly ash and bottom ash from power generation could serve as acceptable final covers for landfills,field revegetation studies and indoor plant elemental bioavailability studies were conducted. It was found that bottom ash alone, or mixed with as much as 10% fly ash performs as well as topsoil in serving as a final cover, in terms of revegetation potential. Based on bioavailability studies, the inhibitory effects of higher proportions of fly ash may be due to boron toxicity.

86   Phosphorus Reduction for Control of Algae

G. K. Young, (Pres., GKY & Associates, Inc., 5411-E Backlick Rd., Springfield, Va. 22151) and K. G. Saunders, (Research Assoc., GKY & Associates, Inc., 5411-E Backlick Rd., Springfield, Va. 22151)

Journal of Environmental Engineering, Vol. 111, No. 5, October, 1985, pp. 574-588

Using time series data of water quality, streamflow records, and mathematical modeling sensitivity results, an analysis of elasticity factors influencing algal water quality is presented. A case study evaluates total regional and local water quality impacts associated with waste treatment; the site is the tidewater Potomac River. Estimates of the partial derivatives of chlorophyll concentrations, with respect to influencing factors, are combined using the total differential to find output elasticities. The elasticities allow a ranking of important factors influencing advanced waste treatment decisions. The method can be transferred to other sites; herein, a local provider's viewpoint is peresented. Results of the analysis indicate that the benefits of phosphorus control are small relative to the effort required to achieve such benefits measured in terms of reducing chlorophyll-α levels. Recognizing the dominant effect of summer streamflow, the significant uncontrollable sources of phosphorus and the unaccounted for factors influencing water quality, reevaluation of highly restrictive phosphorus control is suggested.

87   Pumped Wastewater Collection Systems Optimization

Vijay S. Kulkarni, (Sr. Research Asst., Environmental Sci. and Engrg. Group, Indian Inst. of Tech., Powai, Bombay 400 076, India) and P. Khanna, (Prof. of Environmental Engrg., Indian Inst., of Tech., Powai, Bombay 400 076, India)

Journal of Environmental Engineering, Vol. 111, No. 5, October, 1985, pp. 589-601

Identifying the need for intermediate and/or end pumping in real life wastewater collection systems, an optimization algorithm is developed for the design of gravity-cum-pumped systems with recourse to dynamic programming. The problem of dimensionality is minimized through cost effective feasible grouping at junction manholes and with division of algorithm in two parts. The first part identifies optimal control variables associated with each link and stores the same while the second part uses these stored values along with input data to prepare detailed hydraulic and cost statements. The effectiveness of the algorithm is tested through two case studies.

88   Hazardous Waste Surface Impoundment Technology

Masood Ghassemi,, (Princ. and Tech. Dir., MEESA, 1651 Laraine Circle, San Pedro, Calif. 90732) and **Michael Haro**, (Proj. Engr., MEESA, 1651 Laraine Circle, San Pedro, Calif. 90732)

Journal of Environmental Engineering, Vol. 111, No. 5, October, 1985, pp. 602-617

The design, construction and performance data for hazardous waste surface impoundments (SIs) at nine facilities were reviewed, and actual and projected performances were compared. Discussions were also held with four design engineering firms, one waste management company, one liner installer/fabricator, and regulatory agencies in three states. The following were identified as being essential in achieving good site performance: (1) Siting in suitable geological formation; (2) continuity of the geotechnical support throughout the project planning, site investigation, design, and construction; (3) construction supervision to ensure adherence to specifications; (4) compaction of clay wet of optimum to eliminate air space; (5) consideration of compatbility with waste in selecting liner material; (6) rigorous QA/QC to ensure adequate design and proper liner installation; and (7) providing and maintaining protective cover for liners.

89   Model for Moving Media Reactor Performance

D. S. Bhargava, Fellow, ASCE, (Reader (Pollution Control), Div. of Environmental Engrg., Dept. of Civ. Engrg., Univ. of Roorkee, Roorkee, 247667, U.P., India) and **D. J. Bhatt**, (Asst. Prof. in Civ. Engrg., Engrg. Coll., Morbi, India)

Journal of Environmental Engineering, Vol. 111, No. 5, October, 1985, pp. 618-633

A predictive model for moving media reactor performance is presented adoping a dimensional analysis of parameters involved. It is demonstrated that the reduced equilibrium concentration is a function or sorbent-solute mass input rate ratio. A value of about 20,000 (for $C_o = 1$ mg/1),2,000 (for $C_o = 10$ mg/1) and 200 (for $C_o = 100$ mg/1), for sorbent-solute mass input rate ratio is observed to be optimum for the reactor operation. A correlation between the predicted and the observed values for the reduced equilibrium concentrations of sorbate exhibited a standard error ranging between 0.0004 and 0.017 and a coefficient of correlation between 0.86 to 0.99.

90   Unified Analysis of Biofilm Kinetics

Makram T. Suidan, (Assoc. Prof. of Environmental Engrg., Dept. of Civ. Engrg., Univ. of Illinois, Urbana, Ill. 61801) and **Yi-Tin Wang**, (Research Assoc., Dept. of Civ. Engrg., Univ. of Illinois, Urbana, Ill. 61801)

Journal of Environmental Engineering, Vol. 111, No. 5, October, 1985, pp. 634-646

A simplified algebraic expression for biofilm kinetics is developed. This mathematical model assumes Monod-type biological kinetics and diffusive mass transport. The algebraic expression related the concentration of subtrate at the biofilm surface, and the thickness of the biofilm to the substrate utilization rate. The algebraic relationship is particularly suited for use in the design of biofilm processes such as trickling filters, rotating biological contactors, anaerobic filters, and fluidized or expanded-bed reactors.

91   Dissolved Oxygen Model for a Dynamic Reservoir

Scott C. Martin, (Engr., AquaComp, Inc., 57 Elm St., Potsdam, N.Y. 13676) and Steven W. Effler, (Engr., AquaComp, Inc., 57 Elm St., Potsdam, N.Y. 13676)

Journal of Environmental Engineering, Vol. 111, No. 5, October, 1985, pp. 647-664

A time-variable one-dimensional mathematical model has been applied to Round Valley Reservoir, New Jersey, to test its credibility for simulating oxygen resources in the hypolimnia of dynamic reservoirs. In particular, the model was used to simulate the documented accelerated depletion of hypolimnetic dissolved oxygen (DO) that accompanied the use of the reservoir to augment flow in a nearby river. The model successfully simulated the reduction in hypolimnetic dissolved oxygen that occurred, and displayed potential as both a valuable research and management tool for the reservoir. For example, the model results suggested that the observed acceleration in DO depletion was mostly due to a decrease in the volume of the hypolimnion. Both a sensitivity analysis and a comparison of individual sources and sinks conducted with the model indicated that sediment oxygen demand (calibration value = 0.43 g$\cdot$m$^{-2}\cdot$d^{-1}) was the most important component of the oxygen budget in the hypolimnion of the reservoir.

92   Generating Designs for Wastewater Systems

Shoou-Yuh Chang, Assoc. Member, ASCE, (Asst. Prof., Dept. of Civ. Engrg., Univ. of Missouri-Rolla, Rolla, Mo. 65401) and Shu-Liang Liaw, (Grad. Student, Dept. of Civ. Engrg., Univ. of Missouri-Rolla, Rolla, Mo. 65401)

Journal of Environmental Engineering, Vol. 111, No. 5, October, 1985, pp. 665-679

Optimization models have been used in the preliminary design of wastewater treatment systems to obtain a least cost design. However, this may not produce the best design if other unmodeled issues and incomplete information of the cost and performance data of the treatment processes are considered. Optimization models may be more useful in the design of wastewater treatment systems, if they can be used to generate °good° but °different° design alternatives for evaluation. A study was conducted to examine two modeling-to-generate-alternatives (MGA) methods (the generating and screening, G&S, method and the efficient random generation, ERG, method) for the purpose of generating good and different preliminary designs for a typical wastewater treatment system design problem. The results showed that various attractive designs, which are good with respect to the objectives specified but widely different with respect to the treatment processes; could be generated by the two MGA methods. In comparing the alternatives produced by the two methods with those derived from the original opotimization model and a constraint method, it was found that the MGR methods were more effective for the development of various preliminary wastewater treatment system designs, which not only met minimal requirements but were also widely different from each other.

93   Phosphogypsum Waste Anion Removal by Soil Minerals

Louis C. Murray, Jr., (Grad. Student, Dept. of Civ. Engrg., Duke Univ., Durham, N.C. 27706) and Barbara-Ann G. Lewis, (Assoc. Prof., Dept. of Civ. Engrg., Northwestern Univ., Evanston, Ill. 60201)

Journal of Environmental Engineering, Vol. 111, No. 5, October, 1985, pp. 681-698

The removal of F-, PO_4^{-3}, and SO_4^{-2} from phosphogypsum leachate (pH 1.7) by soil minerals was investigated using a serial batch procedure. pH was the controlling factor in removal of orthophosphate and fluoride, but had less effect on sulfate. Calcium carbonate precipitated major fractions of fluoride and orthophosphate, and smaller amounts of sulfate, during the early stages of leaching; with continued leaching, redissolution of fluoride and orthophosphate occurred, resulting in release of these anions at concentrations exceeding those in the original phosphogypsum leachate. Removal of fluoride and orthophosphate by kaolinite and montmorillonite was strongy affected by pH, due to complex formation, degree of ionization, and charge on the clays. Iron (III) hydroxide had little effect on fluoride or orthophosphate removal due to formation of Ca-, Al-, and Fe-complexes which possibly interfered with adsorption. The iron hydroxide was important to sulfate removal at all pH values; the removal was markedly enhanced by the presence of calcium, presumably due to formation of calcium phosphate complexes that prevented orthophosphate competition with sulfate for adsorption sites.

94 Nitrification in Water Hyacinth Treatment Systems

A. Scott Weber, (Dept. of Civ. Engrg., State Univ. of New York at Buffalo, Buffalo, N.Y. 14260) and George Tchobanoglous, Member, ASCE, (Dept. of Civ. Engrg., Univ. of California, Davis, Calif.)

Journal of Environmental Engineering, Vol. 111, No. 5, October, 1985, pp. 699-713

Factors affecting the nitrification rate in post-secondary water hyacinth treatment systems are investigated. Parameters studied include ammonia concentration, dissolved oxygen concentration, system mixing, and light attenuation. The nitrification rate was found to be influenced by all of the foregoing parameters, and was most sensitive to dissolved oxygen concentration. Based on the results of this study, nitrification in full-scale water hyacinth treatment systems is rate-limited by dissolved oxygen concentration.

95 Water Division at Low-Level Waste Disposal Sites

Edward C. Davis, (Regional Staff Member, Environmental Sci. Div., Oak Ridge National Lab., P.O. Box X, Oak Ridge, Tenn. 37831) and Robert C. Stansfield, (Research Assoc.., Environmental Sci. Div., Oak Ridge National Lab., P.O. Box X, Oak Ridge, Tenn. 37831)

Journal of Environmental Engineering, Vol. 111, No. 5, October, 1985, pp. 714-729

Shallow depth to groundwater, surface drainage, and subsurface flow during storm events can cause major environmental problems in low-level radioactive waste management operations in humid regions. These three problems were encountered at two waste disposal sites on the Oak Ridge Reservation. In September 1983, two similarly designed, engineered drainage projects were initiated at the disposal sites. the SWSA 4 (solid waste storage area four) project was designed to divert surface runoff and shallow subsurface flow originating upslope of the site away from the disposal area. The second project, a passive French drain constructed in SWSA 6, was directed strictly toward suppressing the site water table, thus preventing its intersection with the bottoms of disposal trenches. Postconstruction monitoring for performance evaluation showed that the water table in the SWSA-6 area is suppressed to a depth >4.9 m below the ground surface over 50% of the site, as compared to a depth of only 2.1 m for certain parts of the same area observed during seasonally wet months. The SWSA 4 project evaluation indicates that 56% of the winter-spring 1984 runoff was diverted around SWSA 4 via the drainage system. As a result of the reduced flow in the SWSA 4 tributary to White Oak Creek, a 44% reduction in ^{90}Sr flux from SWSA 4 was calculated.

96   **A Critique of Camp and Stein's RMS Velocity Gradient**

Mark M. Clark, Assoc. Member, ASCE, (Consulting Scientist, Tetra Tech, Inc., 1911 Fort Meyer Drive, Suite 601, Arlington, VA 22209)

Journal of Environmental Engineering, Vol. 111, No. 6, December, 1985, pp. 741-754

The analysis of relative fluid motion by Camp and Stein, which led to their well-known formulation of particle collision rate in flocculation, is examined using standard tools from continuum mechanics. As a result of certain apparent conceptual errors, including the notion that a three-dimensional relative velocity field can be represented in general by a single velocity gradient, several recommendations are offered, including the abandonment of Camp and Stein's terms "absolute velocity gradient" and "root mean square (RMS) velocity gradient." Revisions in nomenclature and an alternate theory of particle collisions are discussed, as well as some implications of inhomogeneity in the turbulent flocculation problem.

97   **Volatilization Rates of Organic Chemicals of Public Health Concern**

T. P. Halappa Gowda, Assoc. Member, ASCE, (Supervisor, Water Quality Studies, Water Resources Div., Gore & Storrie Ltd., Toronto, Ontario, Canada M4G 3C2) and **John D. Lock**, (Organic Chemist, Water Treatment Div., Gore & Storrie Ltd., Toronto, Ontario, Canada M4G 3C2)

Journal of Environmental Engineering, Vol. 111, No. 6, December, 1985, pp. 755-776

Liquid film coefficients of ethylene gas, K_L, are determined from data on ethylene and rhodamine WT dye concentration distributions, which were collected in shallow streams and rivers located in Southern Ontario, Canada. Relationships among K_L, river channel hydraulic characteristics and chemical properties of organic compounds have been developed through dimensional analysis using the Buckingham π-Theorem. A sensitivity analysis of various parameters on K_L has been carried out. A statistical relationship developed by Rathbun and Tai (1982) has been found to underestimate K_L values for the Southern Ontario streams and rivers. An evaluation of the theoretical relationships between K_L and molecular properties (critical volumes and molecular weights) using experimental data reported in the literature has revealed that the relation between K_L and critical volume is suitable to calculate K_L values for highly volatile organic chemicals using the known K_L values of ethylene gas or other compounds (e.g., propane, benzene); however, for moderate and low volatile compounds, norrelationships could be obtained from the available data. The computation of K_L for a given organic compound using the relationships developed in this study is outlined in a step-by-step procedure. An example illustrates the various computations involved. In general, the method is applicable to volatile compounds with Henry's law constants greater than 10^{-3} atm-m^3/mol.

98   **Spatial Estimation of Hazardous Waste Site Data**

J. Zirschky, (Environmental Engr., ERM-Southeast, Inc., Suite 201, 2623 Sandy Plains Road, Marietta, GA 30066), **G. Phil Keary**, (Geologist, U.S. Environmental Protection Agency, Region VII, Kansas City, KS), **Richard O. Gilbert**, (Staff Scientist, Statistics Section, Battelle Pacific Northwest Lab., Richland, WA) and **E. Joe Middlebrooks**, Fellow, ASCE, (Provost and Vice Pres. for Academic Affairs, Tennessee Tech. Univ., Cookeville, TN)

Journal of Environmental Engineering, Vol. 111, No. 6, December, 1985, pp. 777-789

Kriging is a technique which can be used to obtain minimum variance, unbiased estimates of the concentration of a pollutant at a point, or the average concentration in an area or volume. In this paper, the use of simple and universal kriging for analyzing the distribution of dioxin (2,3,7,8-TCDD) in the sediments of a creek is demonstrated. The resulting point and block estimates obtained with both techniques were then compared. Although the universal kriging point estimates were slightly better than the simple kriging estimates, no clear advantage to using universal kriging was apparent in this example. In general, however, universal kriging should be considered whenever a significant drift is present in the data.

99   Suspended Sediment—River Flow Analysis

William J. Grenney, Member, ASCE, (Prof., Dept. of Civ. and Environmental Engrg., Utah State Univ., Logan, UT) and **Edward Heyse**, (Research Environmental Engr., Air Force Engrg. and Services Center, Tyndall Air Force Base, FL)

Journal of Environmental Engineering, Vol. 111, No. 6, December, 1985, pp. 790-803

There is increasing interest in suspended sediment as a water quality parameter because of its potential as a transport mechanism for pollutants and because of its possible effects on fish habitat. The application of bivariate probability distribution functions to represent the suspended sediment concentration—water discharge data for the San Juan River at Bluff, Utah is presented. This approach has several advantages over more traditional methods, including preservation of sediment concentration information and convenience for computer implementation. A bivariate log-normal density function was found to adequately represent the San Juan River data. Average sediment yields were calculated using the closed form solution of the moment generating function, and compared to the results from the flow duration-sediment rating curve method. The bivariate log-normal density function was found to be a convenient, accurate method for parameterizing the frequency distribution of sediment concentrations and water discharges for the San Juan River.

100   Stochastic Optimization/Simulation of Centralized Liquid Industrial Waste Treatment

J. H. Ellis, (Asst. Prof., Dept. of Geography and Environmental Engrg, Johns Hopkins Univ., Baltimore, MD 21218), **E. A. McBean**, (Prof., Dept. of Civ. Engrg., Univ. of Waterloo, Waterloo, Ontario, Canada N2L 3G1) and **G. J. Farquhar**, (Prof., Dept. of Civ. Engrg., Univ. of Waterloo, Waterloo, Ontario, Canada N2L 3G1)

Journal of Environmental Engineering, Vol. 111, No. 6, December, 1985, pp. 804-821

A stochastic optimization-simulation method is presented for delineating least-cost treatment sequences for a centralized liquid industrial waste treatment facility. A dynamic programming model performs the optimization. The function of the model is to delineate least-cost treatment sequences that will produce an acceptable effluent stream quality given a probabilistically-generated influent waste regime. The model is structured to permit the following user-determined options: waste types and respective volumes in the waste inventory; specific contaminants within each waste type; contaminant-specific probability density functions for waste strength; unit treatment processes including performance efficiencies and related costs; and individual contaminant effluent standards. The stochastic dynamic programming model served as a screening device, identifying unit treatment processes and sequences of processes with favorable cost-effectiveness attributes. The treatment paths thus identified were further analyzed and refined using stochastic simulation techniques.

101   **Impact of Lake Acidification on Stratification**

Steven W. Effler, (Engr., Upstate Freshwater Inst., P.O. Box 506, Syracuse, NY 13214) and **Emmet M. Owens**, Assoc. Member, ASCE, (Engr., Upstate Freshwater Inst., P.O. Box 506, Syracuse, NY 13214)

Journal of Environmental Engineering, Vol. 111, No. 6, December, 1985, pp. 822-832

The effects of documented acidification-based increases in transparency, i.e., decreases in diffuse light attenuation, $k_d(m^{-1})$, on the occurrence and character of thermal stratification in lakes is evaluated with a mathematical mixed layer (integral energy) stratification model. Predicted changes in the character of stratification in deep lakes include: deeper epilimnia, reduced density gradients in metalimnia, increased hypolimnetic heating, and reduced stability. These changes, brought about as a result of a reduction in k_d from $0.75 - 0.15$ m^{-1}, were generally as great as or greater than those associated with extremes in meteorological conditions in a north temperate climate. The changes in stratification may have important implications with respect to the vertical cycling of dissolved constituents and the oxygen resources of hypolimnia. Model predictions indicate that lakes of maximum depth less than 25 m located in the Adirondack Region of New York State may change in character from exhibiting strong summer stratification to stratifying only weakly or not at all as a result of similar decreases in k_d. Shallower lakes are more susceptible to less extreme reductions in k_d. A number of lakes in the Adirondack Region have probably been eliminated as cold-water fisheries as a result of this effect.

102   **Separation of Solid Waste With Pulsed Airflow**

Richard I. Stessel, (Research Assoc., Dept. of Civ. and Environmental Engrg., Duke Univ., Durham, NC 27706) and **J. Jeffrey Peirce**, (Assoc. Prof., Dept. of Civ. and Environmental Engrg., Duke Univ., Durham, NC 27706)

Journal of Environmental Engineering, Vol. 111, No. 6, December, 1985, pp. 833-849

The concept and development of a new type of air classifier, the pulsed-flow air classifier, are described. Field experience with air classifiers has indicated the need for re-design. Careful examination of shredded material in light of the literature enables a coherent description of solid waste components. Analyses of fall times show why conventional air classifiers fail to separate adequately. The objective is to achieve separation based more on density and less on aerodynamic characteristics than is possible with current classifier technology. Based on these results and drawing from separation technology, the concept of pulsed-flow air classification is developed. The theory is briefly examined and compared with previous work. Carefully controlled experimentation with specially constructed equipment shows that pulsed flow air classification is capable of separations by density of which conventional classifiers are not capable. The model is shown to aid design of such classifiers.

103   **Removal of Organic Matter in Water Treatment**

Michael R. Collins, Member, ASCE, (Dept. of Civ. Engrg., Univ. of New Hampshire, Durham, NH 03825), **Gary L. Amy**, Member, ASCE, (Assoc. Prof., Environmental Engrg. Program, Dept. of Civ. Engrg., Univ. of Arizona, Tucson, AZ 85721) and **Paul H. King**, Fellow, ASCE, (Prof. and Head, Environmental Engrg. Program, Dept. of Civ. Engrg., Univ. of Arizona, Tucson, AZ 85721)

Journal of Environmental Engineering, Vol. 111, No. 6, December, 1985, pp. 850-864

The performance of several types of water treatment plants in removing various molecular weight (MW) fractions of naturally occurring aquatic organic matter and humic substances is described. An assessment was made of the performance of direct filtration, conventional treatment, and softening in removing trihalomethane (THM) precursors from a diverse array of water sources. In addition, a comparison was made between conventional treatment and direct filtration in removing THM precursors from a common water source, the Colorado River. As a general rule, THM reactivity (ug THM/mg C) generally increased with MW although the $<$ 10,000 MW range was found to be the most consistently reactive fraction of aquatic organic matter. All of the various treatments preferentially removed the most reactive fraction of precursor present in each molecular weight range. None of the various treatments proved to be very effective in removing precursor material below a MW of $<$ 500. The ability to remove THM precursors appears to be related to both the source of humic substances as well as the type of treatment employed.

104   Nitrogen Removal in Rapid Infiltration Systems

Ronald W. Critea, Member, ASCE, (Engrg. Mgr., George S. Nolte and Associates, 1700 L St., Sacramento, CA 95814)

Journal of Environmental Engineering, Vol. 111, No. 6, December, 1985, pp. 865-873

Rapid infiltration systems for wastewater treatment and disposal can be designed rationally based on measured infiltration rates and treatment requirements. The hydraulic loading rate is determined using a percentage of the design infiltration rate. The percentage varies from 2 – 10%, depending on the nature of the field test for infiltration rate, the variability of the site soils, and the ratio of application period to drying period. The hydraulic loading also affects the treatment efficiency for nitrogen and phosphorus. The basis for designing rapid infiltration systems for nitrogen and phosphorus removal and treatment performance data are provided.

105   Particles, Pretreatment, and Performance in Water Filtration

Charles R. O'Melia, Member, ASCE, (Prof., Dept. of Geography and Environmental Engrg., Johns Hopkins Univ., Baltimore, MD 21218)

Journal of Environmental Engineering, Vol. 111, No. 6, December, 1985, pp. 874-890

Relationships among raw water quality, pretreatment facilities, and the design of packed bed filters are presented and applied. The particle size, particle concentration, particle surface characteristics, and solution chemistry in the raw water supply have important and predictable effects on filter design. An integrative approach to water treatment plant design, from raw water quality to filter bed performance, will facilitate process evaluation and has the potential for providing a basis for optimal plant design.

106   External Mass-Transfer Rate in Fixed-Bed Adsorption

Paul V. Roberts, Member, ASCE, (Prof., Dept. of Civ. Engrg., Terman Engrg. Center, Stanford Univ., Stanford, CA 94305), Peter Cornel, (Postdoctoral Research Affiliate, Dept. of Civ. Engrg., Terman Engrg. Center, Stanford Univ., Stanford, CA 94305) and R. Scott Summers, Member, ASCE, (Research Asst., Dept. of Civ. Engrg., Terman Engrg. Center, Stanford Univ., Stanford, CA 94305)

Journal of Environmental Engineering, Vol. 111, No. 6, December, 1985, pp. 891-905

External mass-transfer coefficients are measured with organic compounds, p-nitrophenol and 2,4-dichlorophenol, under conditions representative of fixed-bed activated-carbon adsorption for water treatment. The superficial velocity was in the range 1.3 – 8.0 mm/s and the Reynolds number range was 0.8 – 5. The observed external transfer coefficients are compared with predicted values from four mass-transfer models; the relationship of Gnielinski bestpredicts the velocity dependence. The observed transfer coefficients in all cases exceeded the predictions based on ideal spherical geometry. Depending on the conditions and the choice of predictive model, the particle size correction factors ranged from 1.44 – 2.04. The data from previously published studies show deviations that are similar in direction, but smaller in magnitude.

107   Development and Analysis of Equalization Basins

Boris M. Khudenko, Member, ASCE, (Assoc. Prof., School of Civ. Engrg., Georgia Inst. of Tech., Atlanta, GA 30332)

Journal of Environmental Engineering, Vol. 111, No. 6, December, 1985, pp. 907-922

Adequate control of wastewater treatment processes with variable input parameters can be achieved through the employment of equalization basins. Variability of input and output parameters are analyzed, and mathematical models of conventional and new configurations of equalization basins describing the propagation of regular and random input signals are presented. Comparison of various designs of equalization basins show that new configurations of equalization basins are more efficient than one-channel tanks (complete mix, m-tanks in series, or diffused flow). A substantial advantage of new basin types is in that they can be tuned to the particular spectrum of incoming fluctuations.

108   Self-Cleansing Slope for Partially Full Sewers

Ronald E. Benson, Jr., Member, ASCE, (Asst. Prof. of Civ. Engrg., Rose-Hulman Inst. of Tech., Terre Haute, IN)

Journal of Environmental Engineering, Vol. 111, No. 6, December, 1985, pp. 925-928

An equation is developed that allows direct calculation of the minimum slope required to maintain a given self-cleansing velocity in partially full sewers. The self-cleansing slope is a function of the Manning constant, pipe diameter, ratio of actual flowrate to full flowrate, and self-cleansing velocity. It is found that current recommended minimum slopes correspond to an actual velocity of 0.45 m/s when the actual flowrate is approximately 15% of the full flowrate.

109   Considering the Future in Public Works Planning

David C. Colony, (Prof. of Civ. Engrg., Univ. of Toledo, Toledo, Ohio)

Journal of Professional Issues in Engineering, Vol. 111, No. 1, January, 1985, pp. 1-11

It can be argued that designers of public works projects, in common with others, have some duty to provide for future generations. Various ideas about the nature of that duty have been advanced, including some discussions of resource conservation and environmental protection, but no philosophical theory about future genrations is available which can be directly adapted to problems of public works design. The principle is proposed that the duty of designers of public works is to minimize restrictions upon options open to future people. This concept leads to a proposed method for quantitative determination of the "design period" for which a public works project ought to be planned. The quantitative method, related to a well known problem in operations analysis, involves computation of the "design life" associated with minimum total project cost per unit time (including construction, maintenance and operation). Further research is needed to establish the numerical parameters needed for application of the proposed method, but the philosophical basis is fully described, and certain advantages of the proposed criterion over the benefit/cost ratio are claimed.

110   **The Unviversity: Eisenhower's Warning Reconsidered**

David A. Bella, Member, ASCE, (Prof., Dept. of Civ. Engrg. Oregon State Univ., Corvallis, Oreg. 97331)

Journal of Professional Issues in Engineering, Vol. 111, No. 1, January, 1985, pp. 12-21

Upon leaving office in 1961, U.S. President Dwight D. Eisenhower deliverd his often-quoted farewell address, in which he warned about the "military-industrial complex." Less well-known is his warning, in the same address, concerning the university. This paper examines the relevance of Eisenhower's warning to current conditions within the university. Examples are drawn from the field of water resources engineering. It is argued that the university as a social institution has characteristics not found in other institutions, and that these enable it to make unique contributions to society. These unique characteristics and contributions, however, may be lost because, contrary to Eisenhower's warning, the quest to obtain outside funding has become a virtual substitute for intellectual curiosity and critical inquiry. The consequences, in Eisenhower's words, are 'Gravely to be regarded."

111   **Demoresearch for Resource and Energy Recovery**

Richard Ian Stessel, (Post-Graduate Research Asst. with the Round Table on Science and Public Affairs, Dept. of Civ. and Environ. Engrg., Duke Univ., Durham, N.C. 27706) and **J. Jeffrey Peirce**, (Assoc. Prof., Dept. of Civ. and Environ. Engrg., Duke Univ., Durham, N.C. 27706)

Journal of Professional Issues in Engineering, Vol. 111, No. 1, January, 1985, pp. 22-32

The concept of mechanical separation of municipal solid waste (MSW) was initially viewed with great optimism, prompting the use of unit operations taken from other industries without adequate modification. Many plants were built that could not perform adequately. As a result of these experiences, there is very little current interest in resource recovery in general and waste-to-energy production in particular by municipalities. The reasons for the lack of success of the resource recovery industry are largely attributable to non-technical policy issues and organizational constraints. Research is needed to perform the significant development of the unit operations required for their implementation in resource recovery. The experience of the authors in resource recovery research is that such research is very likely to be fruitful. The ability of research to produce devices capable of separating a variety of MSW streams and producing a variety of products must, however, be demonstrated to a wide audience. To solve such problems, the concept of a research-oriented demonstration facility is suggested. Such demoresearch projects, funded by the federal government, are suggested to be essential to the success of resource recovery.

112   **Motivating and Managing Engineers**

Ralph Aronberg, Assoc. Member, ASCE, (Pres., Florida Engineering Consultants, Ft. Lauderdale, Fla.)

Journal of Professional Issues in Engineering, Vol. 111, No. 1, January, 1985, pp. 33-38

The writer has used his own engineering experience in conjunction with recent research done by others, to develop a guide for the motivation and management of engineers. The intelligence and self–motivating behavior of engineers are best fostered in a nonstructured, participative environment. Engineering managers have to be trained to use their perception to address the needs of their subordinates. Job enrichment is needed in the organizational engineering environment to optimize productivity.

113   User-Pay Funding and Creative Capital Financing

Graham S. Toft, (Visiting Assoc. Prof., Inst. for Interdisciplinary Engrg. Studies, Purdue Univ., West Lafayette, Ind. 47907)

Journal of Professional Issues in Engineering, Vol. 111, No. 2, April, 1985, pp. 39-47

Heightened activity in creative capital financing is an outward sign of drastic alterations taking place in public capital markets. Spurred by the property tax limitation movement, two trends in the structure of state and municipal finance are observable: Increased use of user fees and charges as a source of revenue and a drastic change in the ratio of revenue to general obligation debt. Both trends, coupled with a reawakening of the theoretical foundations of user fee economics, make for enhanced prospects for user-pay funding in infrastructure revitalization. User-pay systems may be a means by which to create more stability and predictability in an otherwise volatile capital finance market. Along with dedicated funds and various forms of public enterprise (utilities, autonomous authorities, special districts), user-pay practices could be used to improve public works organization and management.

114   Estimating Rural and Urban Infrastructure Needs

Enno Koehn, Member, ASCE, (School of Civ. Engrg., Civ. Engrg. Building, Purdue Univ., West Lafayette, Ind. 47907), **John Fisher**, Member, ASCE, (Lawson-Fisher Assoc., South Bend, Ind.) and **James McKinney**, Member, ASCE, (Dept. of Civ. Engrg., Rose-Hulman Inst. of Tech., Terre Haute, Ind.)

Journal of Professional Issues in Engineering, Vol. 111, No. 2, April, 1985, pp. 48-56

The results of an investigation designed to determine the infrastructure needs of a typical midwestern state such as Indiana are presented. One general finding is that the infrastructure condition of small communities appears to be worse than that of larger cities or metropolitan areas. These results should challenge conventional thinking that infrastructure is a "big city" problem. In fact, since Indiana is a state of small to medium sized communities and since the U.S. population is moving in the direction of non-metropolitan areas, considerable attention should be given to the management, technology and financing of infrastructure for small and medium sized communities.

115   Professional Liability of the Architect and Engineer

James P. Holland, Member, ASCE, (Capt., U.S. Air Force, Headquarters Strategic Air Command, Offutt Air Force Base, Omaha, Neb.)

Journal of Professional Issues in Engineering, Vol. 111, No. 2, April, 1985, pp. 57-65

One of the most disturbing issues of the medical profession has begun to disturb those in the engineering community—malpractice. Although professional engineering liability made its debut in the 1950's, it has become an item of concern for every engineer. The threat of a liability frequently has replaced the idea of creativity with ideas of conservation. There are several actions an engineer can take to reduce or eliminate, or both, the potential for a liability lawsuit: (1) use of AEPIC for structural failure data; (2) avoidance of liability-prone projects; and (3) purchase of liability insurance. However, there is no guarantee that an engineer will not be named in a lawsuit, regardless of the precautions he or she takes.

116   **Civil Engineering and Engineering Enrollments**

George K. Wadlin, Fellow, ASCE, (Administrator, Education Services, ASCE, 345 E. 47 St., New York, N.Y. 10017)

Journal of Professional Issues in Engineering, Vol. 111, No. 3, July, 1985, pp. 67-80

A declining undergraduate enrollment in civil engineering is shown both in terms of absolute numbers and as a percentage of all engineering undergraduates. The decline also appears among women and black undergraduate civil engineering students. Increasing graduate enrollments in civil engineering occur at both the master's and doctoral levels. Comparisons are made at all educational levels among civil, electrical, mechanical and chemical engineering enrollments. The largest civil engineering enrollments at the bachelor's, master's and doctoral levels are listed for U.S. universities. Historical comparisons are made from 1968 to 1983. The statistics were extracted from the annual reports on engineering enrollments of the Engineering Manpower Commission of the American Association of Engineering Societies.

117   **Professional Registration of Government Engineers**

Thomas J. Buchanan, Fellow, ASCE, (Asst. Chf. Hydro., U.S. Geological Survey, 441 National Center, Reston, Va. 22092)

Journal of Professional Issues in Engineering, Vol. 111, No. 3, July, 1985, pp. 81-87

Government engineers often are not required to be registered in the state in which they practice, even though their counterparts in the private sector must be registered. Government engineers have a responsibility to safeguard the public health and safety, and must develop in the public a sense of confidence that the public trust is in good hands. One way to generate this confidence is by engineering registration. The American Society of Civil Engineers views professional registration as an appropriate requirement for engineers, including those in government. The National Society of Professional Engineers makes registration a requirement for the grade of member and full privileges in the society. Some Federal agencies require engineering registration for certain positions in their agencies. Engineers in government service should consider the value of engineering registration to themselves and to their agencies and take pride in their professions and in their own capabilities by becoming registered engineers. They should also take steps to encourage their agencies to give more attention to engineering registration, particularly as a qualification for certain positions and as a requirement for senior positions in the organization.

118   **One Plus One Makes Thirty**

Richard H. McCuen, (Prof., Dept. of Civ. Engrg., Univ. of Maryland, College Park, Md. 20742)

Journal of Professional Issues in Engineering, Vol. 111, No. 3, July, 1985, pp. 88-99

A typical engineering program includes 30 credit hours of general educational requirements (GER), which is over 20% of the requirements. Studies suggest that the GER are not meeting the Accreditation Board for Engineering and Technology (ABET) intent"...of making young engineers fully aware of their social responsibilities...." A possible solution to the problem in the form of a one credit hour freshman foundations course and a one credit hour senior capstone course is recommended. The freshman foundations course is intended to make the entering engineering student aware of the social responsibility of the engineer, the value of the GER in meeting this responsibility, and recommended courses that will best fulfill this intent. The senior capstone course is intended to show the relationship between the social value issues analyzed in the courses that the engineering student has used to fulfill the GER and the technical issues covered in engineering courses. The sum of the two one credit hour courses should make the 30 credit hour program of GER more effective in developing a social sensitivity in the engineering student.

119   How Designers Can Avoid Construction Claims

Harvey A. Kagan, Fellow, ASCE, (Professor, Dept. of Civ. Engrg., Rutgers Univ., Coll. of Engrg., P.O. Box 909, Piscataway, N.J. 08854)

Journal of Professional Issues in Engineering, Vol. 111, No. 3, July, 1985, pp. 100-107

The sharp increase in the number of claims associated with construction projects now involves designers in claims brought by construction contractors. Designers were formerly insulated against such claims due to a lack of contractual relationship with the contractors. Broader interpretations by courts have changed this situation. The article reviews some situations in which a design firm can find itself the target of delay claims and offers solutions. These solutions include paying attention to coordination of design documents to minimize document conflicts, use of objective criteria for approval or rejection of "equal" products, establishing written schedules and priorities for shop drawing review, and avoiding use of shop drawings to modify designs. Though not exhaustive, the cases presented should help designers in avoiding or greatly reducing exposure to construction claims.

120   Issues in Engineering Education

Murray A. Muspratt, Member, ASCE, (Visiting Fellow, Dept. of Civil Engrg., Princeton Univ., Princeton, N.J. 08544)

Journal of Professional Issues in Engineering, Vol. 111, No. 3, July, 1985, pp. 108-116

Engineering education today is education for a new millenium. Revolutionary changes are in progress as to the number and quality of students and faculty available, the status of campus facilities, the range of courses offered, the impact of computers, and the correct interface between teaching and research. The government plays a decisive role in cultivating the climate for achieving education goals, while engineering societies and licensing boards have important input. These issues are addressed in an attempt to promote debate and canvass opinions as to where engineering education is heading, or should be heading.

121   LOGO as an Introduction to FORTRAN Programming

W. Craver, Jr., (Assoc Prof., Dept. of Mech. Engrg., Univ. of Texas, El Paso, Tex.), **D. C. Schroder**, (Prof., Dept. of Electrical Engrg., Univ. of Texas, El Paso, Tex.), **A. J. Tarquin**, (Assoc. Prof., Univ. of Texas, El Paso, Tex.) and **Po-Wen Hu**, (Assoc. Prof., Dept of Industrial Engrg., Univ. of Texas, El Paso, Tex.)

Journal of Professional Issues in Engineering, Vol. 111, No. 3, July, 1985, pp. 119-123

An experiment was conducted in a sophomore level computer science course to determine if knowledge of LOGO would be beneficial for students studying FORTRAN programming. The performance of students who were taught six hours of LOGO was compared to that of others who did not know LOGO. The results showed rather conclusively that LOGO was beneficial for learning FORTRAN for students who had no previous computer programming experience.

122   **1984 Civil Engineering Educators Salary Survey**

George K. Wadlin, (Dir., Education Services, ASCE, New York, N.Y. 10017-2398)

Journal of Professional Issues in Engineering, Vol. 111, No. 4, October, 1985, pp. 129-136

The results of a survey of all civil engineering departments taken to determine the 1984-85 academic year salaries are presented. About 80% of the schools responded. average and individual salaries are grouped regionally by ASCE Districts. Separate tables show salaries for all schools plus public, private, Ph.D. and non-Ph.D. granting schools. Nine months' salaries are shown for all four academic ranks, plus the chairman's annual salary and the entry level salary for new assistant professors. A comparison is made between the average salaries reported in 1983 and in 1984. This report is a condensed version of a more comprehensive report by the writer.

123   **A Common Sense Approach to the Fundamentals Exam**

James F. McDonough, (Head and William Thoms Prof. of Civ. Engrg., Univ. of Cincinnati, Cincinnati, Ohio 45221)

Journal of Professional Issues in Engineering, Vol. 111, No. 4, October, 1985, pp. 137-140

A large portion of the graduating civil engineering students take the Fundamentals of Engineering (F.E.) Examination, formally known as the EIT Exam, or Engineering in Training Examination, as they complete their senior year. Proper preparation for, and a complete understanding of, the F.E. Exam and how it is graded is essential to maximizing performance. The Exam format was recently changed, and many available references do not have the latest content and grading schemes. A process for both long-term and short-term preparation is outlined using the most recent format and a general description of how the Exam is graded.

124   **Computer Programming in the Civil Engineering Curriculum**

William J. Rasdorf, (Asst. Prof., Dept. of Civ. Engrg. and Computer Sci., North Carolina State Univ., Box 7908, Raleigh, N.C. 27695)

Journal of Professional Issues in Engineering, Vol. 111, No. 4, October, 1985, pp. 141-148

The rapid advances occurring incomputer science have provided the engineer with a powerful means of processing, storing, retrieving, and displaying data thereby increasing the role of computer science in nearly every engineering discipline. One of the dilemmas in engineering education today is how future engineers can best assimilate the advanced, yet fundamental, knowledge in computer methods and technology appropriate for their specific engineering discipline. This paper suggests that the effective use of such technology in engineering processes and applications is the key to increased individual, company, and national productivity. In the future, an integrated combination of computer-aided analysis and design problems. The implications of this development for the academic community are clear: Students must be prepared to use computer methods and applications as a part of their fundamental education. It is the responsibility of colleges and universities to incorporate contemporary computing fundamentals into their academic curriculum to improve the professional qualifications of their engineering graduates. These graduates will in turn be able to provide their increasingly important expertise to both the engineering profession and the academic community.

125   Ethics of Professionalism

Murray A. Muspratt, Member, ASCE, (Chisolm Inst. of Tech., P.O. Box 197, Calufield East, Victoria, Australia 3145)

Journal of Professional Issues in Engineering, Vol. 111, No. 4, October, 1985, pp. 149-160

As engineering activity becomes increasingly complex, ethical dilemmas without precedent are arising, straddling the technological and legal fields, and evoking emotions that previous generations of engineers have not had to deal with. Some of these issues are addressed against the framework of the political, economic and social value systems.

126   Sewage-Related Wastes and Oceans: A Problem?

Richard C. Kolf, Member, ASCE, (Head, Tech. and Commercial Development, National Sea Grant College Program, NOAA, 6010 Executive Boulevard, Rockville, Md. 20852)

Journal of Professional Issues in Engineering, Vol. 111, No. 4, October, 1985, pp. 161-175

An examination of the case study and other scientific papers concerning the effects of sewage-related wastes on the oceans shows a growing consensus that ocean disposal may, in some cases, be the best option for environmental as well as economic reasons. Environmental issues such as this, however, are complicated by widely varying perceptions and values within the population as a whole; conflict is inevitable, and regulatory reform initiatives seem required by the severity and complexity of the issues involved. While engineers can profitably continue the development of models which predict the fate of sewage particulates, it is equally important that social scientists be encouraged to direct their attention to this issue. Fundamental studies are needed to clarify who bears the economic and social burdens in each disposal option (land, air, water), and to define methods of approaching a fair and efficient decision.

127   Dynamic Characterization of Two-Degree-of-Freedom
                              Equipment-Structure Systems

Takeru Igusa, Assoc. Member, ASCE, (Asst. Research Engr., Dept. of Civ. Engrg., Univ. of California, Berkeley, Calif.) and **Armen Der Kiureghian**, Assoc. Member, ASCE, (Assoc. Prof., Dept. of Civ. Engrg., Univ. of California, Berkeley, Calif.)

Journal of Engineering Mechanics, Vol. 111, No. 1, January, 1985, pp. 1-19

A two-degree-of-freedom equipment-structure system is studied to find its intrinsic properties which are needed for analysis of more general secondary systems. Perturbation theory is used to find closed form expressions for the modal properties of the system in terms of the properties of the individual subsystems. Three important characteristics of the system are identified: tuning, interaction, and nonclassical damping. Mathematical expressions are defined for each of these characteristics and criteria are developed to measure their influences on the response of the equipment. The expressions for the modal properties and the criteria for tuning, interaction, and nonclassical damping are new results for the 2-degree-of-freedom system. These results form the bases for analysis of multiply supported multi-degree-of-freedom secondary systems.

128   **Dynamic Response of Multiply Supported Secondary Systems**

Takeru Igusa, Assoc. Member, ASCE, (Asst. Research Engr., Dept. of Civ. Engrg., Univ. of California, Berkeley, Calif.) and **Armen Der Kiureghian**, Assoc. Member, ASCE, (Assoc. Prof., Dept. of Civ. Engrg., Univ. of California, Berkeley, Calif.)

Journal of Engineering Mechanics, Vol. 111, No. 1, January, 1985, pp. 20-41

An accurate and efficient method for analysis of multiply supported, multi-degree-of-freedom secondary systems is developed. The method accounts for the effects of tuning, interaction, nonclassical damping, and spatial coupling that are intrinsic dynamic characteristics of composite primary-secondary systems. The method consists of two steps: (1) synthesis of the modal properties of the composite primary-secondary system in terms of the known properties of the individual subsystems, and (2) evaluation of secondary system response through modal combination. Concepts from modal synthesis, perturbation theory, and random vibrations are used to produce accurate results. The method is efficient since it avoids large eigenvalue solutions, time-history analysis and floor spectra computations. It is also general, as it is applicable to a variety of stochastic or deterministic inputs. The responses of an example secondary system to seismic input are obtained directly in terms of the ground response spectrum.

129   **State Variable Model for Volumetric Creep of Clay**

Paul R. Dawson, (Asst. Prof., Sibley School of Mech. and Aerospace Engrg., Cornell Univ., 254 Upson Hall, Ithaca, N.Y. 14853), **Joel Lipkin**, (Member of Tech. Staff, Seabed Programs Div., Sandia National Laboratories) and **Herrick S. Lauson**, (Member of Tech. Staff, Computational Physics and Mech. Div., Sandia National Laboratories)

Journal of Engineering Mechanics, Vol. 111, No. 1, January, 1985, pp. 42-61

A state variable constitutive model for the volumetric creep behavior of fine-grained marine clay is presented. The form of the model is motivated by laboratory data for undistrubed illite. Model parameters have been evaluated by least squares methods using data from single-step and double-step loading conditions. The model has been implemented in a saturated creeping porous media finite element code and has been used to simulate the volumetric deformations of clay under laboratory conditions.

130   **Triple Power Law for Concrete Creep**

Zdenek P. Bazant, Fellow, ASCE, (Prof. of Civ. Engrg. and Dir., Center for Concrete and Geomaterials, The Technological Inst., Northwestern Univ., Evanston, Ill. 60201) and **Jenn-Chuan Chern**, (Postdoctoral Research Assoc., Div. of Reactor Analysis and Safety, Argonne National Lab., Argonne, Ill. 60439)

Journal of Engineering Mechanics, Vol. 111, No. 1, January, 1985, pp. 63-83

An improved creep law for concrete at constant temperature and water content is proposed. It gives the creep rate as a product of power functions of the load duration, the age at loading and the current age of concrete. This law exhibits a gradual smooth transition from the double power law for very short load durations to the logarithmic law for very long load durations. The higher the age of loading, the longer the load duration at the transition. The determination of creep compliance requires evaluation of a binomial integral, which can be carried out either with the help of a truncated power series or by replacement of certain integrals with sums. A table of values from which interpolation is possible is also given. Extensive fitting of creep data from the literature reveals only a models improvement in the overall coefficient of variation of the deviations from test data; however, the terminal slopes of creep curves are significantly improved, which is especially important for extrapolation of creep measurements. Compared to the previous double power-logarithmic law, the present formulation has an advantage of continuity in curvature, and compared to the log-double power law, the present formulation has a greater range of applicability involving also very short creep durations, including the dynamic range. The new formulation also significantly limits the occurrence of divergence of creep curves, and permits even a complete suppression of this property, although at the cost of a distinct impairment in data fits.

131 Shear Lag Analysis and Effective Width of Curved Girder Bridges

Kaoru Hasebe, (Asst. Prof., Dept. of Civ. Engrg., Akita Univ., Akita, Japan), Seizo Usuki, Assoc. Member, ASCE, (Assoc. Prof., Dept. of Civ. Engrg., Akita Univ., Akita, Japan) and Yasushi Horie, (Instructor, Dept. of Civ. Engrg., Akita Tech. College, Akita, Japan)

Journal of Engineering Mechanics, Vol. 111, No. 1, January, 1985, pp. 87-92

The effective width of curved girder bridges is formulated by substituting the flange stress derived from present theory into the equation of effective width definition for the curved girders. The required information in formulating the effective width rule for design of curved girder bridges are provided. The actual longitudinal stress distributions for the curved girders are evolved from the present theory for shear lag in order to determine the effective width. The thin-walled curved girders used in this investigation are based on box and channel cross sections, and are analyzed for a uniform lateral load and for a concentrated load. Numerical examples are shown for several problems to investigate the effect on the effective width of curved girder bridges. The values of the effective width obtained by the present theory are compared with those of the straight girder bridges. According to the results, the values of the effective width of curved girder bridges can be regarded as the values of the straight girder bridges approximately.

132 Wake Displacement as Cause of Lift Force on Cylinder Pair

Alireza Bokaian, Member, ASCE, (Sr. Engr., Earl and Wright Ltd., Victoria Station House, 191 Victoria St., London SW1E 5NE, U.K.) and Farhad Geoola, (Grad. Research Student, Dept. of Civ. Engrg., Univ. College London, Gower St., London WC1E 6BT, U.K.)

Journal of Engineering Mechanics, Vol. 111, No. 1, January, 1985, pp. 92-99

When to parallel circular cylinders are in close proximity, mutual interference effects generate large lift forces on both cylinders. A great deal of effort has gone into the task of determining a physical explanation of the lift force. This note describes a series of experiments aimed at further clarifying the nature and source of lift forces on two parallel circular cylinders with a smooth surface positioned at a right angle to approaching flow direction. It is found that changing the relative size of two bodies had a marked influence on the force profiles. This is attributed to displcement of the wake of the upstream cylinder by the downstream one. It is suggested that any flow parameters of the wake, measured without the presence of the rear cylinder when attempting to assess the forces thereon, is an oversimplification of the situation, as far as the lift is concerned.

133   Energy Fluctuation Scale and Diffusion Models

Steven R. Winterstein, (Acting Asst. Prof., Dept. of Civ. Engrg., Stanford Univ., Stanford, Calif. 94305) and **C. Allin Cornell**, Member, ASCE, (Prof., Dept. of Civ. Engrg., Stanford Univ., Stanford, Calif. 94305)

Journal of Engineering Mechanics, Vol. 111, No. 2, February, 1985, pp. 125-142

The energy fluctuation scale, θ_E, of a narrow-band Gaussian response is introduced. It is shown that θ_E and associated bandwidth measures (e.g., $1/\nu_o\theta_E$) are: (1) Simply related to both the spectral density and correlation functions of the response; (2) less sensitive to high frequency information and precise envelope definition than traditional bandwidth measures; and (3) consistent with diffusion models of various response quantities of interest. These diffusion models are found to simplify the practical analysis of a wide range of reliability problems (e.g., various first-passage problems, statistics of extrema in narrow-band responses, etc.). Further, because θ_E is related to statistics of envelope time averages, it has direct application to the analysis of fatigue damage and crack growth under stochastic loading.

134   Model for Flexible Tanks Undergoing Rocking

Medhat A. Haroun, Assoc. Member, ASCE, (Assoc. Prof., Civ. Engrg. Dept., Univ. of California, Irvine, Calif. 92717) and **Hamdy M. Ellaithy**, Assoc. Member, ASCE, (Grad. Research Asst., Civ. Engrg. Dept., Univ. of California, Irvine, Calif. 92717)

Journal of Engineering Mechanics, Vol. 111, No. 2, February, 1985, pp. 143-157

An analytical mechanical model for flexible cylindrical tanks is developed taking into consideration the effect of rigid base rocking motion and lateral translation. Explicit analytical expressions for the parameters of the model are given, and numerical values of these parameters are displayed in charts. The model can be used to evaluate the maximum dynamic response of a rigid or a flexible cylindrical tank subjected to earthquake loading with and without rigid base rocking motion.

135   Impact in Truss Bridge Due to Freight Trains

Kuang-Han Chu, Fellow, ASCE, (Prof. Emeritus, Civ. Engrg. Dept., Illinois Inst. of Tech., Chicago, Ill. 60616), **Vijay K. Garg**, Fellow, ASCE, (Assoc. Prof., Engrg. Dept., Univ. of Maine, Orono, Maine 04469) and **Majeed H. Bhatti**, Member, ASCE, (Engrg. Consultant, GDS and Assoc., Consulting Engrs., Chicago, Ill. 60602)

Journal of Engineering Mechanics, Vol. 111, No. 2, February, 1985, pp. 159-174

The dynamic responses of a railway bridge members due to vehicle-track-bridge interaction were investigated for the effects of vertical and lateral track irregularities, approach track quality, and bridge damping. In each case, a train consisting of three vehicles moving at constant speed was simulated traveling on the bridge. It was found that: (1) Greater approach irregularities produce higher impact factors and dynamic forces in bridge members; (2) impact factors in members with low static stress are high, but the dynamic stresses produced are low; (3) in general, impact factors reduce slightly due to bridge damping; and (4) the dynamic forces in lower lateral bracing members are small as compared to their allowable values.

136   Nonlinear Static and Dynamic Analysis of Plates

Ren-Jye Yang, (Grad. Student, Univ. of Iowa, Iowa City, Iowa) and **M. Asghar Bhatti**, Member, ASCE, (Asst. Prof., Dept. of Civ. Engrg., Univ. of Iowa, Iowa City, Iowa)

Journal of Engineering Mechanics, Vol. 111, No. 2, February, 1985, pp. 175-187

An efficient element for static and dynamic analysis of plates including geometric effects is presented. A formulation of the "heterosis element" presented by Hughes and Cohen for linear static analysis of Mindlin plates is given. This formulation is then extended to include large displacement effects using Von-Karman assumptions and updated Lagrangian formulation. Several numerical examples for both static and dynamic loads are presented.

137   General Failure Criterion for Isotropic Media

Jerzy Podgórski, (Asst. Lect., Inst. of Civ. and Sanitary Engrg., Technical Univ., Lublin, Poland)

Journal of Engineering Mechanics, Vol. 111, No. 2, February, 1985, pp. 188-201

A general failure criterion dependent on three stress-tensor invariants is proposed. It is applicable to a rather large class of materials including, e.g., metals, rocks, concrete and soils. The classical failure criteria and some recently proposed criteria are particular cases of this general criterion. The condition presented permits the uniform description of different groups of materials for which quite different forms of the failure criteria have been applied to date. A general form is applied to formulate failure criteria for plain concrete and sand. These criteria, to which smooth (conical for sand or paraboloidal for concrete) surfaces correspond, provide good agreement between predicted values of failure stresses and experimental results. This was possible due to the introduction of a new two-parameter function describing the deviatoric cross section of the failure surface. Two cross sectional shape characteristic ratios, λ and θ, defined in this paper, make possible the systematic analysis of different criteria and allow prediction of the failure surface features, which can be helpful in interpretation of the experimental results.

138   Stress Intensity Factor Using Quarter Point Element

Viriyawan Murti, (Research Asst., Dept. of Civ. Engrg. Materials, Univ. of New South Wales, Kensington, Australia), **Somasundaram Valliappan**, Fellow, ASCE, (Assoc. Prof., Dept. of Civ. Engrg. Materials, Univ. of New South Wales, Kensington, Australia) and **Ian Kenneth Lee**, Member, ASCE, (Prof. and Head,, Dept. of Civ. Engrg. Materials, Univ. of New South Wales, Kensington, Australia)

Journal of Engineering Mechanics, Vol. 111, No. 2, February, 1985, pp. 203-217

The finite element analysis to determine stress intensity factors (SIF's) is complex but useful. It is complex due to the stress singularity that exists at the crack tip, which requires a large number of conventional elements to model satisfactorily. Its usefulness lies in its generality because of the limited capabilities of analytical tools to calculate SIF's for general crack geometries, complicated boundary conditions and material nonlinearity. The necessity of a very fine mesh of conventional elements in the vicinity of crack tip regions can be overcome by using special elements which incorporate or can generate the required stress singularity. The accuracy and efficiency achieved by the use of these elements will vary. The mathematical expressions are often rigorous, and usually share the common disadvantages, such as the lack of interelement continuity and the constant strain term, excessive programming and error-prone determinations of SIF. The quarter point element (QPE) is free from these weaknesses, is simple, efficient and requires no additional programming. Due to the accuracy of QPE results, QPE is preferred over any other special crack-tip element.

139   Dynamic Stresses and Displacements in a Buried Tunnel

Ken C. Wong, (Grad. Student, Dept. of Civ. Engrg., Univ. of Manitoba, Winnipeg, Canada R3T 2N2), Arvind H. Shah, Member, ASCE, (Prof., Dept. of Civ. Engrg., Univ. of Manitoba, Winnipeg, Canada R3T 2N2) and Subhendu K. Datta, (Prof., Dept. of Mech. Engrg., Univ. of Colorado, Boulder, Colo. 80309)

Journal of Engineering Mechanics, Vol. 111, No. 2, February, 1985, pp. 218-234

Dynamic response of a cylindrical tunnel embedded in a semi-infinite elastic medium is analyzed. The tunnel is assumed to be infinitely long, noncircular in cross section, and lying parallel to the plane-free surface of the medium. The problem considered is one of plane strain, in which it is assumed that the waves are propagating perpendicular to the axis of the tunnel. Since this problem cannot be solved exactly except when the tunnel is circular, a numerical technique that combines the finite element method with the eigenfunction expansions is used. Numerical results are presented for the cases in which the tunnel is disturbed by plane longitudinal (P), vertically polarized shear (SV) and Rayleigh (R) waves. It is shown that the dynamic amplifications of the displacements and stresses induced in the tunnel depend crucially on the properties of the surrounding soil, the depth of embedment and the frequency of the incident simple harmonic disturbance.

Errata: EM Sept. '85, pp. 1213.

140   Model for Beam-Mode Buckling of Buried Pipelines

Heedo Yun, (Grad. Research Asst., Dept. of Aerospace Engrg. and Engrg. Mechanics, Univ. of Texas, Austin, Tex. 78712) and Stelios Kyriakides, Assoc. Member, ASCE, (Asst. Prof., Dept. of Aerospace Engrg. and Engrg. Mechanics, Univ. of Texas, Austin, Tex. 78712)

Journal of Engineering Mechanics, Vol. 111, No. 2, February, 1985, pp. 235-253

An attempt at modeling the so-called "beam mode buckling" exhibited under compression in pipelines is presented. The line is modeled as a long heavy beam on a contacting surface. The reacting surface is modeled first as an elastic and subsequently as a rigid foundation, with the additional constraint that it only reacts to compressive loads. The problem is assumed to possess a localized imperfection. Under compressive axial load, a section of the beam lifts off the foundation. The problem is studied through a large deflection extensional beam nonlinear formulation. The large deflection response of the beam is found to exhibit a limit load which is shown to be imperfection sensitive. A parametric study of the problem as well as a number of examples with actual pipeline parameters are presented.

141   Mathematical Model to Predict 3-D Wind Loading on Building

Giovanni Solari, (Researcher, Instituto di Scienza delle Costruzoni, Univ. of Genova, Genova, Italy)

Journal of Engineering Mechanics, Vol. 111, No. 2, February, 1985, pp. 254-276

A mathematical model to predict 3-D wind loading on buildings with rectangular geometry and wind acting normally to a face is formulated. This methodology allows the evaluation of force distribution in the frequency domain including mean wind loads and fluctuating loads due to alongwind and acrosswind atmospheric turbulence and wake excitation; self-excited forces are not taken into consideration. The reliability of the proposed technique is verified by comparing predicted results with experimental data available in the literature. The satisfactory agreement between theoretical and experimental results demonstrates the good applicability of this methodology, but also emphasizes the necessity of carrying out extensive wind tunnel and full-scale experiments in order to improve the knowledge of the most relevant aerodynamic parameters on which the correctness of the predicted results strongly depends.

142   Shallow Trenches and Propagation of Surface Waves

Kayumars Emad, (Grad. Student, Dept. of Civ. Engrg., State Univ. of New York, Buffalo, N.Y. 14260) and **George D. Manolis**, Assoc. Member, ASCE, (Asst. Prof., Dept. of Civ. Engrg., State Univ. of New York, Buffalo, N.Y. 14260)

Journal of Engineering Mechanics, Vol. 111, No. 2, February, 1985, pp. 279-282

The effect of the placement of shallow trenches of semi-circular or rectangular shape in the path of propagation of surface waves is studied. The ground is represented as a linear elastic halfplane, the vehicle as a point load vibrating in a wide range of frequencies, and an open air trench is interposed between the vehicle and the receiver. A two-dimensional direct boundary element method approach defined in the frequency domain is used to numerically investiate this problem. The results obtained indicate that trenches are beneficial in reducing the amplitude of vibration only in a narrow band around the mid-frequency range and in all other cases a large amplification of the dynamic signal is obtained. The actual shape of the trench is of minor importance.

143   Post-Buckling Equilibrium of Hyperstatic Lattices

S. J. Britvec, (Prof. of Engrg. Mech. and Civ. Engrg., Univ. of Maine, Orono, Me.) and **M. D. Davister**, (Sr. Engr., Applied Research Assoc., Inc., Capital Area Div., Alexandria, Va.)

Journal of Engineering Mechanics, Vol. 111, No. 3, March, 1985, pp. 287-310

Post-buckling equilibrium paths of complex, elastic, hyperstatic, pin-jointed lattices composed of slender members are studied under the aspect of finite deformation using a new simplified theory, which is presented in matrix formulation. A general numerical procedure is developed on a digital computer for the solution of the reduced equilibrium equations and the kinematic admissibility conditions admitting post-buckling equilibrium modes. A method for a direct evaluation of the most degrading mode based on the solution of a quadratic minimization problem, subject to linear constraints, is also presented. The procedure is illustrated on the lattice of a model reticulated shell of a 40 ft base diameter, designed and optimized to carry a head of some 420 ft of sea water. The effect of initial geometrical imperfections on the post-buckling paths is analyzed by means of an imperfection-sensitivity parameter. The method permits an estimation of the load-displacement curves of the imperfect lattice, if the paths for the perfect structure and this parameter are known.

144   Boundary Element Calculations of Diffusion Equation

A. E. Taigbenu, (Grad. Student, Dept. of Environmental Engrg., Cornell Univ., Ithaca, N.Y.) and **J. A. Liggett**, Member, ASCE, (Prof., Dept. of Environmental Engrg., Cornell Univ., Ithaca, N.Y.)

Journal of Engineering Mechanics, Vol. 111, No. 3, March, 1985, pp. 311-328

The boundary element method has become popular for solving elliptic equations and time-dependent problems where time appears only in the boundary conditions. Its most often-cited attribute is efficiency, although user convenience, its ability to solve singular problems, and the ease of solution in infinite regions probably outweigh the efficiency aspects. The boundary element method has been used in parabolic problems, but its advantages in that case are not as apparent. In this paper, three closely related techniques for solving the diffusion equation are explored. Although the method does not retain the feature of a strict boundary technique, since domain integrations are required, the user interface can retain the advantages of a boundary method. Because of the need for repeated integrations, or the need to integrate transcendental functions over a domain, depending on the solution method, the efficiency advantage of boundary elements is lost, at least in simple problems without singularities or infinite regions. A comparison with the finite element method illustrates that fact. Nevertheless, the boundary element remains a viable option for the solution of parabolic equations.

145   Axisymmetrical Vibrations of Tanks—Numerical

Medhat A. Haroun, Member, ASCE, (Assoc. Prof., Civ. Engrg. Dept., Univ. of California, Irvine, Calif. 92717) and **Magdy A. Tayel**, (Grad. Research Asst., Civ. Engrg. Dept., Univ. of California, Irvine, Calif. 92717)

Journal of Engineering Mechanics, Vol. 111, No. 3, March, 1985, pp. 329-345

With few exceptions, current seismic design codes for ground-based cylindrical tanks neglect the effect of vertical ground acceleration. Such motion can be transmitted, in a flexible tank, into radial pulsations of the tank wall resulting in additional stresses. A numerical study of the axisymmetrical dynamic characteristics of partly-filled tanks is carried out. Natural frequencies and mode shapes are evaluated by means of a discretization scheme in which the shell is modeled by finite elements and the liquid region is treated analytically. The distribution of the hydrodynamic pressure along the inner surface of the shell as well as the sidtribution of shell stresses are displayed. For practical applications, a simplified formula is developed to calculate the fundamental natural frequency of full tanks.

146   Axisymmetrical Vibrations of Tanks—Analytical

Medhat A. Haroun, Member, ASCE, (Assoc. Prof., Civ. Engrg. Dept., University of California, Irvine, Calif. 92717) and **Magdy A. Tayel**, (Grad. Research Asst., Civ. Engrg. Dept., University of California, Irvine, Calif. 92717)

Journal of Engineering Mechanics, Vol. 111, No. 3, March, 1985, pp. 346-358

An analytical method for the computation of the axisymmetrical dynamic characteristics of partly-filled cylindrical tanks is presented. The liquid is assumed to be inviscid and incompressible. The tank shell is assumed to be of constant thickness and its material to be linearly elastic. Under these assumptions, two coupled partial differential equations govern the vibrations of the shell. Because the tank is partly-filled with liquid, two different solutions are obtained for the lower (wet) and upper (dry) portions of the shell. A system of linear homogeneous algebraic equations is obtained by satisfying the boundary conditions at the bottom and top of the tank and the compatibility equations at the junction of the wet and dry parts of the shell. The determinant of coefficients of this system leads to the frequency equation. The natural frequencies, mode shapes and stress distributions showed excellent agreement with those obtained from a numerical solution.

147   Bounding Surface Plasticity Model for Concrete

Bing-Lin Yang, (Lect., Wuhan Inst. of Hydr. and Electric Engrg., Wuhan, China), **Yannis F. Dafalias**, Member, ASCE, (Prof. of Civ. Engrg., Univ. of California, Davis, Calif.) and **Leonard R. Herrmann**, Member, ASCE, (Prof. Civ. Engrg., Univ. of California, Davis, Calif.)

Journal of Engineering Mechanics, Vol. 111, No. 3, March, 1985, pp. 359-380

A macroscopic plasticity constitutive model employing the concept of a contracting Bounding Surface is formulated for plain concrete. The concrete strength f_c is the only necessary input material constant. The model is used to predict the stress-strain relations for uniaxial, biaxial, and triaxial compression states for both monotonic and cyclic loading conditions including post-failure states for different strength concretes. The results are successfully compared with available experimental data.

Errata: EM Sept. '85, pp. 1213.

148   Wave Propagation in a Strain-Softening Bar: Exact Solution

Zdenek P. Bazant, Fellow, ASCE, (Prof. of Civ. Engrg., Center for Concrete and Geomaterials, The Technological Inst., Northwestern Univ., Evanston, Ill. 60201) and **Ted B. Belytschko**, Member, ASCE, (Prof. of Civ. Engrg., Center for Concrete and Geomaterals, The Technological Inst., Northwestern Univ., Evanston, Ill. 60201)

Journal of Engineering Mechanics, Vol. 111, No. 3, March, 1985, pp. 381-389

A closed-form solution is given for a one-dimensional bar which undergoes strain softening (i.e., a gradual decline of stress to zero at increasing strain). It is shown that strain softening can occur in the interior of a body, and that the length of the strain-softening region tends to localize into a point, which agrees with what was previously shown by stability analysis for static situations. The stress in the strain-softening cross section drops to zero instantly, regardless of the shape of the strain-softening diagram, and the total energy dissipated in the strain-softening domain of the bar is found to vanish. Despite these unpleasant features, the problem apparently possesses a solution for certain boundary and initial conditions. However, the fact that the energy dissipation in the strain-softening process vanishes is not representative of the experimentally observed behavior of real strain softening materials such as concrete or geomaterials.

149   Strain Softening with Creep and Exponential Algorithm

Zdenek P. Bazant, Fellow, ASCE, (Prof. of Engrg. and Dir., Center for Concrete and Geomaterials, The Technological Inst., Northwestern Univ., Evanston, Ill. 60201) and **Jenn-Chuan Chern**, (Postdoctoral Research Assoc., Div. of Reactor Analysis and Safety, Argonne National Lab., Argonne, Ill. 60439)

Journal of Engineering Mechanics, Vol. 111, No. 3, March, 1985, pp. 391-415

A constitutive relation that can describe tensile strain softening with or without simultaneous creep and shrinkage is presented, and an efficient time-step numerical integration algorithm, called the exponential algorithm, is developed. Microcracking that causes strain softening is permitted to take place only within three orthogonal planes. This allows the description of strain softening by independent algebraic relations for each of three orthogonal directions, including independent unloading and reloading behavior. The strain due to strain softening is considered as additive to the strain due to creep, shrinkage and elastic deformation. The time-step formulas for numerical integration of strain softening are obtained by an exact solution of a first-order linear differential equation for stress, whose coefficients are assumed to be constant during the time step but may vary discontinuously between the steps. This algorithm is unconditionally stable and accurate even for very large time steps, and guarantees that the stress is always reduced exactly to zero as the normal tensile strain becomes very large. This algorithm, called exponential because its formulas involve exponential functions, may be combined with the well-known exponential algorithm for linear aging rate-type creep. The strain-softening model can satisfactorily represent the test data available in the literature.

150   **Timoshenko Beams with Rotational End Constraints**

Timothy J. Ross, Member, ASCE, (Sr. Research Struct. Engr., Civ. Engrg. Research Div., Air Force Weapons Lab., Albuquerque, N. M. 87111-6008) and **Felix S. Wong**, Member, ASCE, (Sr. Assoc., Wiedlinger Assocs., 620 Hansen Way, Suite 100, Palo Alto, Calif. 94304)

Journal of Engineering Mechanics, Vol. 111, No. 3, March, 1985, pp. 416-430

Recent experimental evidence shows that the roof elements of reinforced concrete box-like structures fail in a direct shear mode when subjected to transverse, uniformly distributed, near impulsive pressures. These failures are typically characterized by excessive shear deformations near the roof supports. The Timoshenko beam theory is applied to study the failure of these one-way slabs by assessing the importance of the shear deformations and the importance of the support constraint. The Timoshenko equations are altered to account for variable rotational end constraint and the resulting normal mode solution is illustrated with a numerical example.

151   **Collapse of SDOF System to Harmonic Excitation**

Shuze Ishida, (Prof., Dept. of Architecture, Kyoto Inst. of Tech., Kyoto, Japan) and **Kiyetaka Morisako**, (Asst., Dept. of Architecture, Kyoto Inst. of Tech., Kyoto, Japan)

Journal of Engineering Mechanics, Vol. 111, No. 3, March, 1985, pp. 431-448

The investigation of the dynamic collapse vehavior of structures during strong wind disturbances provides a rational foundation for estimating the safety factor of structures. In order to discuss the essential features of dynamic collapse behavior,analytic and numerical studies are carried out on the dynamic behavior of single-degree-of-freedom (SDOF) inelastic systems, also taking into consideration the effect of gravity under harmonic perturbation conditions with a mean static force. This model disturbance, which may be applied to wind disturbances, will be briefly called the harmonic excitation. The dynamic collapse condition of a system with respect to a harmonic excitation can be given by considering the elastic steady-state response of the system. If an allowable displacement of a system is specified, the harmonic excitation which induces the system to collapse can be found. Inversely, if a harmonic excitation is specified, the tolerable displacement of the system can also be found.

152   **ARMA Representation of Random Processes**

Elias Samaras, Assoc. Member, ASCE, (Asst. Prof. of Civ. Engrg., Columbia Univ., New York, N.Y. 10027), **Masanobu Shinozuka**, Member, ASCE, (Renwick Prof. of Civ. Engrg., Columbia Univ., New York, N.Y. 10027) and **Akira Tsurui**, (Assoc. Prof. of Applied Mathematics and Physics, Kyoto Univ., Kyoto, Japan)

Journal of Engineering Mechanics, Vol. 111, No. 3, March, 1985, pp. 449-461

Auto-regressive moving-average (ARMA) models of the same order for AR and MA components are used for the characterization and simulation of stationary Gaussian multivariate random processes with zero mean. The coefficient matrices of the ARMA models are determined so that the simulated process will have the prescribed correlation function matrix. To accomplish this, the two-stage least squares method is used. The ARMA representation thus established permits one, in principle, to generate sample functions of infinite length and with such a speed and computational mode that even real time generations of the sample functions can be easily achieved. The numerical example indicates that the sample functions generated by the method presented herein reproduce the prescribed correlation function matrix extremely well despite the fact that these sample functions are all very long. This is seen from the closeness between the analytically prescribed auto- and cross-correlation functions and the corresponding sample correlations computed from the generated sample functions.

153   Ambient Vibration Studies of Golden Gate Bridge I: Suspended Structure

Ahmed M. Abdel-Ghaffar, Member, ASCE, (Assoc. Prof., Dept. of Civ. Engrg., Princeton Univ., Princeton, N.J. 08544) and **Robert H. Scanlan**, Member, ASCE, (Prof., Dept. of Civ. Engrg., Princeton Univ., Princeton, N.J. 08544)

Journal of Engineering Mechanics, Vol. 111, No. 4, April, 1985, pp. 463-482

Extensive experimental investigations were conducted on the Golden Gate Bridge in San Francisco, California, to determine, using ambient vibration data, parameters of major interest in both wind and earthquake problems, such as effective damping, the three-dimensional mode shapes, and the associated frequencies of the bridge vibration. The paper deals with the tests that involved the simultaneous measurement of vertical, lateral, and longitudinal vibration of the suspended structure; a subsequent paper addresses the measurement of the tower vibration. Measurements were made at selected points on different cross sections of the stiffening structure: 12 were on the main span and 6 on the side span. Good modal identification was achieved by special deployment and orientation of the motion-sensing accelerometers and by summing and subtracting records to identify and enhance vertical, torsional, lateral, and longitudinal vibrational modes. In all, 91 modal frequencies and modal displacement shapes of the suspended span were recovered: 20 vertical, 18 torsional, 33 lateral, and 20 longitudinal, all in the frequency range 0.0–1.5 hz. These numbers include symmetric and antisymmetric modes of vibration. Finally, comparison with previously computed two- and three-dimensional mode shapes and frequencies shows good agreement with the experimental results, thus confirming both the accuracy of the experimental determination and the reliability of the methods of computation.

154   Ambient Vibration Studies of Golden Gate Bridge II: Pier-Tower Structure

Ahmed M. Abdel-Ghaffar, Member, ASCE, (Assoc. Prof., Civ. Engrg. Dept., Princeton Univ., Princeton, N.J. 08544) and **Robert H. Scanlan**, Member, ASCE, (Prof., Civ. Engrg. Dept., Princeton Univ., Princeton, N.J. 08544)

Journal of Engineering Mechanics, Vol. 111, No. 4, April, 1985, pp. 483-499

Dynamic characteristics such as natural frequencies, mode shapes, and damping ratios of the Golden Gate Bridge tower were determined using ambient vibration data. The ambient vibration tests involved the simultaneous measurement of longitudinal and lateral vibrations of the main tower (San Francisco side). Measurements were made at different elevations of the tower and on the pier, at a total of 10 stations. Good modal identification was achieved by special deployment and orientation of the motion-sensing accelerometers and by summing and subtracting records to identify and enhance definition of longitudinal, torsional, and lateral vibration modes of the tower. A total of 46 modal frequencies and modal displacement shapes of the tower were identified: 20 longitudinal, 15 torsional, and 11 lateral, all in the frequency range of 0.0–5.0 Hz. Finally, comparison with previously computed two- and three-dimensional mode shapes and frequencies shows good agreement with the experimental results, thus confirming both the accuracy of the experimental determination and the reliability of the methods of computation.

155   Lateral Stability of Beams with Elastic End Restraints

John J. Zahn, Member, ASCE, (Research Engr., USDA Forest Service, Forest Products Lab., Madison, Wis. 53705)

Journal of Engineering Mechanics, Vol. 111, No. 4, April, 1985, pp. 500-511

In the analysis of the lateral buckling of simply supported beams, the ends are assumed to be rigidly restrained against tip. Real supports are, of course, never perfectly rigid. This report examines the relation of the stiffness of the end axial rotation restraint to the buckling load when the stabilizing effect of an attached deck is taken into account. For the case of uniform load, it is found that as the restraint stiffness approaches zero, the buckling load also approaches zero. This has implications in the design of large roof systems where end restraint on one member is provided by the torsional rigidity of another member connected in-line. Families of design curves are presented which show the effects of restraint stiffness, span-depth ratio, and shear stiffness of attached roof deck. It is concluded that periodic bracing against axial rotation is essential for stability of long roof systems with several beams spliced together in-line.

156   Torsional Instability in Hysteretic Structures

O. A. Pekau, Member, ASCE, (Assoc. Prof., Dept. of Civ. Engrg., Concordia Univ., Montreal, Quebec, Canada H3G 1M8) and **Pradip K. Syamal**, Member, ASCE, (Research Assoc., Dept. of Civ. Engrg., Concordia Univ., Montreal, Quebec, Canada H3G 1M8)

Journal of Engineering Mechanics, Vol. 111, No. 4, April, 1985, pp. 512-528

The occurrence of inelastic instability in the response of idealized eccentric building structures exhibiting various forms of bilinear hysteretic behavior is investigated. The Kryloff-Bogoliuboff method of averaging provides the response to harmonic ground excitation, with results examined in amplitude-frequency parameter space. The variables involved in the parametric portion of the work are the bilinearity coefficient, the degree of viscous damping, the torsional-to-translational frequency ratio and the magnitude of the eccentricity. it is found that pinched elastoplastic behavior representing steel frames with diagonal tension bracing leads to a high degree of instability, whereas ductile moment resisting frames possess full hysteresis loops and thus remain torsionally stable.

157   Buckling Analysis of FRP-Faced Anisotropic Cylindrical Sandwich Panel

Koganti Mohana Rao, (Asst. Prof., Dept. of Mech. Engrg., I.I.T., Kharagpur, India 721 302)

Journal of Engineering Mechanics, Vol. 111, No. 4, April, 1985, pp. 529-544

Force-deformation relations of a cylindrically curved symmetric anistropic sandwich plate are derived using Castigliano's theorem of minimum complementary energy. With these constitutive relations, the Rayleigh-Ritz method is applied for the buckling analysis of an FRP-faced curved sandwich plate under combined axial and bending loads. The buckling load coefficients of a sandwich plate with a typical fiberglass reinforced face sheet are presented by varying the parameters, e.g., aspect ratio, core to face thickness ratio, fiber orientation angle, and bending load coefficient. The results show that: (1) The buckling strength is maximum when the fiber orientation is about 40° with respect to circumferential direction; and (2) the buckling response is very sensitive to the combined effect of low aspect ratio, low radius, and bending load coefficient.

158   **Third-Variant Plasticity Theory for Low-Strength Concrete**

Howard L. Schreyer, Member, ASCE, (Prof., Dept. of Mech. Engrg. and Research Sci., New Mexico Engrg. Research Inst., Univ. of Mexico, Albuquerque, N.M. 87131) and **Susan, M. Babcock**, Assoc. Member, ASCE, (Research Engr., New Mexico Engrg. Research Inst., Univ. of New Mexico, Albuquerque, N.M. 87131)

Journal of Engineering Mechanics, Vol. 111, No. 4, April, 1985, pp. 545-558

A theory of plasticity for frictional materials is developed in which first and third invariants of stress and strain are used instead of the more conventional second invariants. The usual concepts of strain-hardening plasticity are used with the exception that a nonassociated flow rule is required to control dilatation. A procedure for determining material parameters is outlined. For a weak concrete, detailed comparisons are made between theoretical and experimental stress-strain data for a large number of three-dimensional paths.

159   **Microplane Model for Progressive Fracture of Concrete and Rock**

Zdenek P. Bazant, Fellow, ASCE, (Prof. of Civ. Engrg. and Dir., Center for Concrete and Geomaterials, The Technological Inst., Northwestern Univ., Evanston, Ill. 60201) and **Byung H. Oh**, Assoc. Member, ASCE, (Asst. Prof. of Civ. Engrg., Seoul National Univ., Seoul, Korea)

Journal of Engineering Mechanics, Vol. 111, No. 4, April, 1985, pp. 559-582

A constitutive model for a brittle aggregate material that undergoes progressive tensile fracturing or damage is presented. It is assumed that the normal stress on a plane of any orientation within the material, called the microplane, is a function of only the normal strain on the same microplane. This strain is further assumed to be equal to the resolved component of the macroscopic strain tensor, while the stress on the microplane is not equal to the resolved component of the macroscopic stress tensor. The normal strain on a microplane may be interpreted as the sum of the elastic strain and of the opening widths (per unit length) of all microcracks of the same orientation as the microplane. An additional volumetric elastic strain is introduced to adjust the elastic Poisson ratio to a desired value. An explicit formula which expresses the tangent stiffness of the material as an integral over the surface of a unit hemisphere is derived from the principle of virtual work. The model can represent experimentally observed uniaxial tensile strain-softening behavior, and the stress reduces to zero as the strain becomes sufficiently large. Due to various combinations of loading and unloading on individual microplanes, the response of the model is path-dependent. Since the tensorial invariance restrictions are always satisfied by the microplane system, the model can be applied to progressive fracturing under rotating principal stress directions. This type of application is the main purpose of the model.

160   **Effects of Vortex-Resonance on Nearby Galloping Instability**

Alireza Bokaian, Member, ASCE, (Sr. Engr., Earl and Wright Ltd., Victoria Station House, 191 Victoria Street, London, England SW1E 5NE) and **Farhad Geoola**, (Grad. Research Student, Dept. of Civ. Engrg., Univ. Coll. London, Gower Street, London, England WC1E 6BT)

Journal of Engineering Mechanics, Vol. 111, No. 5, May, 1985, pp. 591-609

Measurements are presented of the response of a rigid smooth rectangular prism, free to oscillate laterally against linear springs in a uniform flow. The prism was of side ratio 1:2 with the broader side facing the flow direction. The experiments also encompassed wake observations behind the prism when fixed, as well as determination of the lift forces on the fixed prism as a function of angle of flow attack. Dynamic tests showed that an increase in structural damping generally causes the instability to begin at a higher flow speed. The behavior of the cylinder showed differing features depending on the level of the response parameter. For low values of the response parameter, the vortex lock-in suppressed the galloping instability in the vicinity of the vortex-resonance speed. In lock-in regime, the vibration frequency showed a considerable variation with flow speed. For high values of the response parameter, the galloping became separated from vortex-resonance. Whereas turbulence was found to have no appreciable effect on vortex lock-in, it had profound effects on galloping oscillation.

---

161   Buckling of Orthotropic Cylinders due to Wind Load

Sukhvarsh Jerath, Member, ASCE, (Asst. Prof., Dept. of Civ. and Environmental Engrg., Washington State Univ., Pullman, Wash. 99164) and **Habib Sadid**, (Research Asst., Dept. of Civ. and Environmental Engrg., Washington State Univ., Pullman, Wash. 99164)

Journal of Engineering Mechanics, Vol. 111, No. 5, May, 1985, pp. 610-622

In many practical problems, e.g., storage bins, oil tanks, missile shells, launch vehicles, etc., cylindrical shells are subjected to unsymmetrical lateral external load. These shells are usually made of thin sheets of metal; therefore, in many instances these structures have failed due to buckling. In this paper, the stability of cylindrical shells under the action of wind load is investigated. Often these shells are constructed of corrugated steel sheets; therefore, the shells are analyzed by considering the material as orthotropic. This analysis can also be used for other orthotropic materials. The principle of minimum potential energy in conjunction with Ritz's approach is used. The buckling loads, as well as the buckling configurations, are obtained for short cylinders made of corrugated steel sheets subjected to wind pressure. The present study is made on cylindrical shells of various dimensions, which are simply supported at the base and are open or closed at the top. For practical use, buckling load curves for these shells are given for different length-to-radius and radius-to-thickness ratios.

---

162   Concrete Fracture in CLWL Specimen

Donald B. Barker, (Assoc. Prof. of Mech. Engrg., Univ. of Washington, Seattle, Wash. 98195), **Neil M. Hawkins**, (Chairman, Dept. of Civ. Engrg., Univ. of Washington, Seattle, Wash. 98195), **Fure-Lin Jeang**, (Grad. Student, Dept. of Civ. Engrg., Univ. of Washington, Seattle, Wash. 98195), **Kyo Zong Cho**, (Asst. Prof., Dept. of Civ. Engrg., Univ. of Washington, Seattle, Wash. 98195) and **Albert S. Kobayashi**, (Prof., Dept. of Civ. Engrg., Univ. of Washington, Seattle, Wash. 98195)

Journal of Engineering Mechanics, Vol. 111, No. 5, May, 1985, pp. 623-638

Results are reported of a series of tests on concrete crack-line wedge loaded double cantilever beams. Those tests provide a data base for fracture parameters characterizing mode I cracking in concrete. The development is reported of a replica technique that provides an easy and accurate method for determining the total extent of cracking in concrete specimens, and it is shown that the fracture process zone, which can be determined with that technique, continues to increase in length with increasing load. From analysis of the results, it is concluded that linear elastic fracture mechanics techniques are not suited to predicting unstable cracking in concrete members of the proportions normally used in buildings.

163   Free Vibration Analysis of Continuous Beams

Toshiro Hayashikawa, (Research Assoc., Dept. of Civ. Engrg., Hokkaido Univ., Nishi 8 Kita 13 Kita-Ku, Sapporo, Japan 060) and **Noboru Watanabe**, (Prof., Dept. of Civ. Engrg., Hokkaido Univ., Nishi 8 Kita 13 Kita-Ku, Sapporo, Japan 070)

Journal of Engineering Mechanics, Vol. 111, No. 5, May, 1985, pp. 639-652

An analytical method for determining eigenvalues of continuous beams is developed by using a general solution for the Bernoulli-Euler differential equation. This method results in an eigenvalue problem in which the solution of a transcendental equation containing trigonometric and hyperbolic functions is obtained, and it leads to an exact solution. Also, the approximate method based on the finite element approach is presented. The mathematical relationship between the exact and the approximate methods is discussed, and the accuracy of the eigenvalues obtained by these methods is investigated. Some typical continuous beams are analyzed to illustrate the applicability of the lumped, consistent, and continuous mass methods, and the computed results are given in tabular form.

164   Large Deflected Plates and Shells with Loading History

Boris Krayterman, Member, ASCE, (Engrg. Specialist, Bechtel Power Corp., Gaithersburg, Md. 20877-1454) and **Gajanan M. Sabnis**, Fellow, ASCE, (Prof. of Civ. Engrg., Howard Univ., Washington, D.C. 20059)

Journal of Engineering Mechanics, Vol. 111, No. 5, May, 1985, pp. 653-663

Among the problems in the design of geometrically nonlinear plates and shells, there is a special class in which the total stress-strain behavior must be considered with loading history. The loading history includes the plate-shell loading, boundary conditions, and material behavior, but assumes only the "discrete steps" in the aforementioned considerations. To indicate the difference between "discrete steps," the axisymmetrical bending of plates and shells is presented along with suitable differential equations for loading history. The supplemental terms are derived and the results illustrated with numerical examples. The arbitrarily loaded large deformed plates and shallow shells are presented along with suitable differential equations for discrete loading history. The importance of variable curvature versus constant curvature consideration is shown for large deformed shallow shells. With consideration of geometrical and statical heredity of every two successive steps of loading history, the linearization can simplify calculations of plates and shells with large deflections.

165   Beam on Generalized Two-Parameter Foundation

Toyoaki Nogami, Member, ASCE, (Assoc. Prof., Dept. of Civ. Engrg., Univ. of Houston, Houston, Tex. 77004) and **Michael W. O'Neill**, Member, ASCE, (Assoc. Prof., Dept. of Civ. Engrg., Univ. of Houston, Houston, Tex. 77004)

Journal of Engineering Mechanics, Vol. 111, No. 5, May, 1985, pp. 664-679

A method for analysis of beams bearing on a ground surface is presented. The method is based on treating the soil medium as a generalized two-parameter model. The inputs required for the model are dimensions and material properties only, which usually are known before the analysis. This contrasts to other two-parameter models in which the parameters are based upon assumed soil displacement distributions that may be difficult to predict before the analysis. Analyses show the new two-parameter model can yield responses of loaded beams reasonably close to those computed by using continuum solutions or a finite element method.

166   Lagrangian Approach to Design Sensitivity Analysis

Ashok D. Belegundu, (Asst. Prof., Mech. Engrg. Dept., GMI Engineering & Management Inst., 1700 West Third Ave., Flint, Mich. 48502)

Journal of Engineering Mechanics, Vol. 111, No. 5, May, 1985, pp. 680-695

A Lagrangian approach to design sensitivity analysis is presented. The final equations obtained by the Lagrangian approach are identical to those obtained by the adjoint method reviewed in the literature. The difference lies in the approach taken to derive these equations. The Lagrangian approach is identical for different categories of design problems, as is demonstrated by considering structural, dynamic, distributed-parameter, and shape optimal design problems. The Lagrangian approach exposes the fact that the "adjoint variables" referred to in the literature are, in fact, the Lagrange multipliers associated with the state equations, and the "adjoint equations" are the classical Euler-Lagrange equations. The clearer understanding that is obtained by this approach leads to some immediate practical advantages, and opens up some new areas for research.

167   Nonlinear Hysteretic Dynamic Response of Soil Systems

Jean-Herve Prevost, Member, ASCE, (Assoc. Prof., Civ. Engrg. Dept., Princeton Univ., Princeton, N.J. 08544), **Ahmed M. Abdel-Ghaffar**, Member, ASCE, (Assoc. Prof., Civ. Engrg. Dept., Princeton Univ., Princeton, N.J. 08544) and **Ahmed-Waeil M. Elgamal**, (Grad. Student, Dept. of Civ. Engrg., Princeton Univ., Princeton, N.J. 08544)

Journal of Engineering Mechanics, Vol. 111, No. 5, May, 1985, pp. 696-713

A simplified analysis procedure for the nonlinear hysteretic dynamic response of soil or structural systems, or both, is presented. The method is based on a Galerkin formulation of the equations of motion in which the solution is expanded using basis functions defined over the spatial domain occupied by the soil system. The basis functions are selected as the normal eigenmodes of the linearized problem. The hysteretic stress-strain behavior is modeled by using elastoplastic constitutive equations based on multi-surface kinematic plasticity theory. Accuracy and versatility of the technique are demonstrated by applying it to analyze the nonlinear dynamic response of an earth dam. The dam is modeled as a one-dimensional hysteretic shear wedge. Parametric studies assessing the influence of the nonlinearities on the response are presented. Finally, comparisons are made with results obtained through a more elaborate finite element representation of the dam.

168   Simplified Earthquake Analysis of Concrete Gravity Dams:
Combined Hydrodynamic and Foundation Interaction
Effects

Gregory Fenves, Assoc. Member, ASCE, (Asst. Prof., Dept. of Civ. Engrg., Univ. of Texas, Austin, Tex. 78712) and **Anil K. Chopra**, Member, ASCE, (Prof. of Civ. Engrg., Dept. of Civ. Engrg., Univ. of Texas, Austin, Tex. 78712)

Journal of Engineering Mechanics, Vol. 111, No. 6, June, 1985, pp. 736-756

A companion paper presented simplified procedures for earthquake analysis of the fundamental mode response of concrete gravity dams including the separate effects of dam-foundation rock interaction and dam-water interaction with reservoir bottom absorption. These procedures are extended to develop a simplified analytical procedure for evaluation of the response of concrete gravity dams to earthquake ground motion including the simultaneous effects of dam-water interaction, reservoir bottom absorption, and dam-foundation rock interaction. Expressions for the parameters of an equivalent SDF system that models the fundamental mode response of dams are derived, a procedure to implement the analytical procedure is outlined, and an extension to consider the response contributions of the higher vibration modes of the dam is briefly mentioned.

169   **Simplified Earthquke Analysis of Concrete Gravity Dams: Separate Hydrodynamic and Foundation Interaction Effects**

Gregory Fenves, Assoc. Member, ASCE, (Asst. Prof., Dept. of Civ. Engrg., Univ. of Texas, Austin, Tex. 78712) and Anil K. Chopra, Member, ASCE, (Prof. of Civ. Engrg., Dept. of Civ. Engrg., Univ. of Texas, Austin, Tex. 78712)

Journal of Engineering Mechanics, Vol. 111, No. 6, June, 1985, pp. 715-735

Simplified procedures are presented for the analysis of the fundamental vibration mode response of concrete gravity dam systems for two special cases: (a) dams with reservoirs of impounded water supported on rigid foundation rock; and (b) dams with empty reservoirs supported on flexible foundation rock. In the first case, the effects of dam-water interaction and reservoir bottom absorption on dam response are included, whereas the effects of dam-foundation rock interaction are included in the second case. In each case, the response of the fundamental vibration mode of a dam monolith is modeled by an equivalent single degree-of-freedom system with frequency-independent properties chosen to represent the effects of complicated, frequency-dependent hydrodynamic terms or foundation-rock flexibility terms, as appropriate. The maximum earthquake-induced deformations and equivalent lateral forces can be computed using the response spectrum for a specified ground motion. The procedures and results presented in this paper are extended in a companion paper to develop a simplified analytical procedure for concrete gravity dams that includes the simultaneous effects of dam-water interaction, reservoir bottom absorption, and dam-foundation rock interaction.

170   **Nonlinear Effects in Creep Buckling Analysis of Columns**

A. M. Vinogradov, (Prof., Dept. of Civ. Engrg., Univ. of Calgary, Calgary, Alberta, Canada)

Journal of Engineering Mechanics, Vol. 111, No. 6, June, 1985, pp. 757-767

A consistent theoretical investigation of geometrically nonlinear creep buckling behavior of structures is presented. The study is based on the analysis of eccentrically compressed viscoelastic columns with the material properties described in terms of linear integral operators of the convolution type. The nonlinear solution is obtained by means of the quasi-elastic method. Subsequently, the linear solution is derived as a special case using the assumptions of the small deformation theory. The analysis involves materials with both limited and unlimited creep behavior. It is observed that in the case of limited creep there is a safe load limit below which the creep buckling process stabilizes in time. The nonlinear and linear creep buckling characteristics are derived and compared for two viscoelastic material models and various magnitudes of the load and the eccentricity parameters. On the basis of these results applicability of the geometrically linear theory is examined and the question is raised as to the usefulness of the critical time criterion in terms of infinite deformations. The derived results are compared with those of previous studies.

171   Laminar Flow in Conduits of Unconventional Shape

Mario F. Letelier S., (Prof., Departmento de Ingenieria Mecanica, Universidad de Santiago de Chile, Casilla, Santiago, Chile 10233) and **Hans J. Leutheusser**, Member, ASCE, (Prof., Dept. of Mech. Engrg., Univ. of Toronto, Toronto, Ontario, Canada M5S 1A4)

Journal of Engineering Mechanics, Vol. 111, No. 6, June, 1985, pp. 768-776

A new analytical solution technique is presented which greatly extends the range of conduit shapes for which a mathematical description of the enclosed laminar flow becomes possible. To this end, a known particular solution u_p of the Poisson equation is linearly combined with some harmonic function u_n in the form $u = u_p + \epsilon\, u_h$. In this, $\epsilon \gtrless 0$ is a parameter which determines the shape of "new" conduits and whose range of possible values is governed by the no-slip condition.

172   Control of Lateral-Torsional Motion of Wind-Excited Buildings

B. Samali, (Asst. Research Prof., Dept. of Civ., Mech. and Environmental Engrg., George Washington Univ., Washington, D.C. 20052), **J. N. Yang**, (Prof., Dept. of Civ., Mech. and Environmental Engrg., George Washington Univ., Washington, D.C. 20052) and **C. T. Yeh**, (Dir., Graduate Inst. of Civ. Engrg., Tamkang Univ., Tamshui, Taipei, Taiwan, R.O.C.)

Journal of Engineering Mechanics, Vol. 111, No. 6, June, 1985, pp. 777-796

An investigation is made of the possible application of an active mass damper control system to tall buildings excited by strong wind turbulence. The effectiveness of active control system, as measured by the reduction of the coupled lateral-torsional motions of tall buildings is studied. The wind turbulence is modeled as a stochastic process that is stationary in time but nonhomogeneous in space. The problem is formulated using the transfer matrices approach, and a closed-loop control law. The random vibration analysis is carried out to determine the statistics of the building responses, the required active control forces, and mass damper displacements. The method of Monte-Carlo simulation is also employed to demonstrate the building response behavior with or without an active control system. A numerical example of a forty-story building under strong wind excitations is given to illustrate the significant reduction of the building acceleration response by use of an active mass damper control system.
Errata: EM Sept. '85, pp. 1213.

173   Constitutive Model for Concrete in Cyclic Compression

En-Sheng Chen, Assoc. Member, ASCE, (Engrg., Brian Watt Associates, Houston, Tex. 77032) and **Oral Buyukozturk**, Member, ASCE, (Prof., Dept. of Civ. Engrg., Massachusetts Inst. of Tech., Cambridge, Mass. 02139)

Journal of Engineering Mechanics, Vol. 111, No. 6, June, 1985, pp. 797-814

A rate-independent constitutive model is proposed for the behavior of concrete in multiaxial cyclic compression. The material composite is assumed to experience a continuous damage process under load histories. The model adopts a damage-dependent bounding surface in stress space to predict the strength and deformation characteristics of the gross material under general loading paths. Reduction in size of the bounding surface as damage accumulates, and the adopted functional dependence of the material moduli on stress and damage permit a realistic modeling of the concrete behavior. Satisfactory prediction is obtained of the generally nonlinear stress-strain response, degradation in stiffness during load cycles, shear compaction-dilatancy phenomena, and post-failure strain softening behavior. Finite element implementation of the proposed model is feasible and computationally efficient.

174   Rigid-Plastic Analysis of Floating Plates

Shankaranarayana U. Bhat, (Inst., Dept. of Ocean Engrg., Massachusetts Inst. of Tech., Cambridge, Mass. 02139) and **Paul C. Xirouchakis**, (Assoc. Prof. of Ocean Engrg., Dept. of Naval Architecture and Marine Engrg., National Tech. Univ. of Athens, Athens, Greece 10682)

Journal of Engineering Mechanics, Vol. 111, No. 6, June, 1985, pp. 815-831

The exact formulation and solution for the static flexural response of a rigid perfectly-plastic freely floating plate subjected to lateral axisymmetric loading is presented. The square yield condition is adopted with the associated flow rule. The plate response is divided into three phases. Initially, the plate moves downwards into the foundation as a rigid body (Phase I). Subsequently, the plate deforms in a conical mode in addition to the rigid body motion (Phase II). At a certain value of the load a hinge-circle forms which may move as the pressure increases further (Phase III). The nature of the solution during the third phase depends upon the parameter $\alpha = a/R$ (ratio of radius of loaded area to the plate radius). When $\alpha = a_s \simeq 0.43505$ the hinge-circle remains stationary under increasing load. For $\alpha < \alpha_s$ the hinge-circle shrinks, whereas for $\alpha > \alpha_s$ the hinge-circle expands with increasing pressure. The application of the present results to the problem of laterally loaded floating ice plates is examined.

175   Plastic Response of Cantilevers with Stable Cracks

H. J. Petroski, Member, ASCE, (Assoc. Prof., Dept. of Civ. Engrg., Duke Univ., Durham, N.C. 27706) and **A. Verma**, Student Member, ASCE, (Research Asst., Dept. of Civ. and Environmental Engrg., Duke Univ., Durham, N.C. 27706)

Journal of Engineering Mechanics, Vol. 111, No. 7, July, 1985, pp. 839-853

Simple analytical models and experiments are employed to demonstrate the structural response of a cantilever beam with a stable crack subjected to impact loading. The mode of plastic deformation is shown to depend very strongly upon the size and location of the crack; and the permanent damage suffered by a cracked beam is found to be significantly different in magnitude and character from that of a correspondingly-loaded uncracked beam.

176   Formulation of Drucker-Prager Cap Model

Luis Resende, (Sr. Research Officer, Nonlinear Struct. Mechanics Research Unit, Univ. of Cape Town, Rondebosch, South Africa 7700) and **John B. Martin**, Member, ASCE, (Dean of Engrg., Univ. of Cape Town, Rondebosch, South Africa 7700)

Journal of Engineering Mechanics, Vol. 111, No. 7, July, 1985, pp. 855-881

The Drucker-Prager cap and similar models for the constitutive behavior of geotechnical materials are widely used in finite element stress analysis. They are multisurface plasticity models, used most frequently with an associated flow rule. The cap may harden or soften, and is coupled to the Drucker-Prager yield surface. As a result of this coupling, plastic deformation in pure shear is possible, after some plastic volume change, for any state of stress on the Drucker-Prager surface. This suggests that for full coupling the constitutive equations for the model can be found consistently; however, the model exhibits unstable behavior under certain conditions. To suppress this instability, some modification of the coupling must be made. Two examples of such modifications which appear in the literature are given; each leads to an inconsistent formulation. Numerical examples are used to illustrate differences and consequences arising from the different assumptions.

177   A Method for Finding Engineering Properties of Sealants

Jon Baxter Anderson, (Research Assoc., Dept. of Civ. Engrg., Texas Tech. Univ., Lubbock, Tex. 79409)

Journal of Engineering Mechanics, Vol. 111, No. 7, July, 1985, pp. 882-892

The increasing use of high modulus polymer sealants, as connectors in structural applications such as glass curtain walls and insulating glass units, has shown that there is a need for better information on the engineering properties of these sealant materials. Currently available information on sealants is often limited to ultimate adhesive and cohesive strengths. More typical engineering properties are required. Stress-strain properties and a number of factors affecting stress-strain properties of polymers are of interest. Factors such as stress-relaxation and strain rate are considered, and the dependence of modulus on the factors are described.

178   Dynamic Instability Analyses of Axially Impacted Columns

Kunitomo Sugiura, (Grad. Student, Dept. of Civ. Engrg., State Univ. of New York at Buffalo, Buffalo, N.Y. 14260), **Eiji Mizuno**, (Research Assoc., Dept. of Civ. Engrg., Nagoya Univ., Nagoya, Japan 464) and **Yuhshi Fukumoto**, Member, ASCE, (Prof., Dept. of Civ. Engrg., Nagoya Univ., Nagoya, Japan 464)

Journal of Engineering Mechanics, Vol. 111, No. 7, July, 1985, pp. 893-908

The dynamic response and critical state of an inelastic simply supported column under an axial impact are studied. This problem is analyzed by solving the dynamic Bernoulli-Euler equation with an axial inertia effect within the framework of finite difference approach. The influence of strain-rate effects on the dynamic response is first examined by using an elastic-viscoplastic theory, and then the critical values of an initial velocity and a mass of striking body for losing stability are evaluated within the context of the numerical results from different initial and boundary conditions. It is found that strain-rate effects are important in the range of post-dynamic instability, and that the mode of the lateral displacement after an impact depends on the initial velocity of the striking body and also on the relationship between the natural period of the first-order lateral mode and that of the first-order axial mode. Present study can give a basic guide to evaluate the dynamic instability of axially impacted columns from a viewpoint of the energy loss of the striking body.

179   Snap-Through and Bifurcation in a Simple Structure

D. A. Pecknold, Member, ASCE, (Prof., Dept. of Civ. Engrg., Univ. of Illinois at Urbana-Champaign, Urbana, Ill. 61801), **J. Ghaboussi**, Member, ASCE, (Prof., Dept. of Civ. Engrg. Univ. of Illinois at Urbana-Champaign, Urbana, Ill. 61801) and **T. J. Healey**, (Visiting Asst. Prof., Dept. of Mathematics, Univ. of Maryland, College Park, Md. 20742)

Journal of Engineering Mechanics, Vol. 111, No. 7, July, 1985, pp. 909-922

Basic characteristics associated with snap-through, bifurcation and post-buckling behavior of structures are illuminated and clarified. This is accomplished by presenting and examining in detail exact closed-form global solutions for the load-deflection response of a simple two-bar planar truss. Three loading conditions are considered: vertical (symmetric), horizontal (anti-symmetric) and combined vertical-horizontal loadings. The symmetric loading case is well-known and has been widely used as a bench mark problem for comparison of numerical solution algorithms. The two-bar planar truss considered involves standard structural and material behavior models which are incorporated in many structural analysis codes, and it can be made to mimic a wide-variety of types of behavior. Concrete examples of behavior which could cause computational difficulties in the numerical analysis of more complex structural models are valuable in the process of developing and improving solution algorithms for nonlinear structural analysis.

180   Instability of Thin Walled Bars

Jerzy W. Wekezer, Member, ASCE, (Visiting Asst. Prof. in Civ. Engrg., Univ. of Southern California, Los Angeles, Calif.)

Journal of Engineering Mechanics, Vol. 111, No. 7, July, 1985, pp. 923-935

Instability of thin walled bars of variable, open cross sections is analyzed. A thin walled bar is treated as a special case of the membrane shell with the internal constraints (Vlasov's and Wagner's assumptions). The geometry of the bar is described by means of coordinates of the discrete points located on the midsurface of the bar. Principal coordinate system is assumed for the cross section. Large deformations of the cross section are analyzed in total Lagrangian formulation. Prebuckling strains are assumed to be small and the Green-Lagrange linearized tensor is used in strain analysis. The general, nonlinear formula for the strain tensor element ge_{22} is obtained for the thin walled bars of variable, open cross sections. Special cases derived from this formula are in agreement with already known solutions. Numerical results are obtained by the use of the finite element approach. Stiffness and geometric matrices are constructed for a thin walled, variable element. Typical shape functions are used and a bifurcation point of stability is determined with the help of the eigenproblem solution.

181   Analytical Theory for Buried Tube Postbuckling

Ian D. Moore, (Lect., Dept. of Civ. Engrg. and Surveying, Univ. of Newcastle, Newcastle, New South Wales, Australia 2308)

Journal of Engineering Mechanics, Vol. 111, No. 7, July, 1985, pp. 936-951

A simplified analytical postbuckling theory has been developed for long, circular, elastically supported tubes. The solution is an extension of linear theories which have previously been developed. Although the analysis is approximate it is found to be in good agreement with a more rigorous solution involving complete finite element analysis. The simplified solution is for both uniform and nonuniform distributions of hoop compression. It performs satisfactorily up to about twice the critical load level. It has been used to briefly examine the effect of geometrical imperfections and nonhydrostatic field stress on buried tube response.

182   Effective Length of a Fractured Wire in Wire Rope

Chi-Hui Chien, (Grad. Research Asst., Dept. of Theoretical and Applied Mechanics, Univ. of Illinois at Urbana-Champaign, 216 Talbot Lab., 104 S. Wright St., Urbana, Ill. 61801-2983) and **George A. Costello**, Member, ASCE, (Prof., Dept. of Theoretical and Applied Mechanics, Univ. of Illinois at Urbana-Champaign, 216 Talbot Lab., 104 S. Wright St., Urbana, Ill. 61801-2983)

Journal of Engineering Mechanics, Vol. 111, No. 7, July, 1985, pp. 952-961

An analytical method is presented for the determination of the length, measured from the fractured end of the wire, in which the wire will be able to carry its appropriate share of the load. The estimate of this effective length is based on the contact loads between the wires, Coulomb type friction, and an invoation of Saint-Venant's principle. The results are applied to a simple strand and a wire rope with a complex cross section.

183   **Acceleration Techniques for Coupled Nonlinear PDE's**

Yehia R. Marmoush, (Research Asst., Dept. of Civ. Engrg., McMaster Univ., Hamilton, Ontario, Canada), **Pulak C. Chakravarti**, (Assoc, Prof., Dept. of Mathematical Sci., McMaster Univ., Hamilton, Ontario, Canada) and **Alan A. Smith**, (Prof., Dept. of Civ. Engrg., McMaster Univ., Hamilton, Ontario, Canada)

Journal of Engineering Mechanics, Vol. 111, No. 7, July, 1985, pp. 962-976

Different acceleration techniques are employed to improve the solution procedure of a set of coupled, nonlinear, partial differential equations. The presented study applies the techniques to an idealized problem of two-dimensional steady laminar flow in an enclosed rectangular cavity with differentially heated end walls. The numerical model is verified by comparing it to analytical solutions of four different cases of the presented problem. An explicit comparison between the alternative acceleration techniques is made to ascertain the relative efficiency.

184   **Periodic Response of Yielding Oscilllators**

Amitabha DebChaudhury, (Asst. Prof., Univ. of Illinois at Chicago, Dept. of Civ. Engrg., Mechanics and Metallurgy, P. O. Box 4348, Chicago, Ill. 60680)

Journal of Engineering Mechanics, Vol. 111, No. 8, August, 1985, pp. 977-994

A new method is employed to obtain the response of a general yielding oscillator to harmonic excitation. This method can be applied to a wide range of hysteretic systems, once the skeleton curve and the loop is defined. In this paper, responses are obtained for four different models, describing the hysteretic behavior of physical system. The analytical expressions for the frequency response curves and the frequency shift at peak responses are obtained. They are extremely simple, even for a general curved hysteresis loop. Results are compared with the numerical solutions and the findings of earlier investigators. Peak responses turn out to be exactly the same as that obtained by earlier researchers, but the frequency shift shows some difference. For a bilinear system the frequency shift at peak response is a function of load level only, and the post yielding slope determines only the lower bound on the frequency shift. Also the frequency response curves are not single valued, contrary to the earlier findings, specially corresponding to lower peak responses.

185   **Continuum Damage Mechanics of Fiber Reinforced Concrete**

D. Fanella, (Grad. Research Asst., Dept. of Civ. Engrg., Univ. of Illinois at Chicago, Chicago, Ill.) and **D. Krajcinovic**, (Prof., Dept. of Civ. Engrg., Univ. of Illinois at Chicago, Chicago, Ill.)

Journal of Engineering Mechanics, Vol. 111, No. 8, August, 1985, pp. 995-1009

A nonlinear analytical model based on the principles of the Continuum Damage Mechanics is developed for the stress-strain behavior of fiber reinforced concrete (or mortar) subjected to monotonic compressive and tensile loading. An equilibrium equation derived from a parallel bar arrangement of the composite material, coupled with the damage laws of the composite and the fibers, form the basis of this model. Stress-strain curves generated from this model are compared to those acquired from experimental data. For both compressive and tensile loading, the analytical expression generates stress-strain quantities well within the range of the scatter of the experimental data.

186   **Random Vibration of Degrading, Pinching Systems**

Thomas T. Baber, Assoc. Member, ASCE, (Asst. Prof., Dept. of Civ. Engrg., Univ. of Virginia, Charlottesville, Va. 22901) and **M. N. Noori**, (Grad. Research Asst., Dept. of Civ. Engrg., Univ. of Virginia, Charlottesville, Va. 22901)

Journal of Engineering Mechanics, Vol. 111, No. 8, August, 1985, pp. 1010-1026

A differential equation model to describe pinching, degrading response of hysteretic elements is presented. The model consists of a nonpinching hysteretic element, in series with a "slip-lock" element. Zero mean response statistics for a single degree of freedom oscillator whose stiffness is described by the series model, computed by equivalent linearization and by Monte Carlo simulation, are compared. The model response statistics are seen to be reasonable estimated by equivalent linearization.

187   **Dynamic Analysis of Orthotropic Plate Structures**

Nabil F. Grace, (Research Asst., Dept. of Civ. Engrg., Univ. of Windsor, Windsor, Ontario, Canada N9B 3P4) and **John E. Kennedy**, Fellow, ASCE, (Prof., Dept. of Civ. Engrg., Univ. of Windsor, Windsor, Ontario, Canada N9B 3P4)

Journal of Engineering Mechanics, Vol. 111, No. 8, August, 1985, pp. 1027-1037

The dynamic response of orthotropic plate structures having fixed-simply supported and free-free boundary conditions is investigated using orthotropic plate theory. The influences of aspect ratio and rigidity ratio on the natural frequencies are examined and compared to those obtained from beam-theory. The analytical results, verified by experimental test results, confirm that for this class of structures the natural frequencies beyond the first cannot be reliably estimated by beam-theory.

188   **Gravity Effects in Consolidation of Layer of Soft Soil**

Chiang C. Mei, Member, ASCE, (Prof., Dept. of Civ. Engrg., Massachusetts Inst. of Tech., Cambridge, Mass. 02139)

Journal of Engineering Mechanics, Vol. 111, No. 8, August, 1985, pp. 1038-1047

The role of gravity in the consolidation of a relatively soft soil layer of large thickness is assessed analytically. Under the assumption of small deformation, Terzaghi's equation is modified and solved for a surface loading. Linearized Biot equations of poro-elasticity are used and solved by Laplace transform. Numerical results are given and examined.

189   **Polyaxial Yielding of Granular Rock**

Paul Michelis, Assoc. Member, ASCE, (Lab. of Reinforced Concrete, National Tech. Univ. of Athens, Greece)

Journal of Engineering Mechanics, Vol. 111, No. 8, August, 1985, pp. 1049-1066

Results from true triaxial testing on a granular dense marble are analyzed with emphasis on the incremental relations during strain-hardening (initial yield to peak strength) at constant intermediate and minor stresses. The strong dependency of the behavior on the value of intermediate principal stress is confirmed. Consistency of experimental flow rule was revealed for the examined yield states corresponding to initial yield, to a given value of plastic strain invariant, and to peak strength. A stress invariant and the corresponding one of plastic strain increment were almost equal, postulating a simple stress strain relation. The vector of plastic strain increment and the normal vector on the yield surface at the relative point formed an angle ranging from 10 to 23 degrees. On the basis of the preceding observations, a work balance equation, the resulting flow rule, and yield equation were extended to model three-dimensional behavior. Reasonable agreement was obtained between the predicted and experimental stress-strain relations.

190   Dynamic Analysis of Short-Length Gravity Dams

Ahmed A. Rashed, Assoc. Member, ASCE, (Asst. Prof. of Civ. Engrg., Johns Hopkins Univ., Baltimore, Md. 21218) and **Wilfred D. Iway**, (Prof. of Civ. Engrg. and Appplied Mechanics, California Inst. of Tech., Pasadena, Calif. 91125)

Journal of Engineering Mechanics, Vol. 111, No. 8, August, 1985, pp. 1067-1083

A simplified and economical procedure is developed to analyze the dynamic behavior of short-length gravity dams. The analysis is based on the Rayleigh-Ritz method and an idealized dam-reservoir geometry. The substructure concept is used, in which the dam is modeled as a thick plate and the water in the reservoir is treated as a continuum. The model accounts for dam-reservoir interaction, water compressibility, and flexibility of the reservoir floor and sides. The natural frequencies and mode shapes of the dam are obtained through a free vibration analysis, and the three-dimensional effects on these properties are illustrated. A forced vibration analysis is carried out in the frequency domain, and the dam response to all three components of ground motion is obtained. The effects of the length to height ratio of the dam on its response, and the significance of the cross-stream component of ground motion are investigated.

191   Experimental Measurement of Multiple-Jet Induced Flow

Joseph H. W. Lee, Assoc. Member, ASCE, (Lect., Dept.of Civ. Engrg., Univ. of Hong Kong, Hong Kong) and **C. W. Li**, (Demonstrator, Dept. of Civ. Engrg., Univ. of Hong Kong, Hong Kong)

Journal of Engineering Mechanics, Vol. 111, No. 8, August, 1985, pp. 1087-1092

Experimental results are confirmed for theoretical predictions of a vortex model for the two-dimensional inviscid flow generated by a line of submerged, turbulent, shallow water jets in a coflowing current. The experiments are performed in a large shallow water basin in which a strong uniform current can be produced and boundary effects are negligible. The vortex model is validated against detailed velocity measurements at and downstream of the momentum source line. Observations of surface flow patterns show that, depending only on the ambient current to ultimate slipstream velocity ratio, distinct flow regimes ranging from a strong inwardly directed sink flow to a smooth contracting slipstream can result.

192   Markov Renewal Model for Maximum Bridge Loading

Michel Ghosn, Assoc. Member, ASCE, (Research Assoc., Dept. of Civil Engrg., Case Western Univ., Cleveland, Ohio 44106) and **Fred Moses**, Member, ASCE, (Prof., Dept. of Civ. Engrg., Case Western Reserve Univ., Cleveland, Ohio)

Journal of Engineering Mechanics, Vol. 111, No. 9, September, 1985, pp. 1093-1104

The prediction of maximum vehicle loadings on a bridge is studied. The stationary distribution of the static response of highway bridges under random truck loading is obtained using a Markov Renewal Model. This model is a generalization of Markov chains and renewal processes and can be used to model the arrival of trucks on a multiline bridge. The model also accounts for random truck characteristics such as axle weights, axle spacings, speed, and the headway distribution between trucks. The stationary distribution of the response is obtained assuming that the bridge (represented by its influence line) acts as a filter to the truck arrival process. The maximum lifetime response is obtained from the stationary distribution using an approximation to Rice's upcrossing rate formula. The results are then compared to a simulation program and acceptable agreement is reported.

193 Uniaxial Cyclic Stress-Strain Behavior of Structural Steel

Nathaniel G. Cofie, (Sr. Consultant, NUTECH Engrs., Inc., 145 Martinvale Lane, San Jose, Calif. 95119) and **Helmut Krawinkler**, Member, ASCE, (Prof., Dept. of Civ.Engrg., Stanford Univ., Stanford, Calif. 94305)

Journal of Engineering Mechanics, Vol. 111, No. 9, September, 1985, pp. 1105-1120

A simple mathematical model for the uniaxial cyclic stress-strain behavior of structural steel is proposed for arbitrary loading histories in the inelastic range. The model uses the monotonic and cyclic stress-strain curves as references and assumes the existence of stress bounds which cannot be exceeded in any given cycle. The movement of the bounds, which is controlled by hardening, softening, and mean stress relaxation, is determined by the strain amplitude of the last excursion and the previous stress-strain history. The rates of hardening, softening, and mean stress relaxation are determined from experimental data. The nonlinear portions of the stress-strain curves are defined by a continuously changing tangent modulus whose magnitude is a function of the distance between the stress bound and the instantaneous stress. A comparison is made of predicted and experimentally obtained stress-strain histories. Although the model is developed specifically for types of histories associated with seismic excitations (small number of cycles), it should be applicable as well to low cycle fatigue problems involving large numbers of cycles since it is based on the stabilized cyclic stress-strain curve.

194 Extreme-Value Statistics for Nonlinear Stress Combination

Henrik O. Madsen, Assoc. Member, ASCE, (Chf. Scientist, Reliability Analysis, A.S. Veritas Research, P.O. Box 300, N-1322 Hovik, Norway)

Journal of Engineering Mechanics, Vol. 111, No. 9, September, 1985, pp. 1121-1129

The extreme-value distribution for the von Mieses stress is determined. The stress components are stationary Gaussian processes. An upcrossing of a stress level by the von Mieses stress is described as he outcrossing of an ellipsoid by the stress component vector process. The extreme-value distribution for the von Mieses stress is expressed in terms of the mean outcrossing rate of the ellipsoid. The mean outcrossing rate is determined by Rice's formula. Code formats for nonlinear stress combinations are compared for an example.

195 Effects of Shear and Normal Strain on Plate Bending

G. Z. Voyiadjis, Member, ASCE, (Assoc. Prof., Dept. of Civ. Engrg., Louisiana State Univ., Baton Rouge, La. 70803), **M. H. Baluch**, (Assoc. Prof., Dept. of Civ. Engrg., Univ. of Petroleum and Minerals, Dhahran, Saudi Arabia) and **W. K. Chi**, (Grad. Student, Dept. of Civ. Engrg., Louisiana State Univ., Baton Rouge, La. 70803)

Journal of Engineering Mechanics, Vol. 111, No. 9, September, 1985, pp. 1130-1143

The equations governing the bending of plates, taking into account the influence of transverse normal strain, are recast into a form involving the average transverse displacement function, $\dot{w}$. The resulting sixth order bending system of equations is solved for the Levy-type plates, with a variety of boundary conditions considered in the direction orthogonal to the simply supported direction. Results are tabulated for the displacement, $\dot{w}$, together with the plate moments M_x and M_y. Comparisons are made to corresponding quantities as obtained from the classical plate theory and the shear deformation theory where available.

196   Elastostatic Infinite Elements for Layered Half Spaces

R. K. N. D. Rajapakse, Assoc. Member, ASCE, (Former Grad. Student, Asian Inst. of Tech., Bangkok, Thailand) and **Pisidhi Karasudhi**, Member, ASCE, (Prof., Asian Inst. of Tech.,Bangkok, Thailand)

Journal of Engineering Mechanics, Vol. 111, No. 9, September, 1985, pp. 1144-1158

The far field behavior of a homogeneous half space and a layered half space is investigated under torsional, vertical, horizontal and moment loadings. Based on the derived far field behavior, three different finite-element based algorithms, i.e., Ordinary Infinite Elements (OIE), Finite Elements by Singular Contraction (FESC) and Exactly Integrable Infinite Elements (EIIE) are developed to model the far field of multilayered half spaces. The coordinate mapping functions and displacement interpolation functions are selected in accordance with the derived far field model. All three schemes satisfy compatibility and completeness. Excellent results are obtained at a very low computational cost by modeling the near field using a small finite-element mesh, together with any of the present schemes modeling the domain exterior to the finite elements. Considering the computational efficiency and ease in incorporating FESC into an existing finite element package, FESC is superior to the other two algorithms.

197   Optimum Building Design for Forced-Mode Compliance

Tsuneyoshi Nakamura, (Prof., Dept. of Architecture, Kyoto Univ., Kyoto, Japan) and **Izuru Takewaki**, (Asst., Dept. of Architecture, Kyoto Univ., Kyoto, Japan)

Journal of Engineering Mechanics, Vol. 111, No. 9, September, 1985, pp. 1159-1174

A new dynamic system response quantity, referred to as forced-mode compliance, is introduced for the forced steady-state vibration of a shear building model subjected to a harmonic ground motion. An optimum design problem subject to the constraints on forced-mode compliance, on fundamental natural frequency and on minimum stiffnesses is formulated and the necessary and sufficient conditions for global optimality are derived. It is shown theoretically and through numerical examples that the closed form optimal solution is useful not only for straightforward optimum design but also for controlling other significant dynamical characteristics.

198   Reliability of Systems under Renewal Pulse Loading

R. Rackwitz, (Dr.-Ing.habil, Technische Universität München, Institut für Bauingenieurwesen III, Lehrstuhl für Massivbau, München, West Germany)

Journal of Engineering Mechanics, Vol. 111, No. 9, September, 1985, pp. 1175-1184

The mean up-crossing rate of simple sums of rectangular-wave renewal load processes frequently used as suitable models for occupancy loading is calculated for processes having Gaussian, auto- and cross-correlated amplitudes. The formulations are extended to simple systems, i.e., series systems and parallel systems and to so-called minimal cut set systems. They involve only elementary manipulations with first and second moments of the variables and evaluations of the multinormal integral. They are believed to be one of the very few analytical solutions to load combination problems available. Therefore, they might be used directly or as reference solutions for the checking of approximations. Possible approximations to the crossing rate of non-normal, dependent load processes out of arbitrarily shaped safe domains of structural states are examined. Also, the formulation for a particular type of dependence between renewals is given.

199 Vibrating Nonuniform Plates on Elastic Foundation

Patricio A. A. Laura, (Dir. and Research Scientist, Inst. of Applied Mechanics, 8111 Puerto Belgrano Naval Base, Argentina) and **Roberto H. Gutierrez**, (Research Engr., Inst. of Applied Mechanics, 8111 Puerto Belograno Naval Base, Argentina)

Journal of Engineering Mechanics, Vol. 111, No. 9, September, 1985, pp. 1185-1196

Circular plates of variable thickness and elastically restrained against rotation along the edge resting on a Winkler-Pasternak medium are addressed. Free and forced vibrations are studied using a very simple polynomial coordinate function and the Ritz method. In the case of simply supported and clamped plates of uniform thickness subjected to static loading the results are in very good agreement with values available in the open literature. The entire algorithmic procedure can be efficiently handled using a microcomputer.

200 Torsion of Composite Bars by Boundary Element Method

J. T. Katsikadelis, (Assoc. Prof. of Struct. Analysis, Dept. of Civ. Engrg., National Tech. Univ. of Athens, 42 Patission St., GR-10682 Athens, Greece) and **E. J. Sapountzakis**, (Grad. Student, Dept. of Civ. Engrg., National Tech. Univ. of Athens, Athens, Greece)

Journal of Engineering Mechanics, Vol. 111, No. 9, September, 1985, pp. 1197-1210

A boundary element (B.E.) solution for the Saint-Venant torsion problem for composite cylindrical bars of arbitrary cross section is presented. The composite bar consists of a cylindrical matrix material surrounding a finite number of inclusions with different shear moduli firmly bonded to it. The problem is formulated in terms of the torsion function, which is established by solving a Neumann-type boundary value problem. Moreover, the torsional rigidity of the composite cross section and the shear stress components are evaluated. Several numerical examples are worked out and the results are compared with those available from analytical or other numerical solutions. The case of the homogeneous cross section with or without holes results in a special case. The efficiency of the B.E. method is also demonstrated and examined.

201 Appropriate Forms in Nonlinear Analysis

Jer-Shi Chen, (Grad. Student, Dept. of Engrg. Mechanics, Univ. of Iowa, Ames, Iowa 50010) and **Tseng Huang**, Member, ASCE, (Prof., Dept. of Civ. Engrg. and Engrg. Mechanics, Univ. of Texas at Arlington, Arlington, Tex. 76019)

Journal of Engineering Mechanics, Vol. 111, No. 10, October, 1985, pp. 1215-1226

In some geometrically nonlinear problems, the effects of external loads are altered significantly by the deformations of structures themself. In finite element analysis, the assumed displacement method is often used to solve this category of problems. But the formulations involve a huge number of unsymmetric matrices in the equilibrium equations and became extremely complex. An efficient algorithm is presented such that only a minimum number of symmetric, differentiation-invariant, repeatable matrices are required in the entire formulation.

202   **Two Parameter Fracture Model for Concrete**

Y. S. Jenq, Student Member, ASCE, (Grad. Research Asst., Dept. of Civ. Engrg., Northwestern Univ., Evanston, Ill. 60201) and **Surendra P. Shah**, (Prof. of Civ. Engrg., Dept. of Civ. Engrg., Northwestern Univ., Evanston, Ill. 60201)

Journal of Engineering Mechanics, Vol. 111, No. 10, October, 1985, pp. 1227-1241

Attempts to apply elastic fracture mechanics (LEFM) to concrete have been made for several years. Several investigators have reported that when fracture toughness, I_{Ic}, is evaluated from notched specimens using conventional LEFM (measured peak load and initial notch length) a significant size effect is observed. This size effect has been attributed to nonlinear slow crack growth occurring prior to the peak load. A two parameter fracture model is proposed to include this nonlinear slow crack growth. Critical stress intensity factory, K_{Ic}, is calculated at the tip of the effective crack. The critical effective crack extension is dictated by the elastic critical crack tip opening displacement, $CTOD_c$. Tests on notched beam specimens showed that the proposed fracture criteria to be size independent. The proposed model can be used to calculate the maximum load (for Mode I failure) of a structure of an arbitrary geometry. The validity of the model is demonstrated by an accurate simulation of the experimentally observed results of tension and beam tests.

203   **Orthotropic Annular Shells on Elastic Foundations**

Y. Nath, (Asst. Prof., Dept. of Applied Mechanics, Indian Inst. of Tech.., New Delhi 110 016, India) and **R. K. Jain**, (Grad. Student, Dept. of Applied Mechnics, Indian Inst. of Tech., New Delhi 110 016 India)

Journal of Engineering Mechanics, Vol. 111, No. 10, October, 1985, pp. 1242-1256

Theoretical nonlinear transient analysis of orthotropic annular shallow spherical shells interacting with Winkler-Pasternak elastic subgrades is performed. The governing nonlinear equations of motion are derived and solved in space and time domains employing Chebyshev polynomials and implicit Houbolt time-marching technique, respectively. The numerical results are obtained for both clamped and simply supported immovable outer edge conditions with free inner edge of the shell. The influence of foundation interaction, material orthotropy, and annular ratio on the response of spherical caps is determined. The results reveal that both foundation interaction and polar orthotropy play a significant role on the response characteristics of these shell structures.

204   **Endochronic Modeling of Sand in True Triaxial Test**

Han C. Wu, (Prof., Dept. of Civ. and Environmental Engrg., Univ. of Iowa, Iowa City, Iowa 52242) and **Zhan K. Wang**, (Grad. Asst., Dept. of Civ. and Environmental Engrg., Univ. of Iowa, Iowa City, Iowa)

Journal of Engineering Mechanics, Vol. 111, No. 10, October, 1985, pp. 1257-1276

The endochronic constitutive equation is written in differential form, which is then used to describe the deviatoric behavior of drained sand in true triaxial test. Explicit equations have been derived. The modified concept of intrinsic time is applied so that no built-in discontinuities exist in the constitutive equation. Triaxial compression test (TC) is used to determine the material constants of the differential form. Other tests considered are the triaxial extension test (TE) and shear tests (SSO with several stress ratios. An equation has been derived that is capable of describing both densification and dilation behaviors. It is shown that the theory does lead to reasonable agreement with experimental results of true triaxial test. Both the deviatoric response and the densification-dilation behavior of Ottawa sand have been investigated.

205 **Lagrangean Continuum Theory for Saturated Porous Media**

P. K. Kiousis, (Asst. Prof., Dept. of Civ. Engrg. and Engrg. Mech., Univ. of Arizona, Tucson, Ariz. 85721) and **George Z. Voyiadjis**, (Assoc. Prof., Dept. of Civ. Engrg., Louisiana State Univ., Baton Rouge, La.70803)

Journal of Engineering Mechanics, Vol. 111, No. 10, October, 1985, pp. 1277-1288

A theory for a two phase material based on the concepts of the theory of mixtures is developed in a Lagrangean reference frame. This theory may be applied for the analysis of flow of water through saturated soils. For this purpose, a new law for the flow of water through porous media is postulated. The material behavior of the soil skeleton is modeled as an elastoplastic, time independent porous solid. The material is assumed to be initially isotropic.

206 **Non-Normal Responses and Fatigue Damage**

Steven R. Winterstein, (Acting Asst. Prof., Dept. of Civ. Engrg., Stanford Univ., Stanford, Calif.)

Journal of Engineering Mechanics, Vol. 111, No. 10, October, 1985, pp. 1291-1295

A simple model of stationary nonnormal responses is proposed. Based on a Hermite series transformation of a normal process, the model requires knowledge of only limited, easily estimated response statistics (i.e., central moments). Applications to fatigue damage problems are considered. Conventional estimates of fatigue damage and life statistics are found to become increasingly unconservative as either the kurtosis coefficient or the stress-law exponent increases. A simple correction factor is developed to account for these effects, and is supported by the simulated behavior of rainflow-counted damage.

207 **Linear Stress Analysis of Torospherical Head**

Phillip L. Gould, Fellow, ASCE, (Harold D. Jolley Prof. and Chmn., Dept. of Civ. Engrg., Washington Univ., St. Louis, Mo. 63130) and **Jhun-Sou Lin**, Student Member, ASCE, (Dept. of Civ. Engrg., Washington Univ., St. Louis, Mo. 63130)

Journal of Engineering Mechanics, Vol. 111, No. 10, October, 1985, pp. 1295-1300

An experimental investigation to determine the buckling and rupture strength of a fabricated torospherical head under internal pressure loading has recently (1984) been carried out under the sponsorship of several industrial and governmental agencies. Preliminary evaluation of the data indicates a larger than expected factor of safety between initial buckling and failure. An important consideration in the investigation is the elastic stress pattern prior to the initiation of buckling. In this paper, the results of a linear elastic stress analysis are presented in order to demonstrate the resulting stress patterns and to compare several different computer codes. This solution is also thought to be suitable as a benchmark problem for the validation of computer codes for thin shell analysis. The analysis demonstrates several interesting aspects of the significance of bending stresses for a situation where there are neither abrupt geometrical discontinuities nor locally concentrated loads, two classical sources of bending in thin shells.

208   Plate Deflections Using Orthogonal Polynomials

Rama B. Bhat, (Assoc. Prof., Dept. of Mech. Engrg., Concordia Univ., Montreal, Quebec, Canada)

Journal of Engineering Mechanics, Vol. 111, No. 11, November, 1985, pp. 1301-1309

Bending deflection of uniform plates under static loading has been determined using beam characteristic orthogonal polynomials in Rayleigh-Ritz formulation. The first member of the orthogonal polynomial set was constructed so as to satisfy the boundary conditions of the corresponding beam problems accompanying the plate problem. The rest of the set was generated using the Gram-Schmidt process. Results are presented for two plate configurations and compared with those obtained by previous methods.

209   Force Spectrum of Horizontal Member

C. C. Tung, Member, ASCE, (Prof., Dept. of Civ. Engrg., North Carolina State Univ., Raleigh, N.C.) and **Beile Yin**, (Grad. Student, Dept. of Civ. Engrg., North Carolina State Univ., Raleigh, N.C.)

Journal of Engineering Mechanics, Vol. 111, No. 11, November, 1985, pp. 1310-1324

Expressions of mean value, mean-square value and spectrum are derived for the vertical force on a horizontal member of square cross section partially immersed in water under the action of a normally incident long-crested Gaussian stationary linear wave train. The derivation takes into account the possibilities that the member may at times be totally submerged or out of the water. Numerical results are obtained and presented graphically to show that, depending on whether the probabilities associated with the events of total submergence and emergence of the members are considered or ignored, the mean value, mean-square value and spectrum of the vertical wave force may be significantly different.

210   Analysis of Plates on a Kerr Foundation Model

Mahmoud C. Kneifati, (Grad. Student, Dept. of Civ. Engrg., Drexel Univ., Philadelphia, Pa. 19104)

Journal of Engineering Mechanics, Vol. 111, No. 11, November, 1985, pp. 1325-1342

This paper studies the foundation model consisting of two spring layers interconnected by a shearing layer that was suggested by A. D. Kerr in 1964. The problem of an infinite plate strip, attached to the Kerr foundation and subjected to a uniform load and boundary forces, is solved. The results are then compared to the corresponding solutions when the base is modeled as: (1) An elastic continuum; (2) a Winkler foundation; and (3) a Pasternak foundation. Finally, paper concludes with a study of the validity of calculating the foundation parameters explicitly from the elastic coefficients (E,ν) of the base material.

211   Oblique Turbulent Jets in a Crossflow

Vincent H. Chu, Member, ASCE, (Assoc. Prof., Dept. of Civ. Engrg. and Applied Mechanics, McGill Univ., Montreal, Canada)

Journal of Engineering Mechanics, Vol. 111, No. 11, November, 1985, pp. 1343-1360

A simple description of oblique turbulent jets in a uniform crossflow is achieved by following the motion of the jet in an oblique coordinate system, which moves with the crossflow. In the moving coordinate system, the ambient is stationary and the motion of the turbulent jet is described by a line-impulse model. The widths and the path of the jet are determined from a series of top- and side-view photographs. The added-mass coefficient and the turbulent entrainment coefficient, each evaluated from independent sets of experimental data, are found to be consistent with the line-impulse formulation and experimental data of related turbulent shear flows.

212   Review of Nonlinear FE Methods with Substructures

Yeon S. Ryu, (Asst. Prof., Dept. of Ocean Engrg., National Fisheries Univ. of Pusan, Korea) and **Jasbir S. Arora**, (Prof., Dept. of Civ. and Environmental Engrg., Univ. of Iowa, Iowa City, Iowa 52242)

Journal of Engineering Mechanics, Vol. 111, No. 11, November, 1985, pp. 1361-1379

Methods of structural design sensitivity analysis are closely related to structural analysis procedures. It is critically important to thoroughly understand the analysis procedures for efficient design sensitivity analysis. This is particularly true for nonlinear finite element analysis methods. This paper reviews some basic solution procedures for nonlinear finite element analysis using the displacement formulation. Solution methods for nonlinear equations are also briefly reviewed. Nonlinear finite element analysis procedures that have been used for geometric as well as material nonlinearities in static and dynamic response problems are described. Use of substructuring for large-scale problems is explained. To take advantage of favorable features of various methods, a procedure to combine them is proposed and analyzed.

213   Impact Tests on Frames and Elastic-Plastic Solutions

J. M. Mosquera, (Research Assoc., Div. of Applied Mathematics, Brown Univ., Providence, R.I.), **H. Kolsky**, (Prof. of Applied Mathematics and Engrg. (Research), Brown Univ., Providence, R.I.) and **P.S. Symonds**, Fellow, ASCE, (Prof. of Engrg. (Research), Brown Univ., Providence, R.I.)

Journal of Engineering Mechanics, Vol. 111, No. 11, November, 1985, pp. 1380-1401

Experimental and analytical work is described in further investigation of a simplified method for estimating final and peak deflection amplitudes of a pulse loaded structure. The method includes effects of elastic response interacting with plastic deformation, but gains simplicity by treating them in artificially separated stages: An initial elastic response stage is followed by a wholly plastic (i.e., rigid-plastic) stage, which in turn is followed by elastic vibrations. Experiments were performed in which portal frames of aluminum alloy and mild steel were impacted at the midpoint of one column by masses projected at various speeds. The contact force pulse was recorded and used as input data for analyses. In addition to the simplified elastic-plastic (SEP) method, numerical analyses were carried out by a finite element code (furnishing nominally "exact" solutions) and by an approach assuming rigid-plastic behavior for the entire response. It is found that with the relatively long pulses produced in the present impact experiments, elastic-plastic interactions may be critically important. Consideration of straing rate sensitive plastic flow in the steel frames requires special treatment. Reasonably good agreement with results of tests and finite element calculations was given by the SEP method, whereas a rigid-plastic solution grossly underestimated final deflections in certain cases.

214   Seismic Floor Spectra by Mode Acceleration Approach

Mahendra P. Singh, Member, ASCE, (Prof., Dept. of Engrg. Sci. and Mechanics, Virginia Polytechnic Inst. and State Univ., Blacksburg, Va.) and **Anil M. Sharma**, Assoc. Member, ASCE, (Engrg. Analyst, Sargent & Lundy, 55 E. Monroe St., Chicago, Ill.)

Journal of Engineering Mechanics, Vol. 111, No. 11, November, 1985, pp. 1402-1419

The truncation of the high frequency modes, so commonly used in dynamic structural analysis, sometimes can cause significant errors in the calculated response; this is caused by the so-called missing mass effect. Such mode truncations can also affect the accuracy of floor spectra generated for stiff structural systems, and also for floors close to the base in even not so stiff structural systems, especially if the mode displacement methods are employed in the analysis. However, this problem due to mode truncation can be alleviated and virtually removed by employing the mode acceleration formulation in the analysis. Here, a direct response spectrum approach for generation of floor response spectra has been developed on the basis of the mode acceleration formulation and random vibration principles. As seismic input, this approach requires the relative acceleration spectra in lieu of the pseudo-acceleration spectra—so commonly used with the mode displacement approaches. Because the relative spectrum values become very small for frequencies higher than the highest frequency component in the design ground motion, such high structural frequencies need not be considered in the analysis. Numerical results demonstrating the effectiveness of the proposed new approach are presented.

215   Symmetric Buckling of Hinged Ring Under External
Pressure

C. Y. Wang, (Michigan State Univ., East Lansing, Mich. 48824)

Journal of Engineering Mechanics, Vol. 111, No. 11, November, 1985, pp. 1423-1427

A ring with equally-spaced hinges is subjected to external pressure. The stability is governed by a fourth order differential equation. The critical pressures for symmetrical buckling are found in the case of one, two, or three hinges.

216   Hydrodynamic Pressure on a Dam During Earthquakes

A. Chakrabarti, (Assoc. Prof., Dept. of Applied Mathematics, Indian Inst. of Sci., Bangalore-560012, India) and **V. N. Nalini**, (Research Scholar, Dept. of Applied Mathematics, Indian Inst. of Sci., Bangalore-560012, India)

Journal of Engineering Mechanics, Vol. 111, No. 12, December, 1985, pp. 1435-1439

A straightforward analysis involving Fourier cosine transforms and the theory of Fourier seies is presented for the approximate calculation of the hydrodynamic pressure exerted on the vertical upstream face of a dam due to constant earthquake ground acceleration. The analysis uses the "Parseval relation" on the Fourier coefficients of square integrable functions, and directly brings out the mathematical nature of the approximate theory involved.

217   Stability of Continuously Restrained Cantilevers

Mahyar Assadi, (Grad. Student, Dept. of Civ. Engrg., Univ. of Washington, Seattle, WA) and **Charles W. Roeder**, Member, ASCE, (Prof., Dept. of Civ. Engrg., Univ. of Washington, Seattle, WA)

Journal of Engineering Mechanics, Vol. 111, No. 12, December, 1985, pp. 1440-1456

The problem of lateral torsional stability of cantilevers with continuous elastic or rigid lateral restraint is examined via a direct variational approach and in view of three major influence parameters: (1) Height of application of lateral restraint; (2) height of application of a point load above the shear center; and (3) stiffness of the elastic restraint. The derived free-end force boundary conditions are examined and a quandary regarding their validity is addressed. Finally, the results of an experimental study are reported, providing some guidance in assessing the buckling capacity of continuously restrained cantilevers.

218   Finite Element Modeling of Infinite Reservoirs

Shailendra K. Sharan, Member, ASCE, (Assoc. Prof. of Civ. Engrg., Laurentian Univ., Sudbury, Ontario, Canada)

Journal of Engineering Mechanics, Vol. 111, No. 12, December, 1985, pp. 1457-1469

A technique is developed to model the effects of radiation damping in the finite element analysis of hydrodynamic pressures on dams subjected to a harmonic horizontal ground motion. The water in the reservoir is treated as being compressible; however, its vibration is assumed to be two-dimensional and of small amplitude. In the finite element modeling, an infinite reservoir must be truncated at a finite distance from the dam, and a suitable boundary condition must be imposed at the truncation surface. Sommerfeld or similar existing boundary conditions are found to be satisfactory for excitation frequencies greater than the fundamental frequency of the reservoir. However, for lower frequencies, which are of greater importance in the seismic response analysis of dams, such boundary conditions require a very large extent of the reservoir to be considered in the analysis. The principal merit of the proposed boundary condition is that the reservoir may be truncated at a very short distance from the dam, resulting in great computational advantages. The effectiveness of the proposed method is demonstrated by analyzing several cases.

219   Interactive Buckling in Thin-Walled Beam-Columns

Srinivasan Sridharan, Member, ASCE, (Assoc. Prof., Dept. of Civ. Engrg., Washington Univ., St. Louis, MO 63130) and **M. Ashraf Ali**, (Doctoral Student, Dept. of Civ. Engrg., Washington Univ., St. Louis, MO 63130)

Journal of Engineering Mechanics, Vol. 111, No. 12, December, 1985, pp. 1470-1486

A new analytical model has been developed for a study of the response of thin-walled beam-columns having doubly symmetric cross sections. The novel features of the model are that: (1) It incorporates the interaction of overall, buckling/bending with two *companion* local modes; (2) it accounts for the phenomenon of amplitude modulation; and (3) it can model any set of realistic end conditions. These features are a result of an innovative combination of the finite strip and finite element techniques. The model has been tested for numerical convergence and against the available results on square box columns. A small number of examples have been studied. These relate to the performance of eccentrically loaded columns and single story frames. Also, the vital necessity of considering amplitude modulation in thin-walled beam columns has been brought out.

220   **The Stability of Cylindrical Air-Supported Structures**

R. Maaskant, (Grad. Student, Dept. of Civ. Engrg., Univ. of Waterloo, Waterloo, Ontario, Canada N2L 3G1) and **J. Roorda**, (Prof. of Civ. Engrg., Solid Mech. Div., Univ. of Waterloo, Waterloo, Ontario, Canada N2L 3G1)

Journal of Engineering Mechanics, Vol. 111, No. 12, December, 1985, pp. 1487-1501

The behavior of cylindrical air-supported membranes subjected to a concentrated line load is studied theoretically and experimentally. Large deflection symmetrical deformations lead eventually to limit point instabilities for high profile initial shapes. The structure may bifurcate from the symmetric shape into a nonsymmetric shape at loads less than the limit load. Some comparisons with available experimental results are included.

221   **On the Reliability of a Simple Hysteretic System**

B. F. Spencer, Jr., Student Member, ASCE, (Asst. Prof., Dept. of Civ. Engrg., Univ. of Notre Dame, Notre Dame, IN) and **L. A. Bergman**, Member, ASCE, (Assoc. Prof., Dept. of Aeronautical and Astronautical Engrg., Univ. of Illinois at Urbana-Champaign, Urbana, IL)

Journal of Engineering Mechanics, Vol. 111, No. 12, December, 1985, pp. 1502-1514

A method to determine statistical moments of time to first passage of a simple oscillator, incorporating the modified Bouc hysteresis model, has been developed. A generalized Pontriagin-Vitt equation is formulated from Markov process theory and solved by a Petrov-Galerkin finite element method. The resulting moments are given as functions of initial displacement and velocity of the oscillator, as well as of the initial hysteretic force in the oscillator. The first two moments are used in conjunction with a maximum entropy distribution to estimate the probability of failure of the oscillator. The accuracy and economy of the method is demonstrated for a particular example drawn from base isolation of a simple structure. A comparison of the finite element results with those obtained by Monte Carlo simulation is then given for this example.

222   **Steady-State Dynamic Analysis of Hysteretic Systems**

Danilo Capecchi, (Research Asst., Istituto di Scienza delle Costruzioni, Facoltà di Ingegneria Monteluco-Roio, 67100 L'Aguila, Italy) and **Fabrizio Vestroni**, (Assoc. Prof., Istituto di Scienza delle Costruzioni, Facoltà di Ingegneria Monteluco-Roio, 67100 L'Aguila, Italy)

Journal of Engineering Mechanics, Vol. 111, No. 12, December, 1985, pp. 1515-1531

A great many hysteretic models have been recently introduced in the analysis of dynamic behavior of structures and structural elements. This paper considers the steady-state oscillations of single-degree-of-freedom systems with different force-deflection relationships. Three types of constitutive laws are covered: bilinear, stiffness degrading, and stiffness-strength degrading. An approximate solution to the equation of motion under sinusoidal excitations is obtained by an analytical procedure and the frequency response curves are drawn. All the models exhibit softening behavior but while the bilinear and the Ramberg-Osgood type models give stable and single-valued response curves, the stiffness and the stiffness-strength degrading models exhibit a multi-valued curve in a certain frequency range. The results obtained are verified by numerically integrating the equation of motion. Numerical solutions are also used to predict the actual response of systems with unstable frequency response curve to general excitation.

223   Flexure of Statically Indeterminate Cracked Beams

Mohammed H. Baluch, (Assoc. Prof., Dept. of Civ. Engrg., Univ. of Petroleum and Minerals, Dharan, Saudi Arabia) and **Abdul K. Azad**

Journal of Engineering Mechanics, Vol. 111, No. 12, December, 1985, pp. 1535-1539

The influence of nonpropagating cracks on redistribution of internal forces in the flexure of statically indeterminate beams is studied by modeling crack as a rotational spring of known compliance, where the compliance is determined by using the relationship between the stress intensity factor for Mode I cracks and the compliance itself. The problem is formulated in terms of conpact slope deflection type equations. Results indicate significance of parameter α, which is defined as the ratio of crack stiffness to beam stiffness. It is found that for values of $\alpha >$ 25, the influence of discrete crack flexibility on force redistribution can be ignored.

224   Buckling with Enforced Axis of Twist

Z. M. Elias, Member, ASCE, (Prof., Dept. of Civ. Engrg., Univ. of Washington, Seattle, WA)

Journal of Engineering Mechanics, Vol. 111, No. 12, December, 1985, pp. 1539-1543

Buckling of a cantilever with a doubly symmetric thin-walled cross section and with one flange laterally restrained is treated by a direct variational method. The critical load is tabulated in terms of a geometric parameter for four cases of loading restraint. The critical load is always increased by flange restraint.

225   Strength of Resin Mortar Tunnel

Tomoaki Nakayama, (Engr., Ibaraki Electrical Communication Lab., Nipon Telegraph and Telephone Public Corp., Tokai, Ibarake-Ken, 319-11, Japan) and **Toshio Takatsuka**

Journal of Engineering Mechanics, Vol. 111, No. 12, December, 1985, pp. 1544-1548

Although the strength of resin mortar tunnel rings is less than 75% of the bending strength of rectangular specimens, the reason for this has not been clarified. In this paper, the resin mortar strength is analyzed statistically using Weibull's theory. Taking size effects and stress distribution into account, the relation between the ring strength and the bending strength of rectangular specimens is determined. Based on this relation, the resin mortar tunnel reliability can then be estimated.

226   Foundation Densification for Fossil Plant Loads

K. A. Kessler, Member, ASCE, (Prin. Geotechnical Engr., Ebasco Services, Inc., Norcross, Ga.) and **J. J. Kuretski, Jr.,** Assoc. Member, ASCE, (Civ.Engr., Florida Power & Light Co., Juno Beach, Fla.)

Journal of Energy Engineering, Vol. 111, No. 1, September, 1985, pp. 1-9

A case history is presented in which foundation soils in excess of 50 ft (15 m) deep, were densified in place to the degree required to support the loads of a coal-fired electric generating plant. The use of drop-weight compaction and compaction grouting techniques resulted in a significant savings when compared to a more conventional pile foundation system. The application of deep densification techniques has gained increased popularity in recent years.

However, even when the soil conditions on a major project are appropriate for in-place densification, the implementation of a deep densification foundation plan may be difficult. A large portion of this difficulty is attributed to the required deviation from the traditional methods of site investigation, contract preparation and construction management. The case history considers how the foundation plan was developed to allow deep densification to be a viable foundation preparation alternative and, ultimately, part of a successful foundation system.

227   Auto-Oxidation Effect on Flue Gas Sludge Systems

Richard W. Goodwin, Member, ASCE, (Consulting Engr., 14 Ramapo Lane, Upper Saddle River, N.J. 07458)

Journal of Energy Engineering, Vol. 111, No. 1, September, 1985, pp. 10-20

Upsets to Flue Gas Desulfurization (FGD) sludge treatment systems may occur under auto-oxidation conditions. A 75-79% auto-oxidized (75-79% $CaSO_4$) FGD sludge exhibits a mean particle diameter of 45 microns—considerably larger than unoxidized ($CaSO_3$) particles. The oxidized sludge reaches terminal settling velocity within 8-10 min, five times faster than unoxidized material. Auto-oxidized waste settles in gravity thickeners to higher densities (0.85 g dry solids/ml) and to thicker underflows (40% solids). Rakearm overload and blanket consolidation can be avoided by proper drive mechanism design and underflow solids/density control. Vacuum filtration of an 80% $CaSO_4$ sludge yields 25% greater solids in the cake as compared to a 30% oxidized material. To avoid cake-cracking, reduce the degree of vacuum during incidents of auto-oxidation and provide dual filtrate receivers. Using predictive curves presented here, a Transportability Analysis was performed on a coal of 0.59-0.91% sulfur and of 10.00-21.50% ash. At an SO/d2 loading of 1.08 lb/MMBTU 38% solids (thickener underflow) or 63% solids (filter cake) could be achieved. Either filter by-pass or adding less fly ash to the cake permits achieving a 75-85% solids transportability blend. Auto-oxidized sludge/ash blends exhibit unconfined compressive strengths from 57-63 psi and permeabilities from $2.2-3 \times 10^5$ cm/sec.

228   Power Production and Air Conditioning by Solar Ponds

Zahra Panahi, (Independent Consultant, Box 44, WSC, Ogden, Utah 84408)

Journal of Energy Engineering, Vol. 111, No. 1, September, 1985, pp. 21-28

The potential of a salt-gradient solar pond for electric power production and thermal energy supply for cooling, using absorption chillers, and heating of a building was investigated by means of computer simulations. The analysis was conducted for the pond and building of the Energy Systems Center (ESC) located in southern Nevada. The experimental pond has a bottom area of about 950 m^2 and 3-m depth. The energy requirement for cooling and heating of the ESC 232-m^2 two-story building was calculated using the TRNSYS model and the ESC facility data. Power production from the pond was evaluated for a solar pond power plant with a low temperature Organic Rankine Cycle (ORC) generator. Several operating strategies were simulated using the salt-gradient solar pond (SGSP) model. The results indicate that the pond can produce 16.8 kWe for 10 h/day, sufficient for ESC cooling demands. Using a pond absorption chiller system for cooling, 27% of the energy available from the pond will meet the cooling and heating demands of the building.

229   Solid Waste Management for Utility Reconversions

Gary K. Welshans, Member, ASCE, (Sr. Environmental Engr., Stone & Webster Engrg. Corp., One Penn Plaza, New York, N.Y. 10001)

Journal of Energy Engineering, Vol. 111, No. 1, September, 1985, pp. 29-39

Since the OPEC oil embargo, several utilities have reconverted or are in the process of reconverting some of their power plants from oil to coal firing. The increased quantity of wastes generated as a result of these projects, coupled with the requirements of the National Environmental Policy Act (NEPA) and the Resource Conservation and Recovery ACt (RCRA), necessitates that the utilities develop comprehensive solid waste management plans. In most cases, the environmental licensing for these projects requires the preparation of a Draft Environmental Impact Statement, in which solid waste management impacts, as well as alternatives to disposal, are discussed. Frequently, other documents, similar to Environmental Reports, containing the results of the subsurface and groundwater investigation and accompanied by a set of conceptual design drawings, must be prepared to satisfy the licensing requirements for the selected solid waste management facility. While air quality impacts remain the single most important environment consideration facing these reconversion projects, solid waste management plans are also vital to successful licensing. Proper attention to the solid waste management issues of a utility's coal reconversion project can expedite the licensing process and, consequently, the overall reconversion timetable. Several of the major aspects which should be considered when selecting and licensing a utility's solid waste management facility are discussed in this paper.

230   Containments for Limerick and Susquehanna Plants

P. S. Sawhney, Member, ASCE, (Engrg. Supervisor, Bechtel Power Corp., San Francisco, Calif.), G. H. Shah, Member, ASCE, (Engrg. Supervisor, Bechtel Power Corp., San Francisco, Calif.) and R. G. Roberts, Member, ASCE, (Structural Engr., Bechtel Power Corp., San Francisco, Calif.)

Journal of Energy Engineering, Vol. 111, No. 1, September, 1985, pp. 40-61

Design and construction of four containment buildings for the Limerick and Susquehanna nuclear power plants has shown that a reinforced concrete containment is feasible. Features of the containment, design crieria, methods of analysis, construction and testing of the containment buildings are presented. Special attention is given to innovations in the design and construction and suggestions are made for improvements in these areas.

231   Field Aging of Fixed Sulfur Dioxide Scrubber Waste

M. S. Aggour, Member, ASCE, (Assoc. Prof., Civ. Engrg. Dept., Univ. of Maryland, College Park, Md.), W. D. Stanbro, (Sr. Staff, Johns Hopkins Univ., Applied Physics Lab., Laurel, Md.) and J. J. Lentz, Assoc. Member, ASCE, (Consultant, William Harrington & Assoc., Inc., Glen Burnie, Md.)

Journal of Energy Engineering, Vol. 111, No. 1, September, 1985, pp. 62-73

A fixed-material test pad was constructed to evaluate the long-term effects of the fixation with lime of blended fly ash and scrubber sludge. The 507.9×10^3 kg test pad was envisioned as a scale model disposal site with average dimensions of 13.72×22.86 m by 0.76 m thick. The fixed material was tested at the time of pad construction in 1979 and again in 1983 when the present investigation was conducted. The study included field and laboratory evaluation of the physical and chemical properties of the four-year-old fixed scrubber sludge. The effect of the exposure of the material to the elements for four years is described, and based on these effects, recommendations on the proper disposal of this type of material in landfilling are made.

232   **MDS Project—Main Dimensions and Economics**

Z. Shalev, (Head, Preliminary Planning Div., TAHAL Consulting Engrs. Ltd., Israel)

Journal of Energy Engineering, Vol. 111, No. 1, September, 1985, pp. 74-88

The conveyance capacity of the tunnel designed to move water from the Mediterranean Sea to a power station on the Dead Sea was calculated to refill the Dead Sea to its historic level at a rate which maximizes the project's net benefits. The generation capacity of the project was calculated to make optimal use of the amounts of water thus conveyed. In calculating the dimensions of these two main components of the scheme, as well as those of other facilities, the long-term economic impact of the project on the entire Israeli national power system was evaluated. These studies formed part of the feasibility study which proceeded in parallel with outline design of the facilities. Both engineering and economic studies have adopted a rather conservative approach and included numerous iterations and sensitivity tests. The feasibility study concludes with a recommendation to implement the scheme on the grounds of direct quantifiable benefits only, even though there are several other benefits to the power system, and to the national economy in general.

233   **Bound Glass in Shredded Municipal Solid Waste**

Jess Everett, (Research Asst., Dept. of Civ. and Environmental Engrg., Duke Univ., Durham, N.C. 27706) and **J. Jeffrey Peirce**, (Assoc. Prof., Dept. of Civ. and Environmental Engrg., Duke Univ., Durham, N.C. 27706)

Journal of Energy Engineering, Vol. 111, No. 1, September, 1985, pp. 91-94

The binding and trapping of glass and organics that occurs during the shredding of municipal solid waste (MSW) is examined experimentally. Such glass is called bound glass. This process produces feed which detrimentally effects air classifier production of refuse derived fuel (RDF). Glass contaminated RDF causes excessive slagging on incinerator walls and abrasion of mechanical parts. Results indicate that the glass prticle size distribution produced by the shredder and the moisture content in the waste steam effect bound glass amounts. Further, bound glass increases with moisture content and increases with decreasing particle size. Dry waste streams and shredders which produce large glass particles are suggested to produce better feeds for air classifiers, and consequentially better RDF.

234   **Cyclic Behavior of Pavement Base Materials**

Michael McVay, Member, ASCE, (Asst. Prof. of Civ. Engrg., Dept. of Civ. Engrg., Univ. of Florida, Gainesville, Fla.) and **Yongyuth Taesiri**, (Research Engr., Asian Inst. of Tech., Bangkok, Thailand)

Journal of Geotechnical Engineering, Vol. 111, No. 1, January, 1985, pp. 1-17

The influence of stress path on the stress-strain behavior of a Florida sand subject to repetitive moving wheel loads is investigated in the laboratory with conventional triaxial equipment. A conventional resilient modulus test with only cyclic varying compressive loads and a moving wheel stress path involving both extension and compressive loads determined from an elastic solution were examined at different initial confining pressures. The tests showed that excursions of applied extension loading followed by compression loading resulted in anisotropic material behavior. This, in turn, influenced both the cyclic permanent strain (rutting) build-up as well as the resilient (distortion) behavior of the material for initial confining pressures below approximately 40 psi (276 kPa). It was concluded that, in order to model both the distortion and rutting characteristics of a sand, improved constitutive relationships over the simple linear elastic theory are warranted. A bounding surface plasticity model was subsequently developed based upon both the plastic shear dilation and volumetric behavior of the sand. The predicted model response agreed both qualitatively and quantitatively for the tests investigated.

235   Effects of Predrilling and Layered Soils on Piles

Pe-Shen Yang, (Grad. Asst., Civ. Engrg., Iowa State Univ., Ames, Iowa), **Amde M. Wolde-Tinasae**, Member, ASCE, (Assoc. Prof. of Civ. Engrg., Univ. of Maryland, College Park, Md.) and **Lowell F. Greimann**, Member, ASCE, (Prof. of Civ. Engrg., Iowa State Univ., Ames, Iowa)

Journal of Geotechnical Engineering, Vol. 111, No. 1, January, 1985, pp. 18-31

A state-of-the-art nonlinear finite element algorithm for pile-soil interaction has been developed and implemented in a modular computer program to study the effects of predrilled oversize holes and layered soils on the vertical load carrying capacity of piles in integral abutment bridges where thermally induced lateral movements occur. A 40-ft (12 m) H pile embedded in very stiff clay is used to study the effect of a predrilled oversize hole with or without sand fill. Numerical results showed that the vertical load carrying capacity of H piles is significantly affected by predrilling. The study of layered soils included the case where compacted soil (50 or more blow counts in standard penetration tests) is used as a fill on top of different layers of natural soil. For such cases, the numerical results indicated that the type of failure is greatly influenced by depth of the compacted soil.

236   Slope Reliability and Response Surface Method

Felix S. Wong, Member, ASCE, (Sr. Assoc., Weidlinger Associates, 3000 Sand Hill Road, Building 4, Suite 155, Menlo Park, Calif. 94025)

Journal of Geotechnical Engineering, Vol. 111, No. 1, January, 1985, pp. 32-53

A method to analyze reliability of soil slopes using the response surface method is described. The soil slope is modelled and analyzed by a finite element code as in prevalent deterministic studies. The simulation which is usually expensive, is repeated a limited number of times to give point estimates of the response corresponding to uncertainties in the model parameters. A graduating function is then fit to these point estimates so that the response given by the finite element code can be reasonably approximated by the graduating function within the region of interest. The approximating function, called the response surface, is used to replace the code in subsequent repetitive computations required in a statistical reliability analysis. The procedure is applied to a sample problem in slope stability involving uncertain soil properties. It is shown that the slope stability statistics from the response surface is within 1-9% of the statistics based on a direct Monte Carlo simulation using the finite element code. The merits of such an approach to reliability analysis are examined.

237   Liquefaction Potential Mapping for San Francisco

Edward Kavazanjian, Jr., Member, ASCE, (Asst. Prof. of Civ. Engrg., Terman Engrg. Center, Stanford Univ., Stanford, Calif. 94305), **Richard A. Roth**, (Staff Engr., Dames and Moore, 455 S. Figuerea St., Suite 3500, Los Angeles, Calif. 90071) and **Heriberto Echezuria**, (Ingeniere Civil, INTEVEP, Caracas, Venezuela)

Journal of Geotechnical Engineering, Vol. 111, No. 1, January, 1985, pp. 54-76

The potential of saturated deposits of cohesionless soils in downtown San Francisco to experience initial liquefaction due to seismically induced pore pressure is evaluated. Initial liquefaction, or the zero effective stress state, is sued as the index of liquefaction potential because it provides the best available index for damage due to seismically induced pore pressures. Liquefaction potential is evaluated by comparing the conditional probability of liquefaction, or liquefaction susceptibility, to the expected intensity of seismic loading, or liquefaction opportunity. The probabilistic evaluation is made using a liquefaction hazard model developed by Chameau. Assuming the water table to be at the ground surface, results indicate that while no liquefaction is expected anywhere for an intensity with an annual probability of exceedance of 0.05, only the most resistant deposits will survive an event with an annual probability of exceedance at 0.02. For an event with an annual probability of exceedance of 0.01, initial liquefaction is expected to occur within all saturated, cohesionless soil deposits in the downtown San Francisco area. It must be emphasized that in dense soil deposits the consequences of initial liquefaction may be minimal because of their limited shear strain potential.

238   Geotechnical Characteristics of Residual Soils

Frank C. Townsend, Member, ASCE, (Prof. of Civ. Engrg., Univ. of Florida, Gainesville, Fla. 32611)

Journal of Geotechnical Engineering, Vol. 111, No. 1, January, 1985, pp. 77-94

Residual soils are products of chemical weathering and thus their characteristics are dependent upon environmental factors of climate, parent material, topography and drainage, and age. These conditions are optimized in the tropics where well-drained regions produce reddish lateritic soils rich in iron and aluminum sesquioxides and kaolinitic clays. Conversely, poorly drained areas trend towards montmorillonitic expansive black clays. Andosols develop over volcanic ash and rock regions and are rich in allophane (amorphous silica) and metastable halloysite. The geological origins greatly affect the resulting engineering characteristics. Both lateritic soils and andosols are susceptible to property changes upon drying, and exhibit compaction and strength properties not indicative of their classification limits. Both soils have been used successfully in earth dam construction, but attention must be given to seepage control through the weathered rock. Conversely, black soils are unpopular for embankments. Lateritic soils respond to cement stabilization and, in some cases, lime stabilization. Andosols should also respond to lime treatment and cement treatments if proper mixing can be achived. Black expansive residual soils respond to lime treatment by demonstrating strength gains and decreased expansiveness. Rainfall induced landslides are typical of residual soil deposits.

239   New Method for Evaluating Liquefaction Potential

Kendiah Arulmoli, Assoc. Member, ASCE, (Staff Engr., Ertec Western, Inc., 3777 Long Beach Blvd., Long Beach, Calif. 90807), Kendiah Arulanandan, Member, ASCE, (Prof. of Civ. Engrg., Univ. of California, Davis, Calif. 95616) and H. Bolton Seed, Fellow, ASCE, (Prof. of Civ. Engrg., Univ. of California, Berkeley, Calif. 94720)

Journal of Geotechnical Engineering, Vol. 111, No. 1, January, 1985, pp. 94-114

A new method of indexing the grain and aggregate properties of sand using electrical parameters is described. Correlations are established between these parameters and relative density, D_r, cyclic stress ratio, τ/σ'_o, and K_{2max}. An electrical probe, used to predict these parameters from ins-situ electrical measurements, is described. Evaluations are made of D_r and τ/σ'_o, which are compared with values measured independently from controlled laboratory tests. Reasonable agreement is found between predicted and measured values. The potential applicability of the electrical probe in the field is shown by evaluation of liquefaction and nonliquefaction at sites affected by the 1906 San Francisco, Niigata and Tangshan earthquakes.

240   Inverted Shear Modulus from Wave-Induced Soil Motion

Ludwig Figueroa, Member, ASCE, (Assoc. Prof., Dept. of Civ. Engrg., Univ. of Miami, Coral Gables, Fla., and Div. of Ocean Engrg., Rosenstiel School of Marine and Atmospheric Science, Univ. of Miami, Miami, Fla.), **Tokuo Yamamoto**, Member, ASCE, (Prof., Div. of Ocean Engrg., Rosenstiel School of Marine and Atmospheric Science, Univ. of Miami, Miami, Fla.) and **Toshihiko Nagai**, (Research Engr., Port and Harbor Research Inst., Ministry of Transportation, Yokosuka, Japan)

Journal of Geotechnical Engineering, Vol. 111, No. 1, January, 1985, pp. 115-132

A previously developed technique, used mainly in seismology and geophysics, has been adopted to determine the shear modulus from soil displacements measured during wave tank experiments. This iterative inversion technique allowed the determination of the shear modulus and Coulomb specific loss (damping) with depth within a bentonite clay profile. It also allowed the observation of both the modulus reduction and damping curves for large shear strain amplitudes (up to 7%) for the first time. There was fairly good agreement between the modulus reduction curve determined by the inversion method and values found in the literature. Refinements to the inversion technique are suggested to improve determination of shear modulus. An alternate procedure to provide fairly reliable and quick estimates of the shear modulus is also presented.

241   Rate of Shear Effects on Vane Shear Strength

Mohsen Sharifounnasab, Assoc. Member, ASCE, (Civ. Engr., Esfahan, Iran) and **C. Robert Ullrich**, Member, ASCE, (Assoc. Prof. of Civ. Engrg., Univ. of Louisville, Louisville, Ky. 40292)

Journal of Geotechnical Engineering, Vol. 111, No. 1, January, 1985, pp. 132-139

A series of laboratory vane shear tests was performed on sedimented samples of kaolinite and slaked Pierre shale to determine the effects of shear rate and vane dimensions on measured strength. Results of the tests were expressed in terms of shear strength on a vertical plane and were plotted versus average shear rate. For kaolinite samples vane shear strength decreased as shear rate increased because of partial drainage occurring at slower shear rates. However, for the highly plastic Pierre shale vane, shear strength increased as shear rate increased because of undrained creep occurring at slower rates of shear. Test results for kaolinite and Pierre shale are compared to published information from other studies.

242   Temperature Effects on Volume Measurements

William Stewart, (Lect. in Civ. Engrg., Univ. of Glasgow, Rankine Building, Oakfield Ave., Glasgow, G12 8LT, Scotland) and **Chung K. Wong**, (Research Student, Univ. of Glasgow, Dept. of Civ. Engrg., Ranking Building, Oakfield Ave., Glasgow, G12 8LT, Scotland)

Journal of Geotechnical Engineering, Vol. 111, No. 1, January, 1985, pp. 140-144

Volume measuring systems for triaxial tests are susceptible to changes in ambient temperature, and the volume change due to temperature can represent 10% or more of the measured value, depending on the magnitude of the total volume change. The automatic volume change logging system showed a clear dependence on temperature, a dependence related to the thermal expansion of the water in the system, the apparatus itself and any entrapped air. By using regression analysis on data accumulated over several tests, a simple temperature correction formula for volume change is derived.

243   Evaluation of Soil Response to EPB Shield Tunneling

Richard J. Finno, Assoc. Member, ASCE, (Asst. Prof. of Civ. Engrg., Illinois Inst. of Tech., Chicago, Ill. 60616) and **G. Wayne Clough**, Member, ASCE, (Prof. of Civ. Engrg., Virginia Polytechnic Inst. and State Univ., Blacksburg, Va. 24061)

Journal of Geotechnical Engineering, Vol. 111, No. 2, February, 1985, pp. 155-173

Earth pressure balance (EPB) shields can be operated so that soil is forced away from its face; this heaving process may cause smaller surface settlements than would result from tunneling, without inducing any initial heave. Field observations made at the first EPB project in th U.S. and finite element analyses results of that project are compared to gain a better understanding of ground behavior around EPB shields. Field observations indicated that the ground response was both three-dimensional and time-dependent. To keep computational costs within reason, the EPB tunneling process was simulated in a finite element analysis using 2 two-dimensional models. Besides approximating EPB shield tunneling effects, the finite element simulations explicitly account for pore pressure mobilization and dissipation in time. Results of parametric studies indicate that intentionally induced heave decreases the net movement into the tail void gap behind the shield at the expense of increasing consolidation settlements at the surface.

244   Transmitting Boundaries and Seismic Response

Albert T. F. Chen, Member, ASCE, (Research Civ. Engrg., U.S. Geological Survey, Menlo Park, Calif.)

Journal of Geotechnical Engineering, Vol. 111, No. 2, February, 1985, pp. 174-180

A parametric study of the seismic response of a chosen site was conducted to demonstrate the obvious inconsistencies in computed ground response as a result of different assumptions made on the transmitting boundary for the site. The cause of these inconsistencies in computed response is the departure from the ideal assumption that the soil deposit below the transmitting boundary is a linear elastic and homogeneous half-space. It is shown that under low-intensity shaking, multiple reflections from the discontinuities below the boundary can be sinificant, and under strong shaking, nonlinear soil response can alter the wave form, as well as reduce the intensity of the motion. Either situation makes correlation between the input motion at the boundary and the control outcrop motion uncertain. For seismic response analyses of sites underlain by very thick soil deposits, it is suggested that consideration also be given to the shear strength profile at the site when deciding on the location of a transmitting boundary.

245   Third-Invariant Model for Rate-Dependent Soils

Howard L. Schreyer, Member, ASCE, (Prof., Dept. of Mech. Engrg. and Research Assoc., New Mexico Engrg. Research Inst., Univ. of New Mexico, Albuquerque, N.M. 87131) and **James E. Bean**, Member, ASCE, (Research Engr., New Mexico Engrg. Research Inst., Univ. of New Mexico, Albuquerque, N.M. 87131)

Journal of Geotechnical Engineering, Vol. 111, No. 2, February, 1985, pp. 181-192

A theory of viscoplasticity for frictional materials is developed in which first and third invariants of stress and strain are used instead of the more conventional second invariants. Rate effects are incorporated directly into the expression for the flow surface so that the numerical algorithm for a plasticity subroutine can be used. The usual concepts of strain- and strain rate-hardening viscoplasticity are used, except that a non-associated flow rule is required to control dilation. For two sandy materials, detailed comparisons are made of theoretical and experimental stress-strain data for both static and dynamic paths.

246   Flow Field Around Cones in Steady Penetration

Mehmet T. Tumay, Member, ASCE, (Prof., Dept. of Civ. Engrg., Louisiana State Univ., Baton Rouge, La.), **Yalcin B. Acar,** Member, ASCE, (Asst. Prof., Dept. of Civ. Engrg., Louisiana State Univ., Baton Rouge, La.), **Murat H. Cekirge,** Member, ASCE, (Research Engr., Water and Environment Div., Research Inst. Univ. of Petroleum and Minerals, Dharan, Saudi Arabia) and **Narayanan Ramesh,** (Grad. Research Asst., Dept. of Civ. Engrg., Louisiana State Univ., Baton Rouge, La.)

Journal of Geotechnical Engineering, Vol. 111, No. 2, February, 1985, pp. 193-204

An analytical solution is presented for the flow field around cones penetrating an inviscid and incompressible fluid. The stream function, velocity field, and strain rates around 18° and 60° cones advancing at a rate of 2 cm/sec are fully determined and presented. It is assumed that this strain rate field will give the first approximation to the in situ strains during penetration of very soft cohesive soils.

247   Performance of a Stone Column Foundation

James K. Mitchell, Fellow, ASCE, (Prof. and Chmn. of Civ. Engrg., University of California, Berkeley, Calif.) and **Timothy R. Huber,** Member, ASCE, (Assoc., Terratech, Inc., Monterey, Calif.)

Journal of Geotechnical Engineering, Vol. 111, No. 2, February, 1985, pp. 205-223

Vibro-replacement stone columns were used to support a large wastewater treatment plant founded on up to 48 ft (15 m) of soft estuarine deposits. Support was required for distributed foundation stresses up to 3,000 psf (145 kPa). The basic design requirement was a loading of 30 tons/ (265 kN) stone column with a settlement of less than 0.25 in. (6 mm) under that load in a test on a single column within a group. Column spacings ranged from a 4 ft × 5 ft (1.2 m by 1.5 m) pattern under the most heavily loaded areas, to a 7 ft × 7 ft (2.1 m by 2.1 m) pattern under lightly loaded areas. Twenty-eight single column load tests were done during the installation of the 6,500 stone columns to evaluate load-settlement behavior. Laboratory tests were done to provide soil property data needed for finite element predictions of both the load test behavior and settlements of the completed structures. Predicted load test settlements were somewhat greater than those recorded during the load tests, but agreement was generally good. The installation of stone columns led to a reduction in settlements to about 30–40% of the values to be expected on unimproved ground. The settlement of a large uniformly loaded area of improved ground was predicted to be about ten times that measured in a load test on a single column within the area. Measured settlements varied from 1.0–2.4 in. (25–60 mm) for a soft sediment thickness of 32–35 ft. (10–11 m). A settlement of about 2.5 in. (64 mm) was predicted by the finite element analysis. Settlement predictions using other, simpler methods gave values which agreed reasonably with both the measured values and the finite element predictions.

248   Real and Apparent Relaxation of Driven Piles

Christopher David Thompson, Member, ASCE, (General Manager, Trow Ltd., Toronto, Ontario, Canada) and **David Elliot Thompson,** Member, ASCE, (Mgr., Oshawa Office, Trow Ltd., Whitby, Ontario, Canada)

Journal of Geotechnical Engineering, Vol. 111, No. 2, February, 1985, pp. 225-237

Relaxation, defined as a decrease in bearing capacity of driven piles with time after initial driving, has been observed to be a rare phenomenon in southern Canadian geotechnical conditions. It has been encountered by the writers with piles driven to end bearing in shale of the Georgian Bay Formation in the Bayfront area of downtown Toronto, Ontario. In this case, the approach of increasing the end area of the piles to reduce the bearing pressure has been counter-productive. In other founding conditions, a reduction in driving resistance between the end of initial driving and the beginning of redrive has not indicated a reduction in bearing capacity, but rather an increase in the developed force from the pile hammer. It appears to be primarily associated with the use of single-acting diesel hammers.

249   Determination of Critical Slope Failure Surfaces

Van Uu Nguyen, Member, ASCE, (Lect., Dept. of Civ. and Mining Engrg., Univ. of Wollongong, P.O.Box 1144, Wollongong, New South Wales, 2500, Australia)

Journal of Geotechnical Engineering, Vol. 111, No. 2, February, 1985, pp. 238-250

The analysis of the stability of slopes using limit equilibrium methods usually requires the determination of the critical failure surface having the minimal factor of safety. This paper presents an optimization technique, adapted from operations research, to determine both the minimal factor of safety and the associated critical failure surface. The technique treats the factor of safety as an optimal function of the N geometrical coordinates defining potential admissible failure surfaces, and searches for the optimum by a reflection algorithm through an N-dimensional space defined by these geometrical coordinates. Application of the technique, as illustrated by examples, revealed that in comparison with the popularly known grid search method, the technique yields lower values of the minimal factor of safety for fewer computations, and is suitable for virtually all types of failure surfaces.

250   Dynamic Centrifugal Modeling of a Horizontal Dry Sand Layer

Philip C. Lambe, Member, ASCE, (Asst. Prof. of Civ. Engrg., North Carolina State Univ., Raleigh, N.C. 27695) and **Robert V. Whitman**, Fellow, ASCE, (Prof. of Civ. Engrg., Massachusetts Inst. of Tech., Cambridge, Mass. 02139)

Journal of Geotechnical Engineering, Vol. 111, No. 3, March, 1985, pp. 265-287

The dynamic response of a 10.7 m (35 ft) thick dry sand layer was modeled on the Cambridge University centrifuge at scale factors of 35 and 80. A stack of teflon-coated aluminum rings and a latex membrane confined a cylindrically shaped sand model that had a height-to-diameter ratio equal to 1. During the scaled sinusoidal base shaking, electronic transducers measured accelerations, transient horizontal displacements and surface settlements. While sensors measuring accelerations and surface settlements worked well, the DC LVDT's used to measure transient horizontal displacements were influenced by frequency-dependent behavior. Measured results were used to evaluate scaling laws governing dynamic centrifugal modeling and to test for consistency with results of numerical methods and laboratory tests reported in the literature. The results of the test program indicate that a simple sand layer can be effectively modeled aboard a centrifuge and that horizontal accelerations, cyclic shear strains, and settlements follow the scaling laws that govern dynamic centrifugal modeling.

251   The Arch in Soil Arching

Richard L. Handy, Member, ASCE, (Prof. of Civ. Engrg., Iowa State Univ., Ames, Iowa 50011)

Journal of Geotechnical Engineering, Vol. 111, No. 3, March, 1985, pp. 302-318

Soil arching action or "bin effect" is usually quantified by use of a horizontal differential element whose support derives in part from Rankine theory. In the 1940's, Krynine mathematically proved this incorrect. The present analysis substitutes a catenary arch describing

Note: "Soil Security Test For Water Retaining Structures" can be found on p. 180 as entry number 500.

the path of the minor principal stress, which thus is complementary to a structural arch, and dips downward instead of upward if supportive. Soil arching action develops in two stages: The first involves rotation of the principal stresses adjacent to a rough wall and causes horizontal wall pressures to significantly exceed those from classical theory, simulating a K_o pressure distribution even in loose backfill soil. The second stage reduces pressures on the lower wall to give a curvilinear distribution typically centered at a height 0.42 times the height of the wall and in close agreement with published data.

252   Loading Rate Method for Pile Response in Clay

Jean-Louis Briaud, Member, ASCE, (Assoc. Prof., Civ. Engrg. Dept., Texas A&M Univ., College Station, Tex. 77843) and **Enrique Garland**, (Geotechnical Engr., Jacento Lara 343, Lima 27, Peru)

Journal of Geotechnical Engineering, Vol. 111, No. 3, March, 1985, pp. 319-335

A method is proposed to predict the behavior of single piles in cohesive soil subjected to vertical loads applied at various rates. The elementary simple shear model is made rate-dependent by introducing the time raised to a negative power n. This viscous exponent n is shown to vary from 0.02–0.08 with a 0.05 average. Correlations with undrained shear strength, water content, plasticity, the liquidity index and overconsolidation ratio are shown based on 152 laboratory tests. Integration of the elementary model leads to rate-dependent elastic-plastic or hyperbolic load transfer curves. A program to predict the head load-head movement response of the pile using these rate-dependent load-transfer curves is described. The validity of the method is first addressed by examining the physical reasons for the rate-dependent properties of clays, then by using the results of 62 rate-dependent pile load tests and by examining a series of applications of the new method and, finally, by presenting the results of a parametric study.

253   Design Strengths for an Offshore Clay

Demetrious C. Koutsoftas, Member, ASCE, (Assoc., Dames and Moore, San Francisco, Calif.) and **Charles C. Ladd**, Fellow, ASCE, (Prof. of Civ. Engrg., Massachusetts Inst. of Tech., Cambridge, Mass.)

Journal of Geotechnical Engineering, Vol. 111, No. 3, March, 1985, pp. 337-355

Stability analyses for stage construction of a breakwater located offshore New Jersey required a detailed evaluation of the undrained stress-strain-strength properties of a plastic marine clay. One set of analyses, representing "conventional" practice, used UU triaxial compression tests to obtain the initial undrained strength c_u and CIU triaxial compression tests to predict strength increases with consolidation. The second approach employed the Stress History And Normalized Soil Engineering Properties (SHANSEP) design procedure. This involved extensive consolidation testing to measure the in-situ preconsolidation pressure and a comprehensive program of CK_oU tests sheared in compression, extension and direct simple shear modes of failure at varying overconsolidation ratios. A strain compatibility technique was applied to the SHANSEP data to account for the effects of anisotropy and progressive failure in circular arc- and wedge-type stability analyses. The SHANSEP approach yielded design strengths significantly smaller than conventional practice, in terms of both the initial c_u profile and the rate-of-strength gain with consolidation.

254 Stress Anisotropy Effects on Clay Strength

Paul W. Mayne, Assoc. Member, ASCE, (Sr. Geotechnical Engr., Law Engrg., P.O. Drawer QQ, Washington, D.C. 22101)

Journal of Geotechnical Engineering, Vol. 111, No. 3, March, 1985, pp. 356-366

For ease and economy, most commercial laboratories perform consolidated undrained triaxial shear tests using an initial isotropic state of stress. The insitu state of stress for most clay soils, however, is anisotropic. How different are the undrained shear strengths and effective stress friction angles measured under "field conditions" (CAU) as opposed to routine laboratory conditions (CIU)? This study reviews available published data from over 40 different clays consolidated under both isotropic and anisotropic conditions before triaxial shear. Most of these clays were normally-consolidated, although one-third were also tested at overconsolidated states. Important factors such as strain rates, soil structure, and differences in laboratories were considered to be beyond the scope of this study. For triaxial compression, it is suggested that the anisotropic undrained strength, on the average, may be estimated as 87% of the isotropic strength. Based on few data, a tentative correction factor of 0.60 is recommended for isotropic extension to account for stress anisotropy. The effective stress friction angle appears little affected by initial stress state.

255 Dynamic Determination of Pile Capacity

Frank Rausche, Member, ASCE, (Pres., Goble, Rausche, Likins and Assoc., Inc., 4423 Emery Industrial Parkway, Warrensville Heights, Ohio 44128), **George G. Goble**, Member, ASCE, (Prof., Environmental and Architectural Engrg., Univ. of Colorado, Boulder, Colo. 80309) and **Garland E. Likins**, (Pres., Pile Dynamics, Inc., 4423 Emery Industrial Parkway, Warrensville Heights, Ohio 44128)

Journal of Geotechnical Engineering, Vol. 111, No. 3, March, 1985, pp. 367-383

A method is presented for determining the axial static pile capacity from dynamic measurements of force and acceleration made under the impact of a large hammer. The basic equation for calculation of the forces resisting pile penetration is derived. The limitations of the basic resistance equation are discussed and illustrative examples of field measurements are given. With the availability of this derivation, it is possible to prove that the Case Pile Wave Analysis Program (CAPWAP) resistance force distribution is unique. Using the assumption that the resistance to penetration can be divided into static and dynamic parts, an expression is developed for calculating the dynamic resistance to penetration. The resulting method requires the selection of a "damping" constant which is shown empirically, to relate to soil size distribution. A correlation of Case Method capacity and the capacity observed in static load tests is given for 69 statically tested piles that were also tested dynamically.

256 Liquefaction Potential of Sands Using the CPT

Peter K. Robertson, (NSERC Univ. Research Fellow, Univ. of British Columbia, Vancouver, British Columbia) and **Richard G. Campanella**, Member, ASCE, (Prof. and Head, Civ. Engrg. Dept., Univ. of British Columbia, Vancouver, British Columbia)

Journal of Geotechnical Engineering, Vol. 111, No. 3, March, 1985, pp. 384-403

Existing methods for the evaluation of liquefaction potential using data obtained from the Cone Penetration Test (CPT) are reviewed. A modified CPT-based method for the liquefaction assessment of sand is presented. Information available on the liquefaction of silty sands is reviewed, and a procedure for considering the influence of silt content is also presented. Field and laboratory data from Canada, Japan, China and the United States are presented to provide a preliminary evaluation of the proposed CPT-based liquefaction assessment methods.

257 Active Earth Pressure Behind Retaining Walls

Sangchul Bang, (Asst. Prof., Dept. of Civ. Engrg., Univ. of Notre Dame, Notre Dame, Ind.)

Journal of Geotechnical Engineering, Vol. 111, No. 3, March, 1985, pp. 407-412

A simple and realistic analytical procedure is described to estimate the eveloped lateral earth pressure behind the rigid retaining wall with cohesionless backfill soil experiencing outward tilt about the base. Included are various stages of wall tilt starting from an initial active state to a full active state. The initial active state is defined as a stage of wall tilt when the soil element at the ground surface experiences a sufficient amount of lateral displacement to reach an active condition, whereas the full active state occurs when the soil elements along the entire depth of the wall are in active condition. The predictions from the developed method of analysis are compared with the model test measurements. The comparison shows very good agreements at various stages of the retaining wall tilt.

258 Interference Between Geotechnical Structures

Kingsley O. Harrop-Williams, Assoc. Member, ASCE, (Asst. Prof., Dept. of Civ. Engrg.,m Carnegie-Mellon Univ., Pittsburgh, Pa. 15213) and **Dimitri Grivas**, Assoc. Member, ASCE, (Assoc. Prof., Dept. of Civ. Engrg., Rensselaer Polytechnical Inst., Troy, N.Y. 12181)

Journal of Geotechnical Engineering, Vol. 111, No. 3, March, 1985, pp. 412-418

The relative safety of neighboring geotechnical structures depends on the degree of dependence between them. This dependence is considered to exist if there is overlapping of their potential failure surfaces. A simple method is presented which defines the interference between two geotechnical structures in terms of a non-dimensional factor α, the distance needed between them to satisfy a certain confidence level that they are independent is obtained.

259 Settlement Analysis of Embankments on Soft Clays

Gholamreza Mesri, Member, ASCE, (Prof. of Civ. Engrg., Univ. of Illinois at Urbana-Champaign, Urbana, Ill.) and **Y. K. Choi**, Assoc. Member, ASCE, (Engr., Geotechnical Engrs. Inc., Winchester, Mass.)

Journal of Geotechnical Engineering, Vol. 111, No. 4, April, 1985, pp. 441-464

Highly realistic analysis of the consolidation of soft clays is possible within the framework of the traditional concepts of soil mechanics. A practical method is described for predicting the magnitude and rate of settlement and pore water pressure dissipation. Detailed interpretation and evaluation of the consolidation properties of soft clays, including the compressibility parameters, preconsolidation pressure and coefficient of permeability are presented. These properties can be determined from the standard oedometer test with direct permeability measurements. Computer procedure ILLICON, using multi-layer analysis, includes any variation of consolidation parameters with depth. The proposed method for evaluating the consolidation parameters, and settlement and pore water pressure analyses is illustrated using case histories of two test fills on soft clays. Excellent agreement is obtained between the computed behavior and field observations.

260 Hydraulic Conductivity Tests on Compacted Clay

Stephen S. Boynton, Assoc. Member, ASCE, (Geotechnical Engr., Trinity Engrg. Testing Corp., Austin, Tex.) and **David E. Daniel**, Member, ASCE, (Asst. Prof. of Civ. Engrg., Univ. of Texas, Austin, Tex.)

Journal of Geotechnical Engineering, Vol. 111, No. 4, April, 1985, pp. 465-478

Permeability tests were performed in the laboratory on compacted clay to study the effects of type of permeameter, hydraulic anistropy, diameter of test specimens, storage time, and desiccation cracking. Essentially identical hydraulic conductivities were measured with compaction-mold, consolidation-cell, and flexible-wall permeameters. With good bonding between lifts, hydraulic conductivity was isotropic. Test specimens having a diameter of 15 cm were twice as permeable as specimens having a diameter of 4 cm, but this difference is too small to be of any consequence. There was no tendency for hydraulic conductivity to increase with increasing storage time. Desiccation cracks in compacted clay close only partially when the clay is moistened and permeated, unless substantial effective stresses are applied to aid in closing the cracks.

261 1-D Pollutant Migration in Soils of Finite Depth

R. Kerry Rowe, Member, ASCE, (Assoc. Prof., Faculty of Engrg. Sci., Univ. of Western Ontario, London, Ontario N6A 5B9) and **John R. Booker**, (Reader, School of Civ. and Mining Engrg., Univ. of Sydney, Sydney, Australia)

Journal of Geotechnical Engineering, Vol. 111, No. 4, April, 1985, pp. 479-499

A technique for the analysis of 1-D pollutant migration through a clay layer of finite depth is presented. This formulation includes dispersive and advective transport in the clay as well as geochemical reactions and permits consideration of the depletion of contaminant in the landfill with time as well as the effect of ground-water flow in a permeable stratum beneath the clay layer. A limited parametric study is presented to illustrate the effect of considering these factors in the analysis. It is shown that for most practical situations the concentration of contaminant within the ground water beneath the landfill will reach a peak value at a specific time and will then decrease with subsequent time. It is shown that the magnitude of this peak concentration and the time required for it to occur are highly dependent upon the mass of contaminant within the landfill and the sorption capacity of the clay. Other important factors which are examined include the thickness of the clay layer, the advection velocity (relative to the dispersivity), and the ground-water flow velocity in any permeable strata beneath the clay layer. The implications of these results for optimizing the design of clay liners is then discussed.

262 Evaluation of Design Methods for Vertical Anchor Plates

Edward A. Dickin, (Lect., Dept. of Civ. Engrg., Univ. of Liverpool, Liverpool, England) and **Chun F. Leung**, (Lect., Dept. of Civ. Engrg., National Univ. of Singapore, Singapore)

Journal of Geotechnical Engineering, Vol. 111, No. 4, April, 1985, pp. 500-520

Design methods for vertical anchors subjected to horizontal pull-out forces are reviewed. Theoretical predictions are compared with results from conventional tests on 2 in. (50 mm) models and centrifugal tests representing 39.4 (1 m) prototypes in dense sand. Good agreement is found between force coefficients from theories for 39.4 in. (1 m) anchors, provided a mobilized friction angle in the mass is used in the analysis. A lack of consistent agreement for 2 in. (50 mm) anchors is attributed to inherent difficulties in obtaining reliable data for small models at low stress levels. Anchor geometry is adequately accounted for by a theoretical shape factor derived by the writers.

263   **Finite Element Analyses of Lock and Dam 26 Cofferdam**

G. Wayne Clough, Member, ASCE, (Prof. and Head, Dept. of Civ. Engrg., Virginia Polytechnic Inst. and State Univ., Blacksburg, Va.) and **Thangavelu Kuppusamy**, Member, ASCE, (Assoc. Prof., Dept. of Civ. Engrg., Virginia Polytechnic Inst. and State Univ., Blacksburg, Va.)

Journal of Geotechnical Engineering, Vol. 111, No. 4, April, 1985, pp. 521-541

Replacement of the old Lock and Dam 26 on the Mississippi River involves construction of one of the largest systems of cellular cofferdams ever built. Preliminary analyses using conventional theories led to sometimes contradictory recommendations about the cofferdam design. To help resolve outstanding issues, instrumentation was placed on the first-stage cofferdam, and finite element procedures were developed for analysis of the cofferdam. Three different two-dimensional finite element procedures were generated, including axisymmetric, vertical slice, and generalized plane-strain models. The alternative approaches allowed prediction of cofferdam response for the most important design conditions and at the critical locations. In all models, allowances were made for nonlinear soil response, slippage on the sheet-pile-soil interfaces, staged construction simulation and seepage effects where appropriate. Further, a means was provided to accommodate the effect of sheet-pile interlock yielding, which led to an orthotropic response of the cellular cofferdam. Predicted behavior from the finite element models was found to be consistent with observed trends. The models appear to have considerable potential for use in future designs.

264   **Punch-Through Instability of Jack-Up on Seabed**

Edmund C. Hambly, Member, ASCE, (Dir., Edmund Hambly Ltd. Consulting Engrg., Berkhamsted, Hertfordshire, England)

Journal of Geotechnical Engineering, Vol. 111, No. 4, April, 1985, pp. 545-550

A punch-through failure of a jack-up platform can occur as a result of an overall instability of the structure on the seabed even when the seabed has bearing capacity increasing with depth. The instability is analogous to the overturning of a top-heavy ship and to the Euler buckling of an inelastic strut. The instability starts when a small tilt of the structure causes an increase in the reaction on a penetrating leg which is greater than the increase in bearing resistance associated with the added penetration. The limiting weight and height to which a jack-up remains stable on a given foundation depends on the sequence of jacking and preloading of the legs.

265   **Plasticity and Constitutive Relations in Soil Mechanics**

Ronald F. Scott, Member, ASCE, (Prof., California Inst. of Tech., 1201 E. California St., Pasadena, Calif. 91125)

Journal of Geotechnical Engineering, Vol. 111, No. 5, May, 1985, pp. 563-605

After early attempts to describe the behavior of beams under load, equations of equilibrium were first formulated correctly in 1827. Isotropic elastic behavior was described shortly after. Plasticity studies were initiated and failure conditions were established in the period 1860 to 1920. By 1900, correct equations of plasticity for soils had been proposed, and solutions had been obtained to a number of practical cases by graphical integration. Punch or footing problems were examined in the 1920s. By 1950, the mathematical basis of plasticity for metals was firmly established. Since that time, they have been extended by many investigators to account for the peculirities of soil behavior, including yielding under hydrostatic stresses. The present state of incremental plasticity theory, necessitating the use of computers for solutions, requires consideration of three basic conditions: A yield criterion, a hardening law, and a flow rule.

266   Latent Heat of Frozen Saline Coarse-Grained Soil

Ted S. Vinson, Member, ASCE, (Prof., Dept. of Civ. Engrg., Oregon State Univ., Corvallis, Oreg. 97331) and Sheldon L. Jahn, Assoc. Member, ASCE, (Asst. Prof., Dept. of Engrg., U.S. Army Military Academy, West Point, N.Y. 10096)

Journal of Geotechnical Engineering, Vol. 111, No. 5, May, 1985, pp. 607-623

In recognition of the need to predict the thermal regime in arctic offshore structures, a laboratory research program was conducted to evaluate the latent heat of fusion of a coarse-grained soil with saline pore water. Latent heat of fusion measurements were made by placing frozen soil specimens with saline pore water in an insulated cylinder and melting the specimens under steady-state heat flow conditions. The specimen temperature and heat transfer through the cylinder as a function of time was recorded. The latent heat of fusion is equal to the heat supplied to the soil specimen in the time interval during which a phase change occurs. The results from the research program indicate that the latent heat released during thawing depends on the initial values of pore–water salinity and subfreezing temperature. The latent heat released increases with decreasing pore–water salinity. For the coarse-grained soil employed in the research program, the latent heat released when thawing from a given subfreezing temperature was found to be equal to the latent heat released in a pure saline solution over a comparable temperature range.

267   Streambank Stability

Floyd M. Springer, Jr., Assoc. Member, ASCE, (Civ. Engr., F.W. Owens & Assoc., Inc., Louisville, Ky. 40208), C. Robert Ullrich, Member, ASCE, (Assoc. Prof., Dept. of Civ. Engrg., Univ. of Louisville, Louisville, Ky. 40292) and D. Joseph Hagerty, Member, ASCE, (Prof., Dept. of Civ. Engrg., Univ. of Louisville, Louisville, Ky. 40292)

Journal of Geotechnical Engineering, Vol. 111, No. 5, May, 1985, pp. 624-640

Erosion of alluvial soils along the Ohio River is a matter of increasing concern to riparian landowners and commercial interests. One mechanism by which erosion of alluvial streambanks takes place is by sliding wedge failures of upper bank layers. Results of a study are presented in which the sliding wedge mechanism was investigated for Ohio River banks. Streambanks were assumed to fail by sliding along sand partings underlying cohesive upper layers. Results of computer wedge stability analyses indicated that bank stability is most sensitive to the depth of water present in tension cracks behind the face of the bank. Other factors found to be important to wedge stability are the effective angle of internal friction of the sand seam underlying the cohesive sliding wedge and the unit weight of the cohesive wedge. The importance of tension crack formation on wedge stability of multi-layered river banks was also evaluated.

268   Load Tests on Tubular Piles in Coralline Strata

James M. Gilchrist, (Sr. Struct. Engr., Britoil Plc, 15 St. Vincent St., Glasgow, Scotland G2 5LJ)

Journal of Geotechnical Engineering, Vol. 111, No. 5, May, 1985, pp. 641-655

At a coral reef site on the Red Sea coast of Saudi Arabia, load tests were performed on 1,422 mm diam tubular steel piles to verify pile design compression and tension capacities predicted by calculations. Although two tests were planned at separate site locations, five were performed (two on test pile A, and three on test pile B) due to the test results disproving some of the calculation assumptions. Test pile A was installed open-ended and the test results concluded: For a coring pile, the measured compression capacities had reasonable agreement with those calculated (calculated overestimates = 5.2% and 16.3%); and the design assumption that a soil plug would form was disproved. Test pile B was installed with a structural plug fitted to the leading end and the test results concluded: (1) The measured skin friction at 11 m penetration = zero; and the unit end bearing capacity assumed in the calculations was considerably larger than that measured in the tests at 11 m and 30 m penetration (calculated overestimates = 181% and 164%).

269   Dimensioning Footings Subjected to Eccentric Loads

William H. Highter, Member, ASCE, (Prof., Dept. of Civ. Engrg., Univ. of Tennessee, Knoxville, Tenn. 37996) and John C. Anders, (Geotechnical Engr., Soil and Material Engrs, Inc., Blountville, Tenn. 37617)

Journal of Geotechnical Engineering, Vol. 111, No. 5, May, 1985, pp. 659-665

An approach often used to determine the bearing capacity of shallow foundations subjected to eccentric loading is based on Meyerhof's concept of a reduced effective area. The effective area is the area of greatest length such that its centroid coincides with the resultant load. The effective width of the reduced area is defined as the effective area divided by its length. The procedure for determining the effective dimensions of rectangular footings with one-way eccentricity is straightforward but the process is complex and time-consuming when the eccentricity is in both directions. Depending on the magnitudes of eccentricities e_l and e_b in the directions of the length, L, and width, B, of the footing, respectively, the effective area is one of four possible shapes. To aid the engineer in the design of eccentrically loaded footings, analysis charts are prepared which yield normalized reduced dimensions as a function of normalized eccentricities, e_l/L and e_b/B, for the four cases. An analysis chart is also prepared for determining normalized reduced dimensions of circular footings as a function of the normalized eccentricity, e_r/R.

270   Seismic Stability of Gentle Infinite Slopes

Tarik Hadj-Hamou, Assoc. Member, ASCE, (Asst. Prof. of Civ. Engrg., Tulane University, New Orleans, La. 70118) and Edward Kavazanjian, Jr., Assoc. Member, ASCE, (Asst. Prof. of Civ. Engrg., Stanford Univ., Stanford, Calif. 94305)

Journal of Geotechnical Engineering, Vol. 111, No. 6, June, 1985, pp. 681-697

Deterministic and probabilistic analyses of the stability of gentle infinite slopes subject to seismically induced excess pore pressures and inertia forces are developed. In the deterministic analysis, classical equations for infinite slope stability are rewritten to explicitly include excess pore pressure and seismic acceleration. Equations for the factor of safety are developed that include these factors. In the probabilistic analysis, the seismic acceleration, excess pore pressure, and effective friction angle are considered random variables. Acceleration peaks are considered Rayleigh distributed. Excess pore pressure is predicted using a model that considers Rayleigh distributed shear stress peaks. The friction angle is modeled with a Beta distribution. Acceleration and pore pressure development within the gentle infinite slope are assumed the same as those in a horizontal deposit of the same average thickness. Finite element analyses are performed to investigate the limits of this assumption. Results from both analyses are compared to documented case histories of lateral spreading.

271   Lateral Stresses Observed in Two Simple Shear Apparatus

Muniram Budhu, Member, ASCE, (Asst. Prof., Dept. of Civ. Engrg., Faculty of Engrg. and Applied Sci., State Univ. of N.Y. at Buffalo, Amherst Campus, Buffalo, N.Y. 14260)

Journal of Geotechnical Engineering, Vol. 111, No. 6, June, 1985, pp. 698-711

Two simple shear apparatus (SSA), the Cambridge University SSA Mk7 and an NGI type, were used to compare the radial stresses developed in the NGI-type SSA with the horizontal normal stresses on the plane perpendicular to the plane of shear deformation and the intermediate principal effective stresses as measured in the Cambridge University SSA Mk7. These stresses are often assumed to be equal. Monotonic and cyclic simple shear tests were conducted on dense and loose Leighton Buzzard sand at constant applied vertical load. Comparison of results from these tests in the two SSA shows that the radial stresses are equal to neither the horizontal normal stresses nor the intermediate principal effective stresses. These radial stresses do not appear to be valuable in calculating the stress state of a specimen in the NGI-type SSA. The values of intemediate principal effective stress ratio and the horizontal stress ratio reached at peak stress ratio (maximum shear stress/vertical) in monotonic tests on sand in its initial densest state represent threshold values for cyclic simple shear tests.

272   Bearing Capacity of Gravity Bases on Layered Soil

Michael Georgiadis, Member, ASCE, (Geotechnical Engr., Inst. of Engrg. Seismology and Earthquake Engrg., Thessaloniki, Greece 54626) and **Alex P. Michalopoulos**, Member, ASCE, (Managing Dir., D'Appolonia, Geneva, Italy)

Journal of Geotechnical Engineering, Vol. 111, No. 6, June, 1985, pp. 712-729

A new numerical method for evaluating the bearing capacity of shallow foundations on layered soil which may contain any combination of cohesive and cohesionless layers is presented. Several potential failure surfaces are analyzed and the minimum material factor for which the foundation is stable is determined. Comparisons between the results obtained with the new method, a number of semi-empirical solutions for uniform and two-layer systems, experiments and other numerical methods including finite elements, provide a valuable assessment of the performance of the various methods currently used and demonstrate the validity of the new method.

273   Strength of Intact Geomechanical Materials

Ian W. Johnston, (Sr. Lect., Dept. of Civ. Engrg., Monash Univ., Clayton, Victoria, Australia 3168)

Journal of Geotechnical Engineering, Vol. 111, No. 6, June, 1985, pp. 730-749

The strength of geomechanical materials can be described by a number of strength criteria. However, each criterion is usually limited to certain material types with a limited range of stress conditions. A new empirical strength criterion is proposed and shown to apply to a wide range of intact materials, from lightly overconsolidated clays through hard rocks, for both compressive and tensile stress regions. It is demonstrated that the strengths of these widely differing materials follow a distinctive progressive pattern.

274   Vertical Response of Arbitrarily Shaped Embedded Foundations

George Gazetas, Member, ASCE, (Assoc. Prof. of Civ. Engrg., Rensselaer Polytechnic Inst., Troy, N.Y. 12180), **Ricardo Dobry**, Member, ASCE, (Prof. of Civ. Engrg., Rensselaer Polytechnic Inst., Troy, N.Y. 12180) and **John L. Tassoulas**, Assoc. Member, ASCE, (Asst. Prof. of Civ. Engrg., Univ. of Texas at Austin, Austin, Tex.)

Journal of Geotechnical Engineering, Vol. 111, No. 6, June, 1985, pp. 750-771

A method to compute the dynamic stiffnesses and damping coefficients of arbitrary-shaped rigid foundations embedded in a reasonably homogeneous and deep soil deposit and subjected to harmonic vertical excitations is developed. The method is based on an improved understanding of the physics of the problem, substantiated by the results of extensive, rigorous parametric studies, including several analytical results compiled from the literature. The results are applicable to both saturated and unsaturated soils, a variety of cross-sectional shapes, and a wide range of embedment depths and types of contact between the vertical sidewalls and the surrounding soil. A numerical example illustrates the applicability of the method. It is also shown that the method constitutes a significant improvement over current practical procedures, such as the "equivalent" circle approximation.

275   Liquefaction Evaluation Procedure

Steve J. Poulos, Member, ASCE, (Princ., Geotechnical Engrs., Inc., Winchester, Mass.), **Gonzalo Castro**, Member, ASCE, (Princ., Geotechnical Engrs., Inc., Winchester, Mass.) and **John W. France**, Member, ASCE, (Proj. Mgr., Geotechnical Engrs., Inc., Winchester, Mass.)

Journal of Geotechnical Engineering, Vol. 111, No. 6, June, 1985, pp. 772-792

     A procedure for evaluating liquefaction susceptibility of a soil mass subjected to shear stress, such as in slopes, embankments, and foundations of structures, is presented. Liquefaction analysis is a stability analysis for which the shear strength in the numerator of the factor of safety equation is the undrained steady-state strength, and the denominator is the driving shear stress. The driving shear stress is the shear stress required to maintain static equilibrium. The undrained steady-state shear strength is a function only of the void ratio. Thus, one critical step of the procedure for liquefaction evaluation is the determination of the in-situ void ratio and the correction of laboratory-measured undrained steady-state strength to account for unavoidable changes in void ratio of the soil during sampling and testing. Dilative soils are not susceptible to liquefaction. Cyclic load tests are not required to evaluate the susceptibility to liquefaction.

276   Cyclic Testing and Modeling of Interfaces

C. S. Desai, (Prof., Dept. of Civ. Engrg. and Engrg. Mechanics, Univ. of Arizona, Tucson, Ariz. 85721), **E. C. Drumm**, (Asst. Prof., Dept. of Civ. Engrg., Univ. of Tennessee, Knoxville, Tenn.) and **M. M. Zaman**, (Asst. Prof., Dept. of Civ. Engrg., Univ. of Oklahoma, Norman, Okla.)

Journal of Geotechnical Engineering, Vol. 111, No. 6, June, 1985, pp. 793-815

     A new device for cyclic testing of large size interfaces between structural and geologic materials and rock joints is described. Test results are reported for a sand-concrete interface, and are used to express shear stress as function of normal stress, relative displacement, number of loading cycles and initial density. Constitutive behavior of the interface is expressed as nonlinear elastic and simulates loading-unloading-reloading response by using a modified Ramberg-Osgood model. The parameters for the models are found from a series of cyclic displacement controlled tests. Predictions of the constitutive model and a finite element analysis are compared with observed test behavior.

277   Experimental Foundation Impedance Functions

C. B. Crouse, Assoc. Member, ASCE, (Sr. Engr., Earth Technology Corp., 3777 Long Beach Blvd., P.O. Box 7765, Long Beach, Calif. 90807), **George C. Liang**, Assoc. Member, ASCE, (Sr. Engr., Earth Technology Corp., 3777 Long Beach Blvd., P.O. Box 7765, Long Beach, Calif. 90807) and **Geoffrey R. Martin**, Member, ASCE, (Vice Pres., Earth Technology Corp., 3777 Long Beach Blvd., P.O. Box 7765, Long Beach, Calif. 90807)

Journal of Geotechnical Engineering, Vol. 111, No. 6, June, 1985, pp. 819-822

Complex foundation impedance functions were determined experimentally from the results of forced vibration tests conducted at an accelerograph station. Two independent vibration experiments were performed to obtain estimates of the translational, K_{11}, and rocking, K_{22}, impedance functions and the impedance functions, K_{12} and K_{21}, representing the cross-coupling between translational and rocking motions. Values of K_{11} and K_{22} computed from the test data were not appreciably affected by K_{12} and K_{21}. This result suggests that, for similar foundations and excitation frequencies, the usual practice of conducting one experiment to obtain estimates of K_{11} and K_{22}, under the assumption that $K_{12} = K_{21} = O$, is valid.

278   Model for Piping-Plugging in Earthen Structures

Kartic C. Khilar, (Asst. Prof. of Chemical Engrg., I.I.T. Bombay, India), **H. Scott Fogler**, (Prof. of Chemical Engrg., Univ. of Michigan, Ann Arbor, Mich.) and **Donald H. Gray**, Assoc. Member, ASCE, (Prof. of Civ. Engrg., Univ. of Michigan, Ann Arbor, Mich.)

Journal of Geotechnical Engineering, Vol. 111, No. 7, July, 1985, pp. 833-846

A capillary model was developed to predict under what conditions piping vs. plugging is likely to occur as a result of clay particle dispersion in soils. In the intermediate size range, i.e., particles too small per se to clog pore constrictions and too large to wash through unimpeded, the outcome depends strongly on the concentration of dispersed particles in the seepage stream. This concentration in turn depends upon the rate at which the particles are eroded from the pore walls. The capillary model was used to ascertain the relationship between key variables at the onset of piping. This relationship involved the initial permeability and porosity of the soil, the critical tractive stress, and the hydraulic gradient. The model was also used to predict under what circumstances clay particle concentrations in a seepage stream will buildup suddenly to very high values, a condition favorable to clogging or particle holdup. Results of the analysis showed that very large particle concentration buildups tend to occur at permeabilities of less than 10^{-9} cm^2 (10^{-4} cm/s for water at 20°C). This prediction is consistent with previous findings reported in the geotechnical literature.

279   Estimating Elastic Constants and Strength of Discontinuous Rock

Pinnaduwa H. S. W. Kulatilake, Assoc. Member, ASCE, (Asst. Prof., Dept. of Mining and Geological Engrg., Univ. of Arizona, Tucson, Ariz. 85721)

Journal of Geotechnical Engineering, Vol. 111, No. 7, July, 1985, pp. 847-864

Methods used to estimate strength and deformability properties of discontinous rock can be categorized into: (1) The direct testing method; (2) the empirical correlation method; and (3) the analytical decomposition model method. In this paper, a numerical method that combines analytical decomposition modeling with statistical simulation is presented. The method consists of: (1) Probabilistic modeling of discontinuity geometry; (2) generation of discontinuities in rock blocks by Monte Carlo simulation; and (3) finite element analysis of simulated rock blocks. Strength and deformability properties of intact rock and discontinuities are needed for the finite element analysis. Results of finite element analysis can be used to obtain statistical properties of the rock mass. The method was used to study the mass properties of a shale that contains slickensides. At a block size of 30 in. × 30 in. (25cm), the mean value of the ratio between shale mass strength and intact shale strength was found to be about 0.6. For the same block size the mean value of the ratio between shale mass modulus and intact shale modulus was found to be around 0.45. At the same block size, Poisson's ratio of shale mass was found to be about two times the Poisson's ratio of intact shale. For ten simulations, the coefficient of variations of all three mass parameters were found to be less than 0.21.

280   Comparison of Natural and Remolded Plastic Clay

J. Graham, (Prof., Dept. of Civ. Engrg., Univ. of Manitoba, Winnipeg, Manitoba, Canada R3T 2N2) and **E. C. C. Li**, (Grad. Student, Dept. of Civ. Engrg., Univ. of Manitoba, Winnipeg, Manitoba, Canada R3T 2N2)

Journal of Geotechnical Engineering, Vol. 111, No. 7, July, 1985, pp. 865-881

Tests have been performed on triaxial samples of natural and remolded plastic clay from Winnipeg, Canada, to examine the way natural samples conform with modern conceptual models of soil behavior based largely on remolded samples. The clay contains mixed illites and montmorillonites. In its field condition, it is layered, fissured, over-consolidated, and contains many inclusions. Natural samples were carefully trimmed from blocks of the clay: Remolded samples were reconsolidated from a slurry. All reconsolidation was one-dimensional. Normal consolidation and critical state lines for the two sets of samples were parallel, with natural values at higher specific volumes than remolded samples. Post-yield compressibilities and pore water pressures were somewhat higher in natural samples, but strengths and moduli were lower. Yield stresses were similar, but the natural yields were again at higher specific volumes. Geotechnical and geological evidence suggests that the natural clay is cemented. Many fundamental features of critical state soil mechanics were found in the natural samples. Differences were in detail and not in principle.

281   Nonlinear Dynamic Analyses of an Earth Dam

Jean H. Prevost, Member, ASCE, (Assoc. Prof., Dept. of Civ. Engrg., Princeton Univ., Princeton, N.J. 08544), **Ahmed M. Abdel-Ghaffar**, Member, ASCE, (Assoc. Prof., Dept. of Civ. Engrg., Princeton Univ., Princeton, N.J. 08544) and **Sara J. Lacy**, (Grad. Student, Dept. of Civ. Engrg., Princeton Univ., Princeton, N.J. 08544)

Journal of Geotechnical Engineering, Vol. 111, No. 7, July, 1985, pp. 882-897

The following investigations are presented: (1) Comparison between the results of 2D nonlinear and 3D nonlinear dynamic finite element analyses of an earth dam subject to two very different input ground motions; and (2) comparison between measured and computed earthquake responses of the dam. The study is based on rigorous nonlinear hysteretic analyses utilizing a multi-surface plasticity theory. The backbone shear stress-strain curve is assumed hyperbolic and symmetrical about the origin. Detailed comparisons of induced stresses, strains, accelerations, and permanent deformations at various locations in the dam are presented. The effects of three-dimensionality on the dynamic response, particularly on resulting permanent deformations, are assessed. The suitability of 2D analyses in determining the dynamic behavior of such structures is evaluated.

282   Chemical Grouting in Soils Permeated by Water

Raymond J. Krizek, Member, ASCE, (Prof. and Chairman, Dept. of Civ. Engrg., Northwestern Univ., Evanston, Ill.) and **Teodoro Perez**, (consulting Engr., Panama City, Republic of Panama)

Journal of Geotechnical Engineering, Vol. 111, No. 7, July, 1985, pp. 898-915

Based on data obtained from seventy-nine large-scale one-dimensional laboratory tests, limiting conditions were established to define the transition zone between retention and elutriation of chemical grout injected into a cohesionless soil permeated by water. Various combinations of four grouts, four sands, and one gravel were included in the experimental program, and the interpretation is presented in terms of seepage velocity versus gel time relationships. The position and extent of the transition zone are strongly dependent on the viscosity and miscibility of the grout and to a lesser extent, on the effective grain size of the soil.

These parameters are largely responsible for the dilution that occurs and that, in turn, lengthens the gel time and decreases the viscosity, thereby governing the distribution of the injected grout and the resulting permeability of the soil. Even if a water seal is established, the grout may be dissolved or eroded, or both, after a relatively short time. Measurements from a limited number of tests demonstrate that the strength is substantially higher in the vicinity of the injection point.

283   Risk Analysis for Seismic Design of Tailings Dams

Steven, G. Vick, Member, ASCE, (Geotechnical Section head, Fox Consultants, Denver, Colo. 80214), Gail M. Atkinson, (Visiting Fellow, Earth Physics Branch, Energy, Mines and Resources, Ottawa, Ontario, Canada K1A 0Y3) and Charles I. Wilmot, (Mill Superintendent, Anaconda Minerals, Nevada Moly Operation, Tonopah, Nev. 89049)

Journal of Geotechnical Engineering, Vol. 111, No. 7, July, 1985, pp. 916-933

Probabilistic seismic risk analysis is a promising method for evaluating design options and establishing seismic design parameters. There have been few examples in the literature to guide practitioners in its use. This paper demonstrates the value of risk analysis for mine tailings dams and provides a case-history application for a seismically active portion of Nevada. Risk analysis provided the basis for selecting among design options having varying liquefaction resistance, and for establishing input parameters for dynamic analysis. Ranges are presented for the quantity and cleanup cost of tailings released is seismic failures to aid in determining expected failure consequences. It is shown that for many tailings dams, accepted lifetime failure probabilities of a few percent may provide a reasonable basis for probabilistic determination of seismic design criteria.

284   Compressibility of Partly Saturated Soils

T. S. Nagaraj, (Assoc. Prof., Dept. of Civ. Engrg., Indian Inst. of Sci., Bangalore, India) and B. R. Srinivasa Murthy, (Asst. Prof., Dept. of Civ. Engrg., Indian Inst. of Sci., Bangalore, India)

Journal of Geotechnical Engineering, Vol. 111, No. 7, July, 1985, pp. 937-942

From the considerations of the truncated diffuse double layer due to partial saturation, an equilibrium equation is written in terms of the soil state parameter viz. void ratio, e, void ratio at liquid limit, $e = wG$, degree of saturation, S and external applied stress, p. The type of fine grained soil is subdued by its liquid limit. The viability of the formulated approach has been experimentally verified. This approach, essentially eliminates the difficulty in the measurement of pore air pressure, u pore water pressure, u and computation of χ parameter.

285   Torsional Response of Piles in Nonhomogeneous Soil

Y. K. Chow, Assoc. Member, ASCE, (Lect., Dept. of Civ. Engrg., Nat. Univ. of Singapore, Kent Ridge, Singapore 0511)

Journal of Geotechnical Engineering, Vol. 111, No. 7, July, 1985, pp. 942-947

A discrete element method is described for the analysis of single piles subjected to torsion. The soil is modelled using the modulus of subgrade reaction, which is expressed in terms of the soil shear modulus and the pile radius. Discrete element matrices are listed for the pile, and the soil that can be readily implemented into a computer program. The accuracy of the discrete element approach is verified by comparison with analytical and finite element solutions. The nonlinear response of the pile-soil system is computed, and compared with two reported torsional load tests in the field.

286 Hydraulic Conductivity of Two Prototype Clay Liners

Steven R. Day, Assoc. Member, ASCE, (Proj. Engr., Geo-Con, Inc., Pittsburgh, Pa.) and **David E. Daniel**, Member, ASCE, (Asst. Prof., Dept. of Civ. Engrg., Univ. of Texas, Austin, Tex.)

Journal of Geotechnical Engineering, Vol. 111, No. 8, August, 1985, pp. 957-970

Two prototype liners were constructed at a site near Austin, Texas, using clays of low and high plasticity. The clays were compacted to 100% of standard Proctor density at a water content slightly wet of optimum using a sheepsfoot roller. The overall hydraulic conductivity (k) of each liner was determined by ponding water on the liners and measuring the rate of seepage. The field-measured k's of the liners were surprisingly high (4×10^{-6} and 9×10^{-6} cm/s). After water in the ponds was removed, laboratory permeability tests were performed on hand-carved samples obtained from the liners, on samples obtained with a thin-walled sampling tube, and on laboratory-compacted samples. Field permeability tests were also performed with ring infiltrometers. The tests showed that: (1) Essentially all of the laboratory tests, even on undistrubed samples, produced a measured k that was approximately 1,000 times less than the field-measured k, and (2) ring infiltration tests showed considerable scatter but the average k was close to the overall field-mesasured k. The findings raise important questions about whether laboratory permeability tests on compacted clay are relevant to clay liners and reinforce previous suggestions that compacted clay liners may contain numerous hydraulic defects such as fissures, slickensides, zones of poor bonding between clods of clay, and zones of relatively poor compaction. The desirability of field permeability tests is evident from the results reported.

287 Microcomputer Based Free Torsional Vibration Test

C. K. Shen, Member, ASCE, (Prof., Dept. of Civ. Engrg., Univ. of California, Davis, Calif. 95616), **X. S. Li**, (Asst. Development Engr., Dept. of Civ. Engrg., Univ. of California, Davis, Calif. 95616) and **Y. Z. Gu**, (Instructor, Dept. of Civ. Engrg., Zhejiang Univ., Hangzhou, China)

Journal of Geotechnical Engineering, Vol. 111, No. 8, August, 1985, pp. 971-986

A microcomputer based free torsional vibration testing system is introduced; its ability to determine the shear modulus and damping ratio of soils is demonstrated. In comparing with the resonant column method, the free torsional vibration testing system seems to offer reliable results, less expensive instrumentation, easier and faster operation, and less disturbance to soil specimens. The system also offers the advantage of conveniently studying the effect of a number parameters, such as the initial shear strain and cycle number on the dynamic modulus of soils. The nondestructive testing system may also be used effectively in the determination of dynamic response of compacted clay soil. The measured maximum shear modulus appears to relate closely to the structure of clay particles and the stiffness of the soil specimen. On that basis, the free torsional vibration test might be used to identify the fabric related stress-strain behavior of both sand and clay with little disturbance to a specimen.

288 Compaction Characteristics of River Terrace Gravel

Vinod K. Garga, Member, ASCE, (Staff Consultant, Klohn Leonoff, Ltd., Consulting Engrs., Richmond, British Columbia, Canada) and **Claudio J. Madureira**, (Geotechnical Engr., Curitiba, Brazil)

Journal of Geotechnical Engineering, Vol. 111, No. 8, August, 1985, pp. 987-1007

Large scale field and laboratory tests were carried out to investigate compaction characteristics of a river terrace gravel soil at the 120 m high Sao Simao Dam in Brazil. The coarse fraction of the deposit was typically 3 in. (7.5 sm) maximum size, while the matrix (defined as material passing No. 4 sieve) was basically a clayey sand with LL = 20-35%, PI = 5-15 and

35-50% passing the No. 200 sieve. The initial objective of the compaction tests was to develop the relationship between maximum density and percentage of gravel fraction which could be used for control of compaction in the field. The scope of the investigation was later enlarged to study the effects of energy of fraction, maximum particle size, mould dimensions, gradation of coarse material and water absorption on compaction of these soils. Field compaction trials were carried out to establish procedures for embankment construction and to evaluate the efficiency of two vibratory rollers.

289 Effect of Underground Void on Foundation Stability

M. C. Wang, Member, ASCE, (Prof., Dept. of Civ. Engrg., Pennsylvania State Univ., University Park, Pa. 16802) and **A. Badie**, Assoc. Member, ASCE, (Sr. Engr., Parsons, Brickerhoff, Quade, and Douglas, Inc., 1625 Van Ness Ave., San Francisco, Calif. 94109-3678)

Journal of Geotechnical Engineering, Vol. 111, No. 8, August, 1985, pp. 1008-1019

The effect of underground void on the stability of shallow foundation supported by a compacted clay was investigated by using a three-dimensional finite element computer program. The clay soil was a commercially available kaolinite which was compacted to 95% compaction of the standard Proctor compactive effort. The analysis was made for different conditions including footing shape (square and strip footings), void shape (continuous circular and cubic voids), orientation of continuous void axis with respect to strip footing axis (parallel and perpendicular directions), and void location. Results of the study indicate that footing stability will be affected by the underground void only when the void is located above the critical depth. The critical depth is not a constant but varies with the shapes of footing and void, void orientation, void size, and soil type. When the void is above the critical depth, the bearing capacity of the footing will decrease with decreasing distance between the footing and the void.

290 Expanded Shale Lightweight Fill: Geotechnical Properties

R. D. Stoll, Member, ASCE, (Prof., Dept. of Civ. Engrg., Columbia Univ., New York, N. Y.) and **T. A. Holm**, Member, ASCE, (Dir. of Engrg., Solite Corp., Mt. Marion, N. Y.)

Journal of Geotechnical Engineering, Vol. 111, No. 8, August, 1985, pp. 1023-1027

Recently, expanded shale aggregates of the kind used to produce lightweight concrete have been used as a substitute for ordinary fill materials in geotechnical applications where the combination of low unit weight and substantial shear strength is important. The results of triaxial compression tests on large-diameter (25 cm) specimens from several different locations in the eastern United States are presented. Depending on the degree of compaction, the angle of internal friction was found to vary from about 40° to 48° in aggregates that weigh about half as much as most naturally occurring fills. Stress-strain curves for the triaxial tests and for some preliminary consolidation tests are included in the paper so that they may be compared with data for ordinary fills when the lightweight aggregate is being considered as a design alternative.

291 Solution of Base Stability Analysis in Layered Soils

Indu Prakash, (Executive Engr., Ministry of Shipping and Transport (Roads Wing), New Delhi, India 110003)

Journal of Geotechnical Engineering, Vol. 111, No. 8, August, 1985, pp. 1027-1032

A solution to analyze stability against base failure in layered soils having slopes and berms of any shape, surcharge and tension crack is presented. The location of critical circle and its factor of safety is obtained through equations rther than by analysis of the number of slip circles. The solution is presented through stability parameters and charts.

292   Uplift Force-Displacement Response of Buried Pipe

Charles H. Trautmann, Assoc. Member, ASCE, (Research Assoc., School of Civ. and Environmental Engrg., Cornell Univ., Ithaca, N.Y.), **Thomas D. O'Rourke**, Member, ASCE, (Assoc. Prof., School of Civ. and Environmental Engrg., Cornell Univ., Ithaca, N.Y.) and **Fred H. Kulhawy**, Member, ASCE, (Prof., School of Civ. and Environmental Engrg., Cornell Univ., Ithaca, N.Y.)

Journal of Geotechnical Engineering, Vol. 111, No. 9, September, 1985, pp. 1061-1076

The design of buried pipelines in areas of vertical ground movement is governed, in part, by the magnitude of the forces imposed on the pipe and the displacements at which they are developed. An experimental study of these effects, dealing in particular with the influence of soil density and depth of burial, is described. The results compare well with several published models for medium and dense sand, but measured values of uplift resistance are much lower than predicted for loose contractive sand. A simplified procedure that can be applied to the design of buried pipelines is given.

293   Lateral Force-Displacement Response of Buried Pipe

Charles H. Trautmann, Assoc. Member, ASCE, (Research Assoc., School of Civ. and Environmental Engrg., Cornell Univ., Ithaca, N.Y.) and **Thomas D. O'Rourke**, Member, ASCE, (Assoc. Prof., School of Civ. and Environmental Engrg., Cornell Univ., Ithaca, N.Y.)

Journal of Geotechnical Engineering, Vol. 111, No. 9, September, 1985, pp. 1077-1092

The results of an experimental program to assess the response of buried pipes to lateral ground movements are presented. The effects of pipe depth, soil density, pipe diameter, and pipe roughness are considered, and test results are compared with published analytical models and experimental data. The results indicate the need to consider vertical equilibrium in predicting the horizontal response of buried pipelines, and the data agree well with several analytical models that include this effect. Pipe surface roughness was found to have little effect on response. Soil density has a large effect on displacements required to mobilize the maximum force but a relatively small effect on the value of the residual force at large displacements for depths typical of transmission pipelines. The study concludes with a simplified design procedure for predicting pipeline response to lateral ground movements.

294   Re-Examination of Slide of Lower San Fernando Dam

Gonzalo Castro, Member, ASCE, (Prin., Geotechnical Engrs., Inc., 1017 Main St., Winchester, Mass. 01890), **Steve J. Poulos**, Member, ASCE, (Prin., Geotechnical Engrs., 1017 Main St., Winchester, Mass. 01890) and **Francis D. Leathers**, Member, ASCE, (Prof. Mgr., Geotechnical Engrs., Inc., 1017 Main St., Winchester, Mass. 01890)

Journal of Geotechnical Engineering, Vol. 111, No. 9, September, 1985, pp. 1093-1107

A major slide occurred in the upstream slope of the Lower San Fernando Dam immediately following the earthquake of February 9, 1971. Previous analyses of this slide have been based on evaluation of seismic shear stresses and the results of cyclic triaxial tests. These previous analyses fail to explain several key features of the slide. The re-analysis presented herein, which is based on the concepts of liquefaction, steady state strength, and flow slide stability analysis, and on the data obtained during the previous analyses, is consistent with the principal features of the observed slide. The estimated in situ undrained steady state strength is substantially lower than the undrained strength measured in conventional laboratory tests due to the sample densification that normally occurs during sampling, handling, setup, and consolidation of loose sandy soils. An interpretation of the limited data on undrained strength available for the San Fernando Dam indicates that the soils that liquefied had very low undrained steady state strengths.

295   Strain Path Method

Mohsen M. Baligh, (Prof. of Civ. Engrg., Massachusetts Inst. of Tech., Cambridge, Mass. 02139)

Journal of Geotechnical Engineering, Vol. 111, No. 9, September, 1985, pp. 1108-1136

The Strain Path Method provides an integrated and systematic framework for elucidating and predicting pile foundation behavior, interpreting in situ tests, assessing sampling disturbance effects and, in general, approaching "deep geotechnical problems" in a consistent and rational manner. This article describes the fundamentals of the method and presents solutions for the effects of quasi-static undrained penetration of piles, cones and samplers on the deformations and strains in saturated isotropic clays. Procedures to determine penetration stresses and pore pressures and extensions of these solutions in an approximate form to more realistic conditions (e.g., anisotropic clays and drained penetration in sands) by means of the Strain Path Method are outlined. Estimates of undrained soil distortions due to sampler penetration indicate the necessity of reevaluating standard sampling and laboratory testing procedures utilized at present to estimate the in situ behavior of foundation soils, especially in the cases of soft clay deposits.

296   Moisture Curve of Compacted Clay: Mercury Intrusion
                              Method

S. Prapaharan, (Grad. Instr. in Research, Purdue Univ., West Lafayette, Ind.), **A. G. Altschaeffl**, Fellow, ASCE, (Prof. of Civ. Engrg., Civ. Engrg. Bldg., Purdue Univ., West Lafayette, Ind. 47907) and **B. J. Dempsey**, Member, ASCE, (Prof. of Civ. Engrg., Univ. of Illinois, Urbana, Ill.)

Journal of Geotechnical Engineering, Vol. 111, No. 9, September, 1985, pp. 1139-1143

This study presents procedures with which one can predict the equilibrium water content, in-service, of a clayey pavement subgrade. Mercury intrusion and pore-size distribution of a specimen of the soil at the in-service fabric are utilized to establish the soil moisture characteristic curve. The procedures are relatively quick, reasonably routine, and reproduceable. Predictions are believed reliable for sites where the soil water suctions are in excess of 500 cm (16 ft) of water depending upon the soil.

297   Dynamic Soil and Water Pressures of Submerged Soils

Hiroshi Matsuzawa, (Assoc. Prof., Dept. of Geotechnical Engrg., Nagoya Univ., Nagoya, Japan) and **Isao Ishibashi**, (Assoc. Prof., School of Civ. and Environmental Engrg., Cornell Univ., Ithaca, N.Y.)

Journal of Geotechnical Engineering, Vol. 111, No. 10, October, 1985, pp. 1161-1176

Current theories and procedures in evaluating dynamic lateral earth and water pressures due to submerged backfill soils against rigid retaining structures are thoroughly reviewed. Available experimental data is gathered and compared to the theories. A new generalized apparent angle of seismic coefficient, which can be easily used to evaluate dynamic soil as well as water pressure for a wide range of backfill soil types, is proposed. The new procedure incorporates the effect of the permeability and the geometry of the backfill soils and the modes of the wall movement.

298   **Crushing of Soil Particles**

Bobby O. Hardin, (Prof. of Civ. Engrg., Dept. of Civ. Engrg., Univ. of Kentucky, Lexington, Ky.)

Journal of Geotechnical Engineering, Vol. 111, No. 10, October, 1985, pp. 1177-1192

In order to understand the physics of the strength and stress-strain behavior of soils and to devise mathematical models that adequately represent such behavior, it is important to define the degree to which the particles of an element of soil are crushed or broken during loading. The amount of particle crushing in a soil element under stress depends on particle size distribution, particle shape, state of effective stress, effective stress path, void ratio, particle hardness, and the presence or absence of water. Data are analyzed for single mineral soils and rockfill-like materials and equations are presented that can be used to estimate the total breakage expected for a given soil subjected to a specified loading.

299   **Trench Effects on Blast-Induced Pipeline Stresses**

Edward D. Esparza, (Sr. Research Engr., Southwest Research Inst., P.O. Drawer 28510, San Antonio, Tex. 78284)

Journal of Geotechnical Engineering, Vol. 111, No. 10, October, 1985, pp. 1193-2010

An experimental program was conducted to obtain data from model tests to evaluate the effects of open trenches on blast-induced circumferential and longitudinal stresses on an underground pipeline. A model pipe was instrumented with five sets of orthogonal strain gages at two longitudinal locations. Five sets of experiments were performed. The first set of experiments, in which the blast-induced stresses covered the range of a stress-predictive equation derived previously, consisted of tests without a trench. Similar blasting tests were then conducted using four different trench geometries. The trenches were all the same width and located the same distance from the model pipe. Two different lengths and depths were used on the four trenches. The measured strains and ground motions from the no-trench experiments showed that the new data compared very well with values-obtained using the predictive equations developed on a previous program. In general, the trenches were more effective in reducing the circumferential pipe stresses than the longitudinal pipe stresses. However, different depths and lengths of the trench affected the pipe stress amplitude variations. The longer and deeper trench was the most effective.

300   **Clay Liner Permeability: Evaluation and Variation**

Kingsley Harrop-Williams, Member, ASCE, (Asst. Prof. of Civ. Engrg., Carnegie-Mellon Univ., Pittsburgh, Pa. 15213)

Journal of Geotechnical Engineering, Vol. 111, No. 10, October, 1985, pp. 1211-1225

The primary criterion used in evaluating the suitability of hazardous waste landfills for containing hazardous wastes is permeability, and many regulatory agencies have adopted regulations requiring clay-lined hazardous waste landfills to have a coefficient of permeability no greater than a fixed value. However, the measurement of in-situ permeability of compacted clay is time-consuming and difficult. If used to monitor construction, it slows the construction rate. Another equally important problem is that clay-liner permeability is extremely variable. Solutions to both of these problems are presented. Firstly, a relationship is developed between permeability and easily measured dry unit weight and moisture content. This would allow for the immediate monitoring of clay liners during construction. Secondly, an alternative is provided to the conventional approach in which permeability is treated as a single-valued quantity. A probabilistic description of the permeability of clay liners is developed from considerations of the heterogeneity of the soil. This would improve the design of clay liners by establishing confidence levels associated with possible ranges of the permeability.

301 Confining Pressure, Grain Angularity, and Liquefaction

Voginder P. Valid, Member, ASCE, (Prof. of Civ. Engrg., Univ. of British Columbia, Vancouver, B.C., Canada), Jing C. Chern, (Doctoral Student, Dept. of Civ. Engrg., Univ. of British Columbia, Vancouver, B.C., Canada) and Hadi Tumi, (Grad. Student, Dept. of Civ. Engrg., Queen's Univ., Kingston, Ontario, Canada)

Journal of Geotechnical Engineering, Vol. 111, No. 10, October, 1985, pp. 1229-1235

Substantial decrease in resistance to liquefaction is shown to occur with increase in confining pressure for a rounded and an angular sand having identical gradation. For a given increase in confining pressure, angular sand suffers a larger loss in resistance than rounded sand. At equal relative densities, angular sand is shown to be more resistant to liquefaction at lower confining pressures but less resistant at higher confining pressures than rounded sand.

302 Thermo-Mechanical Behavior of Seafloor Sediments

Sandra L. Houston, Assoc. Member, ASCE, (Asst. Prof. of Civ. Engrg., Arizona State Univ., Tempe, Ariz. 85287), William N. Houston, Member, ASCE, (Prof. of Civ. Engrg., Arizona State Univ., Tempe, Ariz. 85287) and Neil D. Williams, Assoc. Member, ASCE, (Asst. Prof. of Civ. Engrg., Georgia Inst. of Tech., Atlanta, Ga. 30332)

Journal of Geotechnical Engineering, Vol. 111, No. 11, November, 1985, pp. 1249-1263

A laboratory investigation of the thermo-mechanical response of deep ocean sediments has been performed. The study encompasses the range of thermal loadings expected for the emplacement of a canister of radioactive waste approximately 30 m below the seafloor in deep ocean waters. The experimental program included mechanical and thermal consolidation, undrained triaxial compressive strength and constant stress level creep tests on seafloor sediments over a temperature range of 4 – 200°C.

303 Settlement Analysis of Skirted Granular Piles

B. Govind Rao, (Sci., Central Building Research Inst., Roorkee, India) and Gopal Ranjan, Member, ASCE, (Prof., Dept. of Civ. Engrg., Univ. of Roorkee, Roorkee, India)

Journal of Geotechnical Engineering, Vol. 111, No. 11, November, 1985, pp. 1264-1283

The estimates of settlements of granular pile foundation are of paramount importance. However, in view of the complexity of the problem and non-availability of the precise mathematical models ensuring compatibility of the granular piles and the surrounding weak soil displacements, the settlement of granular piles in weak sub-soil deposits is generally estimated by empirical methods only. Attempts have been made to utilize numerical approaches. These methods require material properties which are difficult to estimate. On the other hand a method based on radial strain measured from in situ pressuremeter tests show promise, but the necessity of specialized equipment for routine use and the skilled personnel for operating the pressuremeter has to be assessed. Besides the computations made by these methods need verification from performance observations and full scale in-situ tests. In the present paper an analytical approach using the concept of equivalent coefficient of volume compressibility of the composite mass of soil pile system is developed to predict the settlements of weak sub-soil deposits reinforced with granular piles for both cohesionless and cohesive soils. The method uses the pile material properties and also the surrounding soil, the size, spacing and depth of piles as well as the pile/soil stiffness ratio and relative pile area. The study is extended to individually and collectively skirted pile groups. The validity of the assumptions and the proposed analytical approach is verified through full scale field tests under design loads at four different sites having different soil conditions. The comparison demonstrates the validity of the proposed method.

304   Stability of Membrane Reinforced Slopes

Dov Leschinsky, Assoc. Member, ASCE, (Asst. Prof., Univ. of Delaware, Dept. of Civ. Engrg., 130 Dupont Hall, Newark, Del. 19716) and **A. J. Reinschmidt**, (Mgr., Track Research Div., Association of American Railroads, 3140 S. Federal St., Chicago, Ill. 60616)

Journal of Geotechnical Engineering, Vol. 111, No. 11, November, 1985, pp. 1285-1300

An analytical approach to membrane reinforced earth is presented. It is based on limit-equilibrium and variational extremization. The results indicate that the potential failure surfaces are either planar or log-spiral. The analysis utilizes a reinforcing membrane sheet that is orthogonal to the radius vector defining its intersection with the slip surface. Results of a closed-form solution imply that: (1) The stronger the membrane the deeper the failure; (2) the membrane's elevation has little effect on the stability or on the location of the slip surface provided that failure is passing through it; (3) the presence of a membrane increases the compressive stress over the critical slip surface; and (4) the presence of a membrane decreases the soil's tensile stress that tends to develop near the crest. The results are presented in a convenient format of stability charts.

305   Soil Tunnel Test Section: Case History Summary

William H. Hansmire, Member, ASCE, (Sr. Professional Assoc., Parsons, Brinkerhoff, Quade & Douglas, Inc., Honolulu, Hawaii) and **Edward J. Cording**, Member, ASCE, (Prof. of Civ. Engrg., Univ. of Illinois at Urbana-Champaign, Urbana, Ill.)

Journal of Geotechnical Engineering, Vol. 111, No. 11, November, 1985, pp. 1301-1320

The strong influence of construction procedures was demonstrated by comprehensive field measurements of initial lining behavior and of ground movements for a soil tunnel. Plowing movements for the first tunnel shield created much of the ground loss and settlement. Much less settlement was observed for the second tunnel construction with a shield of an improved design. Volume changes in the soil played an important role in the entire pattern of ground movement. During first tunnel construction, significant increases in soil volume resulted in less surface settlement than the actual amount of lost ground about the shield. The opposite was observed for the second shield where significant compression of soils took place, particularly in the pillar between the two tunnels and in the previously disturbed soils over the first tunnel. The rib and lagging initial lining was demonstrated to be truly flexible. A ground reaction curve was established on the basis of analysis and measurement of load. The initial lining was observed to be carrying less than half of the full overburden load.

306   Rammed Earth House Construction

Surjya Maiti, (Asst. Prof., Dept. of Mech. Engrg., IIT Bombay-400 076. India) and **Jnanendra Mandal**, (Asst. Prof., Dept. of Civ. Engrg., IIT Bombay-400 076, India)

Journal of Geotechnical Engineering, Vol. 111, No. 11, November, 1985, pp. 1323-1328

An elementary analysis for calculation of stresses in a rammed earth house is presented. The analysis permits calculation of safe wall height, and an examination of safety of these structures. It is observed that the usual wall thickness is quite safe; the thickness can be reduced by at least a third without endangering safety and this can lead to a substantial savings in material and labor costs and an increase in the availability of space.

307   Flexural Behavior of Reinforced Soil Beams

A. P. Chaudhari, (Formerly, Postgrad. Student, Dept. of Civ. Engrg., Indian Inst. of Tech., Kharagpur, India) and **A. N. R. Char**, (Prof., Dept. of Civ. Engrg., Indian Inst. of Tech., Kharagpur, India)

Journal of Geotechnical Engineering, Vol. 111, No. 11, November, 1985, pp. 1328-1333

To investigate the flexural behavior of reinforced soil beams, tests were performed on rectangular and trapezoidal soil beams of varying span to depth ratios and varying amounts of reinforcements. Local silty clay was used in the study and simple supports for the beams were employed. The results indicate that strains are linearly distributed across the depth of the beam up to cracking and become nonlinear thereafter. Improvement due to reinforcement is smaller in trapezoidal and shallow beams. The increase in failure compressive strains is more for beams of smaller span to depth ratio and with larger amounts of reinforcements.

308   On the Determination of Foundation Model Parameters

Arnold D. Kerr, (Prof., Dept. of Civ. Engrg., Univ. of Delaware, Newark, Del. 19716)

Journal of Geotechnical Engineering, Vol. 111, No. 11, November, 1985, pp. 1334-1340

Various methods for the determination of the foundation model parameters are reviewed. It is then shown that one of the four known methods is conceptually of questionable validity because according to it the model parameters may be expressed analytically in terms of the elastic constants (E, v) of the actual subgrade. As part of the presentation, the Pasternak foundation response is derived from the equations of an elastic continuum. The paper concludes with a discussion of the other methods published in the literature for the determination of these parameters.

309   Dynamic Behavior of Pile Groups in Inhomogeneous Soil

Trevor G. Davies, Assoc. Member, ASCE, (Asst. Prof., Dept. of Civ. Engrg., State University of New York, Buffalo, NY 14260), **Rajan Sen**, (Asst. Prof., Dept. of Civ. Engrg. and Engrg. Mechanics, Univ. of South Florida, Tampa, FL 33612) and **Prasanta K. Banerjee**, Member, ASCE, (Prof., Dept. of Civ. Engrg., State University of New York, Buffalo, NY 14260)

Journal of Geotechnical Engineering, Vol. 111, No. 12, December, 1985, pp. 1365-1379

An efficient method of analysis has been developed to determine the steady-state dynamic response of vertically loaded piles and pile groups embedded in a soil stratum in which the stiffness increases linearly with depth. The piles have been represented as compressible columns and the soil as a hysteretic semi-infinite medium. The results of the analysis have been plotted in nondimensional form to demonstrate the effects of key nondimensional parameters on the vertical response of the pile-soil system. From the results of analyses of two pile groups, dynamic interaction factors have been calculated. These factors are very different from the interaction factors for homogeneous soils, particularly if the piles are widely spaced. These interaction factors reveal that piles in groups are subject to our-of-phase motions, which may result in substantially reduced pile group settlements at certain frequencies. Illustrative results for four and nine pile groups are given, as well as a practical design example.

310   Liquefaction Resistance of Thickened Tailings

Steve J. Poulos, Member, ASCE, (Princ., Geotechnical Engineers Inc., Winchester, MA), **Eli I. Robinsky**, Member, ASCE, (Prof. of Civ. Engrg., Univ. of Toronto, and Prin., E. I. Robinsky Assoc. Ltd., Toronto, Canada) and **Thomas O. Keller**, Member, ASCE, (Proj. Mgr., Geotechnical Engineers Inc., Winchester, MA)

Journal of Geotechnical Engineering, Vol. 111, No. 12, December, 1985, pp. 1380-1394

A new method is presented for analysis of the potential for triggering liquefaction, i.e., a flow slide, in liquefiable soil masses. The method is based on the principle of steady state deformation. In conventional mine tailings disposal operations, the tailings are pumped as a water suspension into extensive flat containment ponds. These ponds generally are formed by construction of large tailings dams. However, if the tailings are thickened substantially before discharge, the tailing slurry will form a sloping deposit. Maximum slope angles of 3.5° are normally recommended. Use of thickened tailings is less costly and has much less environmental impact than conventional tailings because tailings dams and their associated slime ponds are eliminated. In this paper, the resistance to liquefaction due to earthquakes of a proposed bauxite tailings deposit placed at a 2.9° slope is analyzed. The water content of the thickened tailings is high enough to make them susceptible to liquefaction even when placed at such gentle slopes. However, due to the clay content and thixotropic nature of these tailings, earthquakes that induce 0.1g peak ground acceleration do not cause enough strain to trigger liquefaction.

311   Effect of Initial Shear on Cyclic Behavior of Sand

Isao Ishibashi, Member, ASCE, (Assoc. Prof., School of Civ. and Environmental Engrg., Cornell Univ., Ithaca, NY), **Makoto Kawamura**, Assoc. Member, ASCE, (Assoc. Prof., Dept. of Civ. and Regional Planning, Toyohashi Univ. of Tech. Toyohashi, Japan (Visiting Asst. Prof., Cornell Univ., May 1984—Feb. 1985).) and **Shobha K. Bhatia**, Assoc. Member, ASCE, (Assoc. Prof., Dept. of Civ. Engrg., Syracuse, NY)

Journal of Geotechnical Engineering, Vol. 111, No. 12, December, 1985, pp. 1395-1410

Drained and undrained cyclic torsional simple shear tests are conducted for saturated Ottawa sand with and without initial static shear applications. It was found from the experiment that when the tests were conducted under uniform cyclic shear strains, both pore water pressure generation in undrained conditions and the volume changes in drained conditions were affected very little by the level of the initial static shear applications. The strain method, which uses shear strain parameters as determinative parameters, has been proven to be very useful for liquefaction and cyclic volume change analyses.

312   Uncertainty of One-Dimensional Consolidation Analysis

Ching S. Chang, Assoc. Member, ASCE, (Assoc. Prof., Dept. of Civ. Engrg., Univ. of Massachusetts, Amherst, MA 01003)

Journal of Geotechnical Engineering, Vol. 111, No. 12, December, 1985, pp. 1411-1424

The measured coefficient of consolidation, c_v, can have a substantial degree of variation even in a uniform clay layer. This paper, through a probabilistic analysis, examines the variability of one-dimensional consolidation solutions. Both the single layer and the multilayer models are evaluated. In the probabilistic analysis, the adequacy of a gamma distribution model for c_v is tested using experimental data. Then both the method of moments and the Monte Carlo method are used to develop solutions for one dimensional consolidation. The results are discussed in terms of confidence level with examples to demonstrate the variability of the solution of consolidation analysis.

313   **The Influence of SPT Procedures in Soil Liquefaction Resistance Evaluations**

H. Bolton Seed, Fellow, ASCE, (Prof. of Civ. Engrg., Univ. of California, Berkeley, CA 94720), **K. Tokimatsu**, (Assoc., Tokyo Inst. of Tech., Yokohama 227, Japan), **L. F. Harder**, Member, ASCE, (Assoc. Civ. Engr., Dept. of Water Resources, Div. of Design Const., Sacremento, CA 95802) and **Riley M. Chung**, Member, ASCE, (Group Leader, Geotechnical Engrg. Group, Structures Div., Center for Bldg. Tech., National Bureau of Standards, Washington, DC 20899)

Journal of Geotechnical Engineering, Vol. 111, No. 12, December, 1985, pp. 1425-1445

The purpose of this paper is to clarify the meaning of the values of standard penetration resistance used in correlations of field observations of soil liquefaction with values of N_1 measured in SPT tests. The field data are reinterpreted and plotted in terms of a newly recommended standard, $(N_1)_{60}$ in SPT tests where the driving energy in the drill rods is 60% of the theoretical free-fall energy. Energies associated with different methods of performing SPT tests in different countries and with different equipment are summarized and can readily be used to convert any measured N-value to the standard $(N_1)_{60}$ value. Liquefaction resistance curves for sands with different $(N_1)_{60}$ values and with different fines contents are proposed. It is believed that these curves are more reliable than previous curves expressed in terms of mean grain size. The results presented are in good accord with recommended practice in Japan and China and should, thus, provide a useful basis for liquefaction evaluations in other parts of the world. Finally, suggestions are made concerning the significance of the term "liquefaction" as it is often used in conjunction with field evidence of this phenomenon.

314   **Comparison of Two Strength Criteria for Intact Rock**

Ian W. Johnston, (Sr. Lect., Dept. of Civ. Engrg., Monash Univ., Clayton, Victoria, Australia 3168)

Journal of Geotechnical Engineering, Vol. 111, No. 12, December, 1985, pp. 1449-1454

Two broadly based empirical strength criteria have recently been proposed for intact rock. The first, which applies to intact rock and rock masses was presented by Hoek and Brown in 1980. The second, which applies to all intact geomechanical materials ranging from slightly overconsolidated clays to hard rocks, was presented more recently by the writer. It is demonstrated that the two different criteria predict essentially the same variations in strength when each is applied to the material of common interest, namely relatively hard intact rocks.

315   **Field Comparison of Three Mass Transport Models**

David A. Hamilton, (Hydr. Engr., Water Management Div., Michigan Dept. of Natural Resources, Lansing, Mich.), **David C. Wiggert**, Member, ASCE, (Prof., Dept. of Civ. Engrg., Michigan State Univ., East Lansing, Mich.) and **Steven J. Wright**, Assoc. Member, ASCE, (Assoc. Prof., Dept. of Civ. Engrg., Univ. of Michigan, Ann Arbor, Mich.)

Journal of Hydraulic Engineering, Vol. 111, No. 1, January, 1985, pp. 1-11

Comparable results are obtained from three mass transport models used to predict the location of a chromium plume in an aquifer. The models are an analytical solution, a finite element formulation, and a method of characteristics formulation. Considering the modeling uncertainties in field applications, costs, and required effort, analytical models are found to be most effective at this site. The finite element and method of characteristics numerical formulations are capable of incorporating many aquifer complexities, but with much greater effort. The Peclet number criterion appears to be critical and must be carefully evaluated before using the finite element model.

316   Model of Dispersion in Coastal Waters

David A. Chin, Member, ASCE, (Asst. Prof., Dept. of Civ. and Architectural Engrg., Univ. of Miami, Coral Gables, Fla. 33124) and **Philip J. W. Roberts**, Member, ASCE, (Asst. Prof., School of Civil Engrg., Georgia Inst. of Tech., Atlanta, Ga. 30332)

Journal of Hydraulic Engineering, Vol. 111, No. 1, January, 1985, pp. 12-28

A mathematical model that predicts the far field dispersion of wastewater discharged from ocean outfalls is presented. The model is a Lagrangian random walk-type simulation and directly uses the data obtained from spatially distributed, continuously recording, current meters. Such data are frequently obtained during the design of major ocean outfalls. The model is tested by using measured current meter data to predict the concentration distribution resulting from instantaneous and continuous point source releases during stratified and unstratified periods. Predictions of the temporal and spatial variations of maximum concentration, diffusing cloud size, and diffusion coefficient were found to be in excellent agreement with previous field studies.

317   Boundary Shear in Smooth Rectangular Ducts

D. W. Knight, Member, ASCE, (Lect., Civ. Engrg. Dept., Univ. of Birmingham, P.O. Box 363, Birmingham, B15 2TT, England) and **H. S. Patel**, (Research Assoc., Civ. Engrg. Dept., Univ. of Birmingham, P.O. Box 363, Birmingham, B15 2TT, England)

Journal of Hydraulic Engineering, Vol. 111, No. 1, January, 1985, pp. 29-47

The results of some laboratory experiments are reported concerning the distribution of boundary shear stresses in smooth closed ducts of a rectangular cross section for aspect ratios between 1 and 10. The distributions are shown to be influenced by the number and shape of the secondary flow cells, which, in turn, depend primarily upon the aspect ratio. For a square cross section with 8 symmetrically disposed secondary flow cells, a double peak in the distribution of the boundary shear stress along each wall is shown to displace the maximum shear stress away from the center position towards each corner. For rectangular cross sections, the number of secondary flow cells increases from 8 by increments of 4 as the aspect ratio increases, causing alternate perturbations in the boundary shear stress distributions at positions where there are adjacent contrarotating flow cells. Equations are presented for the maximum, centerline, and mean boundary shear stresses on the duct walls in terms of the aspect ratio.

318   Geometry of Ripples and Dunes

Muhammad I. Haque, Member, ASCE, (Sr. Research Sci., Civ., Mech., and Environmental Engrg. Dept., George Washington Univ., Washington, D. C. 20052) and **Khalid Mahmood**, Member, ASCE, (Div. of Water Resources Program and Prof. of Civ. Engrg., Civ., Mech., and Environmental Engrg. Dept., George Washington Univ., Washington, D. C. 20052)

Journal of Hydraulic Engineering, Vol. 111, No. 1, January, 1985, pp. 48-63

A kinematic theory for lower-regime bedform shapes is presented and examined in the light of extensive bedform data observed in large prototype channels. Depending upon the extent of bedform maturity, the theory predicts uniquely defined nondimensional shapes for upstream faces. These shapes are borne reasonably well by the observed data in spite of large variation in bedform size. The length of upstream face, relative to the wavelength, is randomly distributed with most probable value of 0.614, which is close to the value 0.667 predicted by the theory for mature ripples and dunes.

319   Stochastic Model of Suspended Solid Dispersion

Wilhelm Bechteler, (Prof. of Civ. Engrg., Inst. fur Hydromechanik and Hydrologie, Hochschule der Bundeswehr Munchen, D-8014 Neubiberg, FRG) and **Kurt Farber**, (Research Eng., Inst. fur Hydromechanik and Hydrologie, HSBw Munchen)

Journal of Hydraulic Engineering, Vol. 111, No. 1, January, 1985, pp. 64-78

A stochastic model of suspended solid dispersion in turbulent open channel flow has been developed that cannot only calculate concentration distributions in equilibrium, but also complex transition cases. The stochastic model gives a statement of probability for the constant movement of a dispersing particle based on measured turbulence parameters. By the Monte Carlo method, many random walks are calculated to determine the concentration of solids at a certain point. It is shown that the description by a diffusion equation is equivalent to that of a stochastic approach. The stochastic model of particle movement is composed from stochastic and deterministic velocity components, a time step, and a weighting function for the influence of concentration effects. A sensitivity analysis of the developed model is carried out. Applications are shown and compared with experiment and other analytical solutions.

320   Modeling of Flow Velocity Using Weirs

Hanna Majcherek, (Dr. Engrg., Inst. of Environmental Engrg., Tech. Univ. of Poznan, Osiedle Wielkiego Pazdziernika 80/36, 61-634, Poznan, Palska, Poland)

Journal of Hydraulic Engineering, Vol. 111, No. 1, January, 1985, pp. 79-92

Basic equations for determining the cross-section shape of the channel, in which the average flow velocity is a specified function of the head, v = v(h), are derived in this work for the case of modeling of flow velocity with proportional weirs. Studies are carried out for one- and two-part weirs with discharge characteristics described by a power function. For both types of weirs, functions that represent the channel cross-section shape are calculated from the flow continuity condition developed for the weir and in the upstream channel. The study on possibility of application of two-part weirs comprises two particular cases: when the weir crest is located at the channel bottom and when the weir crest is located above the channel bottom. The work also presents particular solutions for flow velocity described with a function in the form of power monomial.

321   Bed Load or Suspended Load

Peter J. Murphy, Assoc. Member, ASCE, (Asst. Prof., Dept. of Civ. Engrg., Univ. of Massachusetts, Amherst, Mass. 01003) and **Eduardo J. Aguirre M.**, (Grad. Student, Dept. of Civ. Engrg., Univ. of Massachusetts, Amherst, Mass. 01003)

Journal of Hydraulic Engineering, Vol. 111, No. 1, January, 1985, pp. 93-107

The difference between the sediment transport processes of bed load and suspended load is explained and quantified. The mean and fluctuating components of the particle motion are studied by examining both components of the particle's dynamics in open channel flows. For sand particles, the time required for particles leaving the bed to attain the free fall velocity is shown to be the critical parameter for the saltation-suspension distinction. An example of the use of the distinction is given.

322 Equilibrium Boundary Condition for Suspension

Peter J Murphy, Assoc. Member, ASCE, (Asst. Prof., Dept. of Civ. Engrg., Univ. of Massachusetts, Amherst, Mass. 01003)

Journal of Hydraulic Engineering, Vol. 111, No. 1, January, 1985, pp. 108-117

An explanation of the bottom boundary condition for the advection-diffusion equation is presented. By introducing a source term to account for the sediment erosion process of open channel flow, the equilibrium between erosion and deposition produces a unique reference concentration. The concentration for open channel flow is predicted and compared with existing data. Comparison indicates that the predicted reference concentration agrees with river data.

323 LDA Measurements in Open Channel

P. M. Steffler, (Asst. Prof., Dept. of Civ. Engrg., Univ. of Alberta, Edmonton, Canada), **Nallamuthu Rajaratnam**, Member, ASCE, (Prof., Dept. of Civ. Engrg., Univ. of Alberta, Edmonton, Canada) and **Allan W. Peterson**, Member, ASCE, (Prof., Dept. of Civ. Engrg., Univ. of Alberta, Edmonton, Canada)

Journal of Hydraulic Engineering, Vol. 111, No. 1, January, 1985, pp. 119-130

This paper presents mean velocity and some turbulence measurements for uniform subcritical flow in a smooth rectangular channel for three apsect ratios equal to 5.08, 7.83, and 12.3 obtained with a Laser Doppler Anemometer. These results include some measurements in the viscous sublayer. The agreement of the velocity measurements with the logarithmic law as well as departures from it are studied in the respective regions. Turbulent shear stress profiles are also measured in the central as well as wall regions.

324 Bed Shear From Velocity Profiles: A New Approach

Subrahmanyam Vedula, (Assoc. Prof., Dept. of Civ. Engrg., Indian Inst. of Sci., Bangalore, India) and **Ramakrishna Rao Achanta**, (Lect., Dept. of Civ. Engrg., Indian Inst. of Tech., Madras, India)

Journal of Hydraulic Engineering, Vol. 111, No. 1, January, 1985, pp. 131-143

A new binary law of velocity distribution has been developed to describe the velocity profile for the entire flow region. The law is a combination of logarithmic law, valid in the wall (inner) region, and parabolic law, valid in the core (outer) region of the flow. The validity of the law has been established based on earlier data on flat plates, rough and smooth pines and experimental data obtained from rigid-walled open channels with plane and beds. A procedure of estimating bed shear stress from the proposed law of velocity distribution using the measured velocity profile has been evolved. Bed shear estimates made according to this procedure are in agreement with the values obtained from uniform flow analysis in the case of open channel flow over a sediment bed. The proposed method of estimating the bed shear stress from the observed velocity profiles is found to be particularly useful in case where it is difficult to determine precisely the true bed level, such as in the case of flow over sediment beds.

325   Flow Over Side Weirs in Circular Channels

Ali Uyumaz, (Asst. Prof., Dept. of Civ. Engrg., Iowa State Univ., Ames, Iowa 50011) and **Yilmaz Muslu**, (Prof., Dept. of Civ. Engrg., Istanbul Tech. Univ., Istanbul, Turkey)

Journal of Hydraulic Engineering, Vol. 111, No. 1, January, 1985, pp. 144-160

Theoretical and experimental investigations of the flow over sharp-edged side weirs in circular channels are reported. A theoretical model can be obtained from energy principles, and it is solved by a finite difference method. The results are presented in diagrammatic form for practical use. The experiments cover sub- and supercritical flow regimes, and water surfaces are observed for both. The discharge coefficients are obtained for sub- and supercritical regimes. Derived expressions for the side weir discharge and water surface profiles for these regimes are compared with experimental results.

326   Submerged Weirs

Hanna Majcherek, (Dr. Engrg., Inst. of Environmental Engrg., Tech. Univ. of Poznan, Osiedle Wielkiego Pazdziernika 80/36, 61-634 Poznan, Palska, Poland)

Journal of Hydraulic Engineering, Vol. 111, No. 1, January, 1985, pp. 163-168

A theoretical method is presented for determining the shape of a submerged weir notch with a specified discharge characteristic. This problem was solved for the combined profile of a weir notch, which is described with the function $f_1(y)$ for $y \leqslant a$ and $f_2(y)$ for $y \geqslant a$, assuming a known function $f_1(y)$ and downstream head $h \leqslant a$. In this paper, $f_2(y)$ was derived for the weir with a rectangular bottom part and a linear discharge characteristic for the upper part. In order to obtain the solution, two simplifying assumptions were made: (1) An approximate relation between the downstream, h, and upstream, H, heads; and (2) a constant discharge coefficient, C_d. Based on the experimental tests, it was found that the effect of these simplifying assumptions on the precision of theoretical solutions is small.

327   Prediction of Combined Snowmelt and Rainfall Runoff

Kazumasa Mizumura, (Prof. of Civ. Engrg., Kanazawa Inst. of Tech., Kanazawa, Ishikawa Pref., Japan) and **Pittsburgh, Pa. 15261 Chiu,Chao-Lin**, Member, ASCE, (Prof. of Civ. Engrg., Univ. of Pittsburgh)

Journal of Hydraulic Engineering, Vol. 111, No. 2, February, 1985, pp. 179-193

Prediction of snowmelt runoff is one of the most important problems in the determination of optimum use of water resources. In this study, the combined snowmelt and rainfall runoff during the snowmelt period is predicted using a combination of the tanks and autoregressive models with parameters identified by the Kalman filtering. The proposed method can accurately predict the snowmelt runoff. The input data used are rainfall, snowfall, runoff and air temperature.

328   Channel Scouring Potential Using Logistic Analysis

Yeou-Koung Tung, Assoc. Member, ASCE, (Asst. Prof., Wyoming Water Research Center and Dept. of Statistics, Univ. of Wyoming, Laramie, Wyo.)

Journal of Hydraulic Engineering, Vol. 111, No. 2, February, 1985, pp. 194-205

Water resource engineers often have to relate qualitative dependent variables to one or more independent variables, which may or may not be quantitative. In such circumstances, the use of conventional regression analysis would encounter a number of difficulties. This paper introduces a statistical method called logistic regression which is specially developed for such conditions. The method is applied to a hydraulic problem of relating scouring potential in a channel to depth and velocity of flow. Whether or not the methodology could become a useful addition in water resources engineering analyses further investigations and applications are necessary.

329   Algorithm for Mixing Problems in Water Systems

Richard M. Males, Member, ASCE, (Owner, RMM Services, 3319 Eastside Ave., Cincinnati, Ohio), **Robert M. Clark**, Member, ASCE, (Chief, Physical and Chemical Contaminant Removal Branch, Drinking Water Research Div., U.S. Environmental Protection Agency, Cincinnati, Ohio), **Paul J. Wehrman**, Member, ASCE, (Systems Analyst, W.E. Gates and Assoc., Batavia, Ohio) and **William E. Gates**, Member, ASCE, (Pres., W.E. Gates and Assoc., Fairfax, Va.)

Journal of Hydraulic Engineering, Vol. 111, No. 2, February, 1985, pp. 206-219

The "Solver" algorithm, developed as an outgrowth of work on cost allocation in water distribution systems, is a simple technique that solves a number of interesting problems in water distribution system analysis. Problems related to mixing water from different sources within the distribution network, travel time from any source to any node of the network, and development of the cost of service to any node in the network can be solved assuming steady-state conditions, given a prior solution of the hydraulics (flow in each link) of the network. All these problems are formulated as the solution of simultaneous linear equations. The Solver algorithm has been coded in Fortran and incorporated within the structure of a multi-purpose system of computer programs that allows for the storage, display and manipulation of data associated with node-link networks describing water distribution systems. These programs are identified by the name Water Supply Simulation Model. The formulation of the Solver algorithm is described in detail, with examples of its use.

330   Free Jet Scour Below Dams and Flip Buckets

Peter J. Mason, (Chief Engr., Sir Alexander Gibb & Partners, Reading, England) and **Kanapathypilly Arumugam**, (Head, Fluids Div., Dept. of Civ. Engrg., The City Univ., Northampton Square, London, England)

Journal of Hydraulic Engineering, Vol. 111, No. 2, February, 1985, pp. 220-235

Formulas proposed to date for calculating ultimate scour depth under jets, such as issue from free dam overfalls and flip buckets, are examined. The accuracies of the formulas are evaluated by using each to process sets of scour data from prototypes and models of such prototypes. It is established that scour depth is as adequately calculable using only unit flow, q, and head drop, H, as using more complex considerations, but that those formulas most applicable for model purposes are not those best for prototypes. It is demonstrated that where bed particle size d is also considered, the use of the mean particle size d_m is more appropriate than the d_{90} size.

Chute friction loss allowances and jet impact angle are also examined in terms of improved accuracy; the relevance of Froude law scaling for this type of scour is verified. Lastly, a new formula for predicting ultimate scour depth, which includes an allowance for tailwater depth h is presented and shown to give improved accuracy for both models and prototypes.

331 Open Boundary Condition for Circulation Models

Alan F. Blumberg, (Sr. Sci., Dynalysis of Princeton, Princeton, N.J.) and **Lakshmi, H. Kantha,** (Research Sci., Dynalysis of Princeton, Princeton, N.J.)

Journal of Hydraulic Engineering, Vol. 111, No. 2, February, 1985, pp. 237-255

A persistant difficulty in modeling continental shelf and estuarine circulations is that associated with the correct specification of conditions on the open boundaries. This paper deals with the application of a new form of radiation condition on the open boundaries, which permits the mean subtidal and tidal forcing to be prescribed and, yet, allow transients generated inside the region to be transmitted outwards. A numerical model of shelf circulation is formulated, using vertically integrated equations, which accounts for tidal and transient wind forcing, as well as the effect of density gradients and the offshore large-scale oceanic circulation. The shelf model is dynamically linked to a storm model. The circulation model is applied to the Middle Atlantic Bight in a study of its response to a migratory storm superimposed on the sub–tidal and tidal circulation. The major estuaries of the region—Chesapeake Bay, Delaware Bay and Long Island Sound—are included in the domain, albeit at a resolution coarser than desirable. The model results are consistent with those of earlier observational and modeling studies.

332 Modeling of Unsteady Flows in Alluvial Streams

Bommanna G. Krishnappan, (Research Scientist, Environmental Hydr. Section, Hydr. Div., National Water Research Inst., Canada Centre for Inland Waters, Burlington, Ontario, Canada)

Journal of Hydraulic Engineering, Vol. 111, No. 2, February, 1985, pp. 257-266

A computer model is described to predict unsteady and nonuniform flows in alluvial streams. The model uses a generalized expression for evaluating the energy slope in mobile boundary flows. Therefore, it is possible to adopt different friction factor relationships into the model without affecting its structure. It is also possible to upgrade the model easily as new expressions for alluvial bed roughness become available. The model has been tested for laboratory and field conditions and shows good agreement with measured data. It is capable of predicting long-term changes in riverbed slope and thus is well-suited for estimating chang in river regime caused by certain modifications to river flow, such as river diversions, meander cutoffs, river training works, dams, and hydroelectric power plants.

333 Salt River Channelization Project: Model Study

Yung Hai Chen, Member, ASCE, (Assoc. Prin. Engr., Simons, Li and Assoc., Inc., Fort Collins, Colo.), **Abbas A. Fiuzat,** Member, ASCE, (Asst. Prof., Dept. of Civil Engrg., Clemson Univ., Clemson, S.C. 29631) and **Benjamin R. Roberts,** Assoc. Member, ASCE, (Vice Pres., Anderson-Nichols Co., Palo Alto, Calif.)

Journal of Hydraulic Engineering, Vol. 111, No. 2, February, 1985, pp. 267-283

Two models were tested in order to determine the adequacy of a channelization scheme for Salt River near Sky Harbor International Airport, Phoenix, Arizona. One fixed–bed model was used to study velocity patterns, flow stages, bank freeboards and bank protection requirements. One distorted movable-bed model was utilized to study erosion and deposition patterns and bank stability in the proposed scheme. Incipient motion of bed particles and tractive force theory for stability of banks were utilized for modeling the various complicated parts of the movable-bed model. Results were compared with mathematical model results, and good agreement was observed. Recommendations for improving the channelization scheme were given.

334 Computer Animation of Storm Surge Predictions

Charles M. Libicki, Member, ASCE, (Grad. Research Asst., Dept. of Civ. Engineering, Ohio State Univ., Columbus, Ohio 43210) and **Keith W. Bedford**, Member, ASCE, (Prof., Dept. of Civ. Engrg., Ohio State Univ., Columbus, Ohio 43210)

Journal of Hydraulic Engineering, Vol. 111, No. 2, February, 1985, pp. 284-299

With the proliferation of low-cost graphics systems and small computers, high-quality computer graphic representation of the modeling of geophysical flow is feasible. In this work, storm surge activity on Lake Erie is simulated by a Leendertse-type model and the output from this model (water surface elevations) are pictured as three-dimensional surfaces in an animated sequence. Four methods for constructing such images are surveyed: A wire-frame plot on a line-drawing device; a representation as a set of flat panels on a polygon-drawing raster display, a smooth shaded rendering on a point-addressable monochrome display; and a full-color rendering with "highlights" on a larger color palette display. Each of these methods are assessed in terms of speed, economy, reliability, the ability to render all relevant scales of activity, and the ability to be comprehended readily and unambiguously.

335 Lower Mississippi Salinity Analysis

Armando Ballofet, Fellow, ASCE, (Consultant, Tippetts-Abbett-McCarthy-Stratton, 655 Third Ave., New York, N.Y.) and **Deva K. Borah**, Assoc. Member, ASCE, (Sr. Hydr. Engr., Tippetts-Abbett-McCarthy-Stratton, 655 Third Ave., New York, N.Y.)

Journal of Hydraulic Engineering, Vol. 111, No. 2, February, 1985, pp. 300-315

Two procedures were developed to simulate salinity intrusion in the Mississippi River. The first one is based on the arrested salinity wedge theory and is applicable to quasi-steady-state flow conditions. It simulated very well the experimental results of Keulegan. It also simulated the positions of salt wedge tips and their interfaces in the Mississippi River with proper adjustment of the interfacial friction factor. The second one is a steady-dynamic routing model, which routes salt water in a cascade of reservoirs in the river bottom. Densimetric critical flow is assumed over a reservoir ridge. Error in storage definition is included in a calibration coefficient for the saltwater flow computations over the reservoir ridges. The model satisfactorily simulated durations of salt wedge presence along the lower Mississippi for the existing conditions and after deepening the channel to 55 ft.

336 River Flow Forecasting Model for the Sturgeon River

Donald H. Burn, (Grad. Student, Dept. of Civ. Engrg., Univ. of Waterloo, Waterloo, Ontario, Canada N2L 3G1) and **Edward A. McBean**, (Prof., Dept. of Civ. Engrg., Univ. of Waterloo, Waterloo, Ontario, Canada N2L 3G1)

Journal of Hydraulic Engineering, Vol. 111, No. 2, February, 1985, pp. 316-333

A forecasting technique for predicting river flow resulting from combined snowmelt and rainfall is presented. The technique incorporates Kalman filtering techniques to reflect uncertainty in the measured data as well as errors in the system model. A primary emphasis is given to how the forecasting algorithm is applied to a case study area, thus demonstrating the utility of the technique when applied to real-world data. Methodologies are presented which can be used to calculate the covariance matrices associated with the Kalman filter algorithm utilized by the forecasting procedure.

337   Circulation Structure in a Stratified Cavity Flow

J. R. Koseff, (Acting Asst. Prof., Dept. of Civ. Engrg., Stanford Univ., Stanford, Calif. 94305) and **R. L. Street,** Member, ASCE, (Prof. of Fluid Mech. and Applied Mathematics, Dept. of Civ. Engrg., Stanford Univ., Stanford, Calif. 94305)

Journal of Hydraulic Engineering, Vol. 111, No. 2, February, 1985, pp. 334-354

Experiments were conducted to study the circulation patterns and mixing processes in a stratified, lid-driven cavity flow. The ratio of the cavity depth to width used was 1:1 and that of span to width was 3:1. The flow is different from those reported in the literature in up to four ways, viz., recirculation is a dominant feature, temperature control is used to obtain the stratification, the upper and lower boundaries are held at different constant temperatures, and the flows considered are, at most, only partially turbulent. For the geometric confguration used, the following may be concluded about the flow. The flow is strongly three-dimensional. The number of circulation cells expected in this stratified flow can be estimated from the value of the bulk Richardson number, R_{ib}. Given a local Richardson number, R_i, the entrainment rate, E, is found to be proportional to $R_i^{-1.4}$, indiating a shear-induced mixing process related to the recirculation. Taylor-Görtler-like vortices above the thermocline are significant contributors to the mixing process.

338   Flow Distribution in Compound Channels

Peter Richard Wormleaton, (Lect., Queen Mary College, London Univ., London, England E1 4NS) and **Panos Hadjipanos,** (Post-Doctoral Research Assoc., National Technical Univ. of Athens, Athens, Greece)

Journal of Hydraulic Engineering, Vol. 111, No. 2, February, 1985, pp. 357-361

The performances of four of the most commonly used methods for discharge calculation in compound channels of varying roughness, were assessed by comparison with laboratory data. The four methods are based on splitting the section into main channel and floodplain subdivisions using either horizontal or vertical interfaces. The accuracy of the four methods in calculating overall section discharge generally increases at greater floodplain depth and with smoother floodplains. However, the distribution of flow within the section is generally poorly modelled, which leads to very erroneous values for momentum and kinetic energy flux.

339   Oil Slicks in Ice Covered Rivers

Brent A. Berry, Member, ASCE, (Hydraulic Engr., Northwest Hydr. Consultants, Vancouver, British Columbia, Canada) and **Nallamuthu Rajaratnam,** Member, ASCE, (Prof., Dept. of Civ. Engrg., Univ. of Alberta, Edmonton, Alberta, Canada)

Journal of Hydraulic Engineering, Vol. 111, No. 3, March, 1985, pp. 369-379

The results of a study on the effect of ice cover on oil slicks contained by booms in rivers are presented. Experimental observations on velocities in the oil slick obtained by laser doppler anemometry are found to agree well with theoretically predicted profiles. The effect of the ice cover is to increase the thickness of the ponded slick by one-third. It also increases the stability of the oil water interface.

340   Generalized Water Surface Profile Computations

Albert Molinas, Assoc. Member, ASCE, (Asst. Prof., Dept. of Civ. Engrg., Colorado State Univ., Fort Collins, Colo.) and **Chih Ted Yang**, Member, ASCE, (Civ. Engr., U.S. Dept. of the Interior, Bureau of Reclamation, Engrg. and Research Center, Denver, Colo.)

Journal of Hydraulic Engineering, Vol. 111, No. 3, March, 1985, pp. 381-397

A computer model based on both the energy and momentum equations is developed. This generalized model can be used for the computation of water surface profiles through hydraulic jumps. It also allows computation of water surface profiles regardless of whether the bed slope is steep, mild, horizontal, adverse or a combination of these. The control section can be a lake, weir, gate or a natural river section. The Manning, Chezy or Darcy-Weisbach equations can be used for head loss computation. A detailed description of methods used and a step-by-step computation procedure is given in this paper. Examples are used to demonstrate the applications of this generalized model for water surface profile computations.

341   Boardman Labyrinth—Crest Spillway

John J. Cassidy, Fellow, ASCE, (Chf. Hydro. Engr., Bechtel Civ. and Minerals, Inc., San Francisco, Calif.), **Christopher A. Gardner**, (Proj. Engr., Bechtel Civ. and Minerals, Inc., San Francisco, Calif.) and **Robert T. Peacock**, Member, ASCE, (Proj. Engr., Bechtel Civ. and Minerals, Inc., San Francisco, Calif.)

Journal of Hydraulic Engineering, Vol. 111, No. 3, March, 1985, pp. 398-416

Labyrinth-crest spillways provide an economical flood-handling structure provided the operating head is small. The design of a spillway with a labyrinth crest for which published discharge characteristics were uncertain for higher heads is described. To verify the design, a model study was conducted which indicated that actual discharge would be at least 20% lower than originally estimated. Details are presented for the design criteria for the spillway crest and for the flow conditions downstream including the chute, the energy dissipator, and the receiving channel.

342   Saltwater Upconing in Unconfined Aquifers

Prakob Wirojanagud, (Research Engrg. Assoc., Univ. of Texas at Austin, Bureau of Economic Geology, Austin, Tex. 78712) and **Randall J. Charbeneau**, (Asst. Prof., Dept. Civ. Engrg., Univ. of Texas at Austin, Austin, Tex. 78712)

Journal of Hydraulic Engineering, Vol. 111, No. 3, March, 1985, pp. 417-434

The need for skimming of fresh groundwater above a saline water body without producing an unacceptable contamination in the pumped water calls for the investigations carried out in this research work. Two axisymmetric models, namely the steady state and the transient upconing models, are developed with the assumption of an abrupt interface. Finite element method with an iterative scheme is employed in solving the steady state model whereas a method of finite element in space and finite differences in time is used with iterative schemes for the transient model. The models are calibrated with the analytical solutions for the case of radial flow to a fully penetrating well in an unconfined aquifer. A method of superimposing the effects of hydrodynamic dispersion on the upconing is discussed. The simulation results of both models, together with the technique of dimensional analysis, provide information for establishing design criteria for optimal pumping or choice of well depth or both.

343   Analytical Diffusion Model for Flood Routing

Tawatchai Tingsanchali, Member, ASCE, (Assoc. Prof., Div. of Water Resources Engrg., Asian Inst. of Tech., P.O. Box 2754, Bangkok, Thailand) and Shyam K. Manandhar, Member, ASCE, (Research Assoc., Div. of Water Resources Engrg., Asian Inst. of Tech., P.O. Box 2754, Bangkok, Thailand)

Journal of Hydraulic Engineering, Vol. 111, No. 3, March, 1985, pp. 435-454

An analytical diffusion model for flood routing which can take into account backwater effect and lateral flows has been developed. The model s applied to route the floods in a hypothetical rectangular channel with different upstream, downstream, and lateral boundary conditions. Different channel characteristics are assumed and the results obtained are found to check well with those obtained by the finite difference method of implicit scheme based on the complete Saint-Venant equations for unsteady open channel flow. The model shows good results when applied to simulate flood flow conditions in 1980 and 1981 in the Lower Mun River, in Northeast Thailand. The model cannot be incorporated with detailed data of cross sections or river bed geometry but requires only their average values. The Chézy, C and the diffusivity, k due to channel irregularities are used in the model and are determined by trial and error during model calibration. The model provides an excellent means to analyze individual or overall effects of the boundary conditions and requires much less effort and time for computation at a particular station.

344   Average Uplift Computations for Hollow Gravity Dams

Amrik S. Chawla, (Prof., Water Resources Development Training Centre, Univ. of Roorkee, Roorkee-247667 (U.P), India) and Akhilesh Kumar, (reader, Water Resources Development Training Centre, Univ. of Roorkee, Roorkee-247667 (U.P), India)

Journal of Hydraulic Engineering, Vol. 111, No. 3, March, 1985, pp. 455-466

A closed-form solution is obtained with the help of the seepage theory and the Schwartz-Christoffel transformation for determination of average uplift pressures for any action of hollow gravity dam. The average uplift pressure is calculated numerically from the derived equations and plotted to facilitate the use of the results for design purposes. It is seen that the average uplift pressures drop almost linearly to about 5–22% near the head of the cavity. The average uplift pressure opposite the cavity head decreases with an increase in the value of l_1/b, and decreases in the value of b_1/b. The average uplift pressures beyond the cavity head drop to approximately negligible value at a distance of about 2b from the head of the cavity. By shifting the cavity head towards downstream, the average uplift pressures increase in the entire upstream portion.

345   Volumetric Approach to Type Curves in Leaky Aquifers

Zekai Sen, (Prof. and Head, Hydrology Dept., King Abdulaziz Univ., Faculty of Earth Sci., P.O. Box 1744, Jeddah, Kingdom of Saudi Arabia)

Journal of Hydraulic Engineering, Vol. 111, No. 3, March, 1985, pp. 467-484

A nonequilibrium formula in terms of the depression cone volume is used to predict the drawdown variations in a fully penetrating large diameter well discharging from a leaky aquifer. The derivation of the formula is based on the continuity and Darcy laws. Sets of type curves are presented to find the aquifer and aquitard hydraulic properties. The asymptotic curves, as the leakage diminishes to zero, approximate the type curves for large diameter wells in confined aquifers. Leaky aquifer type curves for drawdowns in the producing well have three distinctive portions. The initial portion is a straightline and it reflects the effect of the well storage. It does not provide any relevant information about either the aquifer or the semipervious layer hydraulic properties. The second portion is a transition from the well storage effect to the complete leakage effect only. The last portion is a horizontal line corresponding to the steady state flow due to leakage. In fact, the second portion is the most significant one from the aquifer parameters determination point of view. Comparison of these type curves with the point source solution indicates that the deviations are larger for small times but large leakages. However, for large times and small leakages, these deviations reduce to practically acceptable limits.

346   Fall Velocity of Particles in Oscillating Flow

Paul A. Hwang, Assoc. Member, ASCE, (Marine Sci., College of Marine Studies, Univ. of Delaware, Lewes, Del. 19958)

Journal of Hydraulic Engineering, Vol. 111, No. 3, March, 1985, pp. 485-502

The paper presents results of harmonic analysis in studying particle motion in oscillating flows. An approximate equation is derived to solve the zeroth harmonic equation, which governs the effective fall velocity, $\bar{v}_s$, of particles in oscillating flows. The equation reveals that the drag force on the particles is significantly modified by the variation of the relative velocity between particles and fluids. Three major factors that govern the variation of $\bar{v}_s$ (as compared to the terminal velocity in still water) are the terminal velocity Reynolds number, the phase lag, and the velocity amplitudes of the flow and particle oscillations. The dimensionless equation is given. Applying to the sediment suspension under wave action, the analysis suggests that an enhancement of suspension should occur.

347   River Morphology and Thresholds

Howard H. Chang, Member, ASCE, (Prof. of Civ. Engrg., San Diego State Univ., San Diego, Calif. 92182)

Journal of Hydraulic Engineering, Vol. 111, No. 3, March, 1985, pp. 503-519

The regime geometry and channel patterns of alluvial rivers are analyzed using an energy approach together with physical relationships of flow continuity, flow resistance, and sediment transport. Because of the discontinuity in flow resistance, and thus in power expenditure, between lower and upper flow regimes, the adjustment in river regime consists of sudden changes in channel geometry, channel pattern, and sometimes silt-clay content, when such a discontinuity is crossed. Thresholds or discontinuities in river morphology are obtained in the analysis. In accordance with such thresholds, rivers of distinct morphological features are classified into four regions based upon the bankfull discharge, channel slope, and median size of bed sediment. Their respective features are described, and certain regime relationships for channel width and depth are established. The predicted channel geometries are compared with river data.

348   Incipient Sediment Motion and Riprap Design

Sany-yi Wang, (Assoc. Prof. of Hydr. Engrg., Tianjin Univ., China) and **Hsieh Wen Shen**, Member, ASCE, (Prof. of Civ. Engrg., Colorado State Univ., Fort Collins, Colo.)

Journal of Hydraulic Engineering, Vol. 111, No. 3, March, 1985, pp. 520-538

The available information on incipient sediment motion criteria for both unidirectional and oscillating flow motions is summarized and analyzed according to the Shields diagram for incipient motion. The Shields diagram is extended for large and small particle sizes on the basis of data collected in China. Design guides prepared by the U.S. Army Corps of Engineers and the California Department of Transportation are compatible with the extended Shields diagram for incipient motion.

349   Equations for Plane, Moderately Curved Open Channel Flows

Willi H. Hager, (Sr. Research Engr., Chaire de Constructions Hydrauliques, EPFL, DGC, CH-1015 Lausanne, Switzerland)

Journal of Hydraulic Engineering, Vol. 111, No. 3, March, 1985, pp. 541-546

Plane flow with moderately curved and involved bounding streamlines is investigated by accounting for the transverse variations of the velocities. The result is a highly nonlinear version of the corresponding Boussinesq relation (valid only for weakly curved and sloped stream lines). Applications include the computation of the cnoidal and solitary wave profiles and indicate a clear advantage of the present formulation over previous ones.

350   Subregion Iteration of Finite Element Method for Large-Field Problems of Ground-Water Solute Transport

Gour-Tsyh Yeh, (Sr. Staff Member, Environmental Science Div., Oak Ridge National Lab., Oak Ridge, Tenn.), **Jack C. Hwang**, (Asst. Prof., Dept. of Civ. Engrg., Drexel Univ., Philadelphia, Pa.) and **Woncheol C. Cho**, (Research Specialist, Dept. of Civ. Engrg., Drexel Univ., Philadelphia, Pa.)

Journal of Hydraulic Engineering, Vol. 111, No. 3, March, 1985, pp. 547-551

To compute groundwater solute transport, the CPU memory requirement can be drastically reduced by using the subregion iteration of the orthogonal-upstream weighting finite element method. The use of an orthogonal-upstream weighting function results in an irreducible and diagonally dominant coefficient matrix. These properties provide sufficient conditions for the convergence of block iteration methods. Computer runs for a test problem were carried out for 10 × 10, 30 × 30, and 50 × 50 grid systems with 2, 3, and 4 subregions. The CPU memory and CPU time were monitored during each run. The number of iterations required for the convergent solution in each time step was also recorded. The results are useful for determining the number of subregions required for a given computer memory capacity, and for estimating the CPU time when using the subregion iteration method.

351   Stage Frequency Analysis at a Major River Junction

Gary B. Dyhouse, Member, ASCE, (Ch., Hydr. Engrg., Dept. of the Army, Corps of Engrs., 210 Tucker Blvd. N., St. Louis, Mo. 63101)

Journal of Hydraulic Engineering, Vol. 111, No. 4, April, 1985, pp. 565-583

A coincident frequency analysis using the total probability theorem was performed for an 80-mile (129 km) reach of the Illinois River to evaluate existing stage frequency relationships and revisions caused by a proposed higher levee system. Frequency and duration relationships were derived to establish the upstream and downstream boundaries. A water surface profile model was applied to calculate stage at 210 points in the reach for given starting backwater and discharge conditions. The analytical process was calibrated to stage data in the reach through iterative adjustments in the boundary duration curves. Comparisons of the adopted results to previous analyses using simplified techniques showed a significant reduction in levee design profile and potential cost savings of $1,600,000 in levee construction.

352 Hydrodynamic Pressures Acting Upon Hinged-Arc Gates

Ryszard Rogala, (Prof., inst. of Geotechnical Science, Technical Univ. of Wroclaw, Poland) and **Jan Winter**, (Dr., Inst. of Geotechnical Sci., Technical Univ. of Wroclaw, Poland)

Journal of Hydraulic Engineering, Vol. 111, No. 4, April, 1985, pp. 584-599

A new method for calculating the mean value of the local hydrodynamic pressure acting upon a hinged gate and a mathematical model of pressure pulsation have been developed. The mean hydrodynamic pressure at an optional point on the surface of a hinged gate was obtained by using the Bernculli equation and model test data. The problem of the pulsation phenomenon of hydrodynamic pressure was solved by considering the random character of the process and by making measurements of pressure fluctuation. Numerical analysis of the results obtained made it possible to determine the autocorrelation function and the power spectral density function and, in consequence, to draw conclusions concerning the character of and reasons for the pulsation phenomenon of the pressure. As a result of the investigations, a normalized power density spectrum equation, which characterized the frequency of the pressure pulsation, was derived.

353 Kinematic Shock: Sensitivity Analysis

Victor Miguel Ponce, (Prof. of Civ. Engrg., San Diego State Univ., San Diego, Calif.) and **Diane Windingland**, (Research Asst., Dept. of Civ. Engrg., San Diego State Univ., San Diego, Calif.)

Journal of Hydraulic Engineering, Vol. 111, No. 4, April, 1985, pp. 600-611

A series of numerical experiments is performed to determine the flow and channel characteristics that are most conducive to kinematic wave steepening and associated kinematic shock phenomena. Relevant flow and channel characteristics are identified at the outset, and a program of 80 computer runs is completed, varying the inflow hydrograph peak Froude number, time-to-peak, and base-to-peak flow ratio, and the channel cross-sectional shape. It is found that all have a definite effect on kinematic shock development. The size of the wave is perhaps mostly responsible for the occurrence or nonoccurrence of the shock.

354 Estimating Mean Velocity in Mountain Rivers

Colin R. Thorne, Affiliate Member, ASCE, (Assoc. Prof., Dept. of Civ. Engrg., Colorado State Univ., Fort Collins, Colo. 80523) and **Lyle W. Zevenbergen**, (Grad. Student, Dept. of Civ. Engrg., Colorado State Univ., Fort Collins, Colo. 80523)

Journal of Hydraulic Engineering, Vol. 111, No. 4, April, 1985, pp. 612-624

It is often necessary to estimate the mean velocity in ungaged mountain rivers, but the flow resistance equations available for this purpose require further testing and development. Such rivers are characterized by coarse bed materials, steep slopes, and low depths. For these conditions, boulders protrude well into or completely through the flow, and bed roughness is said to be large-scale. Recently, three equations specifically intended for large-scale roughness have been developed to address this problem. Data from a mountain river were used to test the equations. All were prone to errors of the order of 30%. The errors were systematic, all the equations tending to overestimate mean velocity compared to observed values. An investigation of the possible sources of error suggested that sampling error in the knowledge of bed material size was a major source of uncertainty in the predicted velocities, but that this and discharge errors could not wholly account for the observed discrepancies. Boulder shape is not represented in the equations, but this was discounted as an important parameter on the basis of a flume study. It was concluded that further research is required to produce a reliable, process based, flow resistance equation for mountain rivers.

355   **Flow Resistance Estimation in Mountain Rivers**

James C. Bathurst, (Sr. Scientific Officer, Inst. of Hydrology, Wallingford, Oxon, U.K.)

Journal of Hydraulic Engineering, Vol. 111, No. 4, April, 1985, pp. 625-643

Examination of the flow resistance of high-gradient gravel and boulder-bed rivers, using data collected in British mountain rivers with slopes of 0.4 - 4%, shows that there are differences in resistance variation between mountain and lowland rivers and that between-site variations do not necessarily reflect at-a-site variations. Comparison of data with the familiar resistance equation relating the Dracy-Weisbach friction factor to the logarithm of relative submergence shows that the equation tends to overestimate the resistance in uniform flow. The equation also tends to underestimate the rate of change of resistance at a site (as discharge varies) with high gradients. The influences of nonuniform channel profile, sediment size distribution, channel slope and sediment transport are reviewed, but the data do not allow any quantification of these effects. Instead an empirical approach based on the available data is presented, allowing the friction factor to be calculated from the relative submergence with an error of up to ±25% to ±35%. A summary of the field data is included.

356   **Water and Sediment Routing Through Curved Channels**

Howard H. Chang, Member, ASCE, (Prof., Dept. of Civ. Engrg., San Diego State Univ., College of Engrg., San Diego, Calif. 92182)

Journal of Hydraulic Engineering, Vol. 111, No. 4, April, 1985, pp. 644-658

A mathematical model for water and sediment routing through curved alluvial channels is developed and applied in a case study. This model, which is for alluvial streams with nonerodible banks, may be employed to simulate stream bed changes during a given flow, thereby providing the necessary information for the design of dikes, levees, or other bank protection. This model incorporates the major effects of transverse circulation, inherent in curved channels, on the flow and sediment processes. In the simulation of the evolution in stream bed profile, the effect of transverse flow is tied in with the aggradation and degradation development. River flow through curved channels is characterized by the changing curvature, to which variations of flow pattern and bed topography are closely related. Simulation of these changing features is based upon the fluid dynamics governing the growth and decay of transverse circulation along the channel.

357   **Transition Zone Width in Ground Water on Ocean Atolls**

Raymond E. Volker, Member, ASCE, (Assoc. Prof., Dept. of Civ. & Systems Engrg., James Cook Univ., Townsville, Australia), **Miguel A. Marino**, Member, ASCE, (Prof., Dept. of Land, Air & Water Resources, Univ. of California, Davis, Calif.) and **Dennis E. Rolston**, (Prof., Dept. of Land, Air & Water Resources, Univ. of California, Davis, Calif.)

Journal of Hydraulic Engineering, Vol. 111, No. 4, April, 1985, pp. 659-676

An analytical solution for the thickness of the transition zone between fresh and seawater in aquifers on ocean islands is presented. The interface is treated as a mixing layer similar to a laminar boundary layer between fluids moving at different velocities. The method requires assumptions of homogeneous isotropic dispersion, uniform hydraulic conductivity and steady flow; all water withdrawal is assumed to occur at the water table. The slope of the sharp interface is used in the solution for the mixing layer and the sharp interface is determined using Dupuit-Forchheimer assumptions of horizontal flow. The method presented efficiently estimates the transition zone thickness, thus providing information not available from sharp interface models. The method is illustrated by application to the freshwater lens on an island of the Tarawa Atoll in the Pacific Ocean, and is used to show the effects of different pumping arrangements on the width of the mixing layer.

358   New Approach to Calibrating Bed Load Samplers

David W. Hubbell, Member, ASCE, (Hydro., U.S. Geological Survey, Box 25046, MS 413, Federal Center, Denver, Colo. 80225), **Herbert H. Stevens**, Member, ASCE, (Hydro., U.S. Geological Survey, Box 25046, MS 413, Federal Center, Denver, Colo. 80225) and **John V. Skinner**, (Hydro., Federal Inter-Agency Sedimentation Project, SAFHL, Univ. of Minnesota, Minneapolis, Minn. 55414)

Journal of Hydraulic Engineering, Vol. 111, No. 4, April, 1985, pp. 677-694

Cyclic variations in bed load discharge at a point, which are an inherent part of the process of bed-load movement, complicate calibration of bed-load samplers and preclude the use of average rates to define sampling efficiencies. Calibration curves, rather than efficiencies, are derived by two independent methods using data collected with prototype versions of the Helley-Smith sampler in a large calibration facility capable of continuously measuring transport rates across a 9 ft (2.7 m) width. Results from both methods agree. Composite calibration curves, based on matching probability distribution functions of samples and measured rates from different hydraulic conditions (runs), are obtained for six different versions of the sampler. Sampled rates corrected by the calibration curves agree with measured rates for individual runs.

359   Radial Turbulent Flow Between Parallel Plates

Girdhari, L. Asawa, (Reader in Civ. Engrg., Univ. of Roorkee, Roorkee, U.P., India), **Pramod K. Pande**, (Prof. of Civ. Engrg., Univ. of Roorkee, Roorkee, U.P., India) and **Pramod N. Godbole**, (Prof. in Civ. Engrg., Univ. of Roorkee, Roorkee, U.P., India)

Journal of Hydraulic Engineering, Vol. 111, No. 4, April, 1985, pp. 695-712

A laboratory investigation was conducted to study the mean flow characteristics of radial turbulent flow. The fluid—air in this case—is supplied from a pipe at the center and spreads out radially between two parallel plates separated by a small distance. The distribution of pressure and shear stress on the front plate and the velocity at different radial distances from the center has been studied. The velocity distribution has been analyzed in a manner similar to that for radial wall jets—with appropriate velocity and length scales to get similarity profiles. The information obtained enables one to predict the velocity if the discharge and geometric characteristics of the flow configuration are known.

360   Scour at Cylindrical Bridge Piers in Armored Beds

Arved J. Raudkivi, (Prof., Dept. of Civ. Engrg., Univ. of Auckland, Auckland, New Zealand) and **Robert Ettema**, (Asst. Prof., Dept. of Civ. and Environmental Engrg., Iowa Inst. of Hydr. Research, Univ. of Iowa, Iowa City, Iowa 52242)

Journal of Hydraulic Engineering, Vol. 111, No. 4, April, 1985, pp. 713-731

Results of laboratory experiments concerning the development of scour around a cylindrical pier placed in a bed of cohesionless sediment overlain by an armoring layer of coarser sediment are presented. Both the armoring layer and the bed beneath it were composed of uniform-sized sediment. It was found that, under these conditions, the depth of scour at a pier may exceed that which occurs at a pier in a nonarmored bed. Local scour due to the local flow structure around a pier may either develop through the armor layer and into the finer, more erodible, sediment, or it may trigger a more extensive localized scour of the bed sediment exposed to the flow by the erosion of the armor layer. Design formulas to estimate the maximum equilibrium depths of scour are proposed.

361　　　　　　　　　Near-Bed Velocity Distribution

Joe C. Willis, Member, ASCE, (Research Hydr. Engr., USDA Sedimentation Lab., Southern Region, ARS, Oxford, Miss. 38655)

Journal of Hydraulic Engineering, Vol. 111, No. 5, May, 1985, pp. 741-753

Velocity distribution equations for fluid flow have previously existed only as limiting relationships for viscous and fully turbulent flow. The turbulent relationships have been further restricted to asymptotic relationships for near-boundary and outer regions of the turbulent boundary layer and even to different forms for rough and smooth boundaries and drag-reducing fluids. This paper presents a continuous representation of the spectrum of equilibrium velocity distributions based on a unified theory of viscous and turbulent shear. This development should aid in describing not only the mechanics of flow, but also other related phenomena such as sediment transportation and chemical and thermal diffusion and transport.

362　　　　　Spatially Varying Rainfall and Floodrisk Analysis

Rafael L. Bras, Member, ASCE, (Prof., Civ. Engrg. Dept., Massachusetts Inst. of Tech., Cambridge, Mass.), **David R. Gaboury**, Assoc. Member, ASCE, (Sr. Project Engr., Woodward-Clyde Consultants, Walnut Creek, Calif.), **Donald S. Grossman**, Assoc. Member, ASCE, (Pres., Coastal Leasing, Inc., Cambridge, Mass.) and **Guillermo J. Vincens**, Assoc. Member, ASCE, (Vice Pres., Camp, Dresser & McKee, Inc., Boston, Mass.)

Journal of Hydraulic Engineering, Vol. 111, No. 5, May, 1985, pp. 754-773

A mathematical simulation model of the spatial and temporal rainfall process was developed. The model is used as input to distributed rainfall-runoff models in order to obtain streamflow series at multiple locations in large river basins. Frequency analysis under different basin development and control alternatives can then be performed. A case study in the Cumberland River Basin (Kentucky and Tennessee) is presented. The basin is 46,397 km² (17,914 mile²) and thus exhibits large rainfall variability ih space. The rainfall model was calibrated with 28 yr of hourly and daily rainfall data in 274 stations in and around the basin. By using generated rainfall as input to a previously calibrated distributed rainfall-runoff model, dischrge frequency curves were obtained at 14 locations. The curves corresponded to both the system with eight operational flood control reservoirs and the natural system witout regulation.

363　　　Transport of Suspended Material in Open Submerged
　　　　　　　　　　　　　　　Streams

Rasmus Wiuff, (Assoc. Prof., Inst. of Hydrodynamics and Hydr. Engrg., Tech. Univ. of Denmark, Bldg. 115, DK-2800, Lyngby, Denmark)

Journal of Hydraulic Engineering, Vol. 111, No. 5, May, 1985, pp. 774-792

The energy exchange in streams carrying suspended material is analyzed using an efficiency concept. The efficiency is defined as the ratio between the gain in potential energy of the suspended material and the turbulent dissipation. It is shown that the efficiency is not constant, but depends linearly on Shields' parameter. The linear relation is used to develop a simple formula, which describes the suspended load as a function of well-known parameters. With the results from open streams an auto-suspension criterion is developed for turbidity currents. This criterion describes the situation when a submerged current carrying suspended material neither deposits nor erodes material.

364   **Surface Buoyant Jets in Steady and Reversing Crossflows**

Dominique N. Brocard, Assoc. Member, ASCE, (Asst. Dir., Alden Research Lab., Worcester Polytechnic Inst., Holden, Mass. 01520)

Journal of Hydraulic Engineering, Vol. 111, No. 5, May, 1985, pp. 793-809

Experimental data on surface buoyant jets in steady and reversing crossflows is presented. The experiments were conducted in a 130 ft (40 m) by 81 ft (25 m) by 1.5 ft (0.45 m) deep basin, which was large enough to ensure negligible boundary effects while allowing controlled ambient conditions. The test results are presented nondimensionally and referenced to controlling length scales. The steady crossflow results are trajectories and longitudinal temperatures, which show the effects of the discharge buoyancy and initial flowrate. Comparing these results with those of other experimental studies leads to the suspicion that boundary effects may be present in some of the latter. The reversing crossflow tests furnish surface temperature patterns and plume depths for a range of conditions. These results can be used to evaluate mathematical models. In addition, the effects of tidal reversals on the offshore extent and depth of the plume are examined. Dispersion tests reveal that buoyancy played an important role in plume spreading. This factor is important as buoyant spreading is frequently omitted in two dimensional farfield mathematical models of thermal plumes.

365   **Dispersion in Anisotropic, Homogeneous, Porous Media**

Qais N. Fattah, Member, ASCE, (Asst. Prof., Coll. of Engrg., Baghdad Univ., Baghdad, Iraq) and **John A. Hoopes**, Member, ASCE, (Prof., Dept. of Civ. and Environmental Engrg., Univ. of Wisconsin, Madison, Wis. 53706)

Journal of Hydraulic Engineering, Vol. 111, No. 5, May, 1985, pp. 810-827

A tensor model for the dispersion coefficient for saturated flow in anisotropic, homogeneous porous media is proposed. The tensor components are evaluated from experiments with an anisotropic, homogeneous porous medium, constructed from thin, alternating layers of two types of sand, having different mean particle sizes. Tests were conducted for flow parallel, perpendicular, and inclined at 60° to the direction of the layers. The hydraulic conductivity and the longitudinal dispersion coefficient were found to be second-rank tensors with equal and constant eccentricities and with major and minor principal axes oriented along and perpendicular to the direction of the sand layers, respectively. The lateral dispersion coefficient was found to be a second-rank tensor whose principal axes were orthogonal to those of the hydraulic conductivity tensor and whose eccentricity increased with increasing seepage velocity. Two values of the off-principal diagonal dispersivities were found to be nearly equal in magnitude but opposite in sign. Experimental results support the dispersion coefficient tensor model proposed in the investigation.

366   **Initial Dilution for Outfall Parallel to Current**

Jon B. Hinwood, (Assoc. Prof., Dept. of Mech. Engrg., Monash Univ., Melbourne, Australia) and **Ian G. Wallis**, (Prin. Engr., Caldwell Connell Engrs., Melbourne, Australia)

Journal of Hydraulic Engineering, Vol. 111, No. 5, May, 1985, pp. 828-845

Rotary tidal currents or site constraints can result in an outfall diffuser being aligned with the current in shallow water. Experiments using a small model laboratory diffuser flume and a much larger scale model in an estuarine channel were used to establish the significant processes causing initial dilution of effluent discharged from a multiport diffuser aligned parallel to the ambient current. From considerations of momentum conservation and mixing, and the results of the flume experiments, an expression for the initial dilution in shallow water was obtained and the conditions defining shallow water were established. The results of the field experiments carried out in an estuarine channel were in agreement with the laboratory results and verified the derived expression. A relationship between the initial dilution and the angle between the diffuser and the current was developed from the field studies.

367   Seawater Circulation in Sewage Outfall Tunnels

D. L. Wilkinson, (Assoc. Prof., Water Research Lab., Univ. of New South Wales, Manly Vale, Australia 2093)

Journal of Hydraulic Engineering, Vol. 111, No. 5, May, 1985, pp. 846-858

The mechanisms of seawater circulation in tunnelled ocean sewage outfalls are investigated. The magnitude of the circulation is determined in terms of the sewage flow and geometric parameters of the outfall. An expression is derived for the sewage flow required to arrest the circulation. The theory is well supported by laboratory experiments. Means by which the risk of seawater intrusion may be minimized are discussed.

368   B-Jumps at Abrupt Channel Drops

Willi H. Hager, (Sr. Research Engr., Chaire de Constructions Hydrauliques, DGC, EPFL, CH-1015, Lausanne, Switzerland)

Journal of Hydraulic Engineering, Vol. 111, No. 5, May, 1985, pp. 861-866

Hydraulic jumps at abrupt bottom drops in rectangular, prismatic channels are either A-jumps (downstream controlled), B-jumps (upstream controlled), or undular jumps (transition between the two). B-jumps are investigated in detail by accounting for the effective pressure distribution at the vertical drop. Traditional description, which assumes strictly hydrostatic pressure distribution at the drop, is modified to yield smaller downstream flow depths for given inflow Froude number, drop height, and inflow depth, a trend suggested by comparing the present formulation with observations. All results are given in terms of design quantities, which enables simple estimation of the major flow characteristics.

369   Water Surface at Change of Channel Curvature

Peter M. Steffler, (Asst. Prof., Dept. of Civ. Engrg., Univ. of Alberta, Edmonton, Alberta, Canada), **Nallamuthu Rajaratnam**, Member, ASCE, (Prof., Dept. of Civ. Engrg., Univ. of Alberta, Edmonton, Alberta, Canada) and **Allan W. Peterson**, Member, ASCE, (Prof., Dept. of Civ. Engrg., Univ. of Alberta, Edmonton, Alberta, Canada)

Journal of Hydraulic Engineering, Vol. 111, No. 5, May, 1985, pp. 866-870

In the study of flow in curved open channels, it is observed that a gradual transition from a flat to a superelevated water surface occurs when the curvature of the channel is abruptly changed. The length of this transition depends on the curvature, channel width and Froude number. An approximate analysis, based on the depth-averaged flow equations with friction neglected, is used to evaluate the water surface configuration and provide a simple expression for the transition length. Comparison of the predicted water surface configuration with a few preliminary experiments is also made, and the agreement is seen to be reasonable.

370   Simplified Design of Contractions in Supercritical Flow

Terry W. Sturm, Member, ASCE, (Assoc. Prof., Georgia Inst. of Tech., School of Civ. Engrg., Atlanta, Ga. 30332)

Journal of Hydraulic Engineering, Vol. 111, No. 5, May, 1985, pp. 871-875

It is shown that minimization of standing waves in supercritical flow in a contraction is equivalent to simultaneous satisfaction of the momentum equation parallel to and perpendicular to the wave front, the continuity equation across the wave front, and continuity through the contraction. Solution of these four governing equations is presented in a graphical form which eliminates the necessity of trial and error in the design procedure. Finally, choking criteria are defined to assist the designer in maintaining supercritical flow through the transition.

371   Fluid and Sediment Interaction over a Plane Bed

Nobuhisa Kobayashi, Member, ASCE, (Asst. Prof., Dept. of Civ. Engrg., Univ. of Delaware, Newark, Del. 19716) and **Seung, Nam Seo**, (Grad. Student, Dept. of Civ. Engrg., Univ. of Delaware, Newark, Del. 19716)

Journal of Hydraulic Engineering, Vol. 111, No. 6, June, 1985, pp. 903-921

A mathematical model is developed for the fluid and sediment interaction in an erodible channel. The model is based on the conservation of mass and momentum for the fluid and sediment. The interaction between the fluid and sediment, as well as the interaction of sediment particles moving in the vicinity of the bed is described in the model. An analysis is performed to derive the vertical distributions of the fluid velocity and sediment concentration over a plane bed in a wide open channel. The analysis yields the friction factor due to the combined effects of grain roughness and sediment movement, as well as the rate of sediment transport. The developed model agrees with existing data for plane beds. The predicted rate of transport of coarse sediment agrees with the Meyer-Peter and Mullen formula. However, detailed measurements of the fluid velocity and sediment concentration over a plane bed are required for a rigorous verification of the developed model.

372   Body-Fitted Coordinates for Flow under Sluice Gates

Jacob H. Masliyah, (Prof., Dept. of Chemical Engrg., Univ. of Alberta, Edmonton, Alberta, Canada), **K. Nandakumar**, (Assoc. Prof., Dept. of Chemical Engrg., Univ. of Alberta, Edmonton, Alberta, Canada), **F. Hemphill**, (Sr. Research Engr., Syncrude Research, Edmonton, Alberta, Canada) and **L. Fung**, (Research Assoc., Dept. of Chemical Engrg., Univ. of Alberta, Edmonton, Alberta, Canada)

Journal of Hydraulic Engineering, Vol. 111, No. 6, June, 1985, pp. 922-933

A boundary-fitted coordinates method was successfully employed in the evaluation of the discharge coefficients, as well as the free-surface profiles, for vertical and radial sluice gates for an ideal fluid. The values of the discharge coefficients and the free upstream and downstream surface profiles are in good agreement with the literature. Some details of the flow, such as the velocity profiles and the bed pressure deficit upstream of the gate, can also be predicted accurately. The boundary-fitted coordinates method was demonstrated to be a viable method in the study of fluid flow with initially unknown discharge and unknown free-surface profiles.

373   Sediment Transport under Ice Cover

Y. Lam Lau, (Head, Environmental Hydraulics Section, Hydraulics Research Div., National Water Research Inst., Canada Centre for Inland Waters, Burlington, Ontario, Canada L7R 4A6) and **Bommanna G. Krishnappan**, (Research Scientist, Environmental Hydraulics Section, Hydraulics Research Division, National Water Research Inst., Canada Centre for Inland Waters, Burlington, Ontario, Canada L7R 4A6)

Journal of Hydraulic Engineering, Vol. 111, No. 6, June, 1985, pp. 934-950

A method for calculating sediment transport in ice-covered flows is developed. Laboratory data on bed form, frictional characteristics and sediment transport are obtained for equivalent free-surface and covered flows. Comparison of the data leads to the conclusion that the lower layer in a covered flow can be treated as a free-surface flow for the purpose of calculating bed load transport. Using the flow characteristics calculated from the k-ε turbulence model and a conventional bed load equation, the sediment transport for covered flows can be obtained. The method was verified using measured data.

374   Cherepnov Water Lifter: Theory and Experiment

H. Liu, Member, ASCE, (Prof., Dept. of Civ. Engrg., Univ. of Missouri-Columbia, Mo. 65211), **M. Fessehaye**, (Asst. Prof., Lincoln Univ., Jefferson City, Mo. 65101) and **R. F. Geekie**, (Resident Asst., Dept. of Civ. Engr., Univ. of Missouri-Columbia, Columbia, Mo. 65211)

Journal of Hydraulic Engineering, Vol. 111, No. 6, June, 1985, pp. 951-969

A theoretical and experimental investigation of the hydraulics of the little-known Cherepnov water lifter was conducted. It was found that as many as 36 algebraic and differential equations are required to predict the cyclic motion of the flow going through the lifter. With these equations, the behavior of the lifter can be conveniently analyzed on any computer having the Continuous System Modeling Program capability. Results of the computer output were compared with experimental data, and good agreement between theory and experiment was found.

375   Simulator for Water Resources in Rural Basins

J. R. Williams, (Hydr. Engr., USDA-ARS, P.O. Box 748, Temple, Tex. 76503), **A. D. Nicks**, (Agricultural Engr., USDA-ARS, P.O. Box 1430, Durant, Okla. 74702) and **J. G. Arnold**, (Hydr. Engr., USDA-ARS, P.O. Box 748, Temple, Tex. 76503)

Journal of Hydraulic Engineering, Vol. 111, No. 6, June, 1985, pp. 970-986

A model called SWRRB (Simulator for Water Resources in Rural Basins) was developed for simulating hydrologic and related processes in rural basins. The objective in model development was to predict the effect of management decisions on water and sediment yields with reasonable accuracy for ungaged rural basins throughout the United States. The three major components of SWRRB are weather, hydrology, and sedimentation. Processes considered include surface runoff, percolation, return flow, evapotranspiration, pond and reservoir storage, and sedimentation. The SWRRB model was developed by modifying the CREAMS (Chemicals, Runoff, and Erosion from Agricultural Management Systems) daily rainfall hydrology model for application to large, complex, rural basins. The major changes were (1) A return flow component was added; (2) the model was expanded to allow simultaneous computations on several sub-basins; (3) a reservoir storage component was added for use in determining the effects of farm ponds and other reservoirs on water and sediment yield; (4) a weather simulation model (precipitation, solar radiation, and temperature) was added to provide for longer-term simulations and more representative weather inputs, both temporally and spatially; (5) a better method was developed for predicting the peak runoff rate; and (6) a simple flood routing component was added. Besides water, SWRRB also simulates sediment yield using the Modified Universal Soil Loss Equation (MUSLE) and a sediment routing model. Tests with data from a 538 km² basin in Oklahoma and a 17.7 km² basin in Texas indicate that SWRRB is capable of simulating water and sediment yield realistically.

376 Bed Material Movement in Hyperconcentrated Flow

Zhaohui Wan, (Research Engr., Inst. of Water Conservancy and Hydroelectri Power Research. P.O. Box 366, Beijing, China)

Journal of Hydraulic Engineering, Vol. 111, No. 6, June, 1985, pp. 987-1002

A series of experiments were made in a closed rectangular conduit with bentonite as fine particles and plastic beads as coarse particles. Compared with that in clear water, in bentonite suspensions coarse particles settle more slowly and start moving at a higher flow intensity. Due to the larger threshold velocity and smaller settling velocity, the bed load is smaller and the suspended load is larger. As a result, the total load, consisting of the bed load and the suspended load, is smaller in the low flow intensity region, but larger in the high flow intensity region. The increase of the suspended load favors the transition from dunes to plane bed. So in clay suspension dunes are lower, flatter, and change to plane bed at lower flow intensity. Correspondingly, the form resistance is smaller in clay suspension.

377 Errors from Using Conservation of Buoyancy Concept in Plume Computations

Frank B. Tatom, (Pres. and Chf. Engr., Engrg. Analysis, Inc., 3109 Clinton Ave. W, Suite 432, Huntsville, Ala. 35805)

Journal of Hydraulic Engineering, Vol. 111, No. 6, June, 1985, pp. 1005-1009

The conservation of buoyancy concept as currently applied to buoyant jets and plumes in water is reviewed with regard to its origin, validity, and accuracy. Although the original buoyant plume model was based on the assumption of a constant thermal coefficient of volumetric expansion, and thus was limited to temperature differences of a few degrees, the original model has been widely applied to cases involving temperature differences as high as 17°C (30.6°F). Such application results in buoyancy forces which are significantly larger than the more accurate values based on models using conservation of energy instead of buoyancy. In the case of the computed trajectories the errors are most noticeable for lower ambient temperatures and lower jet exit velocities. The basic point is that conservation of buoyancy is an approximation and not a fundamental conservation principle. Because of the current use of numerical solutions to the governing equations there is no real justification for its use in place of the more rigorous conservation of energy equation.

378 Negative Outflows from Muskingum Flood Routing

A. T. Hjelmfelt, Jr., Member, ASCE, (Hydr. Engr., North Central Watershed Research Unit, Agricultural Research Service, USDA, 207 Business Loop 70 East, Columbia, Mo. 65203)

Journal of Hydraulic Engineering, Vol. 111, No. 6, June, 1985, pp. 1010-1014

The Muskingum flood routing method was developed in studies made for the Muskingum Conservancy District Flood-Control Project during the mid-1930s. Since its development, investigations into the nature of the method have experienced surges of activity.

Application of linear system theory led to one form of interpretation. Development of the routing equation from kinematic wave theory provided an additional interpretation. Many investigators have transformed the Muskingum hypothesis into a continuous equivalent. The resulting differential equation was then investigated. The inevitability of negative discharges in the routed outflow using the continuous equivalent has drawn comment from these investigators and is a cause for concern among practicing engineers. A definitive solution to this anomalous result is still unavailable. In practice, however, the continuous approach is seldom practical. The inflow is given as discrete values occurring at discrete times. The result of the routing is a table of discrete outflows at discrete times. For this reason, discrete system methods are used in this study and it is shown that the classical limits on routing parameters ensure nonnegative outflows.

379   Head-Discharge Relation for Vortex Shaft

Willi H. Hager, Member, ASCE, (Sr. Research Engr., Chaire de Constructions Hydrauliques, Dept. de Genie Civil, EPFL, Lausanne, Switzerland CH-1015)

Journal of Hydraulic Engineering, Vol. 111, No. 6, June, 1985, pp. 1015-1020

Hydraulic flow characteristics for vortex shafts are investigated for subcritical approach conditions. The analysis is based on Pica's approach by simplifying considerably his procedure. Using appropriate scalings the head-discharge relation may be expressed as a unique function. Using the stability condition that the air core along the shaft axis is not below a certain minimum, explicit expressions for the maximum upstream flow depth and the maximum discharge in terms of the shaft radius, the upstream channel width and the distance of the channel axis from the shaft center are derived. These relations permit a simple and rapid determination of the vortex shaft capacity. Analytical results are compared with observations on model and prototype evaluations. Agreement is favorable in the complete range of the respective parameters. Also, no scale effects could be detected. The present approach, therefore, can be regarded as a universal head-discharge relation for arbitrary, subcritical inflow conditions. Moreover, a graphical evaluation of the relation in question enables a simple and rapid application to cases in practice.

380   Relative Accuracy of Log Pearson III Procedures

James R. Wallis, (Hydro., Inst. of Hydr., Wallingford, England) and Eric F. Wood, (Hydro., Inst. of Hydr., Wallingford, England)

Journal of Hydraulic Engineering, Vol. 111, No. 7, July, 1985, pp. 1043-1056

The U.S. Water Resources Council (WRC) has suggested that the log-Pearson III distribution, fitted by the method of moments, should be used in flood frequency analysis. A Monte Carlo simulation assessment of the WRC procedures shows that the flood quantile estimates obtainable by these procedures are poorer than those obtainable by using an index flood type approach with either a generalized extreme value distribution or a Wakeby distribution fitted by probability weighted moments. It is suggested that the justification for using the WRC Bulletin 17B guidelines is in need of reevaluation.

381   Similarity Solution of Overland Flow on Pervious Surface

A. Osman Akan, Assoc. Member, ASCE, (Assoc. Prof., Dept. of Civ. Engrg., Old Dominion Univ., Norfolk, Va. 23508-8546)

Journal of Hydraulic Engineering, Vol. 111, No. 7, July, 1985, pp. 1057-1067

Similarity solutions for the overland flow-infiltration problem are obtained by use of a physically-based mathematical model. The mathematical model is founded on the kinematic equations of overland flow and the Green-Ampt formulation of the infiltration process. The governing equations of surface flow and infiltration are coupled and rewritten in terms of four physically-based nondimensional parameters. Constant values of these parameters imply hydraulic similarity of an infinite number of different overland flow-infiltration situations. An implicit finite difference technique is employed to solve the governing equations. The similarity approach is verified by comparisons of similarity solutions with the results of a theoretically sophisticated mathematical model for the cases of Columbia sandy loam, Yolo light clay, and Guelph loam. Also, a group of peak discharge prediction charts are developed based on the similar solutions of the overland flow-infiltration problem and presented to emphasize the feasibility of the similarity approach. The use of these charts is illustrated by a practical application section included in the paper.

382   Comparison of Two River Diffuser Models

Joseph Hun-wei Lee, Assoc. Member, ASCE, (Lect., Dept. of Civ. Engrg., Univ. of Hong Kong, Hong Kong)

Journal of Hydraulic Engineering, Vol. 111, No. 7, July, 1985, pp. 1069-1078

The effect of lateral boundaries on the mechanics of mixing multiple shallow water jets in a coflowing current is studied. Two exact inviscid solutions of the momentum-induced flow are examined: A model that accounts for the finite width of the receiving water is compared with a model that neglects lateral confinement. A parametric study reveals remarkably little difference in the key predictions of the velocity and contraction of the ultimate slipstream—especially with reference to a related recent study of boundary effects on multiple submerged jets. The conclusions are corroborated by detailed velocity measurements in a systematic set of experiments covering a wide range of induced flows.

383   Analysis and Stability of Closed Surge Tanks

M. Hanif Chaudhry, Member, ASCE, (Assoc. Prof., Dept of Civ. and Environmental Engrg., Washington State Univ., Pullman, Wash. 99164), **Mostafa A. Sabbah**, (Dir., Dept. of Engrg., City of Newport News, Newport News, Va.) and **John E. Fowler**, (Engr., Dept. of Public Works, City of Virginia Beach, Virginia Beach, Va.)

Journal of Hydraulic Engineering, Vol. 111, No. 7, July, 1985, pp. 1079-1096

The stability of closed surge tanks (tanks with compressed air at the top) is investigated using the phase plane method, which allows inclusion of nonlinear effects in the analyses. All singularities are analyzed and stability criteria are developed. Phase portraits are plotted using the method of isoclines. Six numerical techniques for integrating the governing equations are compared. The effect of the value of the polytropic gas constant on the surge amplitudes is investigated. Several important conclusions are: (1) For the stability of large oscillations, it is neither necessary to provide more tank area than critical area, as is presently done, nor to satisfy the second stability condition presented earler by the writers; (2) the damping rate of oscillations is higher if the limit on the maximum gate opening is included in the analysis; (3) presently used first-order methods for numerically integrating the governing equations may yield incorrect and sometimes unstable results; (4) the second-order modified Euler method yields results comparable to higher-order methods, and is recommended for practical applications; and (5) the polytropic gas law exponent, n, equal to unity (isothermal behavior) produces a larger amplitude of water surface oscillation than $n = 1.4$ (adiabatic).

384   Subdivision Froude Number

David H. Schoellhamer, Member, ASCE, (Research Civ. Engr., U.S. Geological Survey, Gulf Coast Hydroscience Center, Bldg. 2101, NSTL Station, Miss. 39529), **John C. Peters**, Member, ASCE, (Hydr. Engr., Hydr. Engr. Center, Davis, Calif. 95616) and **Bruce E. Larock**, Member, ASCE, (Prof., Civ. Engrg. Dept., Univ. of California, Davis, Calif. 95616)

Journal of Hydraulic Engineering, Vol. 111, No. 7, July, 1985, pp. 1099-1104

A Froude number that is applicable to subdivisions of a cross section is developed assuming one-dimensional flow. The subdivision Froude number is shown to determine accurately the flow regime in subdivisions of a cross section. An imaginary subdivision Froude number in a floodplain indicates that the velocity is increasing with depth and that a one-dimensional flow assumption may be invalid. Mixed flow regimes at a cross section which invalidate the standard step method can be recognized with the subdivision Froude number. A total of 193 cross sections were tested with a modified version of HEC2; eleven had a mixed flow regime and 36 had at least one imaginary Froude number.

385   Effect of Pier Spacing on Scour Around Bridge Piers

Keith R. Elliott, (Student, Dept. of Civ. Engrg., Univ. of Nottingham, Nottingham, U.K.) and **Christopher J. Baker**, (Lect., Dept. of Civ. Engrg., Univ. of Nottingham, Nottingham, U.K.)

Journal of Hydraulic Engineering, Vol. 111, No. 7, July, 1985, pp. 1105-1109

Small scale experiments were designed to investigate effects of lateral spacing between bridge piers on scour depth. The scouring process is effected by two processes: (1) The interaction of the horseshoe vortices around the bases of the piers; and (2) the acceleration of flow between the piers. Equations are derived for clear water scour, for one set of pier geometrics, one water depth and one sediment type.

386   Numerical Determination of Aquifer Constants

S. P. Rai, Member, ASCE, (Sr. Lect., Civ. Engrg. Dept., Nat. Univ. of Singapore, Kent Ridge, Singapore 0511)

Journal of Hydraulic Engineering, Vol. 111, No. 7, July, 1985, pp. 1110-1114

Aquifer constants (transmissibility and storage coefficients) of infinite homogeneous and isotropic aquifers are normally evaluated by graphical procedures using a sufficient amount of pumping test data. A method is proposed which can calculate these constants numerically without involving any curve fitting. The method can be used if only a few sets of the data are available. The accuracy of the method is of the same order as of other graphical methods.

387   Compilation of Alluvial Channel Data

William R. Brownlie, Member, ASCE, (Assoc, Dir., Tetra Tech, Inc., 630 North Rosemead Blvd., Pasadena, Calif. 91107)

Journal of Hydraulic Engineering, Vol. 111, No. 7, July, 1985, pp. 1115-1119

A computer-based collection of alluvial channel data from both laboratory and field investigations has been prepared. The collection contains 5,263 records of laboratory observations and 1,764 records of field observations. Each record contains 10 alluvial channel variables presented in a standard format in SI (International System) units. The paper presents a discussion of the contents of the data collection, a review of some of the data sources, and comparisons with earlier data collections. Information for users of the data collection is also presented.

388   Inflow Seepage Influence on Straight Alluvial Channels

J. R. Richardson, Assoc. Member, ASCE, (Research Asst., Dept. of Civ. Engrg., Colorado State Univ., Fort Collins, Colo. 80523), **S. R. Abt**, Member, ASCE, (Assoc. Prof., Dept. of Civ. Engrg., Colorado State Univ., Fort Collins, Colo. 80523) and **E. V. Richardson**, Fellow, ASCE, (Prof. and Hydr. Program Leader, Dept. of Civ. Engrg., Colorado State Univ., Fort Collins, Colo. 80523)

Journal of Hydraulic Engineering, Vol. 111, No. 8, August, 1985, pp. 1133-1147

A flume study was conducted to identify and analyze the effects of inflow seepage on flow in straight alluvial open channels. The effect of inflow seepage on energy slope, water surface depth, velocity, sediment transport, scour, bed forms, and resistance to flow was investigated. The results of this investigation indicated that inflow seepage increases localized mean channel velocity, energy slope, and stream power in the zone of inflow. The water surface depth decreased for subcritical flow and remained nearly constant for supercritical flow during inflow conditions. Sediment transport was slightly increased when inflow was introduced in the bed. Inflow seepage did not appear to enhance channel scour. Inflow seepage significantly enhanced bed form transformation and the bed roughness. Inflow seepage also caused dunes to become longer, flatter, and move more erratically in the reach where inflow occurred. It was concluded that inflow seepage could significantly influence the channel hydraulics, stream power, bed form and bed roughness in the localized zone of inflow.

389   On the Determination of Ripple Geometry

Mahmet Selim Yalin, Member, ASCE, (Prof., Dept. of Civ. Engrg., Queens Univ., Kingston, Ontario, Canada)

Journal of Hydraulic Engineering, Vol. 111, No. 8, August, 1985, pp. 1148-1155

Laboratory measurements were carried out on the mobile bed covered by ripples generated by an open channel flow. The bed materials used were cohesionless and reasonably uniform; the steady-state subcritical flume flow was in equilibrium and it was nearly two-dimensional. In addition to the conventional runs conducted with water, special runs were conducted with a water and glycerine mixture. Using the results of these measurements, as well as the data of other reliable sources, a series of experimental curves were determined which can be used to predict the length and height of ripples. The analysis of the data indicates that the dimensionless quantities related to the geometry of ripples are functions of two dimensionless variables. One of them must be a combination reflecting the intensity of sediment-transporting flow, the other must be an arrangement of parameters characterizing the physical nature of the liquid and solid phases involved.

390   Simulation of Two-Fluid Response in Vicinity of Recovery Wells

Gerard P. Lennon, Assoc. Member, ASCE, (Asst. Prof., Dept. of Civ. Engrg., Lehigh Univ., Bethlehem, Pa. 18015)

Journal of Hydraulic Engineering, Vol. 111, No. 8, August, 1985, pp. 1156-1168

A dense fluid in an unconfined aquifer is usually removed by pumping a recovery well screened in the region of the aquifer occupied by the fluid. For many problems, the transition zone separating the dense fluid from the overlying water can be modeled as a sharp interface. When pumping begins, the interface is drawn downward, allowing water to enter the well. The recovery process is improved if a second well is drilled and screened in the zone occupied by the water. The simultaneous pumping of water tends to cause the interface to move upward, allowing the dense fluid to be recovered at an increased rate without water entering the recovery well. The boundary integral equation method (BIEM) is used to solve the problem of the response of a dense fluid near a recovery well in a groundwter aquifer. The axisymmetric problems considered here are time-dependent and involve a nonlinear boundary condition along the free surface, as well as a nonlinear, moving interfacial boundary. The entire boundary, including the moving boundaries, are discretized into axisymmetric finite elements for evaluating the required boundary integrations. Numerical examples include pumping both fluids simultaneously as well as separately. The model is used to illustrate the design of a recovery system for the efficient removal of the dense fluid.

391   Prediction of 2-D Bed Topography in Rivers

Nico Struiksma, (Proj. Engr., Delft Hydr. Lab., P. O. Box 152, Emmeloord, Netherlands)

Journal of Hydraulic Engineering, Vol. 111, No. 8, August, 1985, pp. 1169-1182

Mathematical models are being developed at the Delft Hydraulics Laboratory for the computation of time-dependent, two-dimensional bed deformation in alluvial rivers. A most promising verification of the predictive capacity of one such model has been obtained for the Waal River in the Netherlands. This model was developed for rivers with dominant bed load transport without significant grain, sorting effects and with approximately constant width. Also, some results of computations with variable boundary conditions are given which simulate the effect of the river regime and the extraction and return of cooling water on the bed topography.

392   Estimation of Hydraulic Data by Spline Functions

Kazumasa Mizumura, (Prof., Dept. of Civ. Engrg., Kanazawa Inst. of Tech., Kanazawa, Ishikawa Pref., 921, Japan)

Journal of Hydraulic Engineering, Vol. 111, No. 9, September, 1985, pp. 1219-1225

In hydraulics, methods are needed to estimate (interpolate) missing data of various processes. This paper presents a method using the "Spline Functions." For illustration, the method was applied to the generation of missing sediment data. With this estimation method, it is possible to reduce the data requirements.

393   Large Basin Deterministic Hydrology: A Case Study

Victor Miguel Ponce, Member, ASCE, (Prof. of Civ. Engrg., San Diego State Univ., San Diego, Calif. 92182), **Zbig Osmolski**, (Mgr., Flood Control Design Section, Pima County Dept. of Transportation and Flood Control Dist., Tucson, Ariz. 85713) and **Dave Smutzer**, (MGr., Flood Control Planning Section, Pima County Dept. of Transportation and Flood Control Dist., Tucson, Ariz. 85713)

Journal of Hydraulic Engineering, Vol. 111, No. 9, September, 1985, pp. 1227-1245

A case study of large basin hydrology is performed. The basin is the Santa Cruz River upstream of Cortaro Farms Bridge near Tucson, Arizona, draining 3,503 sq. miles. The evaluation uses novel techniques of deterministic hydrologic modeling to calculate frequency-based floods at proposed bridge improvement sites. A computer model capable of simultaneously handling the complex topology of the basin is driven by 100-yr frequency National Weather Service rainfall events of 24-, 48- and 96-hour durations. The model is calibrated using recorded rainfall-runoff data for the flood of October 1983, which produced record flows throughout southeastern Arizona. Once calibrated, a series of general and local storms are simulated. General storms cover the entire basin with low-intensity rainfall events. Local storms cover selected portions of the basin with high intensity rainfall events. Critical peak flows at the bridge sites are shown to be associated with a combination of general 24-hr and local 48-hr storms. Regulatory and design discharges are adopted, based on the findings of the computer model.

394   Structure of Turbulence in Compound Channel Flows

Panagiotis Prinos, (Postdoctoral Fellow, Dept. of Civ. Engrg., Univ. of Ottawa, Ottawa, Canada), **Ron Townsend**, (Prof., Dept. of Civ. Engrg., Univ. of Ottawa, Ottawa, Canada) and **Stavros Tavoularis**, (Assoc. Prof., Dept. of Mech. Engrg., Univ. of Ottawa, Ottawa, Canada)

Journal of Hydraulic Engineering, Vol. 111, No. 9, September, 1985, pp. 1246-1261

The structure of turbulence in compound channel flows is examined in a laboratory study. Shear stresses and turbulence intensities are measured in a channel comprised of a deep central section flanked on either side by wide shallow berms (flood plains). The study concerns the nature of turbulence in the mixing regions separating the deep and shallow zones. Also studied is the mixing region's effect on the compound flow field for both "wide" and "narrow" channel conditions. Under "narrow" channel conditions the mixing process extends to the center of the main channel flow field; however, under "wide" channel conditions, the central region is not affected and observed turbulence levels at the center of the main channel are in close agreement with theoretical values for a two-dimensional flow field. Apparent shear stress at the vertical main channel-flood plain interface was measured directly and compared favorably with estimated values based on momentum considerations.

395   Overbank Flow with Vegetatively Roughened Flood Plains

Erik Pasche, (Visiting Research Engr., Dept. of Civ. Engrg., Univ. of California, Davis, Calif. 95616) and **G. Rouve**, Member, ASCE, (Prof., Head of Inst. for Hydr. Engrg. and Water Resources Development, Technical Univ. of Aachen, Federal Republic of Germany)

Journal of Hydraulic Engineering, Vol. 111, No. 9, September, 1985, pp. 1262-1278

An experimental and theoretical investigation into the flow characteristics of channels with complex cross sections was undertaken. In this study, particular attention was given to the probleem of non-submerged flood-plain roughnesses. This differs from previous studies in which compound-channel flow was primarily investigated for more or less uniform boundary roughnesses. In order to achieve more conformity with natural rivers, the flood plain and the main channel were separated in the model by a sloping bank. Two cross sections with varying aspect ratios and flood-plain roughnesses were investigated. The necessary measurements were carried out by applying LDV and Preston-tube techniques. On the basis of simple turbulence assumptions methods are presented by which the flow resistance in vegetatively roughened flood plains and main channels can be properly predicted as a function of independent and directly determinable basic flow parameters. A first verification of this model was accomplished using field measurements from an actual flood channel.

396   Mixed-Layer Deepening in Lakes after Wind Setup

C. Kranenburg, (Sr. Scientific Officer, Lab. of Fluid Mechanics, Dept. of Civ. Engrg., Delft Univ. of Technology, Delft, Netherlands)

Journal of Hydraulic Engineering, Vol. 111, No. 9, September, 1985, pp. 1279-1297

Laboratory experiments on wind-driven mixed-layer deepening in a wind flume are described for the situation where the wind shear stress has become balanced by a streamwise pressure gradient so that the mean mixed-layer velocity has vanished. The results were obtained at two lengths of the flume, and include visual observations, entrainment rates, density profiles, velocity profiles and profiles of turbulence intensities. An intermediate upwind wedge formed by accumulation of mixed water was observed in all experiments. The entrainment law obtrained agrees with Kraus and Turner's relationship, and is almost independent of the length of the flume.

397   Spillage over an Inclined Embankment

Helge I. Andersson, (Sr. Lect., Div. of Applied Mechanics, Dept. of Physics and Mathematics, Norwegian Inst. of Tech., N-7034 Trondheim NTH, Norway)

Journal of Hydraulic Engineering, Vol. 111, No. 10, October, 1985, pp. 1299-1307

Accidental release of liquid from a storage tank, and the subsequent spillage over the surrounding embankment is considered. The present analysis is based on two simplifying assumptions: (1) The supercritical flow over the inclined embankment can be considered quasi-steady, and thus treated by a classical model for gradually varied flow in open-channels; (2) it is assumed that the overflow quenches when the level in the reservoir has reached a certain critical limit, so that the remaining liquid in the tank will be trapped by the dike. An analytical solution for the spill fraction is derived, showing that the spill fraction is a decreasing function of the ratio of the dike height to the initial tank level. The spill fraction is furthermore found to depend on a critical Froude number, the slope of the embankment, and a Chezy coefficient. The latter quantity, which accounts for ground friction and turbulence, is the only empirical parameter in the analysis. Predictions based on this simple model compare favorably with experimental observations from model inclines.

398   Erosion of Soft Cohesive Sediment Deposits

Trimbak M. Parchure, (Chief Res. Officer, Central Water and Power Res. Sta., Pune, 411024 India) and **Ashish J. Mehta**, Member, ASCE, (Assoc. Prof., Dept. of Coastal and Oceanographic Engrg., Univ. of Florida, Gainesville, Fla. 32611)

Journal of Hydraulic Engineering, Vol. 111, No. 10, October, 1985, pp. 1308-1326

Erosion behavior of soft cohesive sediment deposits has been investigated in laboratory experiments. Such deposits are representative of the top, active layer of estuarial beds. An experimental procedure involving layer by layer erosion under a range of bed shear stresses, τ_b, of successively increasing magnitude was utilized. Interpretation of the resulting concentration-time data together with bed density profiles yielded a description of the variation of the bed shear strength, τ_s, with depth as well as an expression for the rate of surface erosion. In general, τ_s increased with depth and was also influenced by the type of sediment, bed consolidation period and salinity. The rate of erosion was found to vary exponentially with $(\tau_b - \tau_s)^{1/2}$. In modeling estuarial bed erosion, it is essential to take these characteristics of τ_s and the rate of erosion into account.

399   Sediment Transport in Shallow Flows

Nadim M. Azia, Assoc. Member, ASCE, (Asst. Prof. of Engrg. Graphics, Clemson Univ., Clemson, S.C. 29631) and **Shyam N. Prasad**, Member, ASCE, (Prof. of Civ. Engrg., Univ. of Mississippi, University, Miss. 38677)

Journal of Hydraulic Engineering, Vol. 111, No. 10, October, 1985, pp. 1327-1343

The sediment transport problem in shallow flows, such as furrow flows, in the upper flow regime is formulated based upon continuum mechanics principles. Considering the dynamics of the water flow as well as the dynamics of sediment flow, a mathematical expression which is a modification of the classical stability criterion of flow over a fixed bed is obtained for the case when a movable sediment layer is present near the bed of the channel. Coincident with this stability criterion, the amount of sediment being moved is defined as the transport capacity of the flow. The results suggest the dependence of the capacity on the sediment properties, active layer thickness, channel gradient and the hydraulic properties of the flow. Comparison of the theoretical results with experimental data revealed a general agreement. In particular, the theory predicted satisfactorily the transport capacity in models of steep crop-row furrows.

400   Modeling of Rainfall Erosion

Pierre Y. Julien, (Asst. Prof., Dept. of Civ. Engrg., Colorado State Univ., Foothills Campus, Fort Collins, Colo. 80523) and **Marcel Frenette**, (Prof., Dept. of Civ. Engrg., Laval Univ., Quebec, Canada)

Journal of Hydraulic Engineering, Vol. 111, No. 10, October, 1985, pp. 1344-1359

A combined stochastic and deterministic method has been developed to evaluate soil erosion from overland flow. The exponential distributions for rainfall duration and intensity are combined with rainfall-runoff relationships for discrete storms. A general equation for sediment discharge is suggested from dimensional analysis with different sets of coefficients representing several existing equations. The expected value of soil erosion during one rainfall event is theoretically derived using hypergeometric series. When applied to the Chaudière watershed near Québec, this equation converges very rapidly and the first term of the series is recommended. Utilizing sediment-delivery ratios, the sediment yield computed from the total soil erosion is in good agreement with the suspended load measured in the Chaudière River.

401   Culvert Slope Effects on Outlet Scour

Steven R. Abt, (Assoc, Prof. and Assoc. Dept. Head, Dept. of Civ. Engrg., Colorado State Univ., Fort Collins, Colo. 80523) and **James F. Ruff**, (Assoc. Prof., Dept. of Civ. Engrg., Colorado State Univ., Fort Collins, Colo. 80523)

Journal of Hydraulic Engineering, Vol. 111, No. 10, October, 1985, pp. 1363-1367

A circular shaped culvert was tested with 0, 2, 5, 7, and 10% slopes. The scour hole characteristics of depth, width, length and volume after 316 minutes of testing were correlated to the discharge intensity for each slope. The results indicated that a sloped culvert can potentially increase the maximum dimensions of scour by 10 to 40 percent over those for a horizontal culvert.

402   Channel Width Adjustment During Scour and Fill

Howard H. Chang, Member, ASCE, (Prof. of Civ.Engrg., San Diego State Univ., San Diego, Calif. 92182)

Journal of Hydraulic Engineering, Vol. 111, No. 10, October, 1985, pp. 1368-1370

The adjustment of stream channel width during streambed scour and fill is explained by the stream's tendency to establish equal power expenditure along the channel, i.e., the straight water-surface profile. In the process of establishing the uniform power expenditure, channel width adjustment is not necessarily toward streamwise uniformity. It is illustrated that the straight water-surface profile can be approached by significant spatial width variation during scour and fill.

403   Local Scour Downstream of an Apron

N. M. K. Nik Hassan, (Postgrad. Student, Dept. of Civ. and Struct. Engrg., Univ. of Manchester Inst. of Sci. and Tech., Manchester M60 1QD, England) and **Rangaswami Narayanan**, (Lect., Dept. of Civ. and Struct. Engrg., Univ. of Manchester Inst. of Sci. and Tech., Manchester M60 1QD, England)

Journal of Hydraulic Engineering, Vol. 111, No. 11, November, 1985, pp. 1371-1385

Measurements have been made of the rate of scour downstream of a rigid apron due to a jet of water issuing through a sluice opening. Experiments are carried out for various sand sizes, sluice openings, efflux velocities, and lengths of apron. Mean velocity profiles in rigid models simulating the shape of the scour hole are studied in detail. They are found to exhibit similarity. Scales of velocity and length that are adopted bring together all the data concerning the decay of maximum velocity in the scour hole with respect to streamwise distance. A simple semi-empirical theory based on a characteristic mean velocity in the scour hole is proposed to predict the time rate of scour. Predicted results are compared with experimental data.

404   Economics of Pumping and the Utilization Factor

B. B. Sharp, (Reader in Civ. Engrg., Univ. of Melbourne, Parkville, 3052, Australia)

Journal of Hydraulic Engineering, Vol. 111, No. 11, November, 1985, pp. 1386-1396

The derivative method of determining the optimum (minimum cost) diameter of a force main is expanded to deal with the economic sizes of any pipeline where pumping is involved. The benefits to be gained from boosting in a gravity main and the notion of a utilization factor provide new approaches for design rules for economic analysis.

405   Bed Topography in Bends of Sand-Silt Rivers

Syunsuke Ikeda, Member, ASCE, (Assoc. Prof., Dept. of Foundation Engrg., Saitama Univ., 255 Shimo-ohkubo, Urawa, Saitama, 338, Japan) and **Tatsuya Nishimura**, (Engr., Kensetsu-Giken K.K., Nihonbashi-Honmachi, Chuoku, Tokyo, Japan)

Journal of Hydraulic Engineering, Vol. 111, No. 11, November, 1985, pp. 1397-1411

A mathematical model for defining the lateral bed topography in bends of sand-silt rivers is presented. The model includes the effects of suspended sediment that have not been considered in the existing models. The major agency for defining the lateral bed profile is found to be the force balance between the fluid force and the gravitational force exerting on bed materials. The theory suggests that the lateral convective transport of suspended sediment induced by secondary flow may considerably affect the bed profile at the outer region of river bends, while the lateral diffusion due to turbulence has negligible effect on the profile everywhere. The new model is applied to an actual river, suggesting the adequacy of the present model within the possible error.

406   Formation of Alternate Bars

Howard H. Chang, Member, ASCE, (Prof. of Civ. Engrg., San Diego State Univ., San Diego, Calif. 92182)

Journal of Hydraulic Engineering, Vol. 111, No. 11, November, 1985, pp. 1412-1420

River channelization works have often resulted in alternate bar formation at low flows. A criterion for alternate bar formation in straight alluvial-bed channels with rigid banks is developed following a rational approach, and is substantiated with experimental data. The formation of alternate bars is attributed to meandering development within the confined channel. Such development becomes possible if the streamflow's stable width, or regime width, is less than the confinement width. The stable width is analytically established as a direct function of water discharge. If, at a high flow, the stable width is greater than the confinement width, then alternate bars are absent due to the lack of freedom for meandering development.

407   Hydrodynamically Smooth Flows Over Surface Material in Alluvial Channels

Jin Wu, Member, ASCE, (H. Fletcher Brown Prof., Air-Sea Interaction Lab., Univ. of Delaware, Lewes, Del. 19958)

Journal of Hydraulic Engineering, Vol. 111, No. 11, November, 1985, pp. 1423-1427

It is a general practice to treat the flow in alluvial channels in two scales: the microscale over bed materials (sand grains), and the macroscale over bed forms. Contrary to common thinking, the flow over bed materials appears to be in the hydrodynamically smooth regime, or at most in the transition region. This explains why the roughness-resistance coefficient of grains still varies with the Reynolds number. Such an identification is helpful in understanding flow and transport in alluvial channels.

408   Analysis and Simulation of Low Flow Hydraulics

Barbara A. Miller, Assoc. Member, ASCE, (Research Assoc., Surface Water Section, Illinois State Water Survey, Champaign, IL) and Harry G. Wenzel, Fellow, ASCE, (Prof., Dept. of Civ. Engrg., Univ. of Illinois, Urbana, IL)

Journal of Hydraulic Engineering, Vol. 111, No. 12, December, 1985, pp. 1429-1446

A one-dimensional mathematical model has been developed to accurately simulate channel characteristics under low flow conditions in alluvial channels. For a given steady discharge, channel geometry, and channel bed particle size distribution, the model predicts the flow depth, the mean velocity, and the flow resistance. Energy losses are assumed to result from flow resistance, as well as from local losses generated by the contractions and expansions occurring through the pool-riffle sequence. Laboratory and field data were used to calibrate and verify the model, as well as to conduct an in-depth analysis of the flow characteristics associated with low discharges.

409   **River Flood Routing by Nonlinear Muskingum Method**

Yeou-Koung Tung, Assoc. Member, ASCE, (Asst. Prof., Wyoming Water Research Center and Statistics Dept., Univ. of Wyoming, Laramie, WY)

Journal of Hydraulic Engineering, Vol. 111, No. 12, December, 1985, pp. 1447-1460

The linear form of the Muskingum model has been widely applied to river flood routing. However, a nonlinear relationship between storage and discharge exists in most actual river systems, making the use of the linear model inappropriate. In this paper, a nonlinear Muskingum model is solved using the state variable modeling technique. Various curve fitting techniques are employed for the calibration of model parameters, and their performances within the model are compared. Both linear and nonlinear models are applied to an example with pronounced nonlinearity between storage and discharge. The results show that the nonlinear Muskingum model is superior to the linear one.

410   **Stepped Spillway Hydraulic Model Investigation**

Robert M. Sorenson, Fellow, ASCE, (Prof. of Civ. Engrg., Lehigh Univ., Bethlehem, PA 18015)

Journal of Hydraulic Engineering, Vol. 111, No. 12, December, 1985, pp. 1461-1472

A physical hydraulic model investigation was conducted to evaluate the performance of a stepped overflow spillway. The spillway has a standard ogee profile with continuous steps cut into the spillway face from just below the crest, to the toe. The steps significantly increase the rate of energy dissipation on the spillway face, thus eliminating or greatly reducing the need for a large energy dissipation basin at the spillway toe. Primary objectives of the investigation were to evaluate the effectiveness of the flow transition from the smooth crest profile to the steps, to quantify the energy dissipation on the spillway face, and to define the flow characteristics on the steps. The investigation demonstrated that this stepped spillway is quite effective at dissipating energy and that smooth flow transition from the spillway crest to the stepped face is easily achieved.

411   **Turbidity Current with Erosion and Deposition**

J. Akiyama, (Grad. Student, St. Anthony Falls Hydraulic Lab., Dept. of Civ. & Mineral Engrg., Univ. of Minnesota, Minneapolis, MN 55414) and **H. Stefan**, (Prof. and Assoc. Dir., St. Anthony Falls Hydraulic Lab., Dept. of Civ. & Mineral Engrg., Univ. of Minnesota, Minneapolis, MN 55414)

Journal of Hydraulic Engineering, Vol. 111, No. 12, December, 1985, pp. 1473-1496

The equations which govern the movement of two-dimensional gradually varied turbidity currents in reservoirs and over beaches are derived and solved numerically. Turbidity currents are sediment-laden gravity currents that exchange sediment with the bed by erosion or deposition as the flow travels over the downslope. Turbidity currents derive this driving force from the sediment in suspension. They experience a resisting shear force on the bed and entrain water from above. Turbidity currents can be eroding or depositive, accelerating or decelerating, dependent on the combination of initial conditions, bed slope, and size of sediment particles. They can be controlled from upstream (supercritical) or downstream (subcritical). Gravity currents with and without erosion and deposition are examined in order to understand the effects of sediment exchange on the flow.

412   Stability of Dynamic Flood Routing Schemes

Jingxiang Huang, (Visiting Scholar at St. Anthony Falls Hydr. Lab., Wuhan Inst. of Hydr. and Electric Engrg., China) and **Charles C. S. Song**, Member, ASCE, (Prof., St. Anthony Falls Hydr. Lab., Dept. of Civ. and Mineral Engrg., Univ. of Minnesota, Minneapolis, MN 55414)

Journal of Hydraulic Engineering, Vol. 111, No. 12, December, 1985, pp. 1497-1505

The characteristics of the second instability related to energy loss for the diffusive explicit method and the characteristic method for one-dimensional unsteady open channel flow is studied by numerical experiments. The Koren stability criterion is found to be valid not only for the diffusive explicit method but also applicable to the characteristic method. This instability severely restricts the allowable grid size Δt and Δx when the Froude number is small. By treating the energy loss term in a semi-implicit manner without complicating the computational procedure, it is possible to significantly improve the stability of the numerical schemes. The accuracy of various numerical schemes is also studied by numerical experiments. It is found that the most stable method is not necessarily the most accurate method.

413   Evaluation of Hydrologic Models Used to Quantify Major
Land-Use Change Effects

Task Committee on Quantifying Land-Use Change Effects of the Watershed Management and Surface-Water Committees of the Irrigation and Drainage Division, Roger P. Betson, chmn.

Journal of Irrigation and Drainage Engineering, Vol. 111, No. 1, March, 1985, pp. 1-17

An evaluation was made of the capabilities of watershed models when applied in the absence of site calibration data and limited validation data to predict the effects of major land-use changes upon hydrology. Members of the task committee were surveyed for representative models and asked to assess the accuracy to be expected from each. A total of 28 surface hydrology models were considered and most were judged capable of providing "good" accuracy. The reasons for this confidence were explored and appear to be based upon personal experience, possibly tempered by belief in the model originators. In view of the limited number of model comparison studies conducted and the less than encouraging results often obtained, it appears that when models are applied at sites without hydrologic data, considerable care must be taken if reasonable results are to be expected. In addition, there appears to be little justification for using a model more complicated than necessary under these conditions.

414   Modified Venturi Channel

W. H. Hager, Member, ASCE, (Sr. Research Engr., Chaire de Constructions Hydrauliques, Dep. Genie Civil, Ecole Polytechnique Federale de Lausanne, CH-1015 Lausanne, Switzerland)

Journal of Irrigation and Drainage Engineering, Vol. 111, No. 1, March, 1985, pp. 19-35

Flumes having a local diminution of channel width are widely used as discharge measurement structures. The present investigation considers the modified version of a cylinder positioned axially in a nearly horizontal, prismatic channel. This particular configuration has several advantages over the usual one: determination of discharge at arbitrary, well-defined channel cross sections needs no supplementary installations, and the device is simple and inexpensive. Moreover, the device enables a mobile registration of discharge at selected locations of a sewer or channel system. Discharge-head relations are determined using both the traditional hydraulic approach and more sophisticated models that account for the stream-line slope and curvature. Considerations include rectangular, trapezoidal and U-shaped channels. Experiments in rectangular channels corroborate the proposed discharge-head relation ($\pm 5\%$) and reveal the internal flow mechanisms in the vicinity of the cylinder. Examples show the computational procedure, and graphs enable a simple application of results to problems.

415 Design of Stable Alluvial Canals in a System

Howard H. Chang, Member, ASCE, (Prof., Civ. Engrg. Dept., San Diego State Univ., San Diego, Calif. 92182)

Journal of Irrigation and Drainage Engineering, Vol. 111, No. 1, March, 1985, pp. 36-43

A graphical method for the design of stable alluvial canals connected in a distributary system is presented. Under a given plan for water distribution, this method may be used to determine the stable width, depth and slope of each canal in the system. It has the advantages of relative simplicity and handiness as a design tool. Because of the sediment problems in systems design, the geometries and slopes of all canals in the same system must be interrelated in order to maintain sediment equilibrium. Without the use of structural sediment controls, sediment equilibrium is maintained by designing canals with an equal sediment concentration. Under such a system design, the power expenditure per unit weight of water stays approximately the same throughout the system; and the channel slope is inversely proportional to the one-sixth power of the discharge. Canals with greater discharges have higher velocities, but flatter slopes.

416 Miscible Displacement in Porous Media: MOC Solution

Raz Khaleel, (Asst. Prof., Hydr. Dept., New Mexico Inst. of Mining and Tech., Socorro, N.M. 87801) and **Donald L. Reddell,** (Prof., Agricultural Engrg. Dept., Texas A&M Univ., College Station, Tex. 77843)

Journal of Irrigation and Drainage Engineering, Vol. 111, No. 1, March, 1985, pp. 45-64

The method of characteristics (MOC) was revised to solve convective-dispersion equations in a two-dimensional, integrated saturated-unsaturated porous medium. Instead of usual bilinear interpolation schemes, the revised procedure utilizes a three-way linear interpolation scheme to assign seepage velocities to moving points in a two-dimensional grid system. The coordinates and concentrations of moving points reintroduced into the two-dimensional system are assigned using a random distribution. The accuracy of the numerical simulator was tested by comparison with available analytical solutions, numerical solutions and field data. Comparisons provided excellentagreement in almost all cases. Considering the nonhomogeneous field conditions encountered in the experimental plot, the simulated results are considered to be very good. The MOC eliminated numerical dispersion in simulation of miscible displacement in saturated, as well as unsaturated soils. A typical two-dimensional drainage problem was solved in a nonhomogeneous, integrated saturated-unsaturated porous media using the revised MOC.

417 Free Flow Discharge Characteristics of Throatless Flumes

A. S. Ramamurthy, (Prof., Civ. Engrg. Dept., Concordia Univ., 1455 De Maisonneuve W., Montreal, H3G 1M8, Canada), **M. V. J. Rae,** (Visiting Prof., Civ. Engrg. Dept., Concordia Univ., Montreal, Canada) and **Dev Auckle,** (Grad. Student, Civ. Engrg. Dept., Concordia Univ., Montreal, Canada)

Journal of Irrigation and Drainage Engineering, Vol. 111, No. 1, March, 1985, pp. 65-75

A theoretical analysis of the free flow discharge characteristics of the throatless flume is presented. A pressure correction factor which accounts for the non-hydrostatic pressure distribution at the control (throat) section is determined experimenally and incorporated in the free flow discharge equations developed from energy principles. The equations are valid for geometrically similar flumes in which the depth of flow in the approach channel is less than twice the channel width. Tests were conducted in three geometrically similar flumes. Both the present data and previously published data are used to validate the exprssions developed empirically.

418 Flood Storage in Reservoirs

W. H. Hager, Member, ASCE, (Dr. and Sr. Research Engr., Ecole Polytechnique Federale de Lausanne, Dept. of Civ. Engrg., Lausanne, Switzerland) and **R. Sinniger**, (Prof., Civ. Engrg. Dept., Chaire de Constructions Hydrauliques, Institut des Travaux Hydrauliques, EPFL, CH-1015 Lausanne, Switzerland)

Journal of Irrigation and Drainage Engineering, Vol. 111, No. 1, March, 1985, pp. 76-85

Flood storage in reservoirs controlled by free-overflowing, rectangular spillways is investigated by accounting for arbitrary resevoir bathymetry and single-peaked inflow hydrographs. The general solution of the storage equation depends only on a storage factor. It is found that adequate substitution of the effective hydrograph by a model hydrograph has only a secondary effect on the maximum reservoir outflow when using a proper shape factor. The proposed method allows an immediate application to realistic cases. In particular, the maximum reservoir outflow and the corresponding maximum reservoir level are represented graphically. Of further interest are cases in which the spillway crest length is varied for cost optimization purposes. Computations are explained in detail by examples, one of which includes a comparison of the present solution with the result of the numerically integrated basic equation.

419 Drainage Due to Gravity under Nonlinear Law

Bogdan Wosiewicz, (Lect., Dept. of Land Reclamation, Agricultural Univ., Pozan, Poland)

Journal of Irrigation and Drainage Engineering, Vol. 111, No. 1, March, 1985, pp. 89-94

A gravity drainage problem for a horizontal soil stratum underlain by a highly pervious material is analyzed. A non-linear relation between velocity and gradient is assumed. Capillary rise is also taken into account. The general differential equation (non-dimensionalized) for the problem is developed and analytically solved. The degree of drainage as a function of time and the real time of complete drainage are calculated and presented for typical values of the non-linear factor. It is shown that the analyzed non-linear drainage occurs faster compared to linear flow.

420 Electronic Clocks for Timing Irrigation Advance

Allan S. Humpherys, (Agricultural Engr., USDA-ARS, Snake River Conservation Research Center, Kimberly, Ind.), **Michael D. Wilson**, (Research Aide, Univ. of Idaho, Coll. of Agricultural Research and Extension Center, Kimberly, Ind.) and **Thomas C. Trout**, Member, ASCE, (Agricultural Engr., USDA-ARS, Snake River Conservation Research Center, Kimberly, Ind.)

Journal of Irrigation and Drainage Engineering, Vol. 111, No. 1, March, 1985, pp. 94-98

Low-cost electronic digital "stick-on" clocks were modified and used with a water sensor to determine irrigation rate-of-advance. The clock-sensors, spaced intermittently in a test furrow or in a border or basin, were used to determine the time that the advancing water front reached a given point. It was demonstrated that obtaining field data with a limited number of personnel can be greatly facilitated by using the clock-sensors.
Errata: IR Dec. '85, pp. 410-412.

421   Drainage Coefficients for Heavy Land

Lambert K. Smedema, (Dept. of Civ. Engrg., Delft Univ. of Tech., 2600 GA Delft, Netherlands)

Journal of Irrigation and Drainage Engineering, Vol. 111, No. 2, June, 1985, pp. 101-112

Due to the impeding nature of the subsoil, drainage discharge from heavy land predominantly takes the forms of lateral interflow through the more permeable toplayer and of overland flow. Evidence is presented that this type of "shallow drainage" discharge can be described by the linear reservoir type rainfall-discharge model. Procedures and guidelines are given, detailing how this model can be used to derive drainage coefficients for heavy land. Results are evaluated by comparison with other methods of deriving drainage coefficients and with commonly used design standards. It is concluded that the presented method is reliable and universally applicable.

422   Agricultural Benefits for Senegal River Basin

G. Leo Hargreaves, Assoc. Member, ASCE, (Prof. Officer, B. Co. 79th Engr. Bn, U.S. Army, APO, New York, N.Y. 09360), **George H. Hargreaves**, Fellow, ASCE, (Dir. of Research, International Irrigation Center, Dept. of Agr. and Irrigation Engrg., Utah State Univ., Logan, Utah 84322) and **J. Paul Riley**, Member, ASCE, (Prof., Dept. of Civ. and Environmental Engrg. and Utah Water Research Lab., Utah State Univ., Logan, Utah 84322)

Journal of Irrigation and Drainage Engineering, Vol. 111, No. 2, June, 1985, pp. 113-124

A multipurpose water resource development project is proposed for construction and development in West Africa. The plans include the irrigation of an area of 274,805 ha by the year 2030 in three countries: Senegal, Mauritania, and Mali. Agricultural benefits are estimated and analyzed by means of a computer program. The factors influencing crop yields are analyzed and summarized including crop selection, water, fertility, and management. Some of the considerations that will produce higher crop yields in future years are presented. It is proposed that the irrigation project management should be responsible for calculating crop water requirements and insuring acceptable irrigation efficiencies. A method is presented for estimating crop water requirements from maximum and minimum air temperatures. The desirability of using irrigation supplemental to rainfall for maximizing benefits from limited water is presented. The maximum contribution of a unit of irrigation water to yield is possible during the rainy season. Projected revenues and costs and years required for each crop to become economically profitable are presented in tables.

423   Irrigation Scheduling Using Crop Indicators

Robert J. Reinato, (Soil Scientist, USDA/ARS, U.S. Water Conservation Lab., Phoenix, Ariz. 85040) and **John Howe**, (Agronomist, Westlake Farms, Stratford, Calif. 93266)

Journal of Irrigation and Drainage Engineering, Vol. 111, No. 2, June, 1985, pp. 125-133

Petiole water content and canopy temperature of cotton were used as indicators of stress in cotton. A crop water stress index (CWSI) using remotely sensed canopy temperature was a better indicator of plant stress than was petiole water content. This point was most evident when using these parameters as a potential tool for scheduling irrigations and for estimating lint cotton yield.

424 **Furrow Irrigation Simulation Time Reduction**

M. Rayej, (Grad. Student, Agricultural Engrg. Dept., Univ. of California, Davis, Calif.) and **Wesley W. Wallender**, (Asst. Prof., Land, Air and Water Resources and Agricultural Engrg. Depts., Univ. of California, Davis, Calif.)

Journal of Irrigation and Drainage Engineering, Vol. 111, No. 2, June, 1985, pp. 134-146

Temporal and spatial changes in the flow profile decrease with time under furrow irrigation, and these slow changes suggest modifications in the simulation methods. Geometrically increasing time steps and node deletion at the upstream boundary are introduced and evaluated for accuracy and CPU time reduction. Increasing the time steps reduced CPU time without sacrificing accuracy significantly in predicting advance. The geometric progression factor is estimated automatically from the first two advance increments, thus incorporating site-specific flow and infiltration characteristics. The moving left boundary method is slightly more accurate, but is less cost-efficient than the varying time-step modification for the example conditions. However, special infiltration conditions may favor the latter method.

425 **Efficient Water Use in Run-of-the-River Irrigation**

Honorato L. Angeles, (Chmn., Agricultural Engrg. Dept., Central Luzon State Univ., Munoz, Nueva Ecija, Philippines 2320) and **Robert W. Hill**, Member, ASCE, (Prof., Agricultural and irrigation Engrg. Dept., Utah State Univ., Logan, Utah 84322)

Journal of Irrigation and Drainage Engineering, Vol. 111, No. 2, June, 1985, pp. 147-159

Irrigation water demand, crop yield and water allocation models were developed and applied to a Philippine run-of-the-river irrigation project. Dry season water supply hydrographs were classified into four levels of availability from high (I) to low (IV). These levels were correlated with the October stream flow volume to provide a prediction of dry season supply. Project net annual benefits could be increased for all four dry season water supply hydrograph levels by growing an upland crop (soybean) and rice, instead of only rice, providing that costs were less than $90/ha for additional drainage facilities. Net benefits for all rice were greater for reduced area planted, but fully irrigated when the available dry season water supply was lower than average as compared to planting the total area with inadequate irrigation. The irrigation system water demand model was calibrated to match actual diversions for the 1977-78 dry season. The project irrigation efficiency, combined effects of conveyance and application, was determined to be 41%. Thus, a savings in available water supply could be realized by increased system efficiency both from physical modifications as well as from improved management.

426 **Reuse System Design for Border Irrigation**

Muluneh Yitayew, Member, ASCE, (Asst. Prof., Soils, Water and Engrg. Dept., Univ. of Arizona, Tucson, Ariz.) and **D. D. Fangmeier**, Member, ASCE, (Prof., Soils, Water and Engrg. Dept., Univ. of Arizona, Tucson, Ariz.)

Journal of Irrigation and Drainage Engineering, Vol. 111, No. 2, June, 1985, pp. 160-174

A procedure is presented for designing a system to reuse runoff from free outflow irrigation borders. Runoff volumes are obtained using four dimensionless variables and runoff curves developed with a zero-inertia mathematical model. An example of the design procedure includes different modes of reuse system operation.

427   **Electrical Resistivity for Estimating Ground-Water Recharge**

William E. Kelly, Member, ASCE, (Prof., Dept. of Civ. Engrg., Univ. of Nebraska, Lincoln, Neb. 68588-0531)

Journal of Irrigation and Drainage Engineering, Vol. 111, No. 2, June, 1985, pp. 177-180

In the unsaturated zone, fine-grained soils tend to exist in situ at higher degrees of saturation than coarse-grained soils. As a result, fine-grained soils generally will exhibit lower resistivities than coarse-grained soils. On this basis, empirical relations between recharge rates and electrical resistivities measured with surface techniques can be developed. An example using published data is given.

428   **Report of Task Committee on Water Quality Problems Resulting from Increasing Irrigation Efficiency**

Water Quality Committee of the Irrigation and Drainage Division

Journal of Irrigation and Drainage Engineering, Vol. 111, No. 3, September, 1985, pp. 191-198

The Task Committee on Water Quality Problems Resulting from Increasing Irrigation Efficiency coordinated preparation of a series of papers that addressed the perceived effects on water quality from increasing irrigation efficiencies. This paper summarizes and integrates results presented in the requested papers, which deal primarily with salinity and nitrogen relationships, and also analyzes other ramifications of increasing irrigation efficiency. Increasing on–farm efficiency usually connotes providing water to meet crop needs with minimal deep percolation with surface runoff. Topics examined include leaching requirements, chemical reactions on the root zone, potential hazards and benefits of reduced leaching, nitrate leaching, basin-wide implications of increasing efficiencies, and effects on groundwater quality. Effects on erosion and the movement of pesticides are also considered. Physical changes to improve irrigation efficiency cannot be considered independently of the associated economic, social, legal, and institutional restraints. It is concluded that from a water quality viewpoint there is no reason to avoid increasing irrigation efficiency, but any such project should be analyzed thoroughly in advance to determine whether local circumstances warrant the improvements.

429   **Drainage Required to Manage Salinity**

Glenn J. Hoffman, (Research Leader, Water Management Research Lab., USDA-ARS, 2021 S. Peach Ave., Fresno, Calif.)

Journal of Irrigation and Drainage Engineering, Vol. 111, No. 3, September, 1985, pp. 199-206

The amount of drainage required to prevent loss in crop productivity from excess soil salinity has been termed as the leaching requirement (L_r). Steady-state models for predicting I_r are reviewed and compared with experimentally measured values. Values of L_r predicted from the exponential model, which computes the average soil salinity in the root zone based upon an exponential pattern of crop water uptake, correlated best with measured values. Acceptable estimates are also obtained with a simplified model that approximates an average soil salinity based on values for the top and bottom of the root zone. The highest correlation coefficient was 0.67, indicating that none of the models is completely satisfactory.

430   **Chemical Reactions within Root Zone of Arid Zone Soils**

J. D. Oster, (Soil and Water Specialist, Cooperative Extension, Univ. of California, Riverside, Calif. 92521) and **K. K. Tanji**, (Prof., Dept. of Water Sci., Dept. of Land, Air, and Water Resources, Univ. of California, Davis, Calif. 95616)

Journal of Irrigation and Drainage Engineering, Vol. 111, No. 3, September, 1985, pp. 207-217

The quality of an irrigation water, in both its total salinity and individual constituents, influences the chemical character of the soil solution and irrigation return flows including drainage waters. Dissolution, precipitation, adsorption and exchange reactions modify the salt load and composition of return flows. The effects of these reactions can be manipulated by varying the leaching fraction. A change from significant increase in total salts to substantial precipitation of calcite (soil lime) and gypsum can occur as the leaching fraction is reduced. Such a reduction in leaching fraction and the concomitant reduction in applied water will also reduce the rate at which native salts in the return flow pathway are displaced into receiving waters. Methodology exists, as illustrated by several samples, to evaluate potential consequences of reduced leaching and chemical reactions on soil solution composition and on the salt loads of irrigation return flows.

431   **Salt Problems from Increased Irrigation Efficiency**

J. D. Rhoades, (Research Leader, Soil and Water Chemistry, U.S. Salinity Lab., United States Dept. of Agr., Agricultural Research Service, Western Region, Riverside, Calif. 92501)

Journal of Irrigation and Drainage Engineering, Vol. 111, No. 3, September, 1985, pp. 218-229

Increasing on-farm irrigation efficiency reduces deep percolation and generally increases concentrations of soluble salts in the rootzone. Since more salt precipitation occurs within the porous soil matrix, the mass of soluble salts is reduced. In addition exchangeable sodium builds up on the cation exchange complex and plant available water decreases. Conventional irrigation scheduling techniques become less adequate and the need for salinity monitoring increases. Generally, however, the potential benefits of increasing irrigation efficiency outweigh the potential problems caused by the aforementioned.

432   **Effect of Irrigation Efficiencies on Nitrogen Leaching Losses**

William F. Ritter, (Prof., Agricultural Engrg. Dept., Univ. of Delaware, Newark, Del. 19717) and **Katherine A. Manger**, (Grad. Asst., College of Education, Univ. of Delaware, Newark, Del. 19717)

Journal of Irrigation and Drainage Engineering, Vol. 111, No. 3, September, 1985, pp. 230-240

The literature was reviewed to determine the effect increasing irrigation efficiencies have had on NO_3-N leaching. The mass of NO_3-N leached is directly related to the drainage volume. By increasing irrigation efficiencies, both the drainage volume and amount of NO_3-N leaching are reduced. Both water management and nitrogen management are important in controlling NO_3-N leaching. By using improved water management practices that control the amount of water applied and using the proper time to apply water, the irrigation efficiency is increased and NO_3-N leaching is reduced. Applying only enough N fertilizer tomeet crop requirements for a realistic yield goal, time of application and use of slow release fertilizers are N management practices that will reduce NO_3-N leaching. However, it is impossible to reduce NO_3-N leaching to zero in the humid region on coarse-textured soils and still maintain adequate crop yields. Nitrogen simulation models cannot be widely used to evaluate N management practices and irrigation water management practices that increase irrigation efficiencies because of the lack of required input data.

433   Basin-Wide Impacts of Irrigation Efficiency

L. S. Willardson, Member, ASCE, (Prof., Dept. of Agri. and Irrigation Engrg., Utah State Univ., Logan, Utah 84322)

Journal of Irrigation and Drainage Engineering, Vol. 111, No. 3, September, 1985, pp. 241-246

Irrigation efficiency is one of the important factors affecting river basin water quality. Questions frequently arise as to whether increasing irrigation efficiency in a basin will have a net positive or a net negative effect on basin water quality. Quality and quantity of return flow from irrigation are related to irrigation efficiencies and leaching fractions. River basins have leaching fractions analogous to irrigation leaching fractions inasmuch as salts generated by mineral weathering and those dissolved by percolating waters are carried out of the basin by a fraction of the precipitation falling on the basin. Reducing the leaching fraction of a basin by increasing irrigation efficiency and consuming more water in the basin will decrease downstream water quality and quantity. Irrigation efficiency, as a salinity control measure, must be considered in the context of total water management to be properly interpreted.

434   Automatic Throttle Hose—New Flow Regulator

Peter U. Volkart, (Sr. Research Engr., Lab. of Hydr., Hydrology and Glaciology, Fed. Inst. of Technology, 8091-Zurich, Switzerland) and Frits de Vries, (Research Engr., Lab. of Hydr., Hydrology and Glaciology, Fed. Inst. of Technology, 8091-Zurich, Switzerland)

Journal of Irrigation and Drainage Engineering, Vol. 111, No. 3, September, 1985, pp. 247-264

Flow regulators in both irrigation and storm water engineering normally must not be very exact, but should be simple to construct and maintain, "fool-proof," and of course inexpensive. A new flow regulator which matches these requirements, the so-called Automatic Throttle Hose, is described. The device delivers an almost constant discharge independently of fluctuations in the upstream water level. No external power supply is needed for running the device, as it takes all required energy from the fluid itself. Following an explanation of the regulating principle, the application of this new regulator to irrigation practice is described and a design chart for application in the low head range presented. Further, modifications are listed which may be used to extend the effective head range. Practical experiences in both irrigation schemes and storm water projects indicate that this flow regulator has a place alongside other proven regulators for rather low discharges.

435   Irrigation Water Requirements for Senegal River Basin

G. Leo Hargreaves, Assoc. Member, ASCE, (Civ. Engrg. Officer, 79th Engineer Bn. U.S. Army, APO, N.Y. 09360), George H. Hargreaves, Fellow, ASCE, (Dir. of Research, International Irrigation Center, Dept. of Agr. & Irrigation Engrg., Utah State Univ., Logan, Utah 84322) and J. Paul Riley, Member, ASCE, (Prof., Dept. of Civ. and Environmental Engrg. and the Utah Water Research Lab., Utah State Univ., Logan, Utah 84322)

Journal of Irrigation and Drainage Engineering, Vol. 111, No. 3, September, 1985, pp. 265-275

The Senegal River is a major natural resource in West Africa where the principal economic resources are agricultural. A proposed irrigation project will provide a significant increase in crop production and will exert a large influence on the economics of Senegal, Mauritania, and Mali. The magnitude of benefits from the project will depend upon the allocation, scheduling and managing of that portion of the water to be used for irrigating agricultural crops. A procedure is recommended for estimating crop water requirements that only requires the measurement of maximum and minimum temperatures. This procedure although calibrated for the Senegal River Basin using climatic data from four representative locations appears to be generally applicable for other areas without calibration. The importance of rainfall in supplying part of crop water requirements is described. Mean, actual dependable and effective precipitation values are compared for one location. Block farming or the planting of a single crop to manageable areas should be made mandatory. Project managers must assume responsibility or implementing procedures to insure good productivity and reasonable efficiencies.

436   Overland Flow Hydrographs for SCS Type II Rainfall

A. Osman Akan, Assoc. Member, ASCE, (Assoc. Prof., Dept. of Civ. Engrg., Old Dominion Univ., Norfolk, Va. 23508-8546)

Journal of Irrigation and Drainage Engineering, Vol. 111, No. 3, September, 1985, pp. 276-286

A mathematical model, which is founded on the kinematic-wave and Green-Ampt equations, is presented for the conjunctive overland flow-infiltration process. Within the context of this model, it is shown that overland flow resulting from a specified time distribution of rainfall intensity can be described in terms of only three physically-based nondimensional parameters. In cases where a less precise geometric shape is adequate, overland flow hydrographs for the 24-hr SCS-Type II rainfall can be approximated by triangular hydrographs. It is possible to evaluate the elements of such a triangular hydrograph in terms of the three governing physically-based parameters of the overland flow-infiltration process without using a computer. A dimensionless standardized triangular hydrograph is proposed that predicts the peak discharge, the time of occurrenceof the peak discharge, and the volume of runoff before and after the hydrograph peak.

437   Plastic Lining on Riverton Unit, Wyoming

Ronald W. Wilkinson, Member, ASCE, (Chf., Office Engrg. Div., Riverton Proj. Office, P.O. Box 31, Riverton, Wyo. 82501)

Journal of Irrigation and Drainage Engineering, Vol. 111, No. 3, September, 1985, pp. 287-298

Since 1975, the Bureau of Reclamation has installed approximately 25.8 miles (41.5 km) of PVC (polyvinyl-chloride) lining in the distribution system of the Riverton Irrigation Project. The lining is part of a $40,000,000 rehabilitation program designed to stop waterlogging of adjacent farmland and conserve water. This paper describes the methods used in investigating, designing, and constructing the PVC lining. It also summarizes the construction costs from 1973 to 1982.

438   Characteristics of Free Surface Flow Over a Gravel Bed

Ashim Das Gupta, (Assoc. Prof., Div. of Water Resources Engrg., Asian Inst. of Technology, P.O. Box 2754, Bangkok 10501, Thailand) and **Guna Nidhi Paudyal**, (Doctoral Student, Div. of Water Resources Engrg., Asian Inst. of Technology, P.O. Box 2754, Bangkok 10501, Thailand)

Journal of Irrigation and Drainage Engineering, Vol. 111, No. 4, December, 1985, pp. 299-318

Measurements of velocity profiles of a free surface flow over a permeable gravel bed indicate that the logarithmic velocity distribution can be preserved if the reference datum is located a small distance equal to about one third of the median diameter of the bed particles below the surface. The observed value of 0.28 for the Karman constant is significantly reduced below the commonly expected value of 0.4 for impervious boundaries which indicates that the boundary resistance of the permeable bed is higher than that of the impermeable bed having identical rugosity. It is also observed that the friction factor increases with the increase in Reynolds number. A method is proposed to predict the amount of seepage flow through the permeable bed by measuring the hydraulic gradient and the velocity profile above the bed; and the bed material properties such as grain size and permeability. Although the measured bed flow showed considerable deviations from the predictions for the present experiments, the proposed method can be usefully applied in practical problems.

439   **Procedure to Select an Optimum Irrigation Method**

Eduardo A. Holzapfel, Member, ASCE, (Assoc. Prof., Agricultural Engrg. Dept., Univ. of Concepción, Chillán, Chile), **Miguel A. Mariño**, Member, ASCE, (Prof. of Water Sci. and Civ. Engrg., Univ. of California, Davis, CA 95616) and **Jesus Chavez-Morales**, Student Member, ASCE, (Prof., Hydrosciences Center, Postgraduate College, Chapingo, Mexico)

Journal of Irrigation and Drainage Engineering, Vol. 111, No. 4, December, 1985, pp. 319-329

A procedure to select an optimum irrigation method for specified field conditions is presented. The procedure is developed on the basis of indices that show the acceptability of the irrigation methods to the selection parameters (crop density, type of sowing or planting, slope of the field, infiltration rate of the soil, etc.), cost, and financial feasibility criteria.

440   **Runoff Probability, Storm Depth, and Curve Numbers**

Richard H. Hawkins, Member, ASCE, (Prof., Dept. of Forest Resources, Dept. of Civ. and Environmental Engrg., Utah State Univ., Logan, Utah 84322), **Allen T. Hjelmfelt, Jr.**, Member, ASCE, (Hydr. Engr., U.S. Dept. of Agr., Agricultural Research Service, Watershed Research Unit, Columbia, MO 65203) and **Adrain W. Zevenbergen**, (Grad. Student, Agricultural Univ., Wageningen, Netherlands)

Journal of Irrigation and Drainage Engineering, Vol. 111, No. 4, December, 1985, pp. 330-340

Using existing antecedent moisture condition (AMC) Curve Number relationships, a general expression of rainfall-runoff is developed for AMC I and AMC III. The probability distribution of event runoff exceeding zero is found to be lognormal. Relative storm size is then proposed to be defined on the ratio P/S, where a "large" storm has P/S>0.46, when 90% of all rainstorms will create runoff. Consequent problems arising in the definition of CN from rainfall-runoff data are discussed.

441   **Evapotranspiration of Small Conifers**

Michael R. Petersen, (Res. Asst., Agricultural and Irrigation Engrg. Dept., Utah State Univ., Logan, UT 84322) and **Robert W. Hill**, Member, ASCE, (Prof., Agricultural and Irrigation Engrg. Dept., Utah State Univ., Logan, UT)

Journal of Irrigation and Drainage Engineering, Vol. 111, No. 4, December, 1985, pp. 341-351

Three lysimeters were established containing different sized Scotch pines (*P. sylvestris*), and the consumptive water use of each tree was monitored during the 1982 and 1983 growing seasons near Logan, Utah. Weather data including maximum, mean, and minimum daily temperatures, solar radiation, and daily precipitation were collected. Consumptive use data of the first season were of limited use due to the transplanting stress experienced by the trees. The results of the second season were consistent with the usual water use of irrigated crops. Mean monthly crop coefficients were calculated based on the modified Blaney-Criddle and the Jensen-Haise methods, assuming water was extracted from only the crown projection area of the tree. A seasonal Blaney-Criddle crop coefficient was estimated to be 1.222. The growing season was long, and the crop coefficient (Blaney-Criddle) during the winter at this site may be as high as 0.85. An equation was developed to find the composite crop coefficient for conifer tree farms relating tree size, tree spacing, and type of ground cover.

442   **Least-Cost Planning of Irrigation Systems**

Kyung H. Yoo, (Asst. Prof., Alabama Agricultural Experiment Station, Auburn Univ., Auburn, AL 36849) and **J. R. Busch,** (Prof., Idaho Water and Energy Resources Research Inst., Univ. of Idaho, Moscow, ID 83843)

Journal of Irrigation and Drainage Engineering, Vol. 111, No. 4, December, 1985, pp. 352-368

A mixed-inter programming model was used to obtain least-cost system rehabilitation plans for a 6,900-ha (17,000-acre) project in southeastern Idaho. Three types of gravity conveyance system components (existing unlined canal, concrete lined canal, and gravity pipe) were considered, along with five types of irrigation application systems (two gravity and three sprinkler applications systems). The mixed-integer programming model that complied with the constraints specified was flexible and effective. The specified constraints used in this study are water charges and water and land availabilities. The quantitative effects of different constraints were easily evaluated. The same modeling procedure can also be used in developing scenarios of alternative system configurations for a new irrigation project development for least-cost system planning. The model gives descriptive scenarios that can assist planners, irrigators, and other interested parties in making multiple-objective planning decisions for developing or rehabilitating irrigation projects.

443   **Evapotranspiration Model for Semiarid Regions**

Pramod Kumar Jain, (Postdoctoral Fellow of Soil and Water Engrg., Faculty of Agricultural Engrg., Technion-Israel Inst. of Technology, Haifa-32000, Israel) and **Gideon Sinai,** (Sr. Lect. of Soil and Water Engrg., Faculty of Agricultural Engrg., Technion-Israel Inst. of Technology, Haifa-32000, Israel)

Journal of Irrigation and Drainage Engineering, Vol. 111, No. 4, December, 1985, pp. 369-379

The Thornthwaite method for calculating evapotranspiration (ET) as used in the water management model, DRAINMOD, is modified for semi-arid conditions since for such conditions the existing method had underestimated the average monthly ET rate by about 50 percent. The following three modifications are proposed: (1) the constant in the Thornthwaite general equation is increased depending upon the min-max range of the annual mean air temperature wave; (2) the daylight hours wave is shifted by 45 days and brought parallel to the mean air temperature wave. This shift is made in order to account for the heat energy of the atmosphere surrounding the plants; and (3) actual evapotranspiration (EAT) is not equated to zero during irrigation. The AET was measured in a drip irrigated banana field with lysimeters. The simulated monthly adjusted values of AET give satisfactory estimates with an average error of within ± 10 percent, when compared to that of the lysimeter data.

444   **Dimensionless Formulation of Furrow Irrigation**

Theodor Strelkoff, (Hydr. Engr., 136 Seventh Avenue, San Francisco, CA 94118)

Journal of Irrigation and Drainage Engineering, Vol. 111, No. 4, December, 1985, pp. 380-394

The equations governing the flow of water in irrigation furrows are put in nondimensional form by expressing each variable therein in ratio to an appropriate reference variable. Two systems of reference varibles are considered, one based on normal depth in the given furrow geometry at the given inflow, the other on the given inflow and cutoff time. The first is especially valuable because of the clear physical significance of normal depth, but it is useless in horizontal furrows. The second system is pertinent to furrows set on any slope. The matter of choosing a system of reference variables is viewed in some generality to allow application to other systems, suitable for special applications, in particular, design applications. A set of formulas is presented for translation from one system to the other. The necessary input for a completely nondimensional treatment of furrow irrigation is outlined, and the concept of hypothetical dimensioned furrows is introduced. This is designed to provide a physical significance to variables entered in dimensionless form as input to a mathematical model of furrow irrigation.

445   **Brink Depth in Non-Aerated Overfalls**

George C. Christodoulou, Assoc. Member, ASCE, (Assoc. Prof., Dept. of Civ. Engrg., National Tech. Univ. of Athens, Greece)

Journal of Irrigation and Drainage Engineering, Vol. 111, No. 4, December, 1985, pp. 395-403

Small drop structures employed in irrigation canals are often not ventilated. In this paper the importance of the residual pressure at the brink of a rectangular overfall is quantified by means of the 1-D momentum equation. The pressure acting below the nappe under nonaerated conditions is studied experimentally and correlated to the drop height in nondimensional terms. Comparison is made with relevant results for ventilated drop structures and a criterion for defining a low drop is proposed. The results of pressure estimates are then used for a theoretical evaluation of the brink depth ratio as a function of the nondimensional drop height. The theoretical predictions are found to be in good agreement with the experimental results.

446   **The Client Relationship: Effective Marketing Steps**

Lloyd H. Bakan, (Vice Pres., Daniel, Mann, Johnson, and Mendenhall, 3250 Wilshire Blvd., Los Angeles, Calif. 90010)

Journal of Management in Engineering, Vol. 1, No. 1, January, 1985, pp. 3-11

Effective marketing steps involved in establishing strong client relationships are presented, along with ways to maintain continuing and favorable client relationships with previous, current and new users. An outline of purchasing design services in the client-consultant relationship process is also given. Nonpersonal image-building sales and marketing activities, such as advertising, direct mail, and publications aimed at newspapers and industry markets, are also reviewed. These factors, along with social and community activities and direct selling, provide the opportunity and forum for the development of lasting client relationships, and help establish the buyer's attitudes about a firm. A selection of questions posed to the buyers of AE services is also provided.
Discussion: **Melville D. Hensey**, (Prin., Hensey Assocs., Engrg. Management Consultants, Cincinnati, Ohio) ME July '85, pp. 175.

447   **Communicating the Company's Operating Performance Data**

Howard G. Birnberg, (Princ., Birnberg & Assoc., Design Professions Business Consultants, Chicago, Ill.)

Journal of Management in Engineering, Vol. 1, No. 1, January, 1985, pp. 12-19

A report format and indices for communicating the company's operating performance data are presented. The process described can be performed with almost any aspect of engineering firm management; however, in financial management, this process is most vital. A major difficulty preventing adequate financial management is the fact that most financial data is presented in a format understandable and useful to accountants, but unclear to the design professionals who manage the firm.

448   **What Do You Look for in a New Engineer?**

Glenn E. Futrell, Member, ASCE, (Pres., Soil & Material Engrs., Inc., 3109 Spring Forest Road, Box 58069, Raleigh, N.C. 27658-8069)

Journal of Management in Engineering, Vol. 1, No. 1, January, 1985, pp. 20-27

The author finds that certain criteria are most helpful in selecting new engineering personnel. Generally, these criteria stress good "people-relating" skills over mere academic achievement. An applicant's education and technical background should be well-rounded and indicate an ability to grow and learn. A personal work ethic can be gleaned from an applicant's home and pre-college environment, while the personal interview is a good indicator of personality interactions. Communications and social skills are necessary to foster good relations with clients, as well as other employees. A demonstrated positive attitude, and a can do approach will also be helpful to the company. The author finds that pre-employment values are likely to be carried into the market place.

449   **Productivity of Construction Professionals**

Donn E. Hancher, Member, ASCE, (Assoc. Prof. of Construction Engrg., Purdue Univ., West Lafayette, Ind. 47907)

Journal of Management in Engineering, Vol. 1, No. 1, January, 1985, pp. 28-35

Productivity of employees is a major concern in all organizations. The optimization of the production of the total organization can be achieved through the coordinated improvement of the performance of the individual employees. For many years the productivity of blue collar workers in the construction industry has been studied and evaluated. However, several studies have revealed that poor management was the cause for poor worker productivity. This indicates that attention is also needed to improve the productivity of professional employees in construciton organiztions. Organizational psychologists have discovered that there are two basic categories of factors which motivate employee performance: "financial" and "non-financial" motivators. Many feel that the two most significant factors are salary and recognition. They also point out that improved or high performance must be rewarded, or it will soon disgress. In order to reward good performance in an organization, there must be a uniform system to evaluate the performance of the employees. The objectives of this paper is to discuss the improvement of the productivity of professional construction employees, the appraisal of employee performance, and the review of performance appraisal with employees.

450   **Development of an Engineering Manager**

John P. Hribar, Fellow, ASCE, (Office Engr., Howard Needles Tammen & Bergendoff, Milwaukee, Wisc.)

Journal of Management in Engineering, Vol. 1, No. 1, January, 1985, pp. 36-41

In the transition from engineer to manager, the individual's attention shifts by necessity, from detail to overview, specific to general and technical to administrative. To make that transition, the responsibilities of both the individual and the organization in interpreting changing career goals and organizational needs must be identified as the individual moves through the organization. There are many paths the individual may follow through an organization. Some paths may simply turn into dead ends while others may diverge to multiple choices and again converge. The variety of paths will be a challenge to him, requiring an assessment of his own situation and an evaluation of the paths that will meet changing goals and objectives as his career matures.

451   Managing Productivity During an Attrition Program

Bill Macaitis, Member, ASCE, (Asst. Chf. Engr., The Metropolitan Sanitary Dist. of Greater Chicago, Chicago, Ill. 60611)

Journal of Management in Engineering, Vol. 1, No. 1, January, 1985, pp. 42-48

Following the expansion of its Engineering Department to accommodate a large capital works program, The Metropolitan Sanitary District of Greater Chicago initiated an attrition program primarily aimed at reducing the staffing of its Engineering Department to approximately a pre-program level. The attrition environment required that special attention be given to maintaining productivity. The motivations and needs of the employees were analyzed. Management actions directed at productivity have been focused at the individual employee, as well as at general work production. The attrition program has been successful, and the objectives of the Engineering Department have been consistently obtained throughout the program.

452   Effective Management of Engineering Design

Kenneth J. Barlow, (Pres., Barlow Assoc., Inc., Toronto, Ontario, Canada)

Journal of Management in Engineering, Vol. 111, No. 2, April, 1985, pp. 51-66

The effectiveness of project management has a significant impact on the success and profitability of the results achieved. Sophisticated project management computer systems with enormous appetites for manhours are often used instead of effective management of the essential components of the project. Excellent project managers recognize that they will have unparalleled opportunities during the next several decades as the market increasingly recognizes the positive impact which they can have on results. These project managers will motivate, plan, schedule and control team members, their productivities and the project with simple, rather than complex systems. They will organize their teams and ensure that the team's interfaces with clients and customers are carefully managed. They will combine their ability to manage people with effective planning, practical control and meaningful reward systems in order to deliver results on budget and on schedule. Experience and effectiveness, increasing productivity, profit planning and control are considered. Getting off to an effective start, and the elements of a check list that project managers could use as a framework to develop and refine their own personal check lists are explained—check lists they can use to ensure that everything that should be done is done on schedule.

453   A Performance Review System

Richard R. Lenz, Member, ASCE, (Vice Pres., KZF Inc., 111 Merchant St., Cincinnati, Ohio 45246)

Journal of Management in Engineering, Vol. 111, No. 2, April, 1985, pp. 67-78

Review of a person's performance can consist of day-to-day coaching, counseling on specific problem areas, and a periodic formal review. An annual review should serve to fill an employee's need for recognition for work done, skills gained, and improvements made. A comprehensive, fair, and consistent system should be used to evaluate performance. KZF Incorporated, of Cincinnati, Ohio, has developed a system which recognizes the specifics of each job, the relative importance of elements of responsibility, performance factors, and special assignments or job functions. The system is much more objective than what was used before and has worked well when implemented. A description of the background, the development process, the system itself, and training is described.

454   **Earned Value Technique for Performance Measurements**

Daniel R. McConnell, Member, ASCE, (Mgr. of Economic Analysis, Parsons Brinckerhoff Construction Services, Inc., McLean, Va. 22102)

Journal of Management in Engineering, Vol. 111, No. 2, April, 1985, pp. 79-94

    Earned Value is a project control technique which provides a quantitative measure of work performance. It involves a crediting (earning) of budget dollars as scheduled work is performed. The earned value technique is a proven method to evaluate work progress in order to identify potential schedule slippage and areas of budget overruns. Value earned for a given task is computed as budgeted cost of work performed and is a function of time, work completed, and budget. Budgeted cost of work performed is compared against actual cost of work performed and budgeted cost of work scheduled to assess cost and schedule variances, respectively. Cost and schedule variances may be identified at the individual cost account level or at any other level up to the overall project for upper management review. Variances may be identified by work element and organizational disciplines. Variances may be reported as a percent of the baseline and presented graphically. The work breakdown structure, detailed schedules and cost account budgets form the foundation for earned value assessment. Project management control points are established by creating a matrix of the work breakdown structure and the project organizational breakdown. This identifies functional managers and subcontractors responsible for work performance. Each control point is represented by a cost account and establishes the lowest level for evaluating cost and schedule performance.

455   **Selecting the Correct Retirement Plan for Your Business**

Paul A. Randle, (Prof. of Finance, Utah State Univ., Logan, Utah 84321)

Journal of Management in Engineering, Vol. 111, No. 2, April, 1985, pp. 95-104

    For many years the use of tax-qualified pension and profit sharing plans has given engineering firms the opportunity to provide an attractive fringe benefit to employees. In addition, such plans have given firm owners the opportunity to accumulate substantial personal wealth. The Tax Equity and Fiscal Responsibility Act of 1982 (TEFRA) enacted many changes in the law which governs pension and profit sharing plans. Since most of the changes effected by TEFRA went into effect on January 1, 1984, it is important for owners of firms with such plans to understand the nature of these changes. Many important changes, including the attempt to create parity between Keogh (retirement plans for unincorporated businesses) plans and plans for incorporated businesses are explained. The economics of retirement plan participation are also explained. The wealth-accumulation potential which can accrue to firm owners is illustrated, together with cost-benefit comparisons which show owner benefits vis-a-vis employee costs. All such comparisons are constructed so as to show the tax effects of plan contributions, accumulations, and benefits.

456   **Engineers as Managers**

Joseph A. Steger, (Pres., Univ. of Cincinnati, Cincinnati, Ohio 45221)

Journal of Management in Engineering, Vol. 111, No. 2, April, 1985, pp. 105-111

    The intellectually based effectiveness is differentiated from the emotionally based effectiveness of the manager. Although they are obviously interrelated, the required outcomes of a manager's behavior are largely not a function of intellect, but a function of emotions (both handling inputs and projecting positive emotional outputs). The academic setting does not, in a curricular sense, handle the emotional based effectiveness. And in reality it should not, since this base of effectiveness is so well-established in the individual by a young age (and is primarily immutable). Selection for line management is the tool to address this facet of management preparedness. On the other hand, the engineering training in evaluation content (technical), if combined with work experience (co-op), does provide a basis for the intellectual basis of management. However, a good engineering education does not predict management success or effectiveness. And it should not be viewed as designed to insure a cadre of management talent to be line or operating managers. Only selection and carefully designed work experience can insure such a cadre.

457　　　　　Managing the Transition to CADD

David E. Weida, (Prin. and Chief Engr., KZF, Inc., 111 Merchant St., Cincinnati, Ohio 45246)

Journal of Management in Engineering, Vol. 111, No. 2, April, 1985, pp. 112-116

　　　The transition to computer aided drafting and design (CADD) must first begin by establishing the need and the desired or expected results. The transition is comprised of three phases: Planning, administration, and evaluation. Upon making the decision to move to CADD, the planning process establishes goals to meet the desired results. Administration is the implementation of the plan utilizing human relation skills to provide staff motivation and redirection toward the goals after evaluation. The evaluation consists of both financial and staff performance. The two performances are related and overlap; however, each must be evaluated independently to properly identify problem areas and good performance.

458　　　　Private versus Public Ownership of Constructed Facilities

Chris Hendrickson, Assoc. Member, ASCE, (Assoc. Prof., Dept. of Civ. Engrg., Carnegie-Mellon Univ., Pittsburgh, Pa. 15213) and **Tung Au**, Fellow, ASCE, (Prof., Dept. of Civ. Engrg., Carnegie-Mellon Univ., Pittsburgh, Pa. 15213)

Journal of Management in Engineering, Vol. 111, No. 3, July, 1985, pp. 119-131

　　　The ownership arrangements for constructed facilities not only generate the capital and requirements for new facilities, but also influence the management of the construction and operation of these facilities. While it is difficult to conclude definitely that one or another organizational or financial arrangement is always superior, different organizations have systematically chosen the ways in which constructed facilities are financed, designed and constructed. Moreover, the selection of alternative investments for constructed facilities is likely to be affected by the type and scope of the decision-making organization. Some of these systematic differences, particularly with regard to public versus private organizations are considered. An extended example shows that tax shields, high public borrowing costs and private financial leverage may be sufficiently advantageous for private firms to overcome the effects of partial capital subsidies and lower required rates of return for public organizations. Thus, private ownership and operation of infrastructure may prove to be a desirable social alternative to public ownership under some circumstances, even without considering differences in the efficiency of operation.

459　　　　　A Gradualist's Approach to Management

James C. Howland, Fellow, ASCE, (Sr. Consultant, CH2M HILL, P. O. Box 428, Corvallis, Oreg. 97339)

Journal of Management in Engineering, Vol. 111, No. 3, July, 1985, pp. 132-137

　　　Management philosophies and various approaches to applying these philosophies are described. To achieve organizational continuity, it is important to reduce conflict to a minimum.

To do this, the goals of the firm and the goals of the individual should parallel each other; benefits in money and satisfaction should be as equally distributed as possible; and perks should be relatively uniform. Special programs, such as programs full time employees and provided for mandatory sale of stock back to the company at age 65 are effective motivators. Efforts should be made to develop the kinds of work people want to do in locations where they want to be. Although many management consultants believe that bonus systems are demotivators, in a professional services firm with an open management style, in which all employees are continually advised of the operating data, a bonus system can be an important motivator. Long range strategic planning should grow with the company. Both planning successes and planning failures should be expected.

460   Elements of Cost and Schedule Management

John P. Hribar, Fellow, ASCE, (Office Engr., Howard Needles Tammen & Bergendoff, 6815 W. Capitol Dr., Milwaukee, Wisc. 53216) and **Gregory E. Asbury**, (Project Scheduler and Cost Engr., Howard Needles Tammen & Bergendoff, 6815 W. Capitol Dr., Milwaukee, Wisc. 53216)

Journal of Management in Engineering, Vol. 111, No. 3, July, 1985, pp. 138-148

Cost and schedule management is an effort that requires extensive use and control of data. A project coding system such as a Work Breakdown Structure is a key element in developing and controlling cost and schedule. Objectives and techniques used to manage cost and schedule are considered at the three basic stages of project development: Planning, design and construction. The greatest opportunity for cost and schedule control is at the early stages of a project. As a project moves through planning to construction, the detail known increases and thus the detail necessary in the cost and schedule system must increase. Bar charting may be acceptable as an exclusive scheduling tool during planning, but later must be replaced by a logic basic technique such as the Critical Path Method. Cost estimates must be trended from original planning estimates as the project moves into design and construction. Exception reporting is the primary vehicle for upper management review of project cost and schedule data.

461   Entering Technical Management

Richard J. Kosiba, Member, ASCE, (Chief Civ./Struct. Engr., Western Power Div., Bechtel Power Corp., Los Angeles, Calif.)

Journal of Management in Engineering, Vol. 111, No. 3, July, 1985, pp. 149-156

The writer presents a number of selected recommendations in career development of professional engineers interested in pursuing a leadership role in technical management. Initial qualifications and preparation form a sound technical foundation. Consistent attention, upgrading, and strengthening of contemporary skills lead to management interest and investment. The direction toward management success becomes initially established. Attention then shifts toward the development of managerial skills which requires new commitment and development. Maintaining high performance levels through self-motivation and individual development requires additional attention to delegation, investment in colleagues, communications and effective decision-making. Pitfalls in straying off course are presented. The writer presents certain personal recommendations which through experience have formed a basis for success in technical management of professionals.

462   Configuration Management as Applied to Engineering
Projects

Richard W. Schenk, Member, ASCE, (Configuration Mgr., North Alabama Coal-to-Methanol Proj., Tennessee Valley Authority, Knoxville, Tenn. 37920)

Journal of Management in Engineering, Vol. 111, No. 3, July, 1985, pp. 157-165

Configuration management came into being in the early 1960's as a disciplined means for reducing cost overruns primarily caused by engineering changes. Consequently, changes are expected and managed in a formal and systematic manner within the project framework (work breakdown structure). Baselines are established as reference points in the progressive development of the engineering project. As reference points, baselines are a vehicle for conveying project requirements in progressively greater detail whereby proposed changes will have a documented basis for uniform evaluation. Therefore, configuration management policy is implemented to identify baseline requirements, control changes to established baselines, and account for the disposition of all changes by maintaining project documents in a current status mode. Design reviews are a prerequisite to the establishment of effective baselines. Configuration control boards provide comprehensive evaluation of proposed changes. In summary, configuration management is a management mechanism for handling changes by integrating technical and administrative decisions with respect to project goals and objectives.

463   Establishing a Professional Practice

James J. Yarmus, Fellow, ASCE, (Pres., Able Building Inspections and J. Yarmus, Inc., Architectural, Engrg., Construction and Building Inspections Consultants, New City, N.Y.)

Journal of Management in Engineering, Vol. 111, No. 3, July, 1985, pp. 166-171

The steps involved in establishing a practice begin with organizational goals. The methodology and concepts of locating clients are evaluated as critical factors in the formative stages. Attracting and retaining clients who can offer repetitive assignments are parts of the foundation for the new firm. Handling "safely" during the formative period is important to the organization's life. As the young practice consolidates into a viable enterprise, the expansion framework is developed in order to maintain growth and avoid stagnation. Advertising must be used with caution, weighing costs against potential benefits, and analyzing impacts on the firm's image. The attitudes of employees and management, the controls used in everyday operations and the direction of personnel outside the office are issues involved in establishing the practice; as is the interrelationship of procedures, plans and services.

464   Specialization as a Strategy for Growth

Joseph E. Heney, Fellow, ASCE, (Chmn. and Chf. Exec. Officer, Camp Dresser & McKee, Inc., One Center Plaza, Boston, Mass. 02108)

Journal of Management in Engineering, Vol. 111, No. 4, October, 1985, pp. 181-187

One response to the issue of specialization versus diversification is to look for many diverse market segments where specialized capabilities give a firm a competitive advantage over less focused, general practice firms. A clearly defined mission, specialized enough to distinguish a firm from other competitors in the marketplace, is essential to adapting successfully to the rapid changes of a highly uncertain and highly competitive marketplace. Several market forces have created this situation: changes in technology, shifting national priorities, aggressive domestic and international competition, and more complex client needs. Each of these areas demands greater up front investments while offering less assurance of uninterrupted long-term returns. A firm must develop a planned reponse to the oportunities and threats created by these external market conditions. Trade-offs between entering markets a firm knows very little about and adapting to the changes occurring in those areas in which it is established must be evaluated. Formal strategic planning directs attention to these trade-offs. This article describes how one international environmental consulting firm has used the strategic planning process to drive continued growth without giving up its specialized focus.

465   Absenteeism in Construction Industry

Jimmie Hinze, Assoc. Member, ASCE, (Assoc. Prof. of Civ. Engrg., Univ. of Washington, Seattle, Wash. 98195), **Maxwell Ugwu**, (Civ. Engrg., Unipetro Nigeria Ltd, Lagos, Nigeria) and **Larry Hubbard**, (Lect. of Civ. Engrg., Univ. of Missouri, Columbia, Mo.)

Journal of Management in Engineering, Vol. 111, No. 4, October, 1985, pp. 188-200

Construction worker absenteeism was studied on several construction projects. The results of one of these representative studies are presented. Results show that absenteeism is noticeably lower in work units that have a strong team spirit or when the group is cohesive. Absenteeism is also lower when management stresses its displeasure of worker absenteeism. Workers who regard their work as being intellectually stimulating and challenging have fewer absences. Absenteeism is adversely affected by an increase in the distance that the workers must travel to get to the workplace. From the findings it is suggested that management can play a vital role in reducing the level of absenteeism experienced in the workplace. It is also imperative that management should carefully monitor the workforce through such measures as turnover and absenteeism rates, which can indicate when conditions in the work environment need further attention.

466   Critical Human Technology Issues for Engineers

Kenneth Lesley, (Field Engr., Univ. of W. Florida, Dept. of Management, Coll. of Business, Pensacola, Fla.) and **Judith F. Vogt**, (Program Dir. for Organizational Studies and Assoc. Prof. of Management and Organizational Behavior, Univ. of West Florida, Pensacola, Fla.)

Journal of Management in Engineering, Vol. 111, No. 4, October, 1985, pp. 201-213

This article identifies the current status of management and team issues including interpersonal relationship skills as viewed by engineers, colleges of engineering, societies of engineering, and employers of engineers. The data was gathered by interviewing members of each sector of the profession and then compiled into general observations. These observations led to a number of specific recommendations. Schools of engineering need to develop within their students an awareness of the business environment by building ties with management departments at the respective schools and with industry. Engineering employers need to continue the engineers' development through the use of mentors and peer groups. Professional societies should provide active forums for the dissemination of management information geared for the practicing engineer, manager, and team member. Finally, a consistent language and applied theory of Human Technology for engineers needs to be devleoped. Engineers should take the responsibility for preparing themselves for their careers and for ongoing personal and professional managerial development. Overall, we see a need for engineers to obtain more managerial, team, and interpersonal skills and currently, no planned/structured way to assuage that need exists.

467   Harmonizing Organizational and Personal Needs

Willy Norup, Member, ASCE, (Pres., Geodex Int'l, Inc., 669 Broadway, Sonoma, Calif. 95476)

Journal of Management in Engineering, Vol. 111, No. 4, October, 1985, pp. 215-226

A system is presented for result-oriented allocation of time, energy, and resources between the three important forces controlling any manager's life: himself, his family and friends, and his organization. A manager is constantly faced with seeking the proper harmony and balance between his own needs and desires, the needs of his family and the objectives of the organization for which he works. Detailed techniques are presented for managing and controlling projects in all areas which cover a manager's life, at work as well as at home. This is a description of a personal management system within which you collect your ideas; set the proper objectives and make step-by-step project plans to meet those objectives; allocate your personal time in suitable blocks; and delegate and control your projects through to completion and evaluation of results. User survey data of managers demonstrate that application of these techniques typically result in 48.8% increase in perceived personal productivity, with comparable reductions in perceived stress level.

468   Liability: Attitudes and Procedures

Robert E. Vansant, Member, ASCE, (Partner and Proj. Mgr., Black & Veatch, Consulting Engrs., P.O. Box 8405, Kansas City, Mo. 64114)

Journal of Management in Engineering, Vol. 111, No. 4, October, 1985, pp. 227-232

The impact of individual and organizational policies concerning liability are reflected in the techniques and procedures followed by an organization. Minimizing liability exposure is the engineering management imperative of the 1980s. Attitudes of engineers and clients must be changed to reduce liability exposure. Attitude changes are reflected in office procedures, contract procedures, including contract documents, and contract administration. Improved office procedures include limitation of practice, peer review, independent review, maintenance of records, disposal of records, and timely responses. Improved construction contract procedures include EJCDC Contract Documents, procedures for changed conditions, delay compenstion, and geotechnical report for construction.

469   Managing Construction of Israeli Air Bases in Negev--A
Personal Perspective

John F. Wall, Jr., (Major General, U.S. Army Corps of Engrs., Dir. of Civ. Works, Office of Chf. of Engrs., Washington, D.C. 20314-1000)

Journal of Management in Engineering, Vol. 111, No. 4, October, 1985, pp. 233-243

In the spring of 1980, the writer became project manager of the Israeli Air Base Program. This called for the U.S. Army Corps of Engineers to build, from scratch, high performance tactical fighter bases at Ramon and Ovda in the Negev Desert to replace two similar bases nearby in the Sinai Peninsula. The project, being carried out in agreement with the Camp David Accord, presented complex and unusual management challenges and brought with it an extraordinary amount of stress and strain. To begin with, Israelis and Americans shared in the management. The U.S. Air Force managed the program, orchestrating and ensuring activation of the two bases. An Israeli program management office also participated. The United States Corps of Engineers was construction agent for the program. Israeli withdrawal from the Sinai depended on the timely completion of the two air bases. The bases themselves were unlike American military airfields, and resembled more land-based aircraft carriers, with underground hardened hangar complexes connected to multiple runways by high-speed taxiways. Construction management was not routine. Separate U.S. design-construct joint ventures carried out the work on each of the two air bases. The contractors, under a cost-plus-a-fixed-fee contract, were responsible for quality control, while the Corps of Engineers was responsible for quality assurance. Despite management problems, the two bases were built well, completed on time, and within budget.

470   Analysis of Two-Way Acting Composite Slabs

Max L. Porter, Member, ASCE, (Prof. of Civ. Engrg., Iowa State Univ., Ames, Iowa)

Journal of Structural Engineering, Vol. 111, No. 1, January, 1985, pp. 1-18

An ultimate strength analysis procedure was formulated for five full-scale, two-way slabs tested to failure. The concrete slabs were reinforced with composite corrugated cold-formed steel decking. Three slabs contained supplementary reinforcing in the form of welded wire fabric. Nominal out-to-out plan dimensions of the slabs were 16 ft by 12 ft (4.88 m by 3.66 m). All slabs were subjected to four symmetrically placed, concentrated loads. Ultimate failure occurred by a shear-bond action initiated by slippage between the steel deck and the concrete. The analysis procedure was founded on the principles of yield-line theory and of shear-bond regression analysis. A collapse mechanism established by yield-line procedures was utilized to determine the effective load-carrying-segment width of the slabs. A shear-bond regression analysis was used on the effective width to predict the total shear force distributed to the reactive edge perpendicular to the deck corrugations. The shear existing along the sides of the effective load-carrying segment was subsequently added to the shear-bond component to give the predicted ultimate load for each slab. The calculated load agreed very closely (within 9%) with the experimental ultimate for all five slabs.

471   Vibration of Tapered Beams

Arvind K. Gupta, Member, ASCE, (Consultant Engr., Quadrex Corp., Campbell, Calif.)

Journal of Structural Engineering, Vol. 111, No. 1, January, 1985, pp. 19-36

Stiffness and consistent mass matrices for linearly tapered beam element of any cross-sectional shape are derived in explicit form. Exact expressions for the required displacement functions are used in the derivation of the matrices. Variation of area and moment of inertia of the cross section along the axis of the element is exactly represented by simple functions involving shape factors. Numerical results of vibration of some tapered beams are obtained using the derived matrices and compared with the analytical solutions and the solutions based upon stepped representation of the beams using uniform beam elements. The significance of the severity of taper within beams upon solution accuracy and convergence characteristics is examined.
Errata: ST Nov. '85, pp. 2616.

472   Wind-Induced Response Analysis of Tension Leg Platforms

A. Kareem, (Assoc. Prof. and Dir., Dept. of Civ. Engrg., Struct. Aerodynamics and Ocean System Modeling Lab., Univ. of Houston, Houston, Tex.)

Journal of Structural Engineering, Vol. 111, No. 1, January, 1985, pp. 37-55

A procedure is presented for estimating the wind-induced response of tension leg platforms (TLP). Spatiotemporal characteristics of the wind velocity field over the ocean are discussed. It is shown that the wind spectra generally used for land-based structures may not adequately represent wind velocity fluctuations at very low frequencies associated with the compliant modes of a TLP. A new spectral description of the longitudinal wind velocity fluctuations over the ocean is proposed. The wind approaching a TLP is treated as a single-point and a multiple-point random field. Expressions for the wind loads in the surge, yaw, and pitch degrees of freedom are formulated in both time and frequency domains for subsequent dynamic analysis. To account for the nonlinear behavior in the frequency domain, the mean response of a TLP is computed using nonlinear stiffness characteristics of the system. For unsteady response due to the wind fluctuations, the TLP is assumed to oscillate linearly above the static equilibrium position produced by the mean wind loading. A numerical example illustrates the methodology outlined for predicting the response statistics of a typical TLP. It is verified that the TLP response estimates based on frequency and time domain analyses show good agreement for the platforms investigated. The frequency domain analysis is recommended for estimating the wind-induced response at the preliminary design stages, to be followed by the time domain analysis in the final stages.

473   Shear Capacity of Stub-Girders: Full Scale Tests

Rick B. Kullman, (Struc. Engr., MacPhedran & Rebb Engrg. Ltd., Saskatoon, Canada) and **M. U. Hosain**, Member, ASCE, (Prof., Dept. of Civ. Engrg., Univ. of Saskatchewan, Saskatoon, Canada, S7N 0W0)

Journal of Structural Engineering, Vol. 111, No. 1, January, 1985, pp. 56-75

This paper briefly summarizes some of the results of tests on three full-size stub-girders. The objectives of the investigation were to study the failure mechanism and to evaluate the effectiveness of three different types of transverse slab reinforcement in increasing the longitudinal shear capacity. Deflection characteristics and general structural behavior were also studied. In all cases, failure involved crushing of the slab over the interior end of the end stub accompanied by a shear/splitting failure in the slab over the length of the end stub. The observed ultimate load on the stub-girders exceeded the design ultimate load by 45%, 14% and 21% respectively for K-1, K-2 and K-3. Straight transverse reinforcing bars in addition to temperature and shrinkage reinforcement, had little effect on the stub-girder capacity. The incorporation of a wide concrete flute along the slab centreline, together with bent transverse bars significantly increased the ultimate capacity and improved the ductility of the stub-girder. The specimens tested were very stiff and would meet normal deflection requirements under service loads.

474   Extreme Winds Simulated from Short-Period Records

Edmond D. H. Cheng, Member, ASCE, (Assoc. Prof., Dept. of Civ. Engrg., Univ. of Hawaii at Manoa, Honolulu, Hawaii 96822) and **Arthur N.L. Chiu**, Fellow, ASCE, (Prof., Dept. of Civ. Engrg., Univ. of Hawaii at Manoa, Honolulu, Hawaii 96822)

Journal of Structural Engineering, Vol. 111, No. 1, January, 1985, pp. 77-94

In order to utilize a limited length of wind records in providing extreme wind speeds for structural design, a stochastic model for generating long-term annual extreme winds in a well-behaved climate, on the basis of short period records, is developed and presented herein. Basically, this method uses historical wind data to establish Markov transition probabilities at an intended project site. These probabilities will be the guide for producing synthesized hourly wind speeds of a desired period, i.e., 100 annual extreme wind speeds would be obtained if hourly wind data for 100 yr were generated. Furthermore, if correlation is observed between the historical annual extreme winds and annual fastest mile wind speeds at a given site, then long-term annual fastest mile wind speeds can be obtained. Applications of this model are demonstrated. The simulated 15 yr, 50 yr and 100 yr wind speeds compared very favorably with those obtained by fitting Type I distribution of the largest values to the 33 historical annual fastest mile wind speeds for the Honolulu International Airport. However, the results from the proposed simulation model are less favorable for generating extreme winds for typhoon-prone regions on the basis of short period historical data.

475   Inelastic Buckling of Steel Plates

J. L. Dawe, Member, ASCE, (Assoc. Prof. of Civ. Engrg., Univ. of New Brunswich, Fredericton, N.B. E3B 5A3) and **G. Y. Grondin**, (Research Asst., Univ. of New Brunswick, Fredericton, N.B., E3B 5A3)

Journal of Structural Engineering, Vol. 111, No. 1, January, 1985, pp. 95-107

An experimental investigation on inelastic buckling of plates simulating simply supported webs and flanges was conducted. The test results are compared with values predicted using the results of standard tensile coupon tests and inelastic orthotropic material properties suggested by various investigators. A set of orthotropic material properties, derived semi-empirically as a result of this study and in good agreement with test results, is suggested. Agreement between predicted values of buckling loads and test results obtained in this investigation and from other researchers validates the use of these material properties for predicting inelastic plate buckling behavior.

476   Limit States Criteria for Masonry Construction

Bruce Ellingwood, Member, ASCE, (Research Struct. Engrg. and Leader, Struct. Engrg. Group, Center for Building Tech., National Bureau of Standards, Washington, D.C. 20234) and **Andrew Talling**, (Grad. Research Asst., Dept. of Civ. Engrg., John Hopkins Univ., Baltimore, Md. 21218)

Journal of Structural Engineering, Vol. 111, No. 1, January, 1985, pp. 108-122

Specifications for masonry and other construction materials are expected to move gradually over the next several years toward the adoption of probability based limit states criteria for design. This paper illustrates how such criteria might be developed for brick and concrete masonry construction using, as an example, masonry walls loaded in combinations of axial compression and out-of-plane flexure. The paper identifies the type of data and analyses that are necessary to develop probability-based resistance criteria.

477   Test Evaluation of Composite Beam Design Method

Allan E. Bessette, (Adams, Hodson, & Bessette, Tacoma, Wash.) and **Robert J. Hoyle, Jr.**, (Prof., Dept. of Civ. and Environmental Engrg., Washington State Univ., Pullman, Wash. 99164)

Journal of Structural Engineering, Vol. 111, No. 1, January, 1985, pp. 123-141

A series of six wood T-beams, with elastomeric adhesive bonded flange-to-web connections, were tested to compare the measured behavior with that anticipated by the design. The parameters measured were deflection at midspan, stresses at the extreme fibers and the flange-web interfaces, and glue line shear stress. The design method overestimated deflection by an average of 6%, underestimated maximum extreme fiber stress by an average of 10%, and overestimated maximum glue line shear stress by about 12%, for T-beams with continuous flanges. T-beams with transverse flange gaps (unconnected butt-joints) of varying frequency were tested to observe the effect on performance. One gap at mdspan reduced the composite action by 77%, two gaps reduced it by 81% and three equally spaced gaps reduced it by 89%. Flange gaps produced extremely high shear stresses in the glue lines in the vicinity of the gaps. The study results for T-beams with continuous flanges followed the design theory quite well. The flange gap effects were also consistent with theoretical expectations.

478   Nonlinear Response to Sustained Load Processes

Karen C. Chou, Assoc. Member, ASCE, (Asst. Prof., Dept. of Civ. Engrg., Syracuse Univ., Syracuse, N.Y. 13210), **Ross B. Corotis**, Member, ASCE, (Hackerman Prof. and Chmn., Dept. of Civ. Engrg., Johns Hopkins Univ., Baltimore, Md. 21218) and **Alan F. Karr**, (Prof., Dept. of Math Sciences, Johns Hopkins Univ., Baltimore, Md. 21218)

Journal of Structural Engineering, Vol. 111, No. 1, January, 1985, pp. 142-157

Reliability analysis of structural members subjected to a stochastic load process is extended to include material nonlinearity. Characteristics of the nonlinear response are computed for a member with a bilinear force-deformation relationship having the unloading range parallel to the initial elastic range. The expected number of loads exceeding a predetermined deformation level is derived for a poisson load process with rectangular pulses. It is found that the derived values agree very well with those obtained from the simulated sustained load processes. The expected damage duration (the total time a process spends above the threshold deformation level) is also derived. For a square wave process, the derivation is theoretically exact once the expected number of exceedances is known.

479   Direct Biaxial Design of Columns

Michael A. Taylor, (Assoc. Prof. of Civ. Engrg., Univ. of California, Davis, Calif. 95616)

Journal of Structural Engineering, Vol. 111, No. 1, January, 1985, pp. 158-173

A direct design method for determining the required column dimensions, concrete strength, and reinforcement arrangement is presented. The user need know only the n triplets $(Pe_xe_y)_1$, $(Pe_xe_y)_2$, ... $(Pe_xe_y)_n$, which the column must resist. A field of acceptable solutions are computed. If the user supplies the in-place costs of steel, concrete, and formwork, then the program computes a relative cost factor for each design to assist the user in selecting an economical solution.

480   Effect of Support Conditions on Plate Strengths

Rangachari Narayanan, (Reader in Civ. and Struct. Engrg., Univ. College, Cardiff, U.K.) and **Fong-Yen Chow**, (Struct. Engr., Promet Consultancy Sdn. Bhd., Jalan Sultan Ismail, Kuala Lumpur, Malaysia)

Journal of Structural Engineering, Vol. 111, No. 1, January, 1985, pp. 175-189

Approximate formulas for assessing the strengths of plates having various edge support conditions and subjected to uniaxial loading are presented; the edges may be free to pull in or constrained to remain straight. Plates having the loaded and unloaded edges subjected to 15 combinations of edge supports (selected from simple supports, encastré and edges free-to-wave) are studied and comparisons made on the effect of the support conditions upon the first-yield strenghts.

481   Silo as a System of Self-Induced Vibration

Jerzy Kmita, (Head, Dept. of Building, Engrg. and Design, Federal Univ. of Tech., P.M.B. 2373, Makurdi, Nigeria)

Journal of Structural Engineering, Vol. 111, No. 1, January, 1985, pp. 190-204

The experimental investigation of dynamic pressure exerted by a granular material (grit) on the walls of a silo is described. Experimental set-up with the pressure measuring technique and recorded investigation results are presented. Using dimensional analysis, a formula for the pressure on the walls of a silo is developed in terms of dimensionless parameters, π_i. The self-induced vibration theory is used to interpret the experimental results. The general scheme of self-induced vibration sytems is examined and a silo with the granular materials is identified as one of the self-induced vibration systems.

482   Empirical Study of Bar Development Behavior

Theodore Zsutty, Member, ASCE, (Prof., Dept. of Civ. Engrg., San Jose State Univ., San Jose, Calif.)

Journal of Structural Engineering, Vol. 111, No. 1, January, 1985, pp. 205-219

A general form of prediction equation is presented for the strength of reinforcing bar development, lapped bar splices, and hooked bar anchorages in reinforced concrete. The form of the prediction equation shows bond strength as a function of the square root of the development length, cover, and transverse steel ratios, as well as the cube root of the concrete strength. Nearly mean value prediction behavior is indicated for all categories of bar development test data. The equation has applications in both research planning and analysis,and in calibration studies for the related understrength factors for strength design. Results of the analysis of bar development data show significant effects of diagonal cracking, multiple bar groups, and moment gradient on development strength.

483   Torsion of Concrete Beams with Large Openings

Velpula Venkappa, (Lect. in Struc. Engrg., M.R. Engrg. College, Jaipur, India) and **Ganpat S. Pandit**, (Sr. Prof. of Struct. Engrg., M.R. Engrg. College, Jaipur, India)

Journal of Structural Engineering, Vol. 111, No. 1, January, 1985, pp. 223-227

The grid frame approach for analyzing reinforced concrete beams with large openings has been developed in this paper. In this approach, the beam with large web openings is visualized as a grid frame comprising two longitudinal stringers connected by posts between consecutive openings. For analyzing the grid frame, the twisting couple at each end of the beam is replaced by a statically equivalent system comprising two equal and opposite forces. The ultimate torsional strength computing using the grid frame approach has been compared with the observed ultimate strength of 25 specimens tested by the writers and other investigators.

484   Monotone Behavior of Trusses Under Two Loadings

Robert Levy, Assoc. Member, ASCE, (Asst. Prof. of Civ. Engrg., Rutgers Univ., New Brunswick, N.J. 08903)

Journal of Structural Engineering, Vol. 111, No. 2, February, 1985, pp. 474-477

This note provides a convergence proof to the iterative procedure of the analysis/redesign type that attains an upper bound (a fully stressed design) on the minimum weight of trusses under two loading conditions. The proof demonstrates decreasing monotone behavior of an objective function that is proportional to the weight at convergence. In addition, it is speculated that both the lower and upper bound can be attained with relative ease through iterations, and may be useful as a "guidance pair" for minimum weight design.

485   **Stiffness Properties of Fixed and Guyed Platforms**

A. Kumar, (Sr. Design Engr., Marine Div., Brown & Root, Inc., P.O. Box 3, Houston, Tex. 77001), V. V. D. Nair, Member, ASCE, (Engrg. Mgr. IV, Marine Div., Brown & Root, Inc., P.O. Box 3, Houston, Tex. 77001) and D. I. Karsan, Member, ASCE, (Dept. Sr. Mgr., Marine Div., Brown & Root, Inc., P.O. Box 3, Houston, Tex. 77001)

Journal of Structural Engineering, Vol. 111, No. 2, February, 1985, pp. 239-256

For deep-water offshore structures, the fundamental period of lateral vibration is a key factor influencing their design feasibility. This period can be controlled by a rational selection of the parameters governing the stiffness of these structures. These parameters and their relative significance in determining the stiffness properties have been identified. A mathematical spring-mass model to portray the stiffness behavior of platforms is introduced. Data are presented to show how stiffness properties can be rationally augmented. The conclusions of the aforementioned parametric studies are verified with reference to a fixed jacket platform in 450 m water depth. The participation of foundation flexibility in influencing stiffness is also investigated. Finally, a concept of insert piles and a strategy for designing deep-water fixed jackets, and guyed towers for stiffness are introduced.

486   **Racking Deformations in Wood Shear Walls**

William J. McCutcheon, Member, ASCE, (Research General Engr., Forest Products Lab., Forest Service, U.S. Dept. of Agr., Madison, Wisc.)

Journal of Structural Engineering, Vol. 111, No. 2, February, 1985, pp. 257-269

The theory presented in this paper predicts racking deformations in wood-stud shear walls. The energy method employed defines the wall performance in terms of the lateral nonlinear load-slip behavior of the nails which fasten the sheathing to the frame. Using power curves to define the nail load-slip relationship, the theory predicts that wall deformation due to nail slip will also be defined by a power curve. The theory also includes linear deformation due to shear distortion of the sheathing material, and provides accurate estimation of wall performance up to moderate load levels. The method presented should be of interest to engineers who design light frame structures, to researchers, and to those who are concerned with building codes.

487   **Shear Modulus of Precracked R/C Panels**

Philip C. Perdikaris, Assoc. Member, ASCE, (Asst. Prof., Civ. Engrg. (Structures), Case Western Reserve Univ., Cleveland, Ohio 44106) and Richard N. White, Fellow, ASCE, (Prof. and Dir., School of Civ. and Environ. Engrg., Cornell Univ., Ithaca, N.Y. 14853)

Journal of Structural Engineering, Vol. 111, No. 2, February, 1985, pp. 270-289

Modeling of the interface shear transfer (aggregate interlock) and dowel action mechanisms as means of transferring shear stresses along preformed cracks is of great importance in determining the stiffness characteristics of reinforced concrete. An expression of an "effective" tangent shear modulus of precracked reinforced concrete panels subjected to combined in-plane shear and biaxial tension is presented based on equilibrium, compatibility, and constitutive relationships for the average stresses and strains in the concrete and steel. The proposed engineering model for shear transfer is compared with experimental results of concrete panels orthogonally reinforced with #6 bars.

488   Drift Snow Loads on Multilevel Roofs

Michael J. O'Rourke, Member, ASCE, (Assoc. Prof. of Civ. Engrg., Rensselaer Polytechnic Inst., Troy, N.Y.), **Robert S. Speck, Jr.**, Member, ASCE, (Design Engr., Ryan and Biggs P.C., Troy, N.Y.) and **Ulrich Stiefel**, (Doctoral Candidate, ETH, Zurich, Switzerland)

Journal of Structural Engineering, Vol. 111, No. 2, February, 1985, pp. 290-306

A database of snowdrift case histories on multilevel flat roofed structures has been established and is statistically analyzed. Drifted snow loads on multilevel roofs account for a large percentage of the roof losses in the U.S. However, little quantitative information is presently available about factors which influence drift formation. The process of drift formation is discussed; and a relationship between drift height and ground snow load, roof lengths, and roof elevation difference, obtained using multiple linear regression is presented. Drift slope and snow density characteristics are also studied. Finally, snowdrift case histories are compared with drift load provisions in building codes and load standards and recommendations for future research are made.

489   Collapse of Square PSC Slabs with Corner Restraint

Simon H. Perry, (Dir. of Concrete Labs., Dept. of Civ. Engrg., Imperial Coll. of Sci. and Tech., London SW7, England) and **Hilary O. Okafor**, (Dar Al-Handasah Consultants, Nigeria)

Journal of Structural Engineering, Vol. 111, No. 2, February, 1985, pp. 307-327

Three simply-supported 1.5m (60in.) square, 60 mm (2.4in.) thick under-reinforced prestressed concrete slabs with corners held down, were tested to failure under centrally applied concentrated loading. Prestressing was applied eccentrically in both directions by straight bonded 7mm (0.24in.) or 5mm (0.20in.) diam. wires. The concrete cube strength (50N/mm² - 7,500 lb/in.²) and diam. of the loading platen (152mm - 6in.) were kept constant while steel ratio was varied. Measurements of concrete strains, steel strains, slab deflections and corner reactions were taken during the tests. Final failure of each slab was by punching either along, or at a distance from, the perimeter of the platen after the development of the yield-line mechanism. It is shown that the collapse load can be predicted reasonably by the load required for fan mechanism development. For a slab with corners held down, but without corner-lever reinforcement, it is shown that the yield load can only be predicted and distinguished from a similar one with corners allowed to lift, by taking into consideration the flexural strength of concrete across the corner lever.

490   Decay, Weathering and Epoxy Repair of Timber

R. Richard Avent, Member, ASCE, (Prof. and Coordinator of Struct. Engrg. and Mechanics, Dept. of Civ. Engrg., Louisiana State Univ., Baton Rouge, La. 70803)

Journal of Structural Engineering, Vol. 111, No. 2, February, 1985, pp. 328-342

Wood used in structural applications can be a durable material. However, conditions in the surrounding environment can cause premature deterioration. Two common and related conditions are decay and weathering. An experimental investigation was conducted to evaluate the effects of weathering and decay on the epoxy repair of timber. In one phase, unprotected epoxy repaired joints were exposed to natural weathering in the southeastern United States. Joints were periodically load tested over a 4½ yr period and correlated to accelerated weathering tests on small epoxy bonded shear block samples. It is recommended that the dry condition shear strength of epoxy repaired Southern pine be reduced by one-third when the repaired member is exposed to natural weathering conditions. Also, normal precautions (including preservative treatments) should be taken if the repair is exposed to weathering over a multi-year period. A second phase consisted of epoxy repairing both lightly and heavily damaged weathered and decayed material. The strength of the repaired joints compared well with that of undamaged material. However, caution is advised in initiating such repairs without careful evaluation.

491   **Wind Load Effects on Flat Plate Solar Collectors**

Robert P. McBean, Member, ASCE, (Proj. Struct. Engr., Power Div., Black & Veatch, Engrs.-Architects, Kansas City, Mo. 64114)

Journal of Structural Engineering, Vol. 111, No. 2, February, 1985, pp. 343-352

Solar collectors comprise the most visible aspect of large photovoltaic central stations, yet structural costs represent only about 15% of total plant costs. Although the design wind speed is the controlling design parameter, the incremental cost of a station with fixed flat plate collectors is increased only about 2% when the design wind speed is increased from 90–110 mph (40–49 m/s). It is equally significant to select a site with stiff soil properties to reduce foundation costs. For one-axis tracking collectors, which typically stow horizontally at 38 mph (17 m/s), the maximum wind speed is unimportant. Use of a simplified wind pressure distribution in place of wind tunnel test data provides a conservative design except for collectors subject to severe turbulence effects near the edge of the field. The aeroelastic behavior of collectors should be investigated in wind tunnels to evaluate the potential for unacceptable vibrations due to vortex shedding or flutter.

492   **Transverse Stirrup Spacing in R/C Beams**

Wayne Hsiung, (Struct. Engr., China Engineering Consultants, Taipei, Taiwan) and **Gregory C. Frantz**, (Assoc. Prof., Dept. of Civil Engrg., Univ. of Connecticut, Storrs, Conn. 06268)

Journal of Structural Engineering, Vol. 111, No. 2, February, 1985, pp. 353-362

Shear tests of 5 approximately one-third scale models of large, reinforced concrete beams with shear reinforcement are examined to determine how varying web widths and different transverse stirrup distributions effect beam shear strength. Load-deflection response, ultimate shear capacities, stirrup stresses, surface inclined crack widths, and internal crack widths are discussed. The results indicate that the shear capacity of a large reinforced concrete beam is not effected by the transverse spacing of stirrups across the web width.

493   **On the Computation of Slab Effective Widths**

Milija N. Pavlovic, (Dept. of Civ. Engrg., Imperial Coll., London, England) and **Steven M. Poulton**, (French Kier Construction Ltd., Sandy, Bedfordshire, England)

Journal of Structural Engineering, Vol. 111, No. 2, February, 1985, pp. 363-377

An approximate closed-form solution is presented for the computation of effective widths of floor slabs in laterally-loaded multi-story buildings. The resulting formula yields answers in good agreement with existing numerical techniques throughout the entire range of slab aspect ratios and column sizes found in practice. Since the necessary computations consist essentially of a summation of two series and thus can readily be performed on a programmable calculator, the proposed method is especially suited for quick estimates of effective width at the preliminary design stage. The proposed method can easily be extended to anisotropic slabs and to boundary conditions other than those usually assumed in analyses of this type.

494   **Collapse of Plate Girders with Inclined Stiffeners**

Giuseppe Guarnieri, (Assoc. Prof., Instituto di Tecnica delle Costruzioni, Politecnico di Torino, Turin, Italy)

Journal of Structural Engineering, Vol. 111, No. 2, February, 1985, pp. 378-399

Extensive results of tests to collapse I-section plate girders with stiffeners at various angles of inclination, along with an analytical interpretation of the results, are presented. A simple model for determining the ultimate shear in the absence of moment is obtained based on the experimentally observed geometry of the tension creases, while the observed shift of the neutral axis toward the tension flange suggests a simple expression for the corresponding ultimate moment in the absence of shear. Finally, at critical sections whose definition is based on the observed collapse mode, an interaction relation between the two is proposed. For each angle of inclination, a simple nondimensional graph is exhibited of the moment-shear interaction at critical sections as a function of stiffener spacing.

495   **Seismic Response of Light Subsystems on Inelastic Structures**

Jon Lin, (Research Asst., Univ. of California, Berkeley, Calif.) and **Stephen A. Mahin**, Member, ASCE, (Assoc. Prof. of Civ. Engrg., Univ. of California, Berkeley, Calif.)

Journal of Structural Engineering, Vol. 111, No. 2, February, 1985, pp. 400-417

Some preliminary analyses are performed to identify the behavioral characteristics of light, nonstructural subsystems supported on systems that yield during severe earthquake ground motions. The effects of the severity of the inelastic deformations of different hysteretic characteristics of the structure and of the amount of viscous damping of the subsystem are considered. To develop possible design guidelines, the study is carried out using the floor response spectrum approach, which is reasonably satisfactory when subsystem-structure interaction can be neglected. The significant observations made from this study are: (1) Inelastic deformations of the structure tend to shift the floor response spectrum down and toward higher periods; and (2) the variations of the ratio of the floor response spectrum on an inelastic structure over the floor response spectrum on an elastic structure—defined as amplification factor—can be characterized by 3 regions with 2 transition regions in between.

496   **Loads Due to Spectator Movements**

Christopher Y. Tuan, Member, ASCE, (Asst. Prof. of Civ. Engrg., Univ. of Nebraska, Omaha, Neb. 68182) and **William E. Saul**, Member, ASCE, (Dean of Engrg., Univ. of Idaho, Moscow, Idaho 83843)

Journal of Structural Engineering, Vol. 111, No. 2, February, 1985, pp. 418-434

Typical spectator movements were simulated by persons individually on a force platform with which dynamic force components were measured. Impulsive loadings are produced by movements such as rising, sitting, or jumping off the floor, while continuous loadings are generated in repetitive motion such as swaying, bending and straightening knees repetitively in place, jumping or dancing. Descriptive parameters derived from statistical and spectral studies of the load samples are presented. In particular, the vertical components of the impulsive loadings are modeled as random forcing functions, and a narrow-band live load spectrum is proposed for checking potential resonant response of a structure to coherent crowd jumping for serviceability and human comfort. Resonant vibration of a structure can be reduced considerably by installation of energy-dissipating devices. A revised design methodology for assembly structures is proposed to insure both structural integrity and functional serviceability.

497   Generalized Finite Element Evaluation Procedure

John O. Dow, (Asst. Prof., Dept. of Civ., Environ., and Arch. Engrg., Univ. of Colorado, Boulder, Colo. 80309), **Thomas H. Ho**, (Grad. Research Asst., Dept. of Civ., Environ., and Arch. Engrg., Univ. of Colorado, Boulder, Colo. 80309) and **Harold D. Cabiness**, (Grad. Research Asst., Dept. of Civ., Environ., and Arch. Engrg., Univ. of Colorado, Boulder, Colo. 80309)

Journal of Structural Engineering, Vol. 111, No. 2, February, 1985, pp. 435-452

A rational procedure for evaluating the performance of two and three dimensional finite elements is developed. This procedure compares the strain energy content and the strain distribution of the finite element model to that of the continuum region it represents for well-defined strain states. The use of Taylor series expansions which relate the displacements, rotations, and strains through strain gradient terms allows the strain energies and the pointwise strains of the finite element model and the continuum to be directly compared. The evaluation procedure is demonstrated by applying it to a six-node isoparametric element undergoing a series of progressive initial distortions. From these results, a simple algorithm to predict the maximum strain energy error as a function of initial geometry is developed for incorporation into finite element codes at the grid generation level.

498   Warping Moment Distribution

Stefan J. Medwadowski, Fellow, ASCE, (Consulting Structural Engr., 111 New Montgomery, San Francisco, Calif. 94105)

Journal of Structural Engineering, Vol. 111, No. 2, February, 1985, pp. 453-466

An iterative procedure is presented for the calculation of continuous structures subjected to torsional loads and experiencing both the St. Venant and warping torsion. The procedure is entirely analogous to the well-known moment distribution method. Structures to which it applies include crane girders, box-girder bridges, and vertical shear elements of high-rise buildings with prismatic, thin-walled, open sections. The formulae for the necessary constants including warping stiffness, warping carry-over factors and fixed-end warping moments are given, and the procedure is illustrated with the aid of an example of a continuous crane girder which demonstrates its efficiency. The equations of mixed torsion are summarized.

499   Stiffness Matrix for Elastic-Softening Beams

Peter LeP. Darvall, (Sr. Lect., Dept. of Civ. Engrg., Monash Univ., Clayton, Victoria, Australia)

Journal of Structural Engineering, Vol. 111, No. 2, February, 1985, pp. 469-473

Softening, the decrease of bending moment capacity of a section at advanced curvature, may occur at locations within a statically indeterminate reinforced concrete structure prior to the attainment of maximum load. The stiffness matrix for an elastic flexural member with a softening portion is derived, employing the necessary assumption of a finite softening or discontinuity length. For positive values of the softening parameter the stiffness values are as for a beam with a prismatic haunch. For negative values of the parameter stiffness coefficients become unstable at the critical softening parameter for a fixed-end beam. The derived stiffness coefficients allow use of standard frame analysis programs without special treatment for softening. An example of softening at the base of one column of a portal frame is given, showing the effect of softening on sway displacement. The frame demonstrates instability at the critical softening parameter for the particular position and length of the softening hinge.

500   Soil Security Test for Water Retaining Structures

Roman Tadanier, (Supervising Geotechnical Engr., Public Works Dept., Geomechanics Lab., St. Peters, Australia) and **Owen G. Ingles**, (Assoc. Prof., School of Civ. Engrg., Univ. of New South Wales, Australia)

Journal of Geotechnical Engineering, Vol. 111, No. 3, March, 1985, pp. 289-301

A simple test applicable to the evaluation of soils for water-retaining soil structures is described. It combines into one test an appraisal for dispersivity, cracking, softening or slaking in the presence of water, measuring at the same time the influence on the soil of divergence from optimum compaction conditions. The latter aspect has particular value for the selection of compaction limits in construction specifications. The test proposed requires only modest quantities of soil, and is thus particularly suitable with variable soils, for which numerous tests are desirable.

501   Extensional Stiffness of Precracked R/C Panels

Philip C. Perdikaris, Assoc. Member, ASCE, (Asst. Prof., Dept. of Civ. Engrg. (Structures), Case Western Reserve Univ., Cleveland, Ohio 44106), **Said Hilmy**, (Grad. Student, Dept. of Civ. Engrg., Cornell Univ., Ithaca, N.Y. 14853) and **Richard N. White**, (Dir., School of Civ. and Environmental Engrg., Cornell Univ., Ithaca, N.Y. 14853)

Journal of Structural Engineering, Vol. 111, No. 3, March, 1985, pp. 487-504

Very little experimental data is available regarding extensional stiffness in uniaxially or biaxially tensioned reinforced concrete elements. In order to improve our understanding of this important subject, an experimental study was performed to determine an estimate for the effective extensional stiffness, K_N, of precracked concrete panels orthogonally reinforced with No. 4 or No. 6 grade 60 reinforcing bars and subjected to uniaxial or biaxial tension. These panels are meant to be representative of a segment of a cracked containment wall subjected to internal pressurization. To obtain a relationship for the extensional stiffness, the total surface axial deformation was measured in the direction of applied tension. The tension stiffening effect of concrete was accounted for by estimating an average effective steel strain corresponding to an equivalent bond-free reinforcing bar.

502   Friction Coefficient of Steel on Concrete or Grout

B. G. Rabbat, Member, ASCE, (Prin. Engr., Bridge Structures, Struct. Experimental Section, Construction Tech. Labs., Portland Cement Assoc., Skokie, Ill.) and **H. G. Russell**, (Dir., Struct. Development Dept., Construction Tech. Labs., Portland Cement Assoc., Skokie, Ill.)

Journal of Structural Engineering, Vol. 111, No. 3, March, 1985, pp. 505-515

An experimental investigation was conducted to determine the coefficient of static friction between rolled steel plate and cast-in-place concrete or grout. Fifteen tests were performed under conditions that represented the interior and exterior bearing surfaces of a containment vessel. Test parameters included concrete blocks or grout blocks, wet or dry interface, and level of normal compressive stress. For conditions tested, the average effective coefficient of static friction varied between 0.57 and 0.70. It is recommended that the coefficient of static friction for concrete cast on steel plate and grout cast below steel plate should be taken as 0.65 for a wet interface with normal compressive stress levels between 20 and 100 psi (0.14 and 0.69 MPa). For dry interface, the coefficient of static friction should be taken as 0.57.

503   **Fiber Reinforced Beams Under Moment and Shear**

Sanat K. Niyogi, (Prof., Dept. of Civ. Engrg., Indian Inst. of Tech., Kharagpur, West Bengal, India) and **G. I. Dwarakanathan**, (Staff Officer (Designs), Central Zone, India)

Journal of Structural Engineering, Vol. 111, No. 3, March, 1985, pp. 516-527

The science and technique of reinforcing Portland cement concrete by incorporating fibers is now well established through research and practice. However, studies on steel fiber reinforced concrete elements have been devoted primarily to observations under unit actions. The present report gives the results of thirty plain and fiber reinforced concrete beams, the majority of which are tested under combined actions of moment and shear. The principal variables are the concrete mix proportion, fiber volume fraction and shear span.

504   **Z-Section Girts Under Negative Loading**

Dimos Polyzois, Assoc. Member, ASCE, (Asst. Prof., Dept. of Civ. Engrg., Univ. of Texas at Austin, Austin, Tex.) and **Peter C. Birkemoe**, Member, ASCE, (Prof., Dept. of Civ. Engrg., Univ. of Toronto, Toronto, Ontario, Canada)

Journal of Structural Engineering, Vol. 111, No. 3, March, 1985, pp. 528-544

As part of a wall system in low-rise industrial buildings, cold-formed Z-section girts are attached along one side to cold-formed steel panels and are often supported along their span by sag rods. Thus, some degree of restraint is present against the lateral and rotational displacement of these girts. The use of sag rods as a bracing device is often rejected in design on the assumption that the wall panels, once erected, provide adequate lateral restraint to the girts. However, if the wall system is under negative pressure, the Z-sections are attached to the wall panels along the tension flange. In this case, the bracing contribution of the wall panels is also neglected in the design calculations. Results from a theoretical and experimental program showed that the omission of the restraint contribution of sag rods and steel panels leads to overconservative designs.

505   **Reservoir Bottom Absorption Effects in Earthquake Response of Concrete Gravity Dams**

Gregory Fenves, Student Member, ASCE, (Grad. Student, Dept. of Civ. Engrg., Univ. of California, Berkeley, Calif.) and **Anil K. Chopra**, Member, ASCE, (Prof. of Civ. Engrg., Univ. of California, Berkeley, Calif.)

Journal of Structural Engineering, Vol. 111, No. 3, March, 1985, pp. 545-562

Utilizing a recently developed analytical procedure, linear responses of the tallest non-overflow monolith of Pine Flat Dam to Taft ground motion are presented for a range of properties for the reservoir bottom materials and various assumptions for the impounded water and foundation rock. In these analyses, the alluvium and sediments at the reservoir bottom, upstream of the dam, are modelled by a reservoir bottom that partially absorbs incident hydrodynamic pressure waves. Based on these response results, it is demonstrated that the earthquake response of dams is increased by dam-water interaction and decreased by reservoir bottom absorption with the magnitude of these effects depending on the flexibility of the foundation rock and on the component of ground motion. It is also shown that the significance of the response of dams to vertical ground motion was overestimated in earlier studies assuming a rigid reservoir bottom.

506   Constitutive Model for Loading of Concrete

Wimal Suaris, Assoc. Member, ASCE, (Asst. Prof., Dept. of Civ. Engrg., Univ. of Miami, Coral Gables, Fla. 33124) and **Surendra P. Shah**, Member, ASCE, (Prof., Dept. of Civ. Engrg., Northwestern Univ., Evanston, Ill. 60201)

Journal of Structural Engineering, Vol. 111, No. 3, March, 1985, pp. 563-576

Constitutive properties of concrete under dynamic loading are necessary for the rational analysis of concrete structures subject to impact and impulsive loads. The constitutive model presented herein models microcracking through the use of a continuous damage parameter for which a vectorial representation is adopted. The rate of increase of the damage is dependent on the state of strain as well as on the time rate of strain. The constitutive equations are derived from the strain energy function which is influenced by the accumulated damage. The constitutive model is calibrated using uniaxial tension (or flexural) and uniaxial compression test data. The calibrated model is then used to predict certain other load responses of concrete.

507   Composite Box Girder Bridge Behavior During Construction

Fernando A. Branco, (Research Asst., Dept. of Civ. Engrg., Technical University of Lisbon (IST), Lisbon, Portugal) and **Roger Green**, (Prof., Dept. of Civ. Engrg., Univ. of Waterloo, Waterloo, Ontario, Canada)

Journal of Structural Engineering, Vol. 111, No. 3, March, 1985, pp. 577-593

Steel-concrete composite box girder bridges may have a flexible open box section prior to the placement of the slab. Excessive twist or distortion can arise under construction loading. Bracing systems are usually installed within the girder during construction to increase the torsional and distortional stiffness of the open section. The influence of these bracing systems on open box behavior is discussed using finite strip analysis results and a one-quarter scale girder model test data. Tie and distortional bracing were found effective in preventing distortion, with web stiffening found to be more effective than interior cross bracing. Horizontal bracing at the flange level was considered to reduce twisting of the section. This bracing can take the form of either torsion boxes or top chord bracing, both being effective. A torsion-bending analysis based on a rigid section behavior is discussed.

508   Conjugate Frame for Shear and Flexure

A. Abdul-Shafi, Member, ASCE, (Assoc. Prof. of Civ. Engrg., South Dakota State Univ., Brookings, S.D.)

Journal of Structural Engineering, Vol. 111, No. 3, March, 1985, pp. 595-608

A method is described for extending the conjugate beam concept to the solution of the combined effect of shear and flexure in frames and nonprismatic elements. A consistent sign convention is adopted. The method distinguishes between, and evaluates, the slope and cross-sectional plane rotation. These two different rotations usually occur under the influence of shear. In structural analysis these rotations have different considerations: the rotational stiffness and rotational flexibility, of an element, are associated with the cross-sectional plane rotation; but deflection of the element is related to the slope of the deflection curve. The method provides an efficient tool for frame analysis, especially when evaluation of displacements is required at several locations in the structure.

509   **Sparsely Connected Built-Up Columns**

Charles Libove, Member, ASCE, (Prof., Mech. and Aerospace Engrg. Dept., Syracuse Univ., Syracuse, N.Y. 13210)

Journal of Structural Engineering, Vol. 111, No. 3, March, 1985, pp. 609-627

An elastic analysis is presented for the initial buckling and the post-buckling behavior of simply supported built-up columns consisting of two slightly separated identical parallel elements that are joined together at their ends and middle by rigid connections of negligible axial extent. Numerical results for initial buckling indicate that a familiar design rule (to make the slenderness ratio of the individual elements between fillers no greater than that of the column as a whole) will not necessarily lead to a buckling strength approximating that of the integral column. The post-buckling numerical results in the form of load-shortening curves, show that the post-buckling behavior is unstable, i.e., the load required to maintain any buckled configuration is less than that required to initiate the buckling. This post-buckling instability, which is in marked contrast to the virtually neutral stability of the solid elastic column, has important implications regarding imperfection sensitivity. Embedded in the general analysis is that of the 2-connection column obtained by omitting the middle connection.

510   **Load History Effects on Structural Members**

Nevis E. Cook, Jr., (Consulting Engr., Boulder, Colo.) and **Kurt H. Gerstle**, (Prof., Dept. of Civ., Environmental and Architectural Engrg., Univ. of Colorado, Boulder, Colo.)

Journal of Structural Engineering, Vol. 111, No. 3, March, 1985, pp. 628-640

Strength determination in usual structural engineering practice is based on failure under monotonically increasing, proportional loadings. The effects of nonproportional load sequences are largely ignored. Specifically, information on column behavior in steel building frames is generally obtained from analyses under proportionally increasing axial loads and bending moment, although the former, largely due to gravity loads, are likely to precede the latter, which are primarily caused by lateral loads. To explore these effects, the investigation in this paper proceeds in the following steps. Based on measurements of strains during uniaxial cyclic tests of mild steel specimens, a cyclic stress-strain law is formulated, and verified against text results. Secondly, this stress-strain law is incorporated into an analysis to predict the response of structural steel sections under arbitrary applied axial load and moment sequences. The results of this analysis are again verified by comparison with test data. Finally, the section analysis is used to explore the effect of different axial load-moment histories on the response of column sections, and it is concluded that column deformations are insensitive to sequence of load application.

511   **Impact Studies on Small Composite Girder Bridge**

Colin O'Connor, Member, ASCE, (Prof. of Civ. Engrg., Univ. of Queensland, St. Lucia, Queensland, Australia) and **Ross W. Pritchard**, (Civ. Engr., Main Roads Dept., Queensland, Australia)

Journal of Structural Engineering, Vol. 111, No. 3, March, 1985, pp. 641-653

Two impact studies on a small span, composite girder highway bridge have given widely scattered impact fractions, with a maximum of 1.32, compared with the AASHTO code value of 0.30 for this bridge. Field strains were used to measure maximum midspan bending moments for 170 trucks in normal traffic, and these were compared with equivalent static values computed from axle weights measured by a weighbridge. Impact fractions varied from -0.08 to + 1.32 for gross vehicle weights from 27–44 ton. Large impact values occurred for both light and heavy vehicles and were repeated in two independent series of tests. The scatter of the results suggests that impact is vehicle dependent, and that it may vary with suspension geometry. Possible causes of high impact are examined.

512 Structural Behavior of Wood Shear Wall Assemblies

Erik L. Nelson, (Grad. Student, Dept. of Civ. Engrg., University of Texas, Austin, Tex. 78712), **Dan L. Wheat**, (Asst. Prof. of Civ. Engrg., Univ. of Texas, Austin, Tex. 78712) and **David W. Fowler**, (Prof. of Civ. Engrg., Univ. of Texas, Austin, Tex. 78712)

Journal of Structural Engineering, Vol. 111, No. 3, March, 1985, pp. 654-666

The paper is a summary of the results of an experimental program conducted at the University of Texas to investigate the structural behavior of 7 shear wall assemblies used in manufactured housing. Variables studied include the size and location of the shear wall within the assembly, the number of glued sides of hardwood paneling, and the number of floor joists beneath the shear wall. It was found that walls located on the windward side of the assembly had higher ultimate strengths. In addition, measurements of slip between the shear wall and side walls indicate that there may be sinificant load transfer into the side wall. Also, the side walls, it was found, do not provide rigid vertical support to the joists beneath the shear wall, as currently assumed in design.

513 Tests of Full-Size Gusset Plate Connections

Reidar Bjorhovde, Fellow, ASCE, (Prof. of Civ. Engrg. & Engrg. Mechanics, Univ. of Arizona, Tucson, Ariz. 85721) and **S. K. Chakrabarti**, (Grad. Student, Univ. of Arizona, Tucson, Ariz. 85721)

Journal of Structural Engineering, Vol. 111, No. 3, March, 1985, pp. 667-684

The paper presents an evaluation of the behavior and strength of gusset plates on the basis of an experimental investigation of full-scale diagonal bracing connections. Such connections commonly occur in heavy industrial structures, for sample. The evaluations are based on analyses of load and deformation data that were generated during the tests, including the failure patterns for the plates. A total of six tests were run, using three bracing member orientation angles. The results are correlated with analytical studies, with special emphasis on recent finite element work. The results are found to be in reasonable agreement. Current design practices are discussed briefly; some of the methods are shown to be in acceptable agreement with the tests. It is shown that plate boundaries, plate buckling, and related out-of-plane bending phenomena have significant effects on the plate behavior. Recommendations for additional work are also made.

514 Dynamic Lateral-Load Tests of R/C Column-Slabs

Denby G. Morrison, (Sr. Lect. and Assoc., Dept. of Civ. Engrg., Univ. of Stellenbosch, Stellenbosch, 7600 Republic of South Africa)

Journal of Structural Engineering, Vol. 111, No. 3, March, 1985, pp. 685-698

The response of interior reinforced concrete plate-column connections in a laterally loaded structure is investigated. Nine specimens were tested, five statically and four dynamically. This paper describes results from dynamically tested specimens. Experimental variables included slab reinforcement ratio. Nominal slab dimensions were 6 × 6 ft (1.8 × 1.8 m). Slab thickness was 6 in. (76 mm). The column was square with plan dimensions of 1 × 1 ft (0.3 × 0.3 m). Reinforcement layout in the slab was isotropic. Dynamically tested specimens provided data on the strength, stiffness and energy-dissipation characteristics. For joint rotations of approximately 0.02 (related to earthquake design) the stiffness of the slabs corresponded to that calculated from a fully cracked section for one fifth of the slab width, and the associated equivalent viscous damping was approximately 5% of critical.

515   Hysteretic Dampers in Base Isolation: Random Approach

Michalakis C. Constantinou, Assoc. Member, ASCE, (Asst. Prof., Dept. of Civ. Engrg., Drexel Univ., 32nd and Chestnut St., Philadelphia, Pa. 19083) and **Iradj G. Tadjbakhsh**, Member, ASCE, (Prof., Civ. Engrg. Dept., Rensselaer Polytechnic Inst., Troy, N.Y. 12181)

Journal of Structural Engineering, Vol. 111, No. 4, April, 1985, pp. 705-721

A method of random vibration analysis of base isolated structures with hysteretic dampers is employed. The hysteretic restoring force is modelled by a nonlinear differential equation. The equations of motion for shear type structures are linearized in closed form. Nonstationary response statistics for evolutionary nonwhite excitation are determined by solving the associated Lyapunov matrix differential equation. An optimization study which is based on the stationary response is also presented.

516   Mechanistic Seismic Damage Model for Reinforced Concrete

Young-Ji Park, (Grad. Research Asst., Dept. of Civ. Engrg., Univ. of Illinois at Urbana-Champaign, Urbana, Ill. 61801) and **Alfred H. -S. Ang**, Fellow, ASCE, (Prof., Dept. of Civ. Engrg., Univ. of Illinois at Urbana-Champaign, Urbana, Ill. 61801)

Journal of Structural Engineering, Vol. 111, No. 4, April, 1985, pp. 722-739

A model for evaluating structural damage in reinforced concrete structures under earthquake ground motions is proposed. Damage is expressed as a linear function of the maximum deformation and the effect of repeated cyclic loading. Available static (monotonic) and dynamic (cyclic) test data were analyzed to evaluate the statistics of the appropriate parameters of the proposed damage model. The uncertainty in the ultimate structural capacity was also examined.

517   Seismic Damage Analysis of Reinforced Concrete Buildings

Young- Ji Park, (Grad. Research Asst., Dept. of Civ. Engrg., Univ. of Illinois—Urbana-Champaign, Urbana, Ill. 61801), **Alfredo H. -S. Ang**, Fellow, ASCE, (Prof., Dept. of Civ. Engrg., Univ. of Illinois—Urbana-Champaign, Urbana, Ill. 61801) and **Y. K. Wen**, Member, ASCE, (Prof., Dept. of Civ. Engrg., Univ. of Illinois—Urbana-Champaign, Urbana, Ill. 61801)

Journal of Structural Engineering, Vol. 111, No. 4, April, 1985, pp. 740-757

A method for evaluating structural damage of reinforced concrete buildings under random earthquake excitations is proposed. Extensive damage analysis of SDF systems and typical MDF reinforced concrete buildings were performed. On the basis of these results, a simple relationship between the destructiveness of the ground motions, expressed in terms of the "characteristic intensity," and the stuctural damage, expressed in terms of the "damage index," is established. Reinforced concrete buildings that were damaged during past earthquakes were used to calibrate the proposed damage measure; on this basis, practical limits of structural damage are defined.

518   Imperfect Columns with Biaxial Partial Restraints

Zia Razzaq, Member, ASCE, (Assoc. Prof., Dept. of Civ. Engrg., Old Dominion Univ., Norfolk, Va. 23508) and **Antoun Y. Calash**, (Doctoral Candidate, Dept. of Civ. Engrg., Univ. of Notre Dame, Notre Dame, Ind. 46556)

Journal of Structural Engineering, Vol. 111, No. 4, April, 1985, pp. 758-776

An inelastic theoretical study of the effect of biaxial partial end restraints on the response of hollow rectangular steel nonsway columns with or without biaxial crookedness and residual stresses is presented. The partial end restraints considered have linear, elastic-plastic, or trilinear moment-rotation characteristics. The fundamental total equilibrium equations for the problem are derived. An algorithm is presented based on a coupling between an iterative tangent stiffness approach and a finite-difference scheme for evaluating the spatial response of the materially nonlinear columns. The effect of partial end restraints, as well as the individual and combined influence of crookedness and residual stresses is also explained by means of load-deflection, and bending stiffness degradation curves. Several interesting conclusions are drawn regarding the behavior of columns as affected by end restraints, imperfections and slenderness.

519   Design of Slender Webs Having Rectangular Holes

Rangachari Narayanan, (Reader, Civ. and Struct. Engrg., University Coll., Cardiff, Wales, U.K.) and **Norire Gara-Verni Der-Avanessian**, (Post-Doctoral Research Fellow, Dept. of Engrg., Univ. of Lancaster, Lancaster, England)

Journal of Structural Engineering, Vol. 111, No. 4, April, 1985, pp. 777-787

The ultimate shear capacity of a plate girder having a central rectangular web-hole is shown to be the sum of three contributing factors: (1) The elastic critical load of the web; (2) the load carried by the membrane tension in the web in the post critical stage; and (3) the load carried by the flanges. The elastic critical load in shear is evaluated by using an approximate formula derived from a finite element analysis. The load carried by membrane tension is calculated using the Von Mises criterion and a suggested mean value for the angle subtended by it with the longitudinal axis of the girder; the contribution of the flanges is calculated from their plastic moment capacities. The proposed method is shown to be satisfactory by comparing the predictions obtained from it with the corresponding test observations. An ultimate load method of designing flat strip reinforcement is proposed and is aimed at restoring the strength of the web to a value that would have been obtained had the hole not been cut. The suggested design procedure is shown by a worked example.

520   Length-Thermal Stress Relations for Composite Bridges

Jack H. Emanuel, Fellow, ASCE, (Prof. of Civ. Engrg., Univ. of Missouri-Rolla, Rolla, Mo.) and **Charles M. Taylor**, Assoc. Member, ASCE, (Engr.-Design, McDonnell Douglas Astronautics Co., St. Louis, Mo.)

Journal of Structural Engineering, Vol. 111, No. 4, April, 1985, pp. 788-804

Computer-assisted analysis was used to study the relation among uniform, linear, and nonlinear stress components thermally induced in a composite bridge section for hypothetical parameters of varying span lengths, number of spans, and support conditions, as well as for actual bridges. The results were verified by conventional methods of analysis. The following was concluded for prismatic (constant) sections (1) For constant proportionality of span lengths, each of the three thermal stress components is independent of span length; (2) variation of the proportionality of span lengths affects only the linear stress component; (3) support reactions and deflections caused by thermal loading are length dependent, but the induced moments and stresses are independent of length; (4) as the number of spans increases, the (thermally induced) moment magnitudes tend to converge; (5) the magnitude of reactions, for constant proportionality of span lengths, varies inversely with span length; and (6) for total end fixity, no exterior or interior vertical support reactions are thermally induced.

521   Dynamic Response of Tall Building to Wind Excitation

Morteza A. M. Torkamani, Member, ASCE, (Asst. Prof., Dept. of Civ. Engrg., Univ. of Pittsburgh, Pittsburgh, Pa. 15261) and **Eddy Pramone**, (Former Grad. Student, Dept. of Civ. Engrg., Univ. of Pittsburgh, Pittsburgh, Pa. 15261)

Journal of Structural Engineering, Vol. 111, No. 4, April, 1985, pp. 805-825

The dynamic responses of tall buildings subject to wind loading are investigated. One of the objectives of this research is to study the importance of the torsional dynamic response, coupled with translational responses. Finite element modelling is used to assemble the stiffness matrix of the structure. Torsional degrees of freedom are considered in the stiffness formulation of elements and systems. Aerodynamic forces on a tall building are calculated assuming a deterministic, pseudo-turbulent approach. These aerodynamic forces are distributed over the height of the building. The equivalent concentrated aerodynamic loads, acting at each floor level, are calculated using the principle of virtual displacements. The governing differential equations are nonlinear. An iterative method of solution is used to calculate the responses. In order to simplify the solution procedure, a method of linearization is applied to the aerodynamic forces and the final result is a set of second order differential equations with constant coefficients. A 15-story building is modelled as an application. One comparative study has been made between the finite element model and an equivalent continuous cantilever beam model. A second comparative study is between nonlinear and linear models. The results are presented as response spectra for different gust frequencies.

522   Wind-Induced Fatigue on Low Metal Buildings

Brian A. Lynn, (Research Asst., Centre for Building Studies, Concordia Univ., Montreal, Province of Quebec, Canada H3G 1M8) and **Theodore Stathopoulos**, Assoc. Member, ASCE, (Assoc. Prof., Centre for Building Studies, Concordia Univ., Montreal, Province of Quebec, Canada H3G 1M8)

Journal of Structural Engineering, Vol. 111, No. 4, April, 1985, pp. 826-829

Presently, fatigue is not considered a critical design factor for low metal buildings exposed to severe wind storms. Fatigue however, has been shown to be the only possible cause of several roof failures which occurred during cyclones. A simple approach for the evaluation of wind-induced fatigue on low buildings is presented. The number and distribution of maxima occurring in wind pressures acting on low buildings are predicted analytically by a hybrid Gaussian-Weibull extremum model. Experimental measurements in a boundary layer wind tunnel also support these analytical predictions. Wind-induced fatigue data obtained for a case study by applying the present approach agree fairly well with full-scale observations carried out after the cyclone Tracy hit Darwin, Australia, in 1974. It is concluded that a repeated loading test criterion should be introduced and design against wind-induced fatigue should be carried out for low buildings in hurricane-prone regions.

523   Bolted Connections in Round Bar Steel Structures

Henning Agerskov, (Assoc. Prof. of Civ. Engrg., Technical Univ. of Denmark, Lyngby, Denmark) and **Jorgen Bjornbak-Hansen**, (Research Civ. Engr., Tech. Univ. of Denmark, Lyngby, Denmark)

Journal of Structural Engineering, Vol. 111, No. 4, April, 1985, pp. 840-856

High-strength bolted connections in round bar steel structures in which the bolts are subjected to direct tension are studied. Special attention is paid to theprying action in the connection. A theory is developed that permits analysis of bolted end plate connections, including determination of the prying forces. The theory includes both the case in which separation in the bolt line occurs before, and the case in which it occurs after the reduced yield moment of the end plate is reached. A comparison of the results obtained from a test series carried out at the Technical University of Denmark with those obtained by means of the suggested design method shows little difference, the maximum deviation in bolt forces being about 8%. The tests show that, due to strain hardening in the end plate, a significant strength reserve exists beyond the yield load of the connection.

524   Mechanics of Masonry in Compression

W. Scott McNary, (Struct. Engrg., Figg and Muller Engineers, Inc., Denver, Colo. 80237) and **Daniel P. Abrams**, Member, ASCE, (Asst. Prof., Civ. and Architectural Engrg. Dept., Univ. of Colorado, Boulder, Colo. 80309)

Journal of Structural Engineering, Vol. 111, No. 4, April, 1985, pp. 857-870

Strength and deformation of clay-unit masonry under uniaxial concentric compressive force were investigated. Biaxial tension-compression tests of bricks and triaxial compression tests of mortar were done to establish constitutive relations for each material. Mortar strengths and brick types were varied. Interaction effects of these two materials were examined using a theory proposed by others. A numerical model based on this theory was used to compute the force-deformation relationship for a stack-bond prism. Results of the analysis were compared with measured strengths and deformations of test prisms. Results of the study indicated that mechanics of clay-unit masonry in compression could be well represented with a relatively simple model, and the most significant parameter to consider was the dilatant behavior of the mortar.

525   Elastic-Plastic-Softening Analysis of Plane Frames

Peter LeP. Darvall, (Reader, Dept. of Civ. Engrg., Monash Univ., Victoria, Australia) and **Priyantha A. Mendis**, (Grad. Student, Dept. of Civ. Engrg., Monash Univ., Victoria, Australia)

Journal of Structural Engineering, Vol. 111, No. 4, April, 1985, pp. 871-888

Softening of reinforced concrete sections at advanced curvatures in flexure is taken into account in collapse load analysis of frames. The analysis employs a trilinear elastic-plastic-softening approximation to real moment-curvature curves, and elastic unloading and reloading from the softening curve is assumed. Formulas for hinge length in reinforced concrete flexural members are reviewed. Elements of a stiffness matrix for elastic elements with softening portions are derived. The stiffness matrix allows extension of the capability of an existing computer program for elastic-plastic analysis to softening or hardening. Several examples demonstrate the application of computer program PAWS (Plastic Analysis With Softening) to concrete frame structures. A steeper softening slope reduces both the number of hinges formed before collapse and the collapse load, compared with elastic-plastic behavior. Values of critical softening parameters for first or later formed hinges may be found using PAWS, and theoretical values are confirmed. Quite small softening parameters lead to large reductions in static collapse and shakedown loads, especially when there are significant residual moments.

526 Axial Strength of Grouted Pile-to-Sleeve Connections

Nat W. Krahl, Member, ASCE, (Mgr., Struct. Design and Naval Architecture Dept., Brown and Root, Inc., Houston, Tex.) and **Demir I. Karsan**, Member, ASCE, (Sr. Mgr., Fixed Offshore Structures Dept., Brown and Root, Inc., Houston, Tex.)

Journal of Structural Engineering, Vol. 111, No. 4, April, 1985, pp. 889-905

A new equation for the calculation of axial load transfer in grouted pile-to-sleeve connections, either with or without shear keys, is proposed. The general form of the equation was obtained from an ultimate strength formulation for a predicted failure made of the connection. The ultimate strength equation states that the axial strength of the connection is equal to the sum of two terms: First, the strength of the adhesion and friction between the pile and grout, and, second, the product of the confined strength of the grout and the ratio of the shear key's height to its spacing. The first term was estimated from test results. The second term was estimated from both theoretical considerations and test results, with good agreement between the two approaches. The design equation proposed was obtained by applying separate factors of safety to the two terms. Comparison of the design equation to test data gave a safety index of 3.96 for connections without shear keys, and 4.80 for connections with shear keys. These values imply efficient and reliable design when using this design equation. The design equation has been written into the latest revision to the code of recommended practice.

527 Simplified Earthquake Analysis of Structures with Foundation Uplift

Anil K. Chopra, Member, ASCE, (Prof., Dept. of Civ. Engrg., Univ. of California, Berkeley, Calif.) and **Solomon C.-S. Yim**, Assoc. Member, ASCE, (Asst. Research Engr., Dept. of Civ. Engrg., Univ. of California, Berkeley, Calif.)

Journal of Structural Engineering, Vol. 111, No. 4, April, 1985, pp. 906-930

Simplified analysis procedures are developed to consider the beneficial effects of foundation-mat uplift in computing the earthquake response of structures, which respond essentially as single-degree-of-freedom systems in their fixed-base condition. These analysis procedures are presented for structures attached to a rigid foundation mat, which is supported on rigid foundation soil or flexible foundation soil modeled as two spring-damper elements, Winkler foundation with distributed spring-damper elements, or a viscoelastic half-space. In these analysis procedures, the maximum earthquake-induced base shear and deformation for an uplifting structure are computed directly from the earthquake response spectrum. It is demonstrated that the simplified analysis procedures provide results for the maximum base shear and deformation to a useful degree of accuracy for practical structural design.
Errata: ST July '85, pp. 1632-1633.

528 Composite Beam Design Using Interaction Diagram

Shan Somayaji, Member, ASCE, (Assoc. Prof., Dept. of Civ. and Environmental Engrg., California Polytechnic State Univ., San Luis Obispo, Calif. 93407)

Journal of Structural Engineering, Vol. 111, No. 4, April, 1985, pp. 933-938

The interaction diagram approach for design of prestressed concrete members is extended for design of composite prestressed beams with one or more than one critical section. A systematic approach for the selection of prestressing force and eccentricity is presented and compared with the conventional approach which is one of trial and error. In addition, the use of the interaction diagram and the resulting valid domain provides a visual description concerning the safety of the cross section as well as clues for modifications in the design when it is found unsafe.

529   Multivariate Distributions of Directional Wind Speeds

Emil Simiu, Member, ASCE, (Research Engr., Center for Building Tech., National Bureau of Standards, Gaithersburg, Md. 20899), **Erik M. Hendrickson**, Assoc. Member, ASCE, (Research Engr., Center for Building Tech., National Bureau of Standards, Gaithersburg, Md. 20899), **William A. Nolan**, (Cooperative Engrg. Student, Center for Building Tech., National Bureau of Standards, Gaithersburg, Md. 20899), **Ingram Olkin**, (Prof., Dept. of Statistics, Stanford Univ., Stanford, Calif. 94305) and **Clifford H. Spiegelman**, (Mathematical Statistician, Center for Applied Mathematics, National Bureau of Standards, Gaithersburg, Md. 20899)

Journal of Structural Engineering, Vol. 111, No. 4, April, 1985, pp. 939-943

The probability of failure of a structure or structural element subjected to wind forces depends, in large part, on the distribution of extreme wind speeds acting on the structure. In the past, distributions of extreme wind speeds were based on extreme wind data without regard to wind direction, and probabilities of failure were computed accordingly. A method is presented herein for calculating failure probabilities by using directional wind speed data as obtained from weather station records. The method takes advantage of the weak correlations found among wind speeds from different directions and the property of extreme value random variables that zero correlation implies statistical independence. The method is applicable to any type of structure, including structures exhibiting aerodynamic amplification or aeroelastic effects. In addition, it is shown that, in practice, the necessary distributions can be estimated almost as accurately from data obtained from readily available published documents as from data obtained from original weather records.

530   Repair Technique for Buckled Wood Truss Members

David J. Wickersheimer, (Assoc. Prof., School of Architecture, Univ. of Illinois, 309 Architecture Bldg., Urbana, Ill. 61801)

Journal of Structural Engineering, Vol. 111, No. 5, May, 1985, pp. 949-960

Many examples of wood roof trusses built in the early 1900's still exist today. The compression members were often composed of built-up, vertically laminated lumber sections inadequately tied together and often laterally unbraced for substantial distances. Delaminations due to seasonal swelling and shrinkage forced the pieces to behave individually. The resulting overstresses initiated lateral buckling which became amplified by creep in the wood over an extended period of time. The roof surface warps and tears; moisture enters, and evidence of a problem finally becomes visible. Serviceability, and more importantly, safety, is threatened. A relatively unknown repair technique to permanently straighten and strengthen bowed and fractured truss compression members utilizes structural steel shapes to clamp and squeeze out the bow. A series of 1911 wood scissors trusses which were fractured and near collapse, were recently saved by this technique. The primary advantages of this repair method are simplicity, economy, avoidance of roof removal, and minimal intrusion on the original esthetics.

531   Behavior of Columns with Pretensioned Stays

Erling A. Smith, Member, ASCE, (Assoc. Prof., Dept. of Civ. Engrg., Univ. of Connecticut, Storrs, Conn. 06268)

Journal of Structural Engineering, Vol. 111, No. 5, May, 1985, pp. 961-972

The capacity of columns can be increased by using crossarms and pretensioned stays. It is shown that the buckling load of a single crossarm stayed column is a function of the stiffness of the stays, and te ability to achieve that buckling load is a function of the residual tension in the stays at buckling. If the initial pretension is small, then the stays can become slack before the buckling load is reached. If the initial pretension is large, the column will buckle before the stays lose their pretension. It is shown that columns in which stays lose their pretension before buckling occurs have a lower buckling load than columns in which stays do not. It also demonstrated that initial imperfections reduce the capacity below the buckling load of the perfect column. Loss of pretension for a low initial pretension does not cause failure, but for a high initial pretension precipitates instability. Simple but accurate equations are presented that allow noncomputerized solutions to be generated. The theoretical results are compared with recently published results.

532 Behavior of Timber Joints with Multiple Nails

Babu Thomas, (Post-Doctoral Fellow, Dept. of Civ. Engrg., Technical Univ. of Nova Scotia, Halifax, Nova Scotia, Canada B3J 2X4) and **Sundershan K. Malhotra**, Member, ASCE, (Prof., Dept. of Civ. Engrg., Technical Univ. of Nova Scotia, Halifax, Nova Scotia, Canada B3J 2X4)

Journal of Structural Engineering, Vol. 111, No. 5, May, 1985, pp. 973-991

An attempt was made to study the effect of the number of nails on the stiffness of laterally-loaded timber joints with interface friction, fabricated with 2–8 nails in a row. A mathematical model was developed to predict the load-slip behavior of the joints. An experimental program investigated the effect of the number of nails, and some 140 joint tests were conducted, in total, in the present research. The current design concept that a joint fastened together with multiple nails in a row carries a lateral load equal to the product of the lateral capacity of single nail joint, and the number of nails in a row of the joint is somewhat of an overestimation of the overall capacity of the joint. Modification factors are developed to account for the effect of the number of nails in a row on the stiffness of joints with multiple nails. Experiments were also conducted to determine the parameters required for the application of the theoretical model. Frictional forces between the constituent members of joints were observed to be constant for joints with 3–8 nails in a row. The load-slip curves of tested joints and the corresponding curves obtained from the theoretical model are compared with each other to verify the extent of the validity of the model. The theoretical predictions yield an upper bound solution for all types of joints—seven different types—tested in this research.

533 Continuous Timber Diaphragms

Thomas S. Tarpy, Jr., Member, ASCE, (Research Assoc. and Prof., Dept. of Civ. Engrg., Vanderbilt Unit., Nashville, Tenn.), **David J. Thomas**, (Former Grad. Student, Vanderbilt Univ., Nashville, Tenn.) and **Lawrence A. Soltis**, Member, ASCE, (Project Leader, Forest Products Lab., Madison, Wis.)

Journal of Structural Engineering, Vol. 111, No. 5, May, 1985, pp. 992-1002

Current design assumptions for diaphragms assume support conditions which are either simple span or fully continuous. The building codes require a design based on the highest values for moment and shear obtained under either of these two support conditions. More practical criteria for assessing continuity conditions at supports for wood diaphragms are needed. This investigation determines experimentally the effects of continuity conditions on timber floor diaphragms with plywood sheathing subject to inplane loads. Previous testing programs have evaluated simply-supported diaphragms subject to uniform loading; this study evaluates the effects of other support conditions and non-uniform loads. Static loading conditions were used to evaluate the response of the diaphragm for both deflection and ultimate strength. Six 8 ft × 16 ft (2.44m × 4.88m) floor diaphragms typical of certain residential construction techniques with three different sets of boundary and loading conditions were tested in accordance with ASTM E72. The tests demonstrated that: (1) Continuity over a rigid support apparently does not increase the unit shear resistance values of the diaphragm; (2) concentrated loads on the diaphragm produce lower load factors than moment-equivalent uniform loads at a given load level; (3) there is not an apparent direct relationship between relative panel displacement and overall diaphragm deflection for the size diaphragms tested; and (4) local panel buckling has a minimal effect on overall diaphragm failure patterns.

534   Eccentrically Loaded High Strength Bolted Connections

John R. Veillette, (Grad. Student, Vanderbilt Univ., Nashville, Tenn.) and **John T. DeWolf**, Member, ASCE, (Assoc. Prof., Dept. of Civ. Engrg., Univ. of Connecticut, Storrs, Conn.)

Journal of Structural Engineering, Vol. 111, No. 5, May, 1985, pp. 1003-1018

Eccentrically loaded high strength bolted tee-to-column connections were investigated experimentally. The tee and column flanges were fastened with A325 high strength bolts, and the eccentric load was concentrated and directed through the tee's web. Thus, the bolts were subject to shear and tension. load transducers were used to obtain the bolt forces. The connection was loaded until bolt fracture occurred. Three parameters were investigated: (1) The number of bolts; (2) the thickness of the tee flange; and (3) the eccentricity of the load. The test results are compared with current design methods. Recommendations are made for design.

535   Construction Load Analysis for Concrete Structures

Xila Liu, (Research Asst., School of Civ. Engrg., Purdue Univ., West Lafayette, Ind.), **Wai Fah Chen**, Member, ASCE, (Prof. and Head of Struct. Engrg., School of Civ. Engrg., Purdue Univ., West Lafayette, Ind.) and **Mark D. Bowman**, Member, ASCE, (Asst. Prof. of Struct. Engrg., School of Civ. Engrg., Purdue Univ., West Lafayette, Ind.)

Journal of Structural Engineering, Vol. 111, No. 5, May, 1985, pp. 1019-1036

A common practice in multistory reinforced concrete building construction is to shore a freshly placed concrete floor on several previously cast floors. The construction loads on the supporting floors may exceed the slab design loads during maturity, especially when the design live load is small compared with the dead load. A few studies have been conducted to analytically model the construction loading process. However, these early models are based on a number of simplifying assumptions. The objective of this paper is to develop a three-dimensional computer model which can be used to evaluate the effect of variations of the foundation rigidity, column axial stiffness, slab aspect ratio, and shore stiffness distribution on the values of the shore loads and slab moments.

536   Laminated Glass Units under Uniform Lateral Pressure

R. A. Behr, Member, ASCE, (Research Assoc./Lect., Dept. of Civ. Engrg., Texas Tech Univ., Lubbock, Tex. 79409), **J. E. Minor**, Fellow, ASCE, (Prof., Dept. of Civ. Engrg., Texas Tech Univ., Lubbock, Tex. 79409), **M. P. Linden**, (Research Asst., Dept. of Civ. Engrg., Texas Tech Univ., Lubbock, Tex. 79409) and **C. V. G. Vallabhan**, Member, ASCE, (Prof., Dept. of Civ. Engrg., Texas Tech Univ., Lubbock, Tex. 79409)

Journal of Structural Engineering, Vol. 111, No. 5, May, 1985, pp. 1037-1050

Laminated flat glass is gaining popularity as an architectural glazing product. Despite its increased use as a cladding material, its structural properties are not well known. Research undertaken to advance understanding of the behavior of laminated glass units under lateral pressure representing wind loads is reported. Laminated glass units are comprised of two layers of glass connected by a thin interlayer of polyvinyl butyral. The material properties of the interlayer are very different from the properties of the glass plates which it joins together; its modulus of elasticity in shear is only about 1/10,000th that of glass. Experimental stress analyses were conducted on several laminated glass units to ascertain whether their behavior was similar to a monolithic glass plate of the same nominal thickness, or to a layered glass unit consisting of two glass plates with no interlayer. At room temperature the laminated glass unit behaves much like a monolithic glass plate of the same nominal thickness. At elevated temperatures [170°F (77°C)] the behavior changes, and approaches that of a layered glass unit with no interlayer. Results of the experimental stress analyses are compared with theoretical stress analyses designed to characterize the behavior of monolithic glass plates and layered glass units.

537   Local Instability Tests of Plate Elements under Cyclic
Uniaxial Loading

Yuhshi Fukumoto, Member, ASCE, (Prof., Dept. of Civ. Engrg., Nagoya Univ., Nagoya, Japan)
and **Haruyuki Kusama**, (Lect., Dept. of Civ. Engrg., Toyota National Tech. Coll., Toyota, Japan)

Journal of Structural Engineering, Vol. 111, No. 5, May, 1985, pp. 1051-1067

An experimental study of the inelastic cyclic load-deformation behavior of welded built-up square box-section short columns subjected to cyclic axial loading is presented. A total of 10 test specimens were fabricated from mild ($o_y = 240 N/mm^2$) and high strength ($o_y = 700 N/mm^2$) steels, having plate elements with the width-thickness ratios of 40, 60 and 80 for mild steel, and 40 and 60 for high strength steel. Furthermore, monotonically increased loading tests were carried out for comparison with the deformation behavior of cyclic loading tests. This paper emphasizes the development of alternating local instability of plate elements associated with cyclic loading sequences. The hysteretic loops of average stress versus axial strain curves are expressed experimentally, and analytical interpretations are added to the results. Successive reduction of the ultimate strength and residual deformation at zero-load are provided experimentally and analytically under cyclic loadings.

538   Design of Diagonal Roof Bracing Rods and Tubes

Sritawat Kitipornchai, (Sr. Lect. in Civ. Engrg., Univ. of Queensland, Queensland, Australia) and **Scott T. Woolcock**, (Dir., Cardno and Davies Australia Pty. Ltd., Queensland, Australia)

Journal of Structural Engineering, Vol. 111, No. 5, May, 1985, pp. 1068-1084

The behavior of rods and tubes as horizontal tension members is investigated. In the design of double diagonal tension bracing, each member must carry the tension forces as well as its own weight either by cable action in the case of rods, or by beam action in the case of tubes. Rod tensioning experiments reveal that pretensioning forces are much higher than expected. However, excessive pretensioning of rods does not affect their ultimate capacity but would influence the design of the compression members in the bracing system. It is found that the tensile capacity of tubes is only marginally reduced by the effect of self weight and the self weight bending stresses need not be considered in combination with axial tensile stresses. Bending under self weight alone need not be checked as deflection always governs the design.

539   Biaxial Stress-Strain Relations for Brick Masonry

M. Dhanasekar, (Postgrad. Student, Dept. of Civ. Engrg. and Surveying, Univ. of Newcastle, New South Wales, Australia 2308), **Peter W. Kleeman**, (Sr. Lect., Dept. of Civ. Engrg. and Surveying, Univ. of Newcastle, New South Wales, Australia 2308) and **Adrian W. Page**, (Sr. Lect., Dept. of Civ. Engrg. and Surveying, Univ. of Newcastle, New South Wales, Australia 2308)

Journal of Structural Engineering, Vol. 111, No. 5, May, 1985, pp. 1085-1100

Simple nonlinear stress-strain relations for brick masonry constructed with solid pressed bricks have been derived from the results of a large number of biaxial tests on square panels with various angles of the bed joint to the principal stress axes. The macroscopic elastic and nonlinear stress-strain relations were determined from displacement measurements over gage lengths which included a number of mortar joints. Although the initial elastic behavior was found to be close, on average, to isotropic, the nonlinear behavior is strongly influenced by joint deformations and is best expressed in terms of stresses and strains referred to axes normal and parallel to the bed joint. A given strain has been related only to the corresponding stress by a power law, except when the ratio of the shear to normal stress on a bed joint is greater than 1. in the latter case, a bilinear relation is more appropriate. Dependence on other stress components is masked by the variability in the results.

540   Local Buckling of Hollow Structural Sections

John L. Dawe, Member, ASCE, (Assoc. Prof. of Civ. Engrg., Univ. of New Brunswick, Fredericton, New Brunswick, Canada E3B 5A3), **Adel A. Elgabry**, (Grad. Student, Univ. of New Brunswick, Fredericton, New Brunswick, Canada E3B 5A3) and **Gilbert Y. Grondin**, (Research Asst., Univ. of New Brunswick, Fredericton, New Brunswick, Canada E3B 5A3)

Journal of Structural Engineering, Vol. 111, No. 5, May, 1985, pp. 1101-1112

An analytical technique for predicting the local buckling behavior of thin-walled sections was used to predict elastic and inelastic local buckling capacities of axially loaded hollow structural sections. The technique, similar to a finite strip method, is based on the principle of virtual work and uses a Rayleigh-Ritz solution procedure. The effects of manufacturing process and interaction between adjacent plates are included in the formulation. The method is verified by comparing predicted loads with available test results and known elastic solutions of plate buckling. Various parameters considered to be significant in affecting local buckling capacities of hollow structural sections are evaluated.

541   Random Creep and Shrinkage in Structures: Sampling

Zdeněk P. Bažant, Fellow, ASCE, (Prof. of Civ. Engrg. and Dir., Center for Concrete and Geomaterials, Northwestern Univ., Evanston, Ill. 60201) and **Kwang-Liang Liu**, (Grad. Student, Northwestern Univ., Evanston, Ill. 60201)

Journal of Structural Engineering, Vol. 111, No. 5, May, 1985, pp. 1113-1134

This paper deals with uncertainty in the prediction of the effects of creep and shrinkage in structures, such as deflections or stresses, caused by uncertainties in the material parameters for creep and shrinkage, including the effect of random environmental humidity. This problem was previously analyzed using two-point estimates of probability moments. Here the same problem is analyzed using latin hypercube sampling. This has the merit that the number of required deterministic structural creep analyses is reduced from 2^n to approximately 2n, where n = number of random parameters. A method of taking into account the uncertainty due to the error of the principle of superposition is also presented. The mean and variance of creep effects in structures are calculated and scatter bands are plotted for seven typical practical examples, and the results are found to be close to those obtained with two-point estimates. Further, it is demonstrated that the distribution of creep effects is approximately normal if the creep parameters are normally distributed. Finally, it is shown that good results may also be obtained with the age-adjusted effective modulus method, which greatly simplifies structural creep analysis.

542   Cyclic Out-of-Plane Buckling of Double-Angle Bracing

Abolhassan Astaneh-Asl, Assoc. Member, ASCE, (Asst. Prof. of Civ. Engrg. and Environmental Sci., Univ. of Oklahoma, Norman, Okla. 73019), **Subhash C. Goel**, Member, ASCE, (Prof. of Civ. Engrg., Univ. of Michigan, Ann Arbor, Mich. 48109) and **Robert D. Hanson**, Member, ASCE, (Prof. of Civ. Engrg., Univ. of Michigan, Ann Arbor, Mich. 48109)

Journal of Structural Engineering, Vol. 111, No. 5, May, 1985, pp. 1135-1153

The behavior of double-angle bracing members subjected to out-of-plane buckling due to severe cyclic load reversals is investigated. Nine full-size test specimens were subjected to severe inelastic axial deformations. Test specimens were made of back-to-back A36 steel angle sections connected to the end gusset plates by fillet welds or high-strength bolts. Five of the test specimens were designed according to current design procedures and code requirements. These specimens experienced fracture in gusset plates and stitches during early cycles of loading. Based on the observations and analysis of the behavior of these specimens, new design procedures are proposed for improved ductility and energy dissipation capacity of double-angle bracing members which buckle out of plane of gusset plates. Tests of four specimens, designed using proposed procedures, showed significant improvement in their performances.

543   Partial Stress Redistribution in Cold-Formed Steel

Muzaffer Yener, Member, ASCE, (Asst. Prof. of Civ. Engrg., Purdue Univ., West Lafayette, Ind. 47907) and **Teoman Pekoz**, Member, ASCE, (Prof. of Civ. Engrg. and Environmental Engrg.,, Cornell Univ., Ithaca, N.Y. 14853)

Journal of Structural Engineering, Vol. 111, No. 6, June, 1985, pp. 1169-1186

On the basis of the failure criteria developed in an earlier paper by the writers, the utilization of partial section plastification in cold-formed steel flexural members is presented. As in the limit design procedures developed for reinforced concrete, failure is defined in terms of the strain capacity of stiffened compression flanges, for specified values of the width-to-thickness ratio of the flange and the yield stress of the structural steel. Accordingly, depending upon the slenderness of the compression flanges and the cross-sectional shape, ultimate moment capacities in excess of yield moment become possible. As design and analysis aids, formulas to compute the ultimate moment capacity of commonly used cold-formed steel shapes are derived. To illustrate the use of these equations and the economy gained by taking into account the inelastic reserve capacity due to stress redistribution, two example problems are solved. The findings are compared with test results and an elastic solution.

544   Partial Moment Redistribution in Cold-Formed Steel

Muzaffer Yener, Member, ASCE, (Asst. Prof. of Civ. Engrg., Purdue Univ., W. Lafayette, Ind.) and **Teoman Pekoz**, Member, ASCE, (Prof. of Civ. and Environmental Engrg., Cornell Univ., Ithaca, N.Y.)

Journal of Structural Engineering, Vol. 111, No. 6, June, 1985, pp. 1187-1203

Post-yielding strength of redundant cold-formed steel flexural members, due to moment redistribution, is presented. The reserve strength due to section plastification is discussed in an accompanying paper. In an earlier paper, the development of the failure criteria was presented, and limit analysis and design procedures, similar to those developed for reinforced concrete members, were proposed. It is shown that the use of the proposed limit design procedure results in considerable savings, due to both stress and moment redistribution. As cold-formed steel structural sections, in the form of floor and wall panels, roof decks, and framing members are increasingly used in building construction, the inclusion of inelastic reserve strength may become an economic necessity. Additionally, the availability of reliable methods to predict behavior beyond the initiation of yielding would help provide realistic safety factors against complete failure.

545   Stochastic Evaluation of Seismic Structural Performance

Robert H. Sues, (Senior Staff Engr., National Technical Systems/SMA Div., Newport Beach, Calif.), **Yi-Kwei Wen**, Member, ASCE, (Prof., Univ. of Illinois at Urbana-Champaign, Urbana, Ill.) and **Alfredo H-S. Ang**, Fellow, ASCE, (Prof., Univ. of Illinois at Urbana-Champaign, Urbana, Ill.)

Journal of Structural Engineering, Vol. 111, No. 6, June, 1985, pp. 1204-1218

A method is presented for determining the probabilities of a structure sustaining various levels of damage due to seismic activity during its lifetime. Uncertainties in the loading and the structural response analysis are considered. The method is based on a nonlinear random vibration analysis and an analytical technique for evaluating the sensitivity of the response to various structural and load parameters. Recently reported data are used to update the parameter values commonly used in the random process representation of earthquakes. The method is illustrated by analyses of a four-story steel frame building and a seven-story reinforced concrete building.

546   Inelastic Cyclic Analysis of Imperfect Columns

Manolis Papadrakakis, Member, ASCE, (Senior Lect., Inst. of Structural Analysis, National Tech. Univ., Athens, Greece) and **Lefteris Chrysos**, (Grad. Student, Dept. of Civ. Engrg., National Tech. Univ., Athens, Greece)

Journal of Structural Engineering, Vol. 111, No. 6, June, 1985, pp. 1219-1234

The hysteretic behavior of a simply supported prismatic steel column with initial out-of-straightness subjected to cyclic loading is presented. The analysis is based on the plastic hinge concept, which under the assumption of perfect plasticity together with the one-dimensional idealization of the column, has led to an elegant closed form solution for any history of axial loading. The exact yield curve of any type of cross-sectional shape is adequately approximated by a piecewise linear interpolation of the fully plastic states in pure bending and in pure tension and compression. The results presented here illustrate the imperfection sensitivity of column cyclic behavior. The effect of various slenderness ratios combined with different initial imperfection sizes as well as the effect of different approximation to the exact yield curve are also presented.

547   End-Bolted Cover Plates

Fateh Wattar, (Structural Engr., Dar-Alhandasah, Beirut, Lebanon), **Pedro Albrecht**, Member, ASCE, (Prof., Dept. of Civ. Engrg., Univ. of Maryland, College Park, Md. 20742) and **Adnan H. Sahli**, Member, ASCE, (Structural Engrg., Blunt and Evans, Consulting Engrs., Lanham, Md.)

Journal of Structural Engineering, Vol. 111, No. 6, June, 1985, pp. 1235-1249

Welding a cover plate to a rolled beam and high-strength bolting the loose ends with a friction-type connection increased the fatigue strength from Category E for end-welded cover plates to Category B. The required number of end bolts is that needed to develop the cover plate's portion of the bending moment. A parametric analysis indicated that the design of two-span continuous bridges with end-bolted cover plates subjected to HS-20 truck loading was not governed by fatigue. The weight savings from shortening the cover plate from its Category E length to the length between the theoretical cut-off points were greater than the added cost of end-bolting. This method of construction has the potential of fatigue proofing cover plates on rolled beams while lowering their cost of fabrication. It utilizes existing technolocy and can be implemented at once.

548   Calibration of Bridge Fatigue Design Model

William E. Nyman, Assoc. Member, ASCE, (Engr., Hardesty and Hanover, New York, N.Y.) and **Fred Moses**, Member, ASCE, (Prof. of Civ. Engrg., Case Western Reserve Univ., Cleveland, Ohio 44106)

Journal of Structural Engineering, Vol. 111, No. 6, June, 1985, pp. 1251-1266

A structural reliability evaluation is performed of the current AASHTO fatigue specification for steel bridges. The reliability model incorporates uncertainties in vehicular loading, analysis, and fatigue life. Field data is obtained from a weigh-in-motion system, which utilizes existing bridges as equivalent static scales. The load data includes truck axle and gross weights, headways, impact, stress range, girder distribution, and volume. A fatigue life model is formulated in terms of a fatigue failure function. Using the loading and fatigue life data a safety index is calculated using a level II reliability program. The study reviews the current specification in order to derive uniform reliability levels over the range of typical designs. The proposed revisions include: (1) A design vehicle model more representative of the current U.S. truck population; (2) changes in allowable stress ranges to eliminate variations in safety index; and (3) a range of load factors to represent site specific truck volume and loadometer (weights) values. These recommended changes lead to more uniform safety levels and fatigue lives for steel bridges.

549   Maximum Strength Design of Structural Frames

Subramaniam Kanagasundaram, (Postgrad. Student, Dept. of Civ. Engrg. and Surveying, Univ. of Newcastle, New South Wales, Australia 2308) and **Bhushan L. Karihaloo**, Member, ASCE, (Prof. and Head, Dept. of Civ. Engrg., Univ. of Newcastle, New South Wales, Australia 2308)

Journal of Structural Engineering, Vol. 111, No. 6, June, 1985, pp. 1267-1287

A method of solving the minimum weight/maximum strength design problem of plane rigid frames under the influence of their self-weight as well as externally applied forces is described. Each member of the frame attains the maximum allowable normal or shear stress, or both, for the material. Concepts from the differential game theory are used to reduce the corresponding mathematical optimization problem to a minimax variational one. In the first instance, analytical expressions for the optimum shape of typical plane frame members are obtained. These are then combined with a modified stiffness matrix analysis routine to obtain the optimum shape of the plane rigid frame as a whole. The unified method is applicable to a wide variety of plane rigid frames. Several examples are given to demonstrate its potential.

550   Ultimate Load Behavior of Beams with Initial Imperfections

Susumu Nishida, (Prof., Dept. of Civ. Engrg., Kanzawa Inst. of Tech., 7-1 Ohgigaoka Nonoichi, Ishikawa, Japan 921) and **Yuhshi Fukumoto**, Member, ASCE, (Prof., Civ. Engrg., Nagoya Univ., Cikusu-ku, Nagoya, Japan)

Journal of Structural Engineering, Vol. 111, No. 6, June, 1985, pp. 1288-1305

More exact expressions of the fundamental equations of a member with initial imperfections subjected to the action of bending and torsional moments are derived. Since beam problems involve moderately large in-plane deformation prior to collapse, only nonlinear terms of in-plane displacement and in-plane resulting force are considered in the numerical analysis using the transfer matrix method. In the formulation of stress resultants in a partially yielded section, the equivalent section properties are used. The proposed idea is obtained by a transformed-section method of analysis of reinforced concrete beams. Finally, the effects of initial deformation, initial lateral deflection, residual stress patterns, strain hardening, and load and support conditions on the ultimate strength and deformation behavior of beams are investigated.

551 Reliability of Ductile Systems with Random Strengths

Tzyy Shan Lin, (Grad. Asst., Dept. of Civ Engrg., Johns Hopkins Univ., Baltimore, Md.) and **Ross B. Corotis**, Member, ASCE, (Willard and Lillian Hackerman Prof. and Chairman, Dept. of Civ. Engrg., Johns Hopkins Univ., Baltimore, Md.)

Journal of Structural Engineering, Vol. 111, No. 6, June, 1985, pp. 1306-1325

A technique is proposed for determining the reliability of redundant ductile framed structures. A nonlinear structural analysis program and incremental load procedure are used to find the limit state function in load space for fixed structural properties. The failure probability of this deterministic structure is then computed by integrating the joint distribution function of loads over the failure region in load space. The procedure is extended to include randomness in the strength of the structure by introducing a random resistance variable unique to each load path. The statistics of this variable can be found by relating magnitudes of the load vector to successive component failure. The load space formulation of system reliability appears applicable to realistic structures of reasonable complexity.

552 Analytical Modeling of Tube-In-Tube Structure

Peter C. Chang, Member, ASCE, (Asst. Prof., Civ. Engrg. Dept., Univ. of Maryland at College Park, College Park, Md. 20742)

Journal of Structural Engineering, Vol. 111, No. 6, June, 1985, pp. 1326-1337

Tube-in-tube structures are analyzed using a continuum approach in which the two beams are individually modeled by a "tube beam" that accounts for flexural deformation, shear deformation, and shear-lag effect. The beams are forced to have equal lateral deflections, and the amount of load carried by each beam is a function of its relative stiffness. The analyses are performed using the Minimum Potential Energy principle, and the results are compared with results of finite element analyses. It is shown that the continuously compatible condition reasonably models tube-in-tube structures with as few as five stories.

553 Tubular Steel Trusses with Cropped Webs

Glenn A. Morris, Member, ASCE, (Assoc. Dean., Faculty of Engrg., Univ. of Manitoba, Winnipeg, Canada R3T 2N2)

Journal of Structural Engineering, Vol. 111, No. 6, June, 1985, pp. 1338-1357

Experimental investigations of the force-deformation behavior of cropped-web tubular Pratt type truss joints and specimens representing truss compression webs are described. Dimensionless empirical expressions are presented for estimating the strength and flexibility of square-chord and round-chord joints with end-cropped webs. Strength and flexibility comparisons are drawn between round-chord and square-chord joints. The behavior of isolated joint specimens is compared to that of similar joints in trusses. The influence of chord axial force and secondary bending moment on joint strength is examined. The influence of end-cropping on effective-length factors of truss compression webs is demonstrated. A procedure is proposed for designing tubular trusses with end-cropped webs.

554   Models for Human Error in Structural Reliability

Maher A. Nessim, (Research Engr., Det norske Veritas (Canada) Ltd., Calgary, Alberta, Canada) and **Ian J. Jordaan**, (Head of Research and Development, Det norske Veritas (Canada) Ltd., Calgary, Alberta, Canada)

Journal of Structural Engineering, Vol. 111, No. 6, June, 1985, pp. 1358-1376

Probabilistic models are proposed for the occurrence, detection, and consequences of human errors in structures. The models are developed in an overall framework of decision theory applied to the problem of allocating control efforts that would narrow the gap between the estimated and actual rates of structural failure. Two models for error occurrence are presented: (1) A binomial distribution for errors in discrete tasks; and (2) a Poisson process for errors occurring in a continuous production interval. Checking is modeled as a sequence of Bernoulli trials. The restrictive independence assumptions implied in the classical use of the preceding distributions are replaced by the weaker and more appropriate assumption of exchangeability, by accounting for the uncertainty regarding distribution parameters. A bivariate probability model is also developed which correlates the performance of a structure to its error content. The aforementioned probabilistic models provide an outline of areas where prior information is needed and a specific format for the collection of additional data. An application to decision-making in the area of nondestructive testing is also presented.

555   Design and Analysis of Silos for Friction Forces

Demetris Briassoulis, (Visiting Prof., Dept. of Agricultural Engrg., Univ. of Illinois, Urbana, Ill.) and **James O. Curtis**, (Prof., Dept. of Agricultural Engrg., Univ. of Illinois, Urbana, Ill.)

Journal of Structural Engineering, Vol. 111, No. 6, June, 1985, pp. 1377-1398

The effect of axial friction forces on the calculated hoop stress of an unstiffened silo, or of the top unstiffened portion of a silo is analyzed. Ignoring the beam-column interaction between axial friction forces and radial pressure due to the grain may result in an error of 20% in some cases. This results in a variable actual factor of safety which depends on the geometric characteristics of the silo. An instability analysis is carried out which sets limits on the maximum unstiffened length of a silo. This is done on a probabilistic basis and an algorithm is suggested for the design of an unstiffened silo or a silo stiffened by rings.

556   Moment Determination for Moving Load Systems

Marius B. Wechsler, Member, ASCE, (Sr. Engr., Bechtel Power Corp., P.O. Box 60860, Terminal Annex, Los Angeles, Calif. 90060)

Journal of Structural Engineering, Vol. 111, No. 6, June, 1985, pp. 1401-1406

An accurate method is described for the determination of the curves of maximum moments for beams loaded with moving loads. This method avoids errors in the range of 15 percent produced by the commonly used approximation techniques. It applies to simple beams and to beams overhanging their supports. The basic idea of the method is to replace each of the moving load systems by a virtual system of stationary loads. The moments and the reactions produced by this virtual system are equal to the maximum moments and to the maximum reactions respectively produced by the actual moving system of loads. Examples of the application of the theory are given for some simple loading conditions.

557   Limit State Analysis of Arch Segment

R. H. Allen, Member, ASCE, (Asst. Prof., Dept. of Mech. Engrg., Univ. of Houston, Houston, Tex. 77004) and **I. J. Oppenheim**, Member, ASCE, (Assoc. Prof., Dept. of Civ. Engrg., Carnegie-Mellon Univ., Pittsburgh, Pa. 15213)

Journal of Structural Engineering, Vol. 111, No. 6, June, 1985, pp. 1406-1409

A simple, yet useful limit state analysis is performed on a statically determinate arch segment. A failure mode is assumed and relationships governing the kinematic failure mechanism are discovered using statics. The fracture is observed to occur at the midpoint and a minimum thickness to length ratio is found to be 0.150 to insure against brittle failure. The least desirable position of the arch segment is at 52° with respect to the horizontal.

558   Unified Plastic Analysis for Infilled Frames

T. C. Liauw, (Reader, Dept. of Div. Engrg. Univ. of Hong Kong, Hong Kong) and **K. H. Kwan**, (Research Assoc., Dept. of Civ. Engrg. Univ. of Hong Kong, Hong Kong)

Journal of Structural Engineering, Vol. 111, No. 7, July, 1985, pp. 1427-1448

Based on tests and nonlinear finite element analyses, a unified plastic analysis for infilled frames with three types of interface conditions is proposed, in which the stress redistribution towards collapse and the conditions at the infill-frame interface are taken into account. The theory is applicable to single and multistory infilled frames, and comparison with experimental results gives good agreement. Design recommendations for multistory infilled frames are given.

559   Vibrational Characteristics of Multi-Cellular Structures

Thambirajah Balendra, Member, ASCE, (Visiting Assoc. Prof., Dept. of Civ. Engrg. and Engrg. Mech., McMaster Univ., Hamilton, Ontario, Canada) and **Nandivararm E. Shanmugam**, Member, ASCE, (Sr. Lect., Dept. of Civ. Engrg., National Univ. of Singapore, Singapore)

Journal of Structural Engineering, Vol. 111, No. 7, July, 1985, pp. 1449-1459

An experimental study is carried out to verify the grillage idealization for dynamic analysis of multi-cellular structures. Perspex material is used to construct two models of same size, one with no web openings and the other with 25% web openings. The natural frequencies and the corresponding mode shapes are determined for two different sets of boundary conditions, namely all four sides simply supported, two opposite sides simply supported, and the remaining sides free. The experimental results are found to compare well with the theoretical results using grillage idealization and finite element methods. As the linear response due to any dynamic load is the superposition of the model components, it is concluded that the grillage idealization could be used to predict the forced vibration response of multi-cellular structures. Since the study is restricted to linear elastic range, the findings of this investigation are applicable to structures made of other material, e.g., steel cellular structures.

560   Glass Strength Degradation under Fluctuating Loads

D. A. Reed, (Asst. Prof., Dept. of Civ. Engrg., Univ. of Washington, Seattle, Wash. 98195) and **E. R. Fuller, Jr.**, (Research Scientist, Center for Materials Research, National Bureau of Standards, Washington, D.C. 20234)

Journal of Structural Engineering, Vol. 111, No. 7, July, 1985, pp. 1460-1467

An alternative approach for estimating the strength degradation, and ultimately, the failure of glass cladding subjected to fluctuating loadings is proposed. This procedure is formulated using fracture mechanics concepts employed previously. This approach is simple computationally and does not require time integration of fluctuating stress or pressure loading time histories.

561   Analysis of Events in Recent Structural Failures

Fabian C. Hadipriono, Member, ASCE, (Asst. Prof., Dept. of Civ. Engrg., Ohio State Univ., 470 Hitchcock Hall, 2070 Neil Ave., Columbus, Ohio 43210)

Journal of Structural Engineering, Vol. 111, No. 7, July, 1985, pp. 1468-1481

A study of nearly 150 recent major collapses and distresses of structures around the world discloses that external events and deficiencies in the areas of construction and design were the principal sources of failures. More than one-third of the surveyed structures were bridges, and the remaining were low-rise, multi-story, plant-industrial, and long-span buildings. The major incidents in the category of external events were the lateral impact forces and other unexpected live loads on bridges, and explosion impact loads in concrete silos. Construction deficiencies included falsework and concreting faults in several concrete structures and inadequate welding operations in steel bridges. Design deficiency related events were found to vary with the type of the observed structures, but a significant number of failures were attributed to lack of knowledge in long-term creep and shrinkage effects on prestressed concrete members. These findings suggest that attention should be directed to three areas: (1) Identification of potential deficiencies from past failure data; (2) enhancement in procedural methods during design and construction operations; and (3) incorporation of risk analyses of structures during their service life and construction phases.

562   Pseudodynamic Method for Seismic Performance Testing

Stephen A. Mahin, Member, ASCE, (Assoc. Prof. of Civ. Engrg., College of Engrg., Div. of Structural Engrg., Univ. of California, Berkeley, Calif. 94720) and Pui-shum B. Shing, Assoc. Member, ASCE, (Assoc. Research Engr., Dept. of Civ. Engrg., Univ. of California, Berkeley, Calif.)

Journal of Structural Engineering, Vol. 111, No. 7, July, 1985, pp. 1482-1503

The pseudodynamic method is a relatively new experimental technique for evaluating the seismic performance of structural models in a laboratory by means of on-line computer controlled testing. During such a test, the displacement response of a structure to a specified dynamic excitation is numerically computed and quasi-statically imposed on the structure, based on analytically prescribed inertia and viscous damping characteristics for the structure and the experimentally measured structural restoring forces. This paper presents the basic approach of the method, describing the numerical and experimental techniques. Based on current studies, the capabilities and limitations of the method are examined, and possible improvement methods are mentioned. In spite of certain numerical and experimental errors, recent verification tests show that the method can be as reliable and realistic as shaking table testing and that it can be readily implemented in many structural laboratories. The capabilities of the method can be further expanded to test specimens under various load and structural boundary conditions.

563   **Direct Model Test of Stub Girder Floor System**

Anthony J. Nadaskay, Assoc. Member, ASCE, (Assoc., Davis Engrs., P.A., Miami, Fla.) and **C. Dale Buckner**, Member, ASCE, (Assoc. Prof., Dept. of Civ. Engrg., Louisiana State Univ., Baton Rouge, La.)

Journal of Structural Engineering, Vol. 111, No. 7, July, 1985, pp. 1504-1516

Tests to failure of two small-scale model stub girders are described. The tests were performed to determine whether or not behavior observed in an isolated test specimen is indicative of behavior of a similar girder incorporated into a complete floor system. Results indicate a significant difference in both the longitudinal shear strength and mode of failure of the slab in an isolated specimen as compared to a similar girder within a complete floor system.

564   **Curved Bridge Decks: Analytical Strip Solution**

Issam E. Harik, Assoc. Member, ASCE, (Asst. Prof., Dept. of Civ. Engrg., Univ. of Kentucky, Lexington, Ky. 40506) and **Sasan Pashanasangi**, (Grad. Asst., Dept. of Civ. Engrg., Univ. of Kentucky, Lexington, Ky. 40506)

Journal of Structural Engineering, Vol. 111, No. 7, July, 1985, pp. 1517-1532

An "exact" solution is presented for the analysis of orthotropic curved bridge decks subjected to patch, uniform, line and concentrated loads. The bridge deck is idealized as an assemblage of radially supported curved plate elements or strips. The deflection of each plate strip is expressed as a Levy type Fourier series and the loads are expressed as a corresponding series. Concentrated loads and line loads in the tangential direction are incorporated in the solution as a discontinuity in the shear force along the load line. The convergence of this procedure is achieved by increasing the number of modes rather than the number of elements. The solution has the advantage over purely numerical or semi-numerical techniques, using grid analogy or finite element techniques, in that the computational difficulties do not increase as the number of terms is increased to give increased accuracy. Tabular and graphical results are presented for bridge decks subjected to various loading conditions.

565   **New Finite Element for Bond-Slip Analysis**

David Z. Yankelevsky, (Civ. Engrg. Dept., Technion-Israel Inst. of Tech., Haifa, Israel 32000)

Journal of Structural Engineering, Vol. 111, No. 7, July, 1985, pp. 1533-1542

A new finite element for bond stress-slip analysis is presented. A one dimensional model which is based on equilibrium and local bond stress-slip law is developed. The relationship between axial force and slip at the elements modes is expressed through a stiffness matrix. The global stiffness matrix is assembled and solution yields slip, strain and stress distributions along the steel bar. The nonlinear bond stress-slip relationship leads to an iterative technique which is found to converge rapidly. The proposed method predictions are compared with experimental results of monotonic and push-pull tests and very good correspondence is found.

566   Gable Frame Design Considerations

Vernon B. Watwood, Member, ASCE, (Prof. and Chairman, Dept. of Civ. Engrg., Michigan Tech. Univ., Houghton, Mich. 49931)

Journal of Structural Engineering, Vol. 111, No. 7, July, 1985, pp. 1543-1558

A gable frame design is reviewed with regard to the modeling, load specification, and strength evaluation. It is found that the approximation used for the boundary conditions at the column bases should adequately account for the behavior of the foundation and more than one approximation may be necessary. Uneven loading is shown to very significant, and it is recommended that gable frames always be designed for some uneven distribution of load. The question of the correct effective length for the rafter is examined, and a nonlinear analysis or an approximate, but conservative, approach is suggested for gables with significant axial load in the rafter. One method of utilizing a frame buckling analysis in a conventional AISC evaluation is presented. It is suggested that if there is significant axial stress in the rafter, caution should be exercised in the use of specification limitations, which are based on girder behavior where there may have been no axial stress considered in the derivation and testing of the specification limitations.

567   Dynamic Analysis of a Forty-Four Story Building

Bruce F. Maison, Member, ASCE, (Struct. Engrg., 10458 S. St. Louis Ave., Chicago, Ill. 60655) and **Carl F. Neuss**, M. M., (Civ. Engrg., One Soldiers Field Park, No. 401, Boston, Mass. 02163)

Journal of Structural Engineering, Vol. 111, No. 7, July, 1985, pp. 1559-1572

Extensive computer analysis of an existing forty-four story steel frame highrise building is performed to study the influence of various modeling aspects on the predicted dynamic properties and computed seismic response behaviors. The predicted dynamic properties are compared to the building's true properties as previously determined from experimental testing. The seismic response behaviors are computerd using the response spectrum (Newmark and ATC spectra) and equivalent static load methods (ATC and UBC). Interpretations of the analysis results are provided. Conclusions are drawn regarding general results that are relevant to the analysis of other highrise buildings.

568   Optimum Design of Trusses with Buckling Constraints

Juan H. Cassis, (Prof., Dept. of Civ. Engrg., Univ. of Chile, Casilla 5373, Santiago, Chile) and **Abdon Sepulveda**, (Asst. Prof., SECOM, Univ. of Santiago de Chile, Santiago, Chile)

Journal of Structural Engineering, Vol. 111, No. 7, July, 1985, pp. 1573-1589

A method is developed for the optimum design of trusses including constraints on tension and compression stresses, displacements and bounds on the design variables. It is intended to solve a practical problem. The method takes into account the efficiencies achieved for simpler mathematical models where the allowable values for compression stresses are constant. In this work, these values depend on the member cross section and are prescribed in the Code AISC-78. All the pecularities introduced by the buckling constraints are identified and solved. They are: (1) consideration of the radius of gyration, in addition to the cross-sectional area, as design variable, (2) introduction of more complicated constraint functions, and (3) computation of a scale factor to the constraint surface for buckling restrictions. The problem is solved by means of a sequence of approximate problems in the dual space. The approximation is basically due to the fact that the constraint functions are replaced by their first order Taylor series expansions to improve the efficiency of the solution. The effectiveness of the method is shown through some examples.

569   Impact Effect on R.C. Slabs: Analytical Approach

Mohamed Abdel-Rohamn, Assoc. Member, ASCE, (Assoc. Prof., Dept. of Civ. Engrg., Kuwait Univ., P.O. Box 5969, Kuwait) and **Jihad Sawan**, Assoc. Member, ASCE, (Asst. Prof., Dept. of Civ. Engrg., Kuwait Univ., P.O. Box 5969, Kuwait)

Journal of Structural Engineering, Vol. 111, No. 7, July, 1985, pp. 1590-1601

Three simple analytical methods to study the impact effect on R.C. slabs are presented. These methods are the impact factor method, the equivalent mass method, and the continuous mass method. Several aspects that affect the slab's response due to impact are investigated. These aspects include the relation between the impact load and its actual static load when acting on the structure, the duration of impact load when acting on the structure, and the dynamic deflections of the reinforced concrete slabs due to impact. The impact is simulated by dropping a steel ball from different heights on the center of simply supported reinforced concrete slabs. Dynamic deflection, impact time, maximum penetration of the ball into the slab, and the maximum dynamic load that the slab will experience are predicted using the simple analytical methods. Results from these methods are compared with similar experimental ones. In general, the analytical results agreed with the experimental ones.

570   Static and Fatigue Tests on Partially Prestressed Beams

M. H. Harajli, (Research Asst., Dept. of civ. Engrg., Univ. of Michigan, Ann Arbor, Mich. 48109) and **A. E. Naaman**, Member, ASCE, (Prof. of Civ. Engrg., Univ. of Michigan, Ann Arbor, Mich. 48109)

Journal of Structural Engineering, Vol. 111, No. 7, July, 1985, pp. 1602-1618

The fatigue behavior of 12 different sets of partially prestressed concrete beams was experimentally investigated. All beams were rectangular, 9x4.5 in. (229x114 mm) in cross section, simply supported on a 9 ft (2.74 m) span and loaded in 4 point bending. Each set consisted of 2 identical specimens designed with the same input parameters. One control beam was tested in cyclic fatigue at a constant load range varying between 40% and 60% of the ultimate load capacity of the static specimen. The main input variables were the partial prestressing ratio PPR and the reinforcing index $\bar{\omega}$. Four different levels of PPR and three different levels of $\bar{\omega}$ covering both fully prestressed and fully reinforced were explored. Typical results and observed trends are described. Throughout the tests, measurements of strains in the reinforcement, deflections, crack widths, curvatures and their variation under static and cyclic fatigue loading were systematically recorded. The six partially prestressed and the three fully reinforced beams survived 5,000,000 cycles without suffering fatigue failure. The three fully prestressed beams which were loaded beyond cracking failed, respectively, at 1,210,000, 2,170,000, and 1,940,000 cycles.

571   Stiffness Matrices of Symmetric Structures

Robert K. Wen, Member, ASCE, (Prof., Dept. of Civ. Engrg. Mich. State Univ., East Lansing, Mich. 48824)

Journal of Structural Engineering, Vol. 111, No. 7, July, 1985, pp. 1621-1625

By considering the physical symmetry of a structure, a number of pairs of the elements in the upper (or lower) triangular part of the stiffness matrix are equal in absolute value. Such relationships are described in terms of degrees of freedom in "intrinsic coordinates." For similar nodes on each side of a plane of symmetry, the intrinsic coordinates of the nodes are mirror images of each other. The stiffness coefficients for corresponding degrees of freedom in the intrinsic coordinates for similar nodes are equal. The number of independent and dependent coefficients are delineated. For one fold symmetry, the number of independent stiffness coefficients is $(n+2)n/4$, and for two and three fold symmetry, it is $(n+4)n/8$ and $(n+8)n/16$, respectively. The relation of the signs of the dependent and independent coefficients is developed. The information may be used to save computation efforts needed for the formulation of stiffness matrices of symmetric structures.

572 Frequency of Railway Bridge Damage

William G. Byers, Fellow, ASCE, (Bridge Engr., The Atchison, Topeka and Santa Fe Railway Co., Amarillo, Tex.)

Journal of Structural Engineering, Vol. 111, No. 8, August, 1985, pp. 1635-1646

Bridge damage frequencies observed on a 5,500 mile (8,800 km) segment of railroad during a 1-yr period indicate that the probability of damage is strongly influenced by bridge type and moderately influenced by location and traffic density over the bridge. Sixty-eight incidents of serious damage were caused by high or shifted loads, fires, floods, collisions and derailments. The risk of fire damage to timber trestles is greater for open-deck than for ballasted-deck trestles and increases with trestle length. The overall annual probability of serious damage was 0.0016. The annual probability of mechanical damage from derailments was 0.00019. That of high load damage to underpasses was 0.010. For shifted load damage to through trusses, without regard to traffic density, it was 0.030. It was 0.00027 for highwater damage to timber trestles and 0.00036 for fire damage to timber trestles of all lengths. The overall lifetime probability of damage in the order of 0.1, for an assumed 75 yr life, appears to be an appropriate lifetime probability for serviceability failures.

573 Offshore Platform Fatigue Cracking Probability

Alphia E. Knapp, Member, ASCE, (Special Research Assoc., Amoco Production, P. O. Box 3385, Tulsa, Okla. 74102) and Bernhard Stahl, Member, ASCE, (Research Assoc., Amoco Production Co., P. O. Box 3385, Tulsa, Okla. 74102)

Journal of Structural Engineering, Vol. 111, No. 8, August, 1985, pp. 1647-1660

A systems approach is presented for fatigue reliability analysis of offshore platforms. The probability of fatigue crack occurrence in individual structural joints is considered. A nomograph relating deterministic fatigue life to probability of fatigue cracking is developed based on the lognormal fatigue reliability format. Joints are grouped according to ranges of fatigue life. Fatigue cracking probabilities of joints in each group are determined as a function of service life. Bernoulli trials are used to evaluate the probabilities of 0,1,2,3,... cracks occurring in each group and also in the entire structure. The effects of fatigue life correlation are included. It is shown that the number of joints, years of service life, and correlation effects are important factors in assessing system fatigue reliability. Analysis of a North Sea platform indicates a high level of reliability against fatigue damage.

574 Impulsive Direct Shear Failure in RC Slabs

Timothy J. Ross, Member, ASCE, (Sr. Research Struct. Engr., Civ. Engrg. Research Div., Air Force Weapons Lab., Albuquerque, N. M. 87117-6008) and Helmut Krawinkler, Member, ASCE, (Prof. of Civ. Engrg., Dept. of Civ. Engrg., Stanford Univ., Stanford, Calif. 94305)

Journal of Structural Engineering, Vol. 111, No. 8, August, 1985, pp. 1661-1677

Direct shear failure in reinforced concrete slabs under impulsive loads is relatively undocumented because of the paucity of data showing failure characteristics. The combined effects of beam action and wave action are likely to be important in developing models to understand the dynamic direct shear phenomenon. The research summarized in this paper makes an initial attempt to understand this phenomenon by considering elastic beam action to describe incipient direct shear failure conditions. The effects of load rate and beam-end restraint are investigated. Failure curves developed from elastic Timoshenko beam models are compared with experimental data on one-way slabs which failed in direct shear.

575   Fiber-Reinforced Concrete Deep Beams with Openings

Somsak Swaddiwudhipong, Member, ASCE, (Sr. Lect., Dept. of Civ. Engrg., National Univ. of Singapore, Singapore 0511) and **Nandivararm E. Shanmugam**, Member, ASCE, (Sr. Lect., Dept. of Civ. Engrg., National Univ. of Singapore, Singapore 0511)

Journal of Structural Engineering, Vol. 111, No. 8, August, 1985, pp. 1679-1690

The experimental investigation of the first crack and ultimate load behavior of steel fiber reinforced concrete deep beams with openings is reported. The effect of fiber content, positions of openings, and different types of loadings on the behavior of such beams is studied. The experimental results are supplemented and compared with established analytical solutions. They are in good agreement and indicate substantial increase in strength with the increase of fiber content. The significance of the natural load path joining the edge of loading and supporting plates is exemplified through a drastic reduction in strength when it is intercepted by the opening.

576   Analogy for Beam-Foundation Elastic Systems

Marcello Arici, (Asst. Prof., Dept. of Struct. and Geotech. Engrg., Univ. of Palermo, Palermo, Italy)

Journal of Structural Engineering, Vol. 111, No. 8, August, 1985, pp. 1691-1702

An analogy between two systems, each consisting of an elastic beam on elastic foundation, is given. Through this analogy, forces and displacements of one of two beams correspond to the displacements and forces, respectively, of the conjugate one, and field equilibrium and mechanical boundary conditions of one beam correspond to field compatibility equations and kinematical boundary conditions of the other beam. As the beam and foundation stiffnesses of the real system correspond to the foundation and beam compliances of the conjugate one, the problems of statically indeterminate beams not lying on foundation, having uniform or nonuniform cross section can be solved by using simple equilibrium equations on the conjugate beam. The analogy can be applied also when the beam on elastic foundation experiences inelastic deformations both distributed and concentrated. This enables us to solve several practical problems, e.g., temperature change effects, influence lines, secondary moments on statically indeterminate prestressed structures, etc. Applications to cylindrical tanks subjected to rise of temperatures and influence lines of nonprismatic shear beams are reported.

577   Sensitivity of Reliability-Based Optimum Design

Dan M. Frangopol, Member, ASCE, (Assoc. Prof., Dept. of Civ. Engrg., Univ. of Colorado, Boulder, Colo. 80309)

Journal of Structural Engineering, Vol. 111, No. 8, August, 1985, pp. 1703-1721

A reliability-based optimization sensitivity analysis technique based on the feasible directions concept is presented for plastic design of redundant structures. This technique is specialized to a formulation of the problem of plastic optimization for minimum weight based on a rational reliability constraint for sizing members. The technique is demonstrated with a structural design example where emphasis is placed on the sensitivity of the reliability-based optimum solution to both the correlation among strengths and the method for evaluating the overall probability of plastic collapse.

| 578 | Behavior of Wood-Framed Shear Walls |

Ajaya K. Gupta, Member, ASCE, (Prof. of Civ. Engrg., North Carolina State Univ., Raleigh, N. C. 27695-7908) and **Pei-Horng Kuo**, (Grad. Student, Dept. of Civ. Engrg., North Carolina State UNiv., Raleigh, N. C. 27695-7908)

Journal of Structural Engineering, Vol. 111, No. 8, August, 1985, pp. 1722-1733

Shear walls play an important role in the wind and seismic resistance of low-rise wood-framed buildings. The behavior of shear walls is primarily governed by the nail force-slip characteristics. The bending stiffness of studs and the shear stiffness of the sheathing play an important role in providing the stiffness. But the studs and the sheathing play a secondary role in defining the load-deformation properties. New models are presented to represent the shear behavior of these walls. It is shown that the proposed models are comparable to the finite element model, and give results which are in good agreement with the test results. The proposed models are particularly suitable for repetitive applications, e.g. in a nonlinear dynamic analysis.

| 579 | Sensitivity Analysis for Structural Errors |

Andrzej S. Nowak, Assoc. Member, ASCE, (Assoc. Prof., Dept. of Civ. Engrg., Univ. of Michigan, Ann Arbor, Mich. 48109) and **Robert I. Carr**, Member, ASCE, (Prof., Dept. of Civ. Engrg., Univ. of Michigan, Ann Arbor, Mich. 48109)

Journal of Structural Engineering, Vol. 111, No. 8, August, 1985, pp. 1734-1746

Uncertainties in structures result from natural hazards, manmade hazards, variations within common practice, and accepted departures from common practice. However, the major source of structural uncertainty and failures is departure from accepted practice, i.e., human error. The purpose of an error control strategy is to lower the frequency of errors which have large consequences. Sensitivity functions which relate structural errors to structural safety are bases for a control strategy. Sensitivity analysis is performed for several practical cases to demonstrate calculation of sensitivity functions and its importance in establishing an error control strategy.

| 580 | New Constitutive Law for Equal Leg Fillet Welds |

Vernon V. Neis, (Prof. of Civ. Engrg., Univ. of Saskatchewan, Saskatoon, Saskatchewan, Canada)

Journal of Structural Engineering, Vol. 111, No. 8, August, 1985, pp. 1747-1759

A new constitutive law for the load resistance versus displacement of equal leg fillet welds is derived from consideration of the mean stresses induced in a fillet weld due to general in-plane loading of the weld element. Plasticity theory with an isotropic hardening rule is used to derive expressions for the ultimate strength and maximum displacement of the fillet weld. A set of curves representative of the constitutive law is presented. The constitutive law contains the fundamental parameters of the constitutive law contains the fundamental parameters of the problem, illustrates the anisotropic nature of the response of fillet welds and is presented in invariant, nondimensional form.

| 581 | Design Pressures in Circular Bins |

Mark E. Killion, Member, ASCE, (Owner, Proj. Engr., Killion Engrg., 504 E. Jasper St., Paris, Ill. 61944)

Journal of Structural Engineering, Vol. 111, No. 8, August, 1985, pp. 1760-1774

Design pressues in circular bins, which have been of concern in recent years, are examined. Shallow bin pressures are distinguished from deep bin pressures. Furthermore, deep bin pressures are examined for funnel flow and mass flow conditions. Pressures due to outside temperature variations are examined. Examples and drawings are provided to clarify the principles described herein.

582　　RC Structural Walls: Seismic Design for Shear

Ahmet E. Aktan, Member, ASCE, (Assoc. Prof. of Civ. Engrg., Louisiana State University, Baton Rouge, La. 70803) and **Vitelmo V. Bertero**, Fellow, ASCE, (Prof. of Civ. Engrg., Univ. of California, Berkeley, Calif. 94720)

Journal of Structural Engineering, Vol. 111, No. 8, August, 1985, pp. 1775-1791

Provisions of 1982 UBC, ACI 318-83, and ATC 3-06 pertaining to seismic shear design of slender walls in mid-rise construction are evaluated. In the event of major ground shaking in regions of high seismic risk, the actual shear strength demand is expected to equal that associated with the axial-flexural supply. Thus, the codes' minimum design requirements ought to insure that flexure, and not shear, will control the seismic response during the expected rare, major seismic event in the western U.S. The codes do not implement this condition. Expressions suggested by design documents for computing the shear strength of walls were evaluated by comparing the predicted and measured strengths of 10 wall specimens tested at Berkeley. Although generally conservative, since code expressions do not incorporate the actual shear resisting mechanisms of walls under seismic effects, it is possible for the expressions to mislead the designer to poor shear design. Recommendations are formulated to improve the current shear design procedures by: (1) relating the shear strength demands to the actual axial-flexural supply; and (2) incorporating the actual shear resisting mechanisms in predicting shear strength supply of walls.

583　　Variability in Long-Term Concrete Deformations

Daniel J. W. Wium, (Engr., Van Wyk and Louw Inc., P. O. Box 905, Pretoria, South Africa) and **Oral Buyukozturk**, Member, ASCE, (Prof., Dept. of Civ. Engrg., MIT, Cambridge, Mass. 02139)

Journal of Structural Engineering, Vol. 111, No. 8, August, 1985, pp. 1792-1809

The deformation properties of concrete significantly influence the behavior of complex concrete structures. These properties are subject to large variabilities and it is therefore not possible to accurately predict the structural response. The sources of these variabilities are first examined and an example is then preesented of the long term deformation calculations in a multiple span bridge. Finally, a procedure using the finite element method is proposed for predicting the variability in the shortening of the bridge. A number of analyses are performed for selected combinations of material properties, and the deformation at different time steps is calculated. Polynomial models are then fitted to the results at each time step. These models are used to calculate the distribution of the response, the confidence limits and estimates of the decomposition of the variance of the reponse. The results indicate that the variability in the response can be reduced by performing selected tests on the actual material, and using that data to adjust the numerical material models.

584　　Transmission Towers: Design of Cold-Formed Angles

Edwin H. Gaylord, Fellow, ASCE, (Prof. Emeritus, Dept. of Civ. Engrg., Univ. of Illinois at Urbana-Champaign, Urbana, Ill. 61801) and **Gene M. Wilhoite**, Fellow, ASCE, (Consulting Engr., 5533 Pinelawn Ave., Chattanooga, Tenn. 37411)

Journal of Structural Engineering, Vol. 111, No. 8, August, 1985, pp. 1810-1825

Cold-formed members enable the engineer to design more cost-effective transmission towers. Members can be fabricated to closely fit the design requirements. Stiffening lips can increase local-buckling resistance of thinner sections and ensure members that can be handled without damage during erection. Recommendations for the design of plain and lipped angles produced by cold-forming are provided. Methods are presented which allow cold-formed members to be designed by the same criteria presently used for hot-rolled angles. These methods can also be extended to more complex shapes. Test data provides verification.

585   Collapse Loads of Continuous Orthotropic Bridges

John B. Kennedy, Fellow, ASCE, (Prof., Dept. of Civ. Engrg., Univ. of Windsor, Windsor, Ontario, Canada N9B 3P4) and **Ibrahim S. El-Sebakhy**, (Asst. Prof., Dept. of Civ. Engrg., Univ. of Alexandria, Alexandria, Egypt)

Journal of Structural Engineering, Vol. 111, No. 8, August, 1985, pp. 1827-1845

The ultimate-load carrying capacity of relatively wide continuous orthotropic slab bridges is estimated using the yield-line theory. Bridges of skew and rectangular planforms under several loading conditions are considered. The influence of the angle of inclination of the interior pier line support on the collapse load is established. The assumed yield-line patterns, based on a parameteric study using a progressive failure analysis, are verified and substantiated by test results from two 2-span continuous prestressed concrete waffle slab bridges of skewed and rectangular planforms. A method by which a concentrated load is made equivalent to an AASHTO HS 20 truck loading is presented. A design example illustrates the use of the equations derived.

586   Partially Prestressed Concrete Structures

Amin Ghali, Member, ASCE, (Prof., Dept. of Civ. Engrg., Univ. of Calgary, Calgary, Alberta, Canada T2N 1N4) and **Maher K. Tadros**, Member, ASCE, (Prof., Dept. of Civ. Engrg., Univ. of Nebraska, Omaha, Neb. 68182-0178)

Journal of Structural Engineering, Vol. 111, No. 8, August, 1985, pp. 1846-1865

A method is presented for prediction of stresses and strains of partially prestressed concrete structures. Members are assumed to be uncracked under permanent loads, but cracking can occur under transient live load. A superposition procedure is employed to calculate the time-dependent effects of various loadings including the initial prestress and the changes in stress in concrete and steel due to creep and shrinkage of concrete and relaxation of prestressed reinforcement. The analysis accounts for the presence of non-prestressed reinforcement. The analysis accounts for the presence of non-prestressed reinforcement and gives the time-dependent values of axial strain, curvature and stress in concrete, steel and non-prestressed steel. Numerical examples are included for calculation of the curvatures in uncracked and cracked partially prestressed beams and the curvature values are employed to determine the deflections.

587   A Moment of Inertia Invariant

Donald L. Dean, Fellow, ASCE, (Struct. Consultant, Costelow Co., Inc., 121 Kansas Ave., Topeka, Kans. 66603)

Journal of Structural Engineering, Vol. 111, No. 8, August, 1985, pp. 1869-1871

A simple formula is derived for the moment of inertia of a structural section consisting of three or more areas evenly spaced along a circular perimeter. The result is independent of the axis of orientation and the number of areas. For a finite number of areas, elements of the calculus of finite differences are used to sum the second moments of the areas and differential calculus is used to confirm the limiting case.

588   Dynamics of Steel Elevated Guideways—An Overview

Subcommittee on Vibration Problems Associated with Flexural Members on Transit Systems, Committee on Flexural Members of the Committee on Metals of the Structural Division

Journal of Structural Engineering, Vol. 111, No. 9, September, 1985, pp. 1873-1898

The most significant advances over the last 15 years relative to the dynamic problems of aerial guideways are summarized. Emphasized are guideways with steel supporting flexural members, designed for the exclusive right-of-way of a particular type of vehicle: Steel wheel-on-rail, rubber tire or air-levitated. Nondimensional parameters characterizing guideway-vehicle interactions are catalogued and related to recent designs and current design practice. The role of structural dynamic analysis and the need for further research in the control of both passenger ride quality and airborne noise radiating from elevated guideways are discussed.

589   Test of Post-Tensioned Flat Plate with Banded Tendon

Ned H. Burns, Member, ASCE, (Prof. of Civ. Engrg., Univ. of Texas at Austin, Austin, Tex.) and **Roongroj Hemakom**, Member, ASCE, (Sr. Engr., DeLeuw, Cather & Co., Washington, D.C.)

Journal of Structural Engineering, Vol. 111, No. 9, September, 1985, pp. 1899-1915

This investigation was undertaken to observe the strength and behavior of a one-half scale, nine panel, flat plate with banded arrangement of unbonded tendons. The test slab had three 10-ft spans in each direction and 2.5 ft overhangs on two edges with the nominal thickness measured at 2.75 in. The slab was designed with a low P/A stress level of 135 psi. The overall performance of the test slab at service load level (50 psf) was quite satisfactory. The slab behaved elastically and deflection was fully recovered upon releasing the applied load. The slab before failure was very ductile, with large deflections observed in all tests to failure. The failure load was observed at 160 psf, which was in excess of the designed factored load. The punching shear failure was secondary to the flexural failure: thus, the punching shear load was equal to flexural failure load. The minimum bonded reinforcement of 0.15% of the column strip area provided very good crack control up to the failure load.

590   Test of Four-Panel Post-Tensioned Flat Plate

Gary M. Kosut, Assoc. Member, ASCE, (Proj. Engr., Texas Dept. of Water Resources, Austin, Tex.), **Ned H. Burns**, Member, ASCE, (Prof. of Civ. Engrg., Dept. of Civ. Engrg., univ. of Texas at Austin, Austin, Tex.) and **C. Victor Winter**, (Proj. Engr., Page, Southerland, Page Engineers, Austin, Tex.)

Journal of Structural Engineering, Vol. 111, No. 9, September, 1985, pp. 1916-1929

A load test of a prestressed concrete, one-half scale model, flat plate with unbonded tendons is reported. The four panel model was 20 ft (6.1 m) square, 2.75 in. (70 mm) thick, with 8 in. (200 mm) or 7 in. (175 mm) columns spaced at 10 ft (3 m) centers. A uniform tendon arrangement was used in one slab direction and a banded arrangement was employed in the other direction. Banded reinforcement was provided in accordance with *ACI Code* (318-77) provisions. In addition, deformed bars were used as slab stirrups at several column connections. Bonded reinforcement effectively controlled cracking. The measured tendon stress at ultimate was 10–16% lower than the ultimate stress predicted by the *ACI Code* (318-77) equation. Yield-line analysis showed good agreement with measured strength. The banded tendon layout successfully supported a load greater than the design ultimate load. Although slab stirrups were ineffective, the shear strength of each slab-column connection tested exceeded the ultimate strength predicted by the *ACI Code*. Use of the *ACI Code* equivalent frame method of analysis was supported for design.

591   Seismic Response of Pile Supported Cooling Towers

Bor-Jen Lee, (Research Assoc., Petroleum and Energy Research Inst., Univ. of Tulsa, Tulsa, Okla. 74104) and **Phillip L. Gould**, Fellow, ASCE, (Harold D. Jolley Prof. and Chmn., Dept. of Civ. Engrg., Washington Univ., St. Louis, Mo. 63130)

Journal of Structural Engineering, Vol. 111, No. 9, September, 1985, pp. 1930-1947

An analytical method is developed to determine the seismic response of rotational shell structures, such as cooling towers, supported by deep foundations in the form of long piles. The substructure deletion method is employed through the development of a dynamic boundary system at the contact area between the superstructure and the substructure. A new mathematical formulation compatible with the shell deformation is developed to deal with the rigid body motions due to the negation of the fixed base assumption. Two pile foundation cases are considered in order to examine the effect of soil-pile-structure interaction on the seismic response of cooling towers.

592   Relative Reliability of Dimension Lumber in Bending

William M. Bulleit, Member, ASCE, (Asst. Prof., Michigan Tech. Univ., College of Engrg., Dept. of Civ. Engrg., Houghton, Mich. 49931)

Journal of Structural Engineering, Vol. 111, No. 9, September, 1985, pp. 1948-1963

The advanced first-order, second-moment method was used to examine the relative reliability of dimension lumber in bending for designs made in accordance with the National Design Specification for Wood Construction. System behavior was not included, and the load duration effect was approximated using the existing time-strength relationship, the "Madison curve." The species of lumber had an effect on the reliability. For a given grade and size, the reliability indices between certain species differed by more than one unit. The grade of lumber also affected the reliability. Generally, higher grade material exhibited higher reliability indices. One notable exception is No. 1 grade material, which often exhibits relatively low reliability. Increased size (depth) of dimension lumber produced a decrease in the reliability index. Considering lognormal and Weibull 2-parameter distributions for ultimate bending stress, the reliability is greater for the lognormal distribution assumption. The difference in the reliability index can be as much as one unit.

593   Inelastic Post-Buckling Behavior of Tubular Members

H. Sugimoto, (Engr., Kawaski Heavy Industries, Ltd., Chiba, Japan) and **Wai-Fah Chen**, Member, ASCE, (Prof. and Head of Struct. Engrg.,, Dept. of Civ. Engrg., Purdue Univ., West Lafayette, Ind. 47907)

Journal of Structural Engineering, Vol. 111, No. 9, September, 1985, pp. 1965-1978

Post-buckling, post-peak, and cyclic behavior of beam-columns and frames are studied using a computer model developed on the basis of the Finite Segment Method, using generalized cyclic stress-strain relationships and an automatic load control technique. Numerical examples of structural members and simple frames are made using the computer model. The results are compared with available experimental data as well as the results of simple rigid-plastic analyses to verify the proposed method.

594   Stability of Frames with Tapered Built-up Members

John C. Ermopoulos, (Asst. Lect., National Tech. Univ. of Athens, Inst. of Struct. Analysis and Steel Bridges, Athens, Greece) and **Anthony N. Kounadis,** (Prof., National Tech. Univ. of Athens, Inst. of Struct. Analysis and Steel Bridges, Athens, Greece)

Journal of Structural Engineering, Vol. 111, No. 9, September, 1985, pp. 1979-1992

This paper deals with simple frames having latticed members of varying moments of inertia. Crtical loads are established in a closed form by means of a bifurcational analysis. The individual and coupling effects of various parameters upon the load-carrying capacity of such frames are fully assessed. The stresses in the chords and lacing bars of the members are also established with satisfactory accuracy. The analysis presented herein is illustrated through a numerical example.

595   Hurricane Damage on Galveston's West Beach

J. K. Nelson, Member, ASCE, (Asst. Prof. of Civ. Engrg., Texas A&M Univ., College Station, Tex. 77843) and **J. R. Morgan,** Member, ASCE, (Asst. Prof. of Civ. Engrg., Texas A&M Univ., College Station, Tex. 77843)

Journal of Structural Engineering, Vol. 111, No. 9, September, 1985, pp. 1993-2007

Hurricane Alicia came onshore during the morning of August 18, 1983, and caused significant damage to wood-framed, elevated, waterfront construction on Galveston Island. Although the hurricane was moderate in nature (category 3 on the Saffir-Simpson Hurricane Scale), significant damage occurred. This paper examines the damage and outlines possible causes of the damage in the wood-framed structures on Galveston Island, in particular those structures in the West Beach area. The primary weakness appears to be the connection details. Similar damage caused by other hurricanes has been reported in the past. Current building codes for this type of construction are explained. Possible supplements to the codes to help minimize similar damage in the future are presented.

596   Effect of Cable Stiffness on Cable-Stayed Bridges

P. Krishna, (Prof. and Head, Dept. of Civ. Engrg., Univ. of Roorkee, Roorkee, India), **A. S. Arya,** (Prof., Dept. of Earthqauke Engrg., Univ. of Roorkee, Roorkee, India) and **T. P. Agrawal,** (Reader, Dept. of Civ. Engrg., Banaras Hindu Univ. Varanasi-221005, India)

Journal of Structural Engineering, Vol. 111, No. 9, September, 1985, pp. 2008-2020

The effect of cable stiffness on the behavior of cable-stayed bridges is presented. Bridges of radiating arrangement, with 12, 20, 28, and 36 cables per plane were investigated. Three cases of side to main span ratios, 0.35, 0.40, and 0.45, were considered for which cable tensions, girder deflections and tower and girder moments were computed. The computational work was done on IBM 360. The stiffness matrix method, treating the bridge as a two-dimensional structure, was used.

597   **Flexural Strength of Masonry Prisms versus Wall Panels**

Clayford T. Grimm, Fellow, ASCE, (Consulting Architectural Engr. and Sr. Lect. in Architectural Engrg., Univ. of Texas at Austin, Austin, Tex. 78712) and **Richard L. Tucker**, Fellow, ASCE, (Prof. of Civ. Engrg., Univ. of Texas at Austin, Austin, Tex. 78712)

Journal of Structural Engineering, Vol. 111, No. 9, September, 1985, pp. 2021-2032

For a given combination of masonry units and mortar, the flexural strength of masonry is dependent on the quality of workmanship, the method of loading, and the number of mortar joints in the span. The weakest link theory is applied to establish the relationship between flexural strength test data obtained for the same materials and workmanship by different loading conditions and sizes of masonry units. The theoretical results are verified by tests on 71 prisms and 23 wall panels.

598   **Response Maxima of a SDOF System under Seismic Action**

Mario Di Paola, (Assoc. Prof., Instit. di Scienza delle Costruzioni, Facolta di Ingegneria, Univ. di Palermo, Viale delle Scienze, I-90128 Palermo, Italy) and **Giuseppe Muscolino**, (Researcher, Instit. di Scienza delle Costruzioni, Facolta di Ingegneria, Univ. di Palermo,Viale delle Scienze, I-90128 Palermo, Italy)

Journal of Structural Engineering, Vol. 111, No. 9, September, 1985, pp. 2033-2046

A semi-empirical method for the evaluation of the statistical characteristics of the mean, variance and the maximum response peak of a single degree of freedom linear system under seismic action is presented. The nonstationary ground motion is modeled filtering a white Gaussian random noise and enveloping the filtered noise by a deterministic shaping function. The linear differential equations for the determination of the spectral moments are examined and a closed-form solution given. Extensive studies by means of digital simulation are made and the proposed semi-empirical formulation is compared with other approximate methods available in the literature within the framework of the first passage problem.

599   **Indicator of Residential Roof Strength in Wind**

James K. Nelson, Member, ASCE, (Asst. Prof., Civ. Engrg. Dept., Texas A&M Univ., College Station, Tex. 77843) and **W. Lynn Beason**, (Asst. Prof., Civ. Engrg. Dept., Texas A&M Univ., College Station, Tex. 77843)

Journal of Structural Engineering, Vol. 111, No. 9, September, 1985, pp. 2049-2053

Much residential construction in the United States is not engineered and is inspected by building officials who often are not engineers. Roof failures in these structures occur during severe windstorms. A simplified method to determine the adequacy of the roof-wall connection is presented in this paper. The procedure utilizes design charts and nomographs.

600   Cold-Formed Steel Farm Structures Part I: Grain Bins

George Abdel-Sayed, Member, ASCE, (Prof., Dept. of Civ. Engrg., Univ. of Windsor, Windsor, Ontario, Canada) and **Frank Monasa**, (Prof., Dept. of Civ. Engrg., Tech. Univ. of Michigan, Houghton, Mich.)

Journal of Structural Engineering, Vol. 111, No. 10, October, 1985, pp. 2065-2089

Cylindrical cold-formed steel structures are widely used for grain storage on the farm, as well as for stock facilities, storages, and utility structures. The information related to their analysis and design is scattered and incomplete; therefore, a task committee was established by the ASCE Committee on Cold-Formed Steel Members, with the following objectives: (1) To define the problems encountered in the analysis and design of these types of structures; and (2) to gather and develop information that could lead to rational methods of design. The state of the art for the upright cylindrical grain bins made of cold-formed steel is discussed. The paper critically examines the existing methods of analysis and, wherever possible, additional information and charts are provided to simplify the design process.

601   Cold-Formed Steel Farm Structures Part II: Barrel Shells

George Abdel-Sayed, Member, ASCE, (Prof., Dept. of Civ. Engrg., Univ. of Windsor, Windsor, Ontario, Canada) and **Frank Monasa**, (Prof., Dept. of Civ. Engrg., Technical Univ. of Michigan, Houghton, Mich.)

Journal of Structural Engineering, Vol. 111, No. 10, October, 1985, pp. 2090-2104

Cylindrical cold-formed steel structures are widely used for grain storage on the farm, as well as for stock facilities, storages, and utility structures. The information related to their analysis and design is scattered and incomplete; therefore, a task committee was established by the ASCE Committee on Cold-Formed Steel Members, with the following objectives: (1) To define the problems encountered in the analysis and design of these types of structures; and (2) to gather and develop information that could lead to rational methods of design. The state of the art for barrell shells is discussed. The paper critically examines the existing methods of analysis and, wherever possible, additional information and charts are provided to simplify the design process.

602   Confinement Effectiveness of Crossties in RC

Jack P. Moehle, Assoc. Member, ASCE, (Asst Prof., Dept. of Civ. Engrg., Univ. of California, Berkeley, Calif.) and **Terry Cavanagh**, (Grad. Student, Dept. of Civ. Engrg., Univ. of California, Berkeley, Calif.)

Journal of Structural Engineering, Vol. 111, No. 10, October, 1985, pp. 2105-2120

An experimental study is made of confinement effectivensss of crossties in reinforced concrete columns subjected to monotonically increasing axial compression. Ten columns were constructed, of which eight were reinforced. The main variable was the type of transverse reinforcement. Comparison is made between strength and ductility obtained by the different types of transverse steel, and analytical moment-curvature studies are used to estimate the influence of different confined concrete behaviors on flexural behavior of columns and structural walls. It is observed that crossties having 180° hooks are as effective in confining concrete as intermediate hoops. Crossties having 135° and 90° hooks are nearly as effective. It is concluded that both types of crossties are acceptable details for confinement of concrete where large inelastic strains will be applied monotonically.

603   **Minimum Weight Sizing of Guyed Antenna Towers**

William H. Greene, (Aerospace Engr., Struct. Concepts Branch, NASA Langley Research Center, Hampton, Va. 23665)

Journal of Structural Engineering, Vol. 111, No. 10, October, 1985, pp. 2121-2139

A procedure is described for automated sizing of the members and selection of pretension values for guyed antenna towers. The towers consist of a triangular cross section, tubular member mast supported by a number of sets of pretensioned guys. The sizing procedure employs a nonlienar finite element analysis coupled to a nonlinear mathematical progamming based optimizer. Numerical studies were performed on a typical VLF antenna configuration to obtain minimum weight tower designs. Various loading conditions, tower heights, and tower materials were considered.

604   **Seismic Response of RC Frames with Steels Braces**

Ashok K. Jain, (Reader in Civ. Engrg., Univ. of Roorkee, Roorkee, India 247667)

Journal of Structural Engineering, Vol. 111, No. 10, October, 1985, pp. 2138-2148

Steel racing members are widely used in steel structures to reduce lateral displacements and dissipate energy during strong ground motions. This concept is extended to concrete frames. Inelastic seismic response of reinforced concrete frames with K and X steel gracing patterns is presented. A two bay, six story frame designed by limit state design procedure was subjected to 1.25 times the North-South component of the 1940 El Centro earthquake and the artificially generated Bl earthquake. Takeda's bilinear hysteresis model was assumed for girders, and elastoplastic model was assumed for columns with axial force effects including, and a multilinear buckling model was assumed for steel bracing members. It is concluded that the inelastic response of both the K and X braced frames are satisfactory under the modified El Centro earthquake. The response of X braced frame is much better compared to that of the K braced frame under the modified Bl earthquake. It is also noticed that there is a considerable increasing in the column axial forces due to the presence of bracing members, which is typical of braced frames.

605   **Lateral Load Response of Flat-Plate Frame**

Jack P. Moehle, Assoc. Member, ASCE, (Asst. Prof., Dept. of Civ. Engrg., Univ. of California, Berkeley, Calif.) and John W. Diebold, (Research Asst., Dept. of Civ. Engrg., Univ. of California, Berkeley, Calif.)

Journal of Structural Engineering, Vol. 111, No. 10, October, 1985, pp. 2149-2162

Lateral-load response of a flat-plate frame under simulated earthquake base motions is examined. The test structure is a three-tenths scale model of a two-story, three-bay flat plate frame, which was designed and detailed according to current procedures for beamless slabs in regions of moderate seismic risk. Overall resistance to low, moderate, and high intensity base motions is examined, and design-oriented methods to evaluate lateral-load stiffness and strength are examined. It is concluded that the design resulted in a flexible but reasonably tough structural system. Relatively simple procedures to evaluate stiffness and strength are presented and found to be acceptably accurate for the test structure.

606   Active Control of Seismic-Excited Buildings

B. Samali, (Asst. Research Prof., Dept. of Civ., Mech. and Environmental Engrg., George Washington Univ., Washington, D.C. 20052), **J. N. Yang**, (Prof., Dept. of Civ., Mech. and Environmental Engrg., George Washington Univ., Washington, D.C. 20052) and **S. C. Liu**, (Professorial Lect., Dept. of Civ., Mech. and Environmental Engrg., George Washington Univ., Washington, D.C. 20052)

Journal of Structural Engineering, Vol. 111, No. 10, October, 1985, pp. 2165-2180

An investigation is made of the possible application of both the active tendon and active mass damper control systems to buildings excited by strong earthquakes. The effectiveness of both active control systems as measured by the reduction of coupled lateral-torsional motions of buildings is studied. The earthquake ground acceleration is modeled as a uniformly modulated non-stationary random process. The problem is formulated using the transfer matrices approach and a closed-loop control law. The random vibration analysis is carried out to determine the statistics of the building response and the required active control forces. The method of Monte Carlo simulation is also employed to demonstrate the building response behavior with or without an active control system. A numerical example of an eight-story building under strong earthquake excitations is given to illustrate the significant reduction of the building response by use of an active tendon or an active mass damper control system.

607   Interaction of Plastic Local and Lateral Buckling

Alan R. Kemp, Member, ASCE, (Prof. of Struct. Engrg., and Dept. Head, Andrew Roberts Chair of Civ. Engrg.,Univ. of Witwatersrand, Johannesburg, 2001, South Africa)

Journal of Structural Engineering, Vol. 111, No. 10, October, 1985, pp. 2181-2196

Attention is given to the interaction between local and lateral buckling of I-sections and their influence on plastic flexural ductility in regions of moment gradient. It is shown by means of simple theoretical formulations and the results of a series of tests to failure that rotation capacity in the plastic range is influenced significantly not only by conventional codified parameters, but also by the ratio of span to flange thickness and coincident axial force. In addition, the interaction between local and lateral buckling in the plastic region is demonstrated, and a model is formulated for quantifying this strain-weakening effect. The results are relevant to the plastic design of structures as well as more general research into methods of improving flexural ductility of I-sections.

608   Wind Induced Lateral-Torsional Motion of Buildings

Andrew Tallin, Member, ASCE, (Asst. Prof., Dept. of Civ. Engrg., Polytechnic Inst. of New York, Brooklyn, N.Y. 11201) and **Bruce Ellingwood**, Member, ASCE, (Research Struct. Engr. and Leader, Struct. Engrg. Group, Center for Building Tech., National Bureau of Standards, Gaithersburg, Md. 20899)

Journal of Structural Engineering, Vol. 111, No. 10, October, 1985, pp. 2197-2213

Fluctuating wind forces on tall buildings can cause excessive building motion that may be disturbing to the occupants. A method to relate dynamic alongwind, acrosswind, and torsional forces acting on square isolated buildings to building accelerations is developed using random vibration theory. Wind tunnel test data are analyzed to determine the spectra of force components and correlations among the different components of force. The effects on building vibration of statistical correlations among components of force and mechanical coupling of components of motion introduced by eccentricities of the centers of mass and rigidity from the building centroid are examined. Comparisons are made with more common building analyses, where the forces are assumed to be statistically uncorrelated and the components of motion are assumed to be uncoupled.

609   Nonlinear Analysis of Plates with Plastic Orthotropy

Manouchehr Gorji, Member, ASCE, (Assoc. Prof. of Civ. Engr., Portland State Univ., Portland, Oreg. 97207)

Journal of Structural Engineering, Vol. 111, No. 10, October, 1985, pp. 2214-2226

The behavior of plates exhibiting different longitudinal and transverse stress-strain relations beyond the elastic limit is investigated. The equivalent load concept is used to assess the influence of plastic orthotropy on the behavior of paltes undergoing large deflections. For plates considered, the lateral deflection is increased only slightly by the orthotropic behavior; the maximum extreme fiber stresses are, however, influenced more significantly as the aspect ratio of the plate approaches unity. The effect of orthotropy on stresses is reduced with increasing the aspect ratio of the plate.

610   Light-Frame Shear Wall Length and Opening Effects

Marcia Patton-Mallory, Assoc. Member, ASCE, (Engr., USDA Forest Products Lab., One Gifford Pinchot Dr., Madison, Wisc. 53705) and **Ronald W. Wolfe**, (Research Engr., USDA Forest Products Lab., One Gifford Pinchot Dr., Madison, Wisc. 53705)

Journal of Structural Engineering, Vol. 111, No. 10, October, 1985, pp. 2227-2239

Standard methods of testing the racking capacity of light-frame walls are inefficient and may give erroneous estimates of shear wall performance. This study is concerned with improving the data base for racking resistance of light frame walls with plywood and gypsum sheathings. The shear resistance of small walls sheathed with gypsum was compared to that of full-size walls. Results of both tests indicated that racking strength was linearly proportional to wall length. Contributions of sheathing layers and panel sections appear to be additive, and length of wall sections containing door and window openings may be neglected in determining racking resistance of conventional walls. One notable difference in the performance of these racking tests involved the effect of panel length on racking stiffness. Stiffness of small walls increased linearly with length while the stiffness of full-size walls increased nonlinearly. Knowledge that hypsum wallboard can contrtibute to shear wall performance and that it is additive to the resistance of plywood sheating should be helpful to building designers and code officials.

611   Inelastic Response of Tubular Steel Offshore Towers

Dept. of Civ. Engrg., Div. of Struct. Engrg., Berkeley, Calif. 94720 Popov. Egor P., Fellow, ASCE, (Prof. Emeritus of Civ. Engrg., Univ. of California-Berkeley), **Stephen A. Mahin**, Member, ASCE, (Assoc. Prof. of Civ. Engrg., Univ. of California-Berkeley, Dept. of Civ. Engrg., Div. of Struct. Engrg., Berkeley, Calif. 94720) and **Ray W. Clough**, Fellow, ASCE, (Nishkian Prof. of Struct. Engrg., Univ. of California-Berkeley, College of Engrg., Dept. of Civ. Engrg., Berkeley, Calif. 94720)

Journal of Structural Engineering, Vol. 111, No. 10, October, 1985, pp. 2240-2258

Three alternative experimental methods of subjecting structural models of steel offshore towers to severe inelastic lateral loadings simulating seismic effects are described. These pertain to either pseudo-static or pseudo-dynamic methods of loading, or to experiments performed using a shaking table. These methods are evaluated with respect to their advantages and disadvantages. Test results are highlighted to provide insight into the inelastic cyclic behavior of this type of structure. Since the seismic problem is by its very nature nondeterministic, seismic experiments primarily provide a data base for verifying the capabilities of available computer programs. Methods for predicting the inelastic seismic response of braced tubular steel platforms are briefly reviewed. Comparisons of experimental with analytical results are then presented to indicate the confidence that can be placed in such analytical predictions.

612   Stability of Plane Frames Omitting Axial Strains

Moshe Eisenberger, (Lect., Dept. of Civ. Engrg., Technion, Haifa, Israel) and Victor Gorbonos, (Grad. Student, Dept. of Civ. Engrg., Technion, Haifa, Israel)

Journal of Structural Engineering, Vol. 111, No. 10, October, 1985, pp. 2261-2265

The number of degrees of freedom for the stability analysis of plane frames may be considerably reduced by introducing axial constraints. Determination of the rank and basis of the constraint matrix by Gauss-Jordan elimination with pivoting leads to automatic selection of the best set of dependent and independent joint translations. Then a transformation of the buckling equations produces a reduced set of equations, that may be solved for the buckling loads of frames. This technique can be applied in conjunction with the elimination of joint rotations, thereby accomplishing a double reduction without significant loss of accuracy. The procedure described has been programmed on a digital computer.

613   Flexural Limit Design of Column Footings

Hans Gesund, Fellow, ASCE, (Prof. of Structural Engineering, Dept. of Civ. Engrg., Univ. of Kentucky, Lexington, Ky. 40506-0046)

Journal of Structural Engineering, Vol. 111, No. 11, November, 1985, pp. 2273-2287

Flexural collapse loads of eccentrically loaded, individual column footings were investigated using yield line theory. It was found that the cantilever failure mechanism recommended by the ACI Building Code does not give the lowest upper bound on the loads. Governing equations were derived for mechanisms that led to flexural collapse loads as low as one-half those predicted by the cantilever mechanism for some column-footing combinations. Suggestions are given for a reasonably simple design procedure that should ameliorate the problem.

614   Structural Optimization Using Reliability Concepts

Dan M. Frangopol, Member, ASCE, (Assoc. Prof., Dept. of Civ., Environmental, and Architectural Engrg., Univ. of Colorado, Bouder, Colo. 80309)

Journal of Structural Engineering, Vol. 111, No. 11, November, 1985, pp. 2288-2301

An overview of the major concepts and methods used in reliability-based optimization is presented. New formulations related to multicriteria optimization are developed. Comparative results are presented when different criteria are used for the optimum design of a structure under service or ultimate reliability constraints, or both. It is concluded that reliability-based optimization is now practicable for structural engineers. The writing of appropriate reliability-based optimum design software is a vital element at present receiving too little attention.

615   Theory for Thermally Induced Roof Noise

C. H. Ellen, (Research Mgr., Research & Tech. Ctr., John Lysaght (Australia) Ltd., P.O. Box 77, Port Kembla, NSW, 2505, Australia), C. V. Tu, (Research Officer, Research Officer, Research & Tech. Ctr., John Lysaght (Australia) Ltd., P.O. Box 77, Port Kembla, NSW, 2505, Australia) and W. Y. D. Yuen, (Research Officer, Research & Tech., John Lysaght (Australia) Ltd., P.O. Box 77, Port Kembla, NSW, 2505, Australia)

Journal of Structural Engineering, Vol. 111, No. 11, November, 1985, pp. 2302-2319

This paper presents a theoretical analysis of thermally induced noise in sheet metal roofs. A quasi-static model of the roof behavior is developed based on roof support batten frictional forces constraining the thermal expansion and contraction of the sheeting. As a result, in-plane stresses, set up in the sheeting, build up until the static frictional force limit is exceeded at one of the fasteners. At this point the sheeting slips over the batten at that fastener, the local frictional force is reduced to the dynamic frictional force limit, and a redistribution of in-plane stresses and fastener forces occurs. The redistribution can cause an overload of an adjacent fastener frictional force to exceed its limit, thereby inducing another slip. The sequence will continue until all frictional forces are below the static frictional force limit. The change in deflection pattern of the roof accompanying each slip is used to estimate the noise generated. With the assumptions that the curvature strain is small relative to the thermal strain, the coefficients of friction at each batten are identical and the batten spacing is uniform, the theory may be used to derive considerable detail of the slip cascade pattern and to determine expressions for the rate of noise bursts, for the necessary temperature change to cause a roof to change from "heating" to "cooling" cycling, and for the impulsive noise level.

616   Investigation of Wind Effects on a Tall Guyed Tower

Reginald T. Nakamoto, Member, ASCE, (Research Struct. Engr., Naval Civ. Engrg. Lab., Port Hueneme, Calif.) and Arthur N. L. Chiu, Fellow, ASCE, (Prof. of Civ. Engrg., Dept. of Civ. Engrg., Univ. of Hawaii at Manoa, Honolulu, Hawaii)

Journal of Structural Engineering, Vol. 111, No. 11, November, 1985, pp. 2320-2332

Full-scale wind velocity and structural response data from a tall guyed tower have been analyzed to obtain information concerning wind characteristics and dynamic response. Anemometers and accelerometers were installed at five stations along the height of the tower, and orthogonal components of wind velocities and tower accelerations were recorded. Design guidelines specify $1/7$ as the power-law exponent for wind profiles in coastal regions. However, data from this project, though limited in number of samples and conditions, indicate a mean exponent of 0.287. Digital correlation and spectral analysis are applied to the tower acceleration data to obtain estimates of resonant frequencies and critical damping ratios.

617   Approximations for Momements in Box Girders

Ishac I. Ishac, Member, ASCE, (Lect., Faculty of Engrg., Zagazig Univ., Zagazig, Egypt) and Tom R. Graves Smith, (Sr. Lect., Civ. Engrg. Dept., Southampton Univ., Southampton, U.K.)

Journal of Structural Engineering, Vol. 111, No. 11, November, 1985, pp. 2333-2342

Simple design approximations are presented for determining the transverse moments in single-span, single-cell concrete box girder bridges. Five representative loading cases are considered which can be used in combination to model a range of vertical dead and live loading and wind loading. The design approximations are based on the analysis of the cross section as an isolated framework. In the case of twisting loads, the longitudinal distributions are expressed in terms of a dimensionless panel length. The design approximations are derived from a finite element parameter study and give moments that are accurate to about 5%.

618   Effects of Weld Configurations on Strain Field

Taiji Inui, (Engr., Engrg. Dept., Ocean Proj. Div., Mitsui Engrg. & Shipbuilding Co., 6-4, Tsukiji 5-Chome, Chuo-ku, Tokyo 104, Japan), **Koichiro Yoshida**, (Prof., Dept. of Naval Architecture, Univ. of Tokyo, Tokyo, Japan) and **Kunihiro Iida**, (Prof., Dept. of Naval Architecture, Univ. of Tokyo, Tokyo, Japan)

Journal of Structural Engineering, Vol. 111, No. 11, November, 1985, pp. 2343-2354

Effects of geometrical configurations of welds on stress and strain distributions are investigated experimentally and theoretically. Three types of weld configurations are analyzed under axial and bending loads. Analyses by using a three-dimensional finite element method are performed, and the results are compared with the experimental ones with good accuracy. Stress and strain concentration factors are also investigated, and the applicability of Neuber's equation is examined.

619   Static Infinite Element Formulation

Prabhat Kumar, (Scientist, Struct. Engrg. Research Ctr., Roorkee, 247667, India)

Journal of Structural Engineering, Vol. 111, No. 11, November, 1985, pp. 2355-2372

In several problems of engineering and sciences, the domain of analysis extends to large distances in one or more directions. Such problems can be analyzed efficiently by the method of finite elements by using a combination of finite and infinite elements to discretize the domain of analysis. The existing approaches to formulating infinite elements have not been exploited to their full extent. In this research, these formulations are used to derive some advanced infinite elements with various forms of decay. The implementation of these elements and general formulation are verified by solving an elastic half-space problem in axisymmetric as well as in plane strain formulations. Also, this method has been applied to the problem of a shallow underground opening in horizontal tensile stress field. The numerical and analytical results are found to be in satisfactory agreement. These infinite elements overcome the difficulties of parametric formulation and numerical integration, but are suitable for the problems of elastostatics only.

620   Multistory Frames Under Sequential Gravity Loads

Chang-Koon Choi, Member, ASCE, (Prof. and Chrmn., Dept. of Civ. Engrg., Korea Advanced Inst. of Sci. and Tech.) and **E-Doo Kim**, (Grad. Student, Dept. of Civ. Engrg., Korea Advanced Inst. of Sci. and Tech.)

Journal of Structural Engineering, Vol. 111, No. 11, November, 1985, pp. 2373-2384

This paper deals with the bending moments and shear forces that are induced in the members of the frame by the differential column shortening, taking into account the construction sequence and the sequential application of dead weight in the analysis. Utilizating the substructuring techniques, the entire frame is analyzed by "one substructure at a time" approach in the reverse order of construction. Numerical examples of two high-rise buildings clearly show the significance of the differential column shortening effects and the effectiveness of the proposed frame analysis technique in coping with the problem.

621   Effective Length of Columns in Steel Radial Gates

Yunlin He, (Visiting Assoc. Prof. of Civ. Engrg., Univ. of Texas at Arlington, Arlington, Tex.) and Tseng Huang, Member, ASCE, (Prof. of Civ. Engrg. and Engrg. Mech.,, Univ. of Texas at Arlington, Arlington, Tex.)

Journal of Structural Engineering, Vol. 111, No. 11, November, 1985, pp. 2385-2401

The stability of steel radial gates used in reservoirs, hydroelectric power stations, etc. is studied by modeling them as rectangular, trapezoidal or triangular frames. The elastic resistance of the rubber seals for preventing leakage is incorporated in the formulation. Results are expressed in terms of the effective length factors of the compression members in the gates.

622   Nonlinear Analysis of Clamped Circular Plates

Boris L. Krayterman, (Engrg. Specialist, Bechtel Power Corp., Gaithersburg, Md. 20877-1454) and Chung C. Fu, Member, ASCE, (Sr. Staff Engr., Bechtel Power Corp., Gaithersburg, Md. 20877-1454)

Journal of Structural Engineering, Vol. 111, No. 11, November, 1985, pp. 2402-2415

The integral collocation method (ICM) is used to analyze large deflected circular plates with clamped (radially held and not held) outer edge under uniform pressure. The ICM has been completely computerized with automatic precision control and exhibits excellent results for the aforementioned plates. Convenient formulas have been derived using the least squares method, following an evaluation of many solutions of Von Kármán's equations. Critical values of fiber stresses and deflections are presented in numerical and graphical form. For confirming the validity and usefulness of the proposed formulation, comparisons are made with existing formulas. The comparative study shows a significant discrepancy in stresses; however, a good agreement is shown in deflections. The proposed computational technique presents lower stresses in the midplate and higher stresses at the end of the plate than the existing formulas. However, results show that the edge stresses are always higher than the midplate stresses in both methods. Further confirmations are made by comparison with the finite element results. Results of the comparison indicate that the proposed formulas produce more exact results than the existing formulas.

623   Thin Glass Plates on Elastic Supports

C. V. Girija Vallabhan, Member, ASCE, (Prof. in Civ. Engrg., Texas Tech. Univ., Lubbock, Tex.), Bob Yao-Ting Wang, (Formerly, Grad. Student, Dept. of Civ. Engrg., Texas Tech. Univ., Lubbock, Tex.), Gee David Chou, (Grad. Student, Dept. of Civ. Engrg., Texas Tech. Univ., Lubbock, Tex.) and Joseph E. Minor, Fellow, ASCE, (Prof. in Civ. Engrg. and Dir. of Glass Research and Testing Lab., Texas Tech. Univ., Lubbock, Tex.)

Journal of Structural Engineering, Vol. 111, No. 11, November, 1985, pp. 2416-2426

Von Karman equations are used for the analysis of rectangular window glass plates subjected to lateral pressures. A previously reported model that used a finite difference method solved the glass plate problem with simply supported boundary conditions. Here, this model is modified to include conditions where the plate is elastically supported on the boundary. Both linear and nonlinear elastic supports are considered. An efficient iterative procedure is used to solve the nonlinear equations.

624   Design of Cold-Formed Latticed Transmission Towers

Adolfo Zavelani, Member, ASCE, (Prof., Dept. of Struct. Engrg., Politecnico di Milano, Piazza Leonardo da Vinci 32, 20133 Milano, Italy) and **Paolo Faggiano**, (Research Engr., SAE S.p.A., Milano, Italy)

Journal of Structural Engineering, Vol. 111, No. 11, November, 1985, pp. 2427-2445

Cold-formed sections allow for very high efficiency since the shape and the size can be fitted to the specific working conditions. The use of these shapes in lieu of hot-rolled angles in latticed transmission towers favorably affects the overall cost. Recommendations for the design of cold-formed members for latticed transmission towers are presented, based upon the AISI "Specification for the Design of Cold-Formed Steel Structural Members" (3) and on significant specific testing experience. These recommendations cover all the shapes except plain angles. Experimental results are presented reflecting the correlation between test results and design recommendations. The fabrication technology calls for special care in the selection of steels which exhibit adequate ductility. Recommendations regarding galvanizing, shape tolerances, and connections are also included to provide guidance for proper design.

625   Floor Spectra for Nonclassically Damped Structures

Mahendra P. Singh, Member, ASCE, (Prof., Dept. of Engrg. Sci. and Mechanics, Virginia Polytechnic Inst. and State Univ., Blacksburg, Va. 24061) and **Anil M. Sharma**, (Engrg. Analyst, Sargent and Lundy Engrs., 55 East Monroe St., Chicago, Ill. 60633)

Journal of Structural Engineering, Vol. 111, No. 11, November, 1985, pp. 2446-2463

An approach is presented for the generation of seismic floor response spectra for structures which cannot be modeled as classically damped. Although the nonclassically damped structures do not possess classical modes, the proposed approach can still employ prescribed ground response spectra directly as input for the generation of floor spectra. The approach requires the solution of a complex eigenvalue problem which provides the information about the equivalent modal frequencies and damping ratios of the structure. The floor response spectrum value is expressed in terms of these equivalent modal parameters. The seismic input is required to be defined in terms of pseudo acceleration and relative velocity ground spectra for the equivalent modal frequencies and damping ratios of the structure and also at the oscillator frequency and its damping ratio. To verify the applicability of the proposed approach in view of simplifying assumptions made in this development, a numerical simulation study has been conducted. The results obtained by the proposed direct approach and the time history approach for an ensemble of acceleration time histories as inputs compare very well with each other, thus validating the applicability of the proposed approach.

626   Minimum Cost Prestressed Concrete Beam Design

Harry L. Jones, (Assoc. Prof., Dept. of Civ. Engrg., Texas A&M Univ., College Station, Tex. 77843)

Journal of Structural Engineering, Vol. 111, No. 11, November, 1985, pp. 2464-2478

An integer programming formulation for the minimum cost design of precast, prestressed concrete simply supported beams is developed. The cross section and gridwork within which strands must be placed are assumed given, and the design problem is to determine the concrete strength and the number, location, and draping of strands that minimizes the total cost of the beam. Constraints limit release and service load stresses, ultimate moment capacity, cracking moment capacity, and release camber. A box girder for a multibeam highway bridge is designed to demonstrate the computer program written to implement the methodology.

627   Lateral-Torsional Motion of Tall Buildings

Ahsan Kareem, Member, ASCE, (Assoc. Prof. and Dir.,, Struct. Aerodynamics and Ocean System Modeling Lab., Civ. Engrg. Dept., Univ. of Houston, Houston, Tex. 77004)

Journal of Structural Engineering, Vol. 111, No. 11, November, 1985, pp. 2479-2496

The lateral-torsional motion of tall buildings is investigated. For a square cross-section building, expressions for the alongwind, acrosswind and torsional loading are developed through the use of spatio-temporal fluctuations in the pressure field around the building. A simplified formulation is used to represent the dynamic behavior of torsionally coupled buildings by considering a class of buildings in which all floors have the same geometry in plan, the eccentricities between the elastic and mass centers are the same for all stories, and the ratio of the story stiffness in lateral directions is about the same for all stories. Methods of random vibration theory are used to estimate the rms and peak values of various components of response. The results indicate that the torsional response contributes significantly towards the overall dynamic response of a symmetric building. Inclusion of coupling between the lateral and torsional degrees-of-freedom further increases the building response. The procedure presented here provides a convenient checking procedure, from a structural reliability viewpoint, for serviceability and survivability limit states, thus ensuring serviceability under normal conditions as well as safety under extreme loads.

628   Wood Floor Behavior: Experimental Study

Ricardo O. Foschi, (Prof., Dept. of Civ. Engrg., Univ. of British Columbia, Vancouver, B.C., Canada V6T 1W5)

Journal of Structural Engineering, Vol. 111, No. 11, November, 1985, pp. 2497-2508

A series of 13 wood floors was tested under uniformly distributed load. The floors were manufactured with 12 (nominal) 2×8 hemlock joists, "No. 2 and better" grade, 5/8 in. (15.8 mm) Douglas-fir plywood, and 8d common nails at 6 in. (152 mm) o.c. In two floors the nail spacing was increased to 12 in. (305 mm) and in one floor, 5/8 in. (15.8 mm) waferboard replaced plywood as the sheathing. Experimental results are presented for floor deflections, first joist failure load, and subsequent failure loads. Test results were compared with theoretical predictions using the computer program FAP. It is concluded that, although first joist failure should be used in reliability analysis, the load carrying capacity of the floor assembly is normally higher and several more joist failures usually occur before total collapse. The variability in floor strength is substantially lower than the variability in bending strength of individual joists. The tests showed that floor behavior was nearly linear, and the results compared well with FAP predictions. In particular, the required nail stiffness may be obtained from a simple nailed joint test in compressions. It was also concluded that the weakest joists are not generally the ones that fall first in a floor, and that reliability predictions of floor strength can only be obtained when the bending strength is known for the weaker as well as the stronger joists in the population.

629   State-of-the-Art Report on Redundant Bridge Systems

The Task Committee on Redundancy of Flexural Systems of the ASCE-AASHTO Committee on Flexural Members of the Committee on Metals of the Structural Division, D. H. Hall, chmn.

Journal of Structural Engineering, Vol. 111, No. 12, December, 1985, pp. 2517-2531

Although most flexural systems are designed as a series of elements, they generally behave as a system in which loads are transferred to the foundation through a combination of members which resist loads in proportion to their relative stiffness. These structures are said to possess multiple load paths. The redundant members may even be small bracing members not designed to resist primary loads. The desirability of multiple load paths in flexural systems is presented. The AASHTO Highway Bridge Specifications penalize nonredundant steel members. The engineer is assigned the responsibility of determining if the structure is sufficiently redundant. Presently available methods of determining the degree of structural redundancy are reviewed. A survey of bridges known to have suffered some distress is presented. A surprising amount of redundancy was observed in most of these structures. Several areas in this field which need considerably more work are identified.

630   Effective Length of Luffing Jibs

Bosheng Qin, (Research Engr., Research Inst. of Construction Machinery, People's Republic of China) and **Chuan C. Feng**, Fellow, ASCE, (Prof. of Civ. Engrg., Univ. of Colorado, Boulder, CO 80309)

Journal of Structural Engineering, Vol. 111, No. 12, December, 1985, pp. 2532-2544

The side-sway effective length of a crane jib is investigated. For jibs with uniform cross section, restraints provided by the derrick, hoist cable and tower are also considered. When the jib end is completely fixed at the pivot, an improved approximate analysis for the effective length is presented. For a varying moment of inertia cross section, the stability equation is derived. Based on the computational results, the approximate formula for effective length is developed with sufficient accuracy.

631   Elastic Analysis of Gusseted Truss Joints

Kazuyuki Yamamoto, (Assoc. Prof. of Struct. Engrg., Shibaura Inst. of Tech., 3-9-14 Shibaura, Minato, Tokyo, Japan), **Narioki Akiyama**, (Prof. of Struct. Engrg., Saitama Univ., 255 Shimo-Okubo, Urawa, Saitama, Japan) and **Toshie Okumura**, Member, ASCE, (Prof. of Struct. Engrg., Emeritus, Univ. of Tokyo, 7-3-1 Hongo, Bunkyoku, Tokyo, Japan)

Journal of Structural Engineering, Vol. 111, No. 12, December, 1985, pp. 2545-2564

A series of experimental investigations was performed to determine the stress distribution in basic two types of gusset plates, the maximum stress intensity and its location, and to propose a simple and acceptable method for rational gusset plates designs. Experiments were performed on eight specimens of Warren- and Pratt-type truss joints with both spliced and monolithic gusset plates. Based on these test results, design formulas for determining the required gusset plates thickness are given. These design formulas are incorporated in "Tentative Design Guidelines of Gusseted Truss Joints" by Honshu-Shikoku Bridge Authority, which give the required gusset plate thickness from the following viewpoints: (1) Transmitting axial and bending forces from web members to gusset plate; (2) transmitting the shear stress resultant along the critical section; and (3) the maximum equivalent stress in the critical section.

632   Dynamics of Trusses by Component-Mode Method

William Weaver, Jr., Member, ASCE, (Prof. of Struct. Engrg., Dept. of Civ. Engrg., Stanford Univ., Stanford, CA) and **C. Lawrence Loh**, (Staff Engrg., Engineering Information Systems, Inc., San Jose, CA)

Journal of Structural Engineering, Vol. 111, No. 12, December, 1985, pp. 2565-2575

For dynamic analysis of trusses, we usually consider only axial deformations in the members. However, it is also known that inertial and body forces occur along the members in their transverse directions, which cause flexure. Therefore, we have devised a component-mode analytical model in which a member in bending constitutes a substructure. Treating the member

1985 ASCE PUBLICATIONS

as a simply-supported beam, we include a limited number of its vibrational mode shapes as displacement shape functions. For plane and space trusses with only a few members, we find by computer analysis that flexure plays an important role in their dynamic responses. However, in trusses with many members, this influence is not very significant, because the flexural deformations in members are localized.

633   Biaxially Loaded L-Shaped Reinforced Concrete Columns

Cheng-Tzu Thomas Hsu, Member, ASCE, (Assoc. Prof., Dept. of Civ. and Environmental Engrg., New Jersey Inst. of Tech., Newark, NJ 07102)

Journal of Structural Engineering, Vol. 111, No. 12, December, 1985, pp. 2576-2595

Results of an experimental and analytical investigation on the strength and deformation of biaxially loaded short and tied columns with L-shaped cross section are presented. The study explores the behavior of reinforced concrete columns under loads monotonically up to failure. A few tests loaded cyclically are also compared with those loaded monotonically. The strength interaction curves and load contours of L-shaped columns based on analysis and some test results are shown in this paper to provide advice for design information. Two design examples are given to provide possible design procedures.

634   Structural Performance and Wind Speed-Damage Correlation in Hurricane Alicia

Ahsan Kareem, Member, ASCE, (Assoc. Pres. and Dir., Struct. Aerodynamics and Ocean System Modeling Lab., Civ. Engrg. Dept., Univ. of Houston, Houston, TX 77004)

Journal of Structural Engineering, Vol. 111, No. 12, December, 1985, pp. 2596-2610

This paper describes the performance of buildings and other constructed facilities in the Houston-Galveston area during Hurricane Alicia on August 18, 1983. Records obtained from 17 anemometer sites in the Houston-Galveston area provide estimates of the fastest mile speed at 10m above ground in open terrain. The wind speed estimates at various locations are compared with codes and standards for wind speed design values for constructed facilities. The analysis indicates that wind speed estimates exceeded the ANSI (American National Standards Institute) Standard A58.1982 recommended values in only the Baytown-Laporte area and, possibly, near the west end of Galveston Island. For performance evaluation purposes, all structures are divided into four categories based on the level of engineering attention given during their design and construction phases: fully engineered, pre-engineered, marginally engineered, and nonengineered structures. Inasmuch as the wind speeds equaled or slightly exceeded ANSI standard recommendations at only a few locations, little building and structural damage should have occurred during Alicia, yet this was not the case. In summary, most of the damage from Alicia in the Houston-Galveston area was caused by lack of hurricane-resistant construction rather than the severity of the storm. Provision of adequate fastenings and anchorage of houses in the Galveston area and control of the availability of windborne debris in the Houston area would have substantially reduced damage caused by Alicia.

635   Design Spectra for Degrading Systems

Ghazi J. Al-Sulaimani, (Asst. Prof., Civ. Engrg., Univ. of Petroleum and Minerals, Dhahran, Saudi Arabia) and **Jose M. Roesset**, (Prof. of Civ. Engrg., Univ. of Texas, Austin, TX)

Journal of Structural Engineering, Vol. 111, No. 12, December, 1985, pp. 2611-2623

The effect of stiffness or strength degradation, or both, under cyclic loading on the seismic response of single-degree-of-freedom inelastic systems with 5% initial damping is investigated. Rules are proposed to construct inelastic design spectra for these systems (typical of reinforced concrete construction) as a function of the desired displacement (or ductility) ratio. These rules complement those used at present based on the response of simple elasto-plastic systems without any degradation.

636   Dynamic Analysis of Structures by the DFT Method

Anestis S. Veletsos, Member, ASCE, (Brown & Root Prof., Dept. of Civ. Engrg., Rice Univ., Houston, TX 77251) and **Carlos E. Ventura**, Student Member, ASCE, (Grad. Student, Dept. of Civ. Engrg., Rice Univ., Houston, TX 77251)

Journal of Structural Engineering, Vol. 111, No. 12, December, 1985, pp. 2625-2642

Following a brief review of the Discrete Fourier Transform (DFT) method of analyzing the dynamic response of linear structures, the limitations and principal sources of potential inaccuracies of this approach are identified, and an evaluation is made of the nature and magnitudes of the errors that may result. Two versions of a modification are then presented which dramatically improve the efficiency of the DFT procedure, and the relative merits of the two techniques are examined. The concepts involved are developed by reference to single-degree-of-freedom systems, and they are illustrated with the aid of judiciously selected numerical solutions.

637   Design Criteria for Box Columns Under Biaxial Loading

Shi-Ping Zhou, (Visiting Scholar, School of Civ. Engrg., Purdue Univ., W. Lafayette, IN 47907) and **Wai-Fah Chen**, Member, ASCE, (Prof. and Head, Struct. Engrg. Dept., School of Civ. Engrg., Purdue Univ., W. Lafayette, IN 47907)

Journal of Structural Engineering, Vol. 111, No. 12, December, 1985, pp. 2643-2658

Several design criteria for isolated, simply supported welded rectangular box-column subjected to compression and combined with biaxial bending are examined. The load carrying capacity of the columns is presented in terms of nonlinear interaction equations. In the present development, four major subjects are considered: (1) Uniaxial bending of columns; (2) biaxial strength of short columns; (3) biaxial strength of long columns; and (4) the role of residual stress.

638   Rapid Exact Inelastic Biaxial Bending Analysis

J. M. Rotter, (Sr. Lect., Univ. of Sydney, School of Civ. and Mining Engrg., Sydney, NSW, 2006, Australia)

Journal of Structural Engineering, Vol. 111, No. 12, December, 1985, pp. 2659-2674

A general analysis of cross sections subjected to axial force and biaxial bending is presented. The method is based on Green's Theorem and allows for exact determination of stress-resultants and tangent stiffnesses for a given set of deformations when the section boundary is rectilinear. The method may be applied to cross-sections of any material provided that the stress-strain relationships are integrable. The method is extremely rapid because stress integrals need only be evaluated at a small number of points on the section boundary. Unlike existing tangent stiffness methods, discontinuous stress-strain relationships may be used.

639   **Nonuniform Bending-Stress Distribution (Shear Lag)**

Luigino Dezi, (Research Asst., Istituto di Scienza e Tecnica delle Costruzioni, Universita degli Studi di Ancona, Via della Montagnola 30, 60100 Ancona, Italia) and **Lando Mentrasti**, (Research Asst., Istituto di Scienza e Tecnica delle Costruzioni, Universita degli Studi di Ancona, Via della Montagnola 30, 60100 Ancona, Italia)

Journal of Structural Engineering, Vol. 111, No. 12, December, 1985, pp. 2675-2690

Nonuniform normal longitudinal stress distribution in a trapezoidal box beam with lateral cantilever is discussed. The problem is solved by the variational method assuming as unknown the displacement of the beam axis and three functions which describe the warping of the horizontal flanges; the unknowns are reduced in case of vertical walls. The state of stress in each structural element is analyzed and an example is given to illustrate the results.

640   **Application of NLFEA to Concrete Structures**

Claude Bédard, (Asst. Prof., Centre for Building Studies, Concordia Univ., 1455 de Maisonneuve Blvd. W., Montréal, Québec, H3G 1M8, Canada) and **Michael D. Kotsovos**, (Lect. of Concrete Structures and Tech., Civ. Engrg. Dept., Imperial Coll., London, SW7 2AZ, U. K.)

Journal of Structural Engineering, Vol. 111, No. 12, December, 1985, pp. 2691-2707

Nonlinear finite element analysis (NLFEA) systems are rarely used in practice for the analysis of concrete structures even though several systems have been developed to date. The main reason appears to be a lack of agreement concerning the numerical description of material behavior characteristics. A successful NLFEA system is developed by incorporating realistic descriptions of material characteristics into a basically standard nonlinear finite element package. The system applies to the analysis of concrete structures under short-term stress and axisymmetric loading conditions. Simple, yet realistic, numerical models are presented to describe important features of material behavior such as failure criteria and deformational properties of concrete and steel, fracture processes of concrete and steel-concrete interaction. Procedures to represent crack propagation in concrete appear to influence significantly the quality of predictions. Three structural configurations of known experimental performance are finally analyzed. The predictions show good agreement with the experiment which is typical of several analyses covering a wide range of concrete structural forms.

641   **Simplified Earthquake Analysis of Multistory Structures With Foundation Uplift**

Solomon C.-S. Yim, Assoc. Member, ASCE, (Research Engineer, Offshore Systems, Exxon Production Research Co., Houston, TX) and **Anil K. Chopra**, Member, ASCE, (Prof. of Civil Engineering, University of California, Berkeley, CA)

Journal of Structural Engineering, Vol. 111, No. 12, December, 1985, pp. 2708-2731

A simplified analysis procedure is developed to consider the beneficial effects of foundation-mat uplift in computing the earthquake response of multistory structures. This analysis procedure is presented for structures attached to a rigid foundation mat which is supported on flexible foundation soil modeled as two spring-damper elements, Winkler foundation with distributed spring-damper elements, or a viscoelastic half space. In this analysis procedure the maximum, earthquake induced forces and deformations for an uplifting structure are computed from the earthquake response spectrum without the need for nonlinear response history analysis. It is demonstrated that the maximum response is estimated by the simplified analysis procedure to a useful degree of accuracy for practical structural design.

642   Optimum Characteristics of Isolated Structures

Michalakis C. Constantinou, Assoc. Member, ASCE, (Asst. Prof., Dept. of Civ. Engrg., Drexel Univ., Philadelphia, PA 19104) and **Iradj G. Tadjbakhsh**, Member, ASCE, (Prof., Dept. of Civ. Engrg., Rensselaer Polytechnic Inst., Troy, NY 12181)

Journal of Structural Engineering, Vol. 111, No. 12, December, 1985, pp. 2733-2750

The effect of frequency content of the ground excitation on the optimum design of a base isolation system and a first-story damping system is investigated. The ground excitation is modeled as a filtered white noise with Kanai-Tajimi power spectral density. The stationary response, including peak response, is derived in closed form. Small "ground frequencies" result in considerable differences of the optimum design as compared with the one based on a white noise model of the excitation. Examples show the manner of applying the results to an ordinary two-story structure. Time-history analyses of the optimized structures confirm the results of the stochastic analysis.

643   Design of a Seismic-Resistant Friction-Braced Frame

Mark A. Austin, (Res. Asst., Dept. of Civ. Engrg., University of California, Berkeley, CA 94720) and **Karl S. Pister**, (Prof. of Engrg. Sci., Dept. of Civ. Engrg., University of California, Berkeley, CA 94720)

Journal of Structural Engineering, Vol. 111, No. 12, December, 1985, pp. 2751-2769

This paper illustrates the design of a ten-story, single bay, earthquake-resistant friction-braced steel frame using a computer-aided design system called DELIGHT.STRUCT. Linear and nonlinear time history analyses are built into the design procedure itself rather than serving as a check at the end of the design process. The frame's performance is assessed on the basis of its response to gravity loads alone, gravity loads plus a moderate earthquake, and finally gravity loads combined with a rare severe earthquake ground motion. A preliminary analysis is conducted first to find the most active frame performance constraints. The method of feasible directions is employed to solve the constrained optimization problem. Objective functions include minimum volume, minimum dissipated energy and minimum sum of squared story drifts. A sensitivity analysis of frame response for perturbed ground motion and modeling parameters is included.

644   Distortional Buckling of Steel Storage Rack Columns

Gregory J. Hancock, (Sr. Lect., School of Civ. and Mining Engrg., Univ. of Sydney, N.S.W., Australia 2006)

Journal of Structural Engineering, Vol. 111, No. 12, December, 1985, pp. 2770-2783

A distortional mode of buckling is described for cold-formed lipped channel columns that have additional flanges attached to the flange stiffening lips. The purpose of the additional flanges, called "rear flanges," is to permit bolting of braces to the channel section so as to form upright frames of steel storage racks. Theoretical and experimental evidence of this mode of buckling is provided for perforated and unperforated columns. A set of design charts is included to permit calculation of the distortional buckling stresses of a range of sizes of practical channel sections with rear flanges.

645   **Functions for Relating Steel-Sections Properties**

Carlos I. Pesquera, (Res. Engr.,, 3D/EYE, Ithaca, NY 14850), **Shlomo Ginsburg**, Assoc. Member, ASCE, (Asst. Prof., Dept. of Civ. Engrg., Univ. of Kansas, Lawrence, KS 66045) and **William McGuire**, Fellow, ASCE, (Prof., Dept. of Struct. Engrg., Cornell Univ., Ithaca, NY 14853)

Journal of Structural Engineering, Vol. 111, No. 12, December, 1985, pp. 2787-2791

Functional relationships between cross-sectional properties of standard sections are useful for preliminary design, and for automated optimization. Expressions which relate the area, moment of inertia, and the torsional constant for steel sections are used for saving efforts associated witn numerous searches for section properties in a table or database. They also provide means for deriving recurrence relations for optimizations. This paper presents a modified set of relations for commonly used steel sections. A distinction is made between beam and column sections, resulting in less scatter of discrete points around the correlation functions, as compared with other expressions published so far.

646   **On Minimum Reinforcement in Concrete Structures**

Franco Levi, (Prof. of Civ. Engrg., Technical Univ. of Turin, 24, Corso Duca degli Abruzzi, Turin, Italy)

Journal of Structural Engineering, Vol. 111, No. 12, December, 1985, pp. 2791-2796

The usual criteria to fix minimum reinforcement in concrete structures require first, that the cross section of reinforcement be dimensioned so that the tensile stresses acting on the "embedment section" of concrete, when concentrated on the steel after cracking, do not exceed its elastic limit, and second, the crack width be kept within established limits. But these rules neglect some important aspects as soon as cracking provokes significant redistribution of stresses. The incidence of redundancy is decisive in the presence of imposed deformations. An interesting example is the case of a well in which thermohygrometric effects generates a diffused state of tension. From a finite elements analysis it appears that, in the mid-vertical section, a limited opening of the crack enduces a marked relaxation, the sum of the tensile stresses along the section being much smaller than before crack formation. This result suggests the idea to replace evenly distributed bars with series of well-reinforced chords (or prestressed prismatic elements) spaced so that they may keep within acceptable limits the opening of cracks in the intermediate zones and allowing to exploit the release of stresses to cut down load effects in restraining elements.

647   **Telecommunications Access to National Geodetic Survey Integrated Data Base**

Dave L. Pendleton, (Asst. Chief, Systems Development Branch, National Geodetic Survey Div., C&GS, NOS, NOAA, Rockville, Md. 20852)

Journal of Surveying Engineering, Vol. 111, No. 1, March, 1985, pp. 3-9

The National Geodetic Survey plans to implement a nationwide telecommunications network to allow on-line access to an integrated geodetic data base currently under development. The network will use one of the commercially available X.25 packet-switched telecommunications services, and the data base will use a back-end data base machine. This paper provides an overview of the data base concept, the network concept, a user session scenario, and outlines long range plans for including interactive computer graphics and local area networking capabilities.

648   Information Systems in Surveying

Clifford D. Cullings, (Chf., Castral Electronics Station, BLM, Div. of Cadastral Survey, 4700 E. 72nd Ave., Anchorage, Alaska 99507)

Journal of Surveying Engineering, Vol. 111, No. 1, March, 1985, pp. 10-13

The information systems of the Bureau of Land Management, Alaska State Office, are highly sophisticated and support many disciplines. One of these disciplines deals with the management of volumes of land-related data that is being collected. This overview examines data collection systems, their products and users, as well as automated land and mineral records systems.

649   USGS Telecommunications Responding to Change

James L. Hott, (Chf., Branch of Telecommunications Management, 804 National Center, Reston, Va. 22092)

Journal of Surveying Engineering, Vol. 111, No. 1, March, 1985, pp. 14-22

The telecommunications industry is undergoing tremendous change due to the court ordered breakup of the monopoly once enjoyed by American Telephone & Telegraph (AT&T). This action has resulted in a plethora of new services and products in all of the communications fields, including traditional voice and data. The new products are making extensive use of computer technology. At the same time, costs of telecommunications services have risen dramatically over the past three years. This article reviews some of the major actions that the Geological Survey has taken in response to these changes.

650   Use of Computerized Databases in Legal Research

Dan J. Freehling, (Law Librarian and Assoc. Prof., Univ. of Maine School of Law, 246 Deering Ave., Portland, Maine 04102) and **Robert F. Seibel**, (Assoc. Prof., Univ. of Maine School of Law, 246 Deering Ave., Portland, Me. 04102)

Journal of Surveying Engineering, Vol. 111, No. 1, March, 1985, pp. 23-35

The use by lawyers of specialized national computer databases in legal research, a task basic to the practice of law, is described. First, traditional legal research tools and methods are described; then the computer-assisted research technique is explained. The two methods are then compared by example and, finally, a general critique of both is presented. The conclusion is that the two approaches complement each other and enhance the profession's ability to serve clients effectively.

651   Role of Telecommunications in Timekeeping

Francis N. Washington, (Astronomer, Time Service Div., U.S. Naval Observatory, Washington, D.C. 20390)

Journal of Surveying Engineering, Vol. 111, No. 1, March, 1985, pp. 36-42

In all aspects of precise timekeeping the efficient and timely acquisition of an increasing amount of data is becoming of vital importance. To effectively handle these data, the U.S. Naval Observatory Time Service Division utilizes modern techniques of telecommunication. By processing the data through international computer networks, such as the General Electric MARK III, by offering a computerized Automatic Data Service, and by setting up remote automatic data acquisition systems, the Naval Observatory is not only ensuring the rapid exchange of data, but is freeing personnel for important analysis and research.

652   Computers and Group Communications: The Law:Forum
Experiment

Jennifer K. Bankier, (Assoc. Prof., Faculty of Law, Dalhousie Univ., Halifax, Nova Scotia, Canada B3H 4H9)

Journal of Surveying Engineering, Vol. 111, No. 1, March, 1985, pp. 43-56

Communications applications for computers include information retrieval, electronic mail, and computer conferencing. Computer conferencing is the use of the computer as a tool to facilitate communication among groups of individuals. The utility of computer conferencing is not restricted to the sciences, but may extend to any group that needs to communicate over long distances, or where regular communication is needed on a day-to-day basis among individuals with conflicting schedules. The article describes the implementation and results of a successful computer conference for lawyers and others who are interested in the relationship between law and computers. The success of a computer conference depends on the presence of an active organizer, the adaptation of the conferencing software to meet the needs of the particular discipline, the contents of the initial discussions in the conference, and an active recruitment strategy.

653   BITNET: Inter-University Computer Network

John Sherblom, (Analyst/Programmer, Computer Center, Univ. of Maine, Orono, Me. 04469)

Journal of Surveying Engineering, Vol. 111, No. 1, March, 1985, pp. 57-70

BITNET is an inter-university network that can facilitate the communication among researchers at colleges and universities throughout the United States, Canada, and some sites in Europe. This network gives users the capability of sending messages, text mail files, computer programs, data, and documents to each other. It also provides a service machine, a user directory indexed by keyword, gateways to other networks, and access to KERMIT, a file transfer profram. BITNET is easy to use and has many applications for the researcher with colleagues at other institutions.

654   Local Area Networks

Edmund M. Sheppard, (Prof., Dept. of Electrical Engrg., Barrows Hall, Univ. of Maine, Orono, Me. 04469)

Journal of Surveying Engineering, Vol. 111, No. 1, March, 1985, pp. 71-74

Local area networks will have a profound effect on the manner in which business is conducted. They will enable businesses, government agencies, and educational institutions to provide a wide variety of services by establishing communication links between many users that are capable of transmitting voice and video, as well as digital data. Some of the basic technologies are examined along with their advantages and disadvantages. Broadband cable systems have considerably greater capabilities than other forms of networks. Fiber optic systems are promising for the future.

655   **KERMIT Protocol for Inter-Computer Data Transfer**

Eloise R. Kleban, (Programmer/Analyst, Computing Center, Univ. of Maine, Orono, Me. 04469)

Journal of Surveying Engineering, Vol. 111, No. 1, March, 1985, pp. 75-78

To make satisfactory use of their machines, computer users must be able to send and receive files of data to and from other computer systems. Standard rules currently exist governing the hardware aspect of computer communication, but there is no generally agreed upon set of rules for the file transfer software. KERMIT, designed and maintained by Columbia University, consists of a software protocol for file transfer together with a set of implementations for various computers. KERMIT is of particular interest in situations where a variety of computers of different sizes must communicate using the telephone lines.

656   **Distributed Data Bases for Surveying**

Andrew U. Frank, (Dept. of Civ. Engrg., Surveying Engrg. Program, Univ. of Maine at Orono, Orono, Me. 04469)

Journal of Surveying Engineering, Vol. 111, No. 1, March, 1985, pp. 79-88

Space related data is gathered in different places and by different services. Considerable duplication of effort to maintain the same data is obvious. The use of a distributed database management system to manage such data collections is proposed. Information and data collection have special characteristics, which must be taken into account. They are expensive to update, but can easily be duplicated or transported from one place to another. Different users may produce or use the data and data may be an important asset for a user group. A distributed database may be adapted to such situations and most adequately suit managerial considerations. Users may maintain control over data of importance to them but nevertheless let other users gain access to it.

657   **Land Information Management: A Canadian Perspective**

John McLaughlin, (Prof., Univ. of New Brunswick, Dept. of Civ. Engrg., P. O. Box 4400, Fredericton, New Brunswick, Canada)

Journal of Surveying Engineering, Vol. 111, No. 2, August, 1985, pp. 93-104

Canadian society is entering an era in which a dominant public issue will be the need for more careful stewardship of the land, and the more intensive management of its limited resource stocks. This in turn will require a re-examination of the procedures and strategies for providing information about the land. This paper provides a conceptual framework for such a re-examination from a land information management perspective.

658   **Mathematical Models within Geodetic Frame**

Alfred Leick, Assoc. Member, ASCE, (Assoc. Prof., Univ. of Maine, Dept. of Civ. Engrg., Surveying Engrg. Program, 103 Bordman Hall, Orono, Me. 04469)

Journal of Surveying Engineering, Vol. 111, No. 2, August, 1985, pp. 105-117

The geodetic frame is defined from the practitioner's point of view, i.e., the deflection of the vertical and the geoid undulation must be available for the reduction of observations. The three-dimensional, the two-dimensional ellipsoid, and the conformal mapping model are identified as the prime mathematical models to be used for the analysis of surveying data. It is pointed out that the three-dimensional model is the most natural one and allows the incorporation of vector observations as derived from modern satellite surveying techniques. Whereas conformal mapping of the ellipsoidal surface may be very suitable for purposes of pictorial display, the ellipsoidal and the conformal model are quite complicated mathematically because the geodesic is required. A minimum requirement for a software package on the three-dimensional model is described.

659   Detection and Localization of Geometrical Movements

Lother Gründig, (Dr. of Engrg., Inst. für Anwendungen der Geodasie im Bauwesen, Univ. Stuttgart, Keplerstrabe 10, D-7000 Stuttgart 1), **Matthias Neureither**, (Dipl. of Engrg., Inst. Für Anwendungen der Geodasie im Bauwesen, Univ. Stuttgart, Keplerstrabe 10, D-7000 Stuttgart 1) and **Joachim Bahndorf**, (Dipl. of Engrg., Inst. Für Anwendungen der Geodasie im Bauwesen, Univ. Stuttgart, Keplerstrabe 10, D-7000 Stuttgart 1)

Journal of Surveying Engineering, Vol. 111, No. 2, August, 1985, pp. 118-132

For the geometric analysis of deformations in survey control networks, a program system has been developed and favorably applied to detect deformations, i.e., of dams, hang slides, buildings, or earthquake faults. The system allows for a one-, two-, or three-dimensional analysis. The basic mathematical theory of deformation analysis is summarized, and its implementation into a modular program system is demonstrated. The analysis is characterized by the application of statistical tests to estimable quantities. The datum definition and the change of datum applying S-transformations play an important role. It is shown how those strategies have been included in a powerful program system. Applications with real and simulated deformation models demonstrate the power and the efficiency of the underlying theoretical concept.

660   Relative Error Analysis of Geodetic Networks

Haim B. Papo, (Prof., Dept. of Civ. Engrg., Technion-Israel Inst. of Technology, Haifa 32 000, Israel) and **David Stelzer**, (Prof., Dept. of Civ. Engrg., Technion-Israel Inst. of Technology, Haifa 32 000, Israel)

Journal of Surveying Engineering, Vol. 111, No. 2, August, 1985, pp. 133-139

Positional accuracy estimators for all the points in a primary control network are derived from the covariance matrix. Relative error ellipses in scale and in orientation are proposed as objective descriptors of the positional accuracy of every point with respect to a ring of its neighbors. The ring size is chosen as a function of the average distance between points in the network. A free-net error ellipse is considered as another alternative for describing the positional accuracy of a point relative to its neighbors. A few examples of the analysis of a primary horizontal network are given as an illustration.

661   Terrain Profiling Using Seasat Radar Altimeter

Paul H. Salamonowicz, Assoc. Member, ASCE, (Cartographer, Office of Geographic and Cartographic Research, National Mapping Div., U. S. Geological Survey, 521 National Center, Reston, Va. 22092) and **Gregory C. Arnold**, (Cartographer, Data Management Branch, Eastern Mapping Center, U. S. Geological Survey, 562 National Center, Reston, Va. 22092)

Journal of Surveying Engineering, Vol. 111, No. 2, August, 1985, pp. 140-154

The ability to remotely establish elevation profiles, especially in inaccessible areas of

Alaska and Antarctica, could decrease the cost and time required to produce topographic maps. One possible means is the use of Seasat radar altimeter data to determine terrain profiles. Prior research has indicated the possibility of determining spot elevations accurate to about 1 m from terrain profiles derived from Seasat radar altimeter data, even though the altimeter was designed only for over-ocean sensing. Over-land use of the altimeter is generally limited to flat or gently rolling terrain. Using the data reduction method developed by the Geoscience Research Corporation, data from the Aleutian Islands, Alaska; Wilkes Land, Antarctica; and the Delmarva Peninsula, Delaware-Maryland-Virginia were investigated to assess the achievable accuracy. Terrain profiles were generated from the Antarctic and Maryland data, however no profiles could be generated from the Alaskan data. The Antarctic profiles were compared with nearby doppler determined elevations; a 50-ft. (approximately 15 m) discrepancy between the two was found. Where the altimeter gathered data in Maryland, a profile with a root-mean-square error of 12 ft (3.7 m) was determined.

662   The Surveyor and Written Boundary Agreements

Andrew C. Kellie, Assoc. Member, ASCE, (Asst. Prof., Dept. of Engrg. Tech., Murray State Univ., Murray, Ky. 42071)

Journal of Surveying Engineering, Vol. 111, No. 2, August, 1985, pp. 155-160

A written boundary agreement is one of several methods—parol agreement, acquiescence and acceptance, and practical location—which can be used to clarify the location of an uncertain boundary line. Unlike the other methods listed, the written agreement requires no litigation to establish the agreed line as a coterminous boundary. A written boundary agreement can also change the location of a line whose position is unambiguous by deed and susceptible to field location. Without a written agreement, a change in a locatable boundary amounts to an unwritten conveyance and, as such, is within the Statute of Frauds. In either case, the execution of the agreement requires an exchange of deeds between adjoiners. The agreed line should be described on the basis of a field survey which ties that line to the lines and corners of the adjoining owners. Once executed, the line agreement appears in the title record of each parcel of land involved.

663   Definition of the Term "Engineering Surveying"

Committee on Engineering Surveying of the Surveying Engineering Division

Journal of Surveying Engineering, Vol. 111, No. 2, August, 1985, pp. 161-164

A definition of the term "engineering surveying" is proposed in an effort to ease the confusion relatively recent developments have created with respect to the role of civil engineers in the practice of surveying. This confusion is further aggravated by a lack of understanding by certain engineering disciplines other than civil of the importance of surveying to the practice of civil, aeronautical, mechanical and mining engineering, among others. To clarify the roles of the various disciplines and professions, dialogue between such professional organizations as the American Congress on Surveying and Mapping and the American Society of Civil Engineers, as well as between these professional organizations and the registration boards of the various states, will be necessary. It is hoped that the definition will elicit a response from Society members so that a final version can be prepared and submitted to the governing board for formal adoption.

664   Seismically Induced Fluid Forces on Elevated Tanks

Medhat A. Haroun, Assoc. Member, ASCE, (Assoc. Prof., Civ. Engrg. Dept., Univ. of California, Irvine, CA 92717) and **Hamdy M. Ellaithy**, Assoc. Member, ASCE, (Grad. Research Asst., Civ. Engrg. Dept., Univ. of California, Irvine, CA 92717)

Journal of Technical Topics in Civil Engineering, Vol. 110, No. 1, December, 1985, pp. 1-15

Seismically induced fluid forces on elevated liquid storage tanks and their supporting systems are evaluated. X-braced as well as pedestal towers are analyzed. The interaction between liquid sloshing modes, tank rotation, tank translation and tank wall flexibility is taken into consideration to assess the relative importance of such effects on the overall earthquake response of elevated tanks. Ideal liquids and elastic towers are assumed throughout the first phase of the study. A comparison between the calculated forces and those provided by applicable codes is made.

665   The Microcomputer in C.E. Education: A Survey

James F. McDonough, Member, ASCE, (William Thoms Prof. of Civ. Engrg. and Head, Dept. of Civ. & Environmental Engrg., Univ. of Cincinnati, Cincinnati, OH)

Journal of Technical Topics in Civil Engineering, Vol. 110, No. 1, December, 1985, pp. 16-19

What is the state of microcomputers in civil engineering education? Which microcomputers are being used? Are program developers willing to exchange programs? The survey conducted by the Civil Engineering Division of the American Society for Engineering Education (ASEE) attempts to answer these questions and determine the next step to take in encouraging the development of usage by faculty and students. The respondents to the survey indicated a desire to participate in a software exchange. Many institutions have software they are prepared to make available. There are many problems of compatibility and transportability of programs. A summary of the most popular micros at the department, college, and university levels is provided. The C.E. division has established a committee to carry forward this effort. A special software demonstration and exchange was conducted at the 1984 ASEE national convention.

666   Heavy Oil Mining—An Overview

Tracy J. Lyman, Assoc. Member, ASCE, (Geotechnical Engrg., Stone & Webster Engrg. Corp., Denver, CO) and **Edwin M. Piper**, (Mgr. Process Proj., Stone & Webster Engrg. Corp., Denver, CO)

Journal of Technical Topics in Civil Engineering, Vol. 110, No. 1, December, 1985, pp. 20-32

Heavy oil mining is a promising technology that may increase ultimate oil recovery from existing reservoirs from 35–40% to 90% of the original oil in place. If successful, mining for heavy oil could effectively double or even triple the quantity of oil recovered from the known resources in the United States and throughout the world. Current estimates of these known world resources are several thousand billion barrels. Three mining methods appear to offer technical feasibility for heavy oil recovery: surface extractive mining, underground extractive mining and underground mining for access. These three methods have been attempted in several projects worldwide during the past century, and have demonstrated the technical feasibility of each method. Each project has failed economically in the past due to competition from oil produced from traditional surface wells. Successful implementation of the theories behind heavy oil production using mining techniques will require the cooperation of several diverse disciplines including petroleum geology, petroleum engineering, mining engineering, civil engineering and environmental sciences.

667   Response of Semi-Submersibles in Ice Environment

M. Arockiasamy, Member, ASCE, (Assoc. Prof., Dept. of Ocean Engrg., Florida Atlantic Univ., Boca Raton, FL 33431), D. V. Reddy, Member, ASCE, (Prof., Dept. of Ocean Engrg., Florida Atlantic Univ., Boca Raton, FL 33431), D. B. Muggeridge, (Prof., Faculty of Engrg. and Applied Sci., Memorial Univ. of Newfoundland, St. John's, Newfoundland, Canada A1B 3X5) and A. S. J. Swamidas, (Assoc. Prof., Faculty of Engrg. and Applied Sci., MemorialUniv. of Newfoundland, St. John's, Newfoundland, Canada A1B 3X5)

Journal of Technical Topics in Civil Engineering, Vol. 110, No. 1, December, 1985, pp. 33-46

The dynamic response of moored semi-submersibles to forces imposed on them by ice floes and bergy-bit/growler impacts is studied. Ice forces on the vertical columns of the semi-submersibles are determined using a strain rate, temperature, and ice-thickness-to-column-diameter ratio dependent force formulation. Three different analytical approaches are suggested for the solution of semi-submersible bergy-bit impact problems, and numerical results are obtained using the initial velocity approach. The semi-submersible chosen for numerical example is similar to the SEDCO 700 Series. The global and local structural responses are obtained for forces imposed on it by ice floes and bergy bits.

668   Are Pedestrians Safe at Right-On-Red Intersections?

Himmat S. Chadda, Member, ASCE, (Asst. Chief, Traffic Surveys, Data Collection and Analysis Branch, Bureau of Traffic Services, D.C. DPW, Washington, D.C.) and Paul M. Schonfeld, Assoc. Member, ASCE, (Assoc. Prof., Civ. Engrg. Dept., Univ. of Maryland, College Park, Md.)

Journal of Transportation Engineering, Vol. 111, No. 1, January, 1985, pp. 1-16

The permission of right-turn-on-red (RTOR) at intersections has operational, fuel consumption and air quality benefits but poses safety hazards, especially for pedestrians. A review and analysis of this pedestrian safety problem suggests a set of countermeasures including engineering (e.g., geometric design, signalization, signing, marking, prohibition, and visibility improvements), educational, and enforcement countermeasures to alleviate the problem.

669   Testing of Cement-Mortar Lined Carbon Steel Pipes

Chang-Ning Sun, Member, ASCE, (Civ. Engr., Geology and Geotechnical Engrg. Group. Tennessee Valley Authority, 400 Summit Hill Drive West, 156 LB-K, Knoxville, Tenn. 37902), James M. Hoskins, Assoc. Member, ASCE, (Civ. Engr., Geology and Geotechnical Engrg. Group. Tenessee Valley Authority, 400 Summit Hill Drive West, 156 LB-K, Knoxville, Tenn. 37902) and R. Joe Hunt, Member, ASCE, (Group Head, Geology and Geotechnical Engrg. Group, Tennessee Valley Authority, 400 Summit Hill Drive West, 156 LB-K, Knoxville, Tenn. 37902)

Journal of Transportation Engineering, Vol. 111, No. 1, January, 1985, pp. 17-32

Corrosion and incrustation in carbon steel raw water pipelines have been found to be problems at many of TVA's fossil and nuclear plants. Lining existing and new carbon steel raw water pipelines with cement mortar conforming to AWWA C602-76 requirements is one way to correct these problems. However, for nuclear applications, the cement-mortar lining under seismic loading conditions must be qualified by either analysis or testing. This paper describes a TVA-1 conducted full-scale testing program to structurally qualify the cement-mortar lined carbon steel pipelines for safety-related nuclear applications. Static and dynamic tests were carried out in the laboratory and field. After analyzing all data, it is concluded that the tests have verified the structural integrity of the cement-mortar lining for both nuclear and non-nuclear applications at TVA facilities.

670　　　　　　　　　**Blowup of Concrete Pavements**

Arnold D. Kerr, (Prof., Dept. of Civ. Engrg., Univ. of Delaware, Newark, Del. 19716) and **W. Alex Dallis**, Assoc. Member, ASCE, (Assoc., Holmes Thomson Logan and Cantrell, Charleston, S.C. 29402)

Journal of Transportation Engineering, Vol. 111, No. 1, January, 1985, pp. 33-53

Blowups of concrete pavements are caused by axial compression forces induced in the pavements by a rise in temperature and moisture. Although numerous papers and reports have been published on this subject, they have not yet resulted in the development of a generally accepted analysis. The purpose of the study is to establish the blowup mechanism for concrete pavements and to provide an analysis for problems of this type. The analysis presented is based on the assumption that blowups are caused by lift-off buckling of the pavement. A "safe" temperature and moisture increase in the pavement was defined; how it is affected by various pavement-subgrade parameters is shown.

671　　　　**Catalytic Modification of Road Asphalt by Polyethylene**

Wladislaw Milkowski, (Prof., Technical Univ. of Gdansk, Poland)

Journal of Transportation Engineering, Vol. 111, No. 1, January, 1985, pp. 54-72

The purpose of this research is to achieve asphaltic concrete of much higher stability and lower thermal susceptibility. Tests results showed that the addition of 5.0 wt% of polyethylene and of 2.0 wt% of a catalyst raised the softening temperature of asphalt D70 by 27.2°C (80.9°F), reduced penetration at 25°C (77°F) by 50.7 pen and increased the shear strength of asphaltic joints by $2.62 \cdot 10^5 \cdot N \cdot m^{-2}$. The stability of mean-grained asphaltic concrete containing 5 wt% of polyethylene in its binder, determined by modified Marshall test, was as much as 2,720 daN (6,118 lbf) higher than that of concrete containing nonmodified asphalt. The examination of rheological properties of modified asphaltic concrete samples, performed at the temperatures of 10°C (50°F) and -20°C (-4°F), proved their increased tensile strength as well as reduced stiffness coefficient. Mineral-asphalt mixes containing polyethylene modified binders, used for road construction and hydrotechnics seals, lead to remarkable savings due to their prolonged service life.

672　　　　**Correction of Head Loss Measurements in Water Mains**

Thomas M. Walski, Member, ASCE, (Research Civ. Engr., U.S. Waterways Experiment Station, Vicksburg, Miss. 39180)

Journal of Transportation Engineering, Vol. 111, No. 1, January, 1985, pp. 75-78

Pipe roughness or carrying capacity are occasionally determined in water mains to assess the internal condition of the mains or determine coefficients for pipe network models. As part of these tests, it it necessary to measure the head loss between two points in the pipe. Two methods are commonly used to determine the head loss: (1) two gage method; and (2) parallel pipe method. The effect of the differing temperatures in the main and parallel pipe is theoreticlly determined and a means for compensating for this effect is suggested. The parallel pipe method can be an extremely accurate method for measuring head loss in water mains. When the tests are conducted in hilly areas, caution must be exercised since the gage reading may not exactly correspond to the head loss, in which case a correction would be necessary to account for the effect of water temperature difference between the main and the parallel pipe.

673   High-Speed Rail Systems in the United States

The Subcommittee on High-Speed Rail Systems of the Committee on Public Transport of the Urban Transportation Division, Murthy V. A. Bondada, chmn.

Journal of Transportation Engineering, Vol. 111, No. 2, March, 1985, pp. 79-94

A great resurgence of interest in High Speed Rail (HSR) systems is occurring in the U.S. Several intercity corridors are under detailed scrutiny for application of various HSR technologies that have been developed or are in use outside the U.S. At present, all the HSR studies are in the physical and financial feasibility stages. The HSR industry is projecting that the U.S. has a potential for spending several billions of dollars on HSR systems. If the studies show that these systems are feasible, especially financially, the feasibility studies will lead to preliminary engineering, detailed design and, eventually, to deployment of HSR systems in the U.S.; in such a case, civil engineers around the country will play a major role in planning, design, construction and operation of the new intercity transportation systems. This paper reviews the ongoing effort in the area of HSR systems in the U.S. and presents an overview of the state-of-the-art HSR and Magnetic Levitation (Maglev) technologies presently considered for the intercity corridors of the U.S. The paper also includes an examination of issues associated with implementing HSR systems in the U.S.

674   Pedestrian Traffic on Cincinnati Skywalk System

Manmohan K. Bhalla, (Superintending Engr., Ministry of Shipping and Transport (Roads Wing), Jaipur, India) and Prahlad D. Pant, Member, ASCE, (Asst. Prof., Dept. of Civ. and Environmental Engrg., Univ. of Cincinnati, Cincinnati, Ohio)

Journal of Transportation Engineering, Vol. 111, No. 2, March, 1985, pp. 95-104

A method for estimating pedestrian trips on the skywalk system in the central business district of Cincinnati, Ohio has been developed. The central business district experiences congestion and conflicting pedestrian-vehicle movements, especially at signalized intersections. Attempts have been made to reduce the complexity of the problem by the construction of the skywalk system. Data on pedestrian trips on the skywalk system was obtained from a manual count conducted simultaneously at twelve strategic points on the skywalk system. Land use data was obtained from published sources and from related public and private agencies. Regression analysis was conducted to estimate pedestrian trips on the skywalk system during lunch peak (11:30 a.m. - 1:30 p.m.) and evening peak (3:30 p.m. - 5:30 p.m.). The results have shown that restaurants (number of seats), floor space area for office use (sq. ft. or m²), floor space area for retail use (sq. ft. or m²) and hotels (number of rooms) are the most significant independent variables. Parking space was not found significant. The analysis shows that simple quantitative models for estimating pedestrian trips on the skywalk system can be conveniently developed on the basis of a relatively simple data base.

675   Railway Route Rationalization: A Valuation Model

David Arditi, Member, ASCE, (Assoc. Prof. of civ. Engrg., Illinois Inst. of Tech., Chicago, Ill. 60616) and Dan Steven Krieter, (Regional Mgr., Barber-Greene Information Systems, Downers Grove, Ill. 60515)

Journal of Transportation Engineering, Vol. 111, No. 2, March, 1985, pp. 105-113

Abandonment of rail branches and secondary lines with low traffic density is an effective means of maintaining railway company profitability in many countries such as Britain,

Canada, Australia and Japan. In the United States, it was not until passage by Congress of the Stagger's Act in 1980, that the industry's abandonment vigor became fully manifested. This is because the Stagger's Act substantially reduced the time required for the Interstate Commerce Commission to act upon abandonment applications. An economic model is developed to predict losses in a railroad branch line without going through the cumbersome and lengthy calculations normally undertaken, thus reducing the time necessary for the decision-making process. To this purpose, regression analysis has been performed between losses incurred in 50 railroad branch lines and independent variables extracted from the abandonment applications made for these lines to the Interstate Commerce Commission. The resulting statistically significant model indicates that losses can confidently be predicted by making use of cost and revenue data that are readily obtainable by railway companies: freight revenues, maintenance costs for way and structures, rehabilitation costs and equipment maintenance costs.

676   Transmission Corridor Location Modeling

Dennis L. Huber, (Engr., Enertech Consultants, P.O. Box 17390, Pittsburgh, Pa. 15235) and **Richard L. Church**, (Prof., Dept. of Geography, Univ. of California at Santa Barbara, Santa Barbara, Calif. 93106)

Journal of Transportation Engineering, Vol. 111, No. 2, March, 1985, pp. 114-130

The location of corridor Rights-of-Way has been approached by a number of investigators dealing with problems of transmission power lines, pipelines, highways, and other transport systems. A commonly used three step approach in computerized approaches for locating a transmission corridor is reviewed. This three step methodology is widely accepted and employed in virtually all computerized corridor models. Unfortunately there can be significant geometric errors in the actual application of these techniques due to the underlying network representation. The sources of error are discussed and an approach is presented that can measurably reduce such errors. Applications to a hypothetical data set and to a statewide geo-based information system data base are presented.

677   3R Standards Implementation

David K. Phillips, Member, ASCE, (Dir., Office of Engrg., Federal Highway Administration, 400 7th Street, S.W., Washington, D.C. 20590)

Journal of Transportation Engineering, Vol. 111, No. 2, March, 1985, pp. 131-137

The Surface Transportation Assistance Act of 1982 increased the federal funding for highway improvements by well over 50%. In that act, the Congress also directed that a large portion of the non-Interstate funds be used for resurfacing, restoring, rehabilitating and reconstructing (3R/4R) existing highways and bridges in a manner which would enhance highway safety. The extent of appropriate safety improvements on 3R/4R projects varies depending on a number of factors, such as the existing highway condition, the scope of the pavement improvement, social or environmental impacts, available right-of-way and cost. Appropriate safety improvements include better skid resistance quality of the pavement; cross-slope changes to improve drainage; superelevation corrections; better signing, marking and delineation; regrading of roadside slopes, removal or upgrading of roadside hardware; and removal or mitigation of hazardous roadside features. Improvements to the geometrics may also be included, such as lane and shoulder widening, and horizontal or vertical curve reconstruction. Designing the rehabilitation of a highway is a greater challenge to the skills of highway designers than is the designing of a new highway. For a 3R project, designers need to use accident and traffic data and all that is known about safety design to custom tailor safety into the remodeled highway. The challenge is there for the highway administrators and designers to take up the enormous task of revitalizing the Nation's highway system.

678   Flyovers and High Flow Arterial Concept

Wilfred W. Recker, (Prof., Civ. Engrg. Dept., Inst. of Transportation Studies, Univ. of California, Irvine, Calif.), Michael G. McNally, (Research Assoc., Inst. of Transportation Studies, Univ. of California, Irvine, Calif.) and Gregory S. Root, (Asst. Prof., Dept. of Civ. Engrg., Univ. of South Florida, Tampa, Fla.)

Journal of Transportation Engineering, Vol. 111, No. 2, March, 1985, pp. 139-154

The feasibility of the development of high flow urban arterials by means of an integration of flyover technology with signal optimization is examined. Simulation results, using a modified version of the TRANSYT model, are presented for a case study arterial in the Greater Los Angeles area. These results indicate that the use of prefabricated flyovers, in conjunction with signal optimization, can effectively reduce travel delays and stops along heavily congested major arterials. The resulting high flow arterials can function effectively as "continuous flow boulevards," even when embedded in relatively dense urban traffic networks.

679   Examining Air Travel Demand Using Time Series Data

Ashish Sen, (Prof., School of Urban Planning and Policy, Univ. of Illinois at Chicago, Chicago, Ill.)

Journal of Transportation Engineering, Vol. 111, No. 2, March, 1985, pp. 155-161

One of the most commonly available types of data available to transportaton public and private properties are monthly data on ridership and revenue. A technique, based almost solely on such data, for determining the effects of various factors on ridership and revenue and for making forecasts under various pricing assumptions is presented. The technique is very simple (the analysis on which this paper is based took less than 2 hr without the use of a computer), and the results are easy to present to a non-technically oriented audience. The approach consists of first smoothing the data using running medians, and then examining the smoothed data for patterns, using data for Indian Airlines.

680   Mathematical Programming for Highway Project Analysis

Frank R. Wilson, (Prof., Transportation Group, Dept. of Civ. Engrg., Univ. of New Brunswick, Fredericton, New Brunswick, Canada) and Hugo Gonzales, (Research Asst., Transportation Group, Dept. of Civ. Engrg., Univ. of New Brunswick, Fredericton, New Brunswick, Canada)

Journal of Transportation Engineering, Vol. 111, No. 2, March, 1985, pp. 162-171

The development of a methodology to assist in the evaluation of alternative highway projects is described. The methodology selected is based on a multiple objective decision making approach and uses a goal programming technique as the evaluation mechanism. To examine the effectiveness of the technique, goal programming was compared to linear programming and to cost-benefits analysis. Proposed improvements to the road system in the Acadian Peninsula of New Brunswick, New Brunswick, Canada were selected as the case study to demonstrate the effectiveness of the technique. Goal programming was considered to be the most suitable technique for evaluating sets of alternative highway projects with multiple policies and objectives. The technique provides an easy method of assessing the impact of different policy structures on the selection of alternative highway improvement programs. The methodology is capable of multiyear analysis as used in programming of highway projects and accepts input from both the public and the decision makers in the highway planning process.

681   **Design, Construction and Maintenance of Torrijos Airport**

Maurice S. Greenberg, Fellow, ASCE, (Mgr. of Geotechnical Services, The Ralph M. Parsons Co., 100 West Walnut St., Pasadena, Calif. 91124), **Vernon A. Smoots**, Fellow, ASCE, (Partner, Dames & Moore, Suite 3500, 445 S. Figueroa St., Los Angeles, Calif. 90071) and **Rudy M. Pacal**, Member, ASCE, (Pres. and Princ. Engr., Gorian and Assoc., Inc., Suite A, 766 Lakefield Road, Westlake Village, Calif. 91361)

Journal of Transportation Engineering, Vol. 111, No. 2, March, 1985, pp. 173-189

In 1970, the Direccion de Aernoautica Civil (DAC) of the Republic of Panama realized that its facilities at Tocumen International Airport could not accommodate the heavy jumbo jet traffic of world airlines using the airport as a stop on routes between cities in North and South America. The existing single runway, built in 1933-34, was of inadequate pavement thickness, lacked modern lighting and instrumentation services, and badly needed improvement in surface and subdrainage to relieve hydroplaning and pavement pumping. In 1972, the DAC implemented the final design and construction of a second parallel runway and new modern passenger terminal in accordance with its master plan, prepared by The Ralph M. Parsons Company. Design and construction problems were successfully surmounted, resulting in extremely low-cost maintenance programs and highly satisfactory service conditions of the new runways, taxiways, apron area and terminal of the now-named General Omar Torrijos International Airport.

682   **Selected Bibliography on Traffic Operations**

Committee on Traffic Operations of the Urban Transportation Division

Journal of Transportation Engineering, Vol. 111, No. 3, May, 1985, pp. 191-195

A select list of references pertaining to traffic operations is presented. The list contains 62 references including handbooks, manuals, conference proceedings, textbooks, and case studies that the committee considers useful sources of technical information for civil engineers involved with urban traffic management. While this list is not all-inclusive, further sources can be found in the bibliographies of most of its entries.

683   **Minimum-Cost Design of Flexible Pavements**

Nagui M. Rouphail, Assoc. Member, ASCE, (Asst. Prof. of Transportation Engrg., Univ. of Illinois, Chicago, Ill.)

Journal of Transportation Engineering, Vol. 111, No. 3, May, 1985, pp. 196-207

The formulation of a mixed inter-linear programming model to determine a minimum-cost flexible pavement design is presented. The model identifies the number, type, and thicknesses of paving materials required to meet the structural strength requirements of the pavement system at a minimum initial cost to the highway agency. Policies regarding the designation of minimum and maximum layer thicknesses are accommodated in the model formulation, including the establishment of variable minimum and maximum thicknesses based on the underlying material. In addition, the structural layer coefficients are allowed to vary with the pavement configuration, including the number and type of constituent layers. The model can be used to calculate a marginal-cost function which has applications in the selection of cost-effective measures designed to strengthen the structural pavement capacity.

684   Statistical Designation of Traffic Control Subareas

James E. Moore, (Grad. Student, Dept. of Civ. Engrg., Program in Infrastructure Planning and Management, Stanford Univ., Palo Alto, Calif. 94305) and **Paul P. Jovanis,** (Asst. Prof. of Civ. Engrg. and Transportation, Northwestern Univ., Evanston, Ill. 60201)

Journal of Transportation Engineering, Vol. 111, No. 3, May, 1985, pp. 208-223

A method for identifying control subareas in traffic signal networks has been developed and tested with data from the Miami, Florida, business district. The procedure combines the use of cluster and discriminant analysis to group signalized intersections. Unlike existing procedures which use degrees of association between adjacent signals, the multivariate procedures used in this method allow for consideration of degrees of association among all signals in the original network. Model tests used the flow ratio for each intersection approach as the vector of intersection attributes. The resulting subareas were compact, well defined and slightly different for morning and midday traffic conditions. The procedure appears to have considerable utility in defining control subgroups in signal system design.

685   Effects of Vehicles on Buried, High-Pressure Pipe

John C. Potter, Member, ASCE, (Research Civ. Engr., USAE Waterways Experiment Station, Vicksburg, Miss.)

Journal of Transportation Engineering, Vol. 111, No. 3, May, 1985, pp. 224-236

The mechanical effects of selected types of traffic on buried, high-pressure steel pipe are examined. A section of the Colorado Interstate Gas pipeline across Fort Carson's Pinon Canyon Maneuver Site, which should be most susceptible to damage, is studied. The relationship between depth of cover and steel pipe deflection is clarified using field test results, with special emphasis on dynamic effects (impact factor). The results presented provide the technical basis for evaluating the feasibility of pipeline protection from anticipated traffic by earth cover.

686   Wave Load-Submarine Pipeline-Seafloor Interaction

M. R. Pranesh, (Prin. Scientific Officer, Ocean Engrg. Centre, Indian Inst. of Tech., Madras, India 600-036) and **G. S. Somanatha,** (Research Scholar, Ocean Engrg. Centre, Indian Inst. of Tech., Madras, India 600-036)

Journal of Transportation Engineering, Vol. 111, No. 3, May, 1985, pp. 237-250

Pipeline stresses due to loss of support caused by depressions or scour below submarine pipelines are analyzed. The sea floor on which the pipe is resting is idealized by Winkler medium. Bending in horizontal plane due to hydrodynamic loads is considered in the analysis, and theoretical equations and general solutions are presented. Theoretical results for vertical deflections due to loss of support are compared with experimental results, and computer results are presented in nondimensional form for a wide range of parameters. It is concluded that the stresses due to hydrodynamic drag will be affecting the resultant stresses considerably and will need consideration in the analysis. The deflection and moment coefficients are found to be linearly proportional to the load coefficient. An illustrative example has also been worked out to show the use of design charts.

687   Microcomputer Data Management System

Lansford C. Bell, Member, ASCE, (Assoc. Prof., Dept. of Civ. Engrg., Auburn Univ., Auburn, Ala. 36849) and **Charles P. Markert**, (County Engrg., Chambers County Alabama, Lafayette, Ala. 36862)

Journal of Transportation Engineering, Vol. 111, No. 3, May, 1985, pp. 251-257

The introduction of inexpensive microprocessor-based computer systems has given county engineers a valuable tool for manipulating, storing and retrieving timely information. A research project funded by the State of Alabama Highway Department has demonstrated the utility of an integrated set of microcomputer programs to a typical county highway department. The microcomputer is well suited to tracking labor, equipment and materials expenditures, comparing actual and budgeted expenditures, and tabulating equipment ownership, repair and gasoline consumption expenditures.

688   Use of NDT and Pocket Computers in Pavement Evaluation

Jacob Greenstein, Member, ASCE, (Chief Soils Engr., Louis Berge Int'l., Inc., 100 Halsted St., P.O. Box 270, East Orange, N.J. 07019)

Journal of Transportation Engineering, Vol. 111, No. 3, May, 1985, pp. 258-267

The profile and clearance heights at seven bridges of the New Jersey approach roads to the George Washington Bridge needed upgrading. To do this, nondestructive test (NDT) techniques were carried out on the pavement surfaces to determine the subgrade properties under the existing pavement. The NDT dynamic load and the measured deflection basin were analyzed to calculate the subgrade parameters, the elastic modulus (E), the CBR, and the modulus of reaction (K). The testing and calculation of 100 points, under heavy traffic volume, were completed in four hours. The subgrade parameters were used to design the reconstructed new pavement structure for local environmental conditions and for 20-yr projected traffic loading.

689   Vehicle Pooling in Transit Operations

Peter G. Furth, (Asst. Prof., Dept. of Civ. Engrg., Northeastern Univ., Boston, Mass. 02115) and **Andrew B. Nash**, (Doctoral Candidate, Dept. of Civ. Engrg., Univ. of California, Berkeley, Calif.)

Journal of Transportation Engineering, Vol. 111, No. 3, May, 1985, pp. 268-279

The benefits of pooling vehicles among routes that emanate from a common focus terminal are examined. In this strategy, trips are still scheduled, but vehicles are not assigned to specific trips. Instead, vehicles belonging to the pool serve all of the round trips leaving that terminal in a first in/first out sequence. Pooling improves schedule adherence, since in a pooled system a bus returning early can "cover" for a bus returning late. Pooling also facilitates interlining (sharing of buses among routes), which reduces the need for slack time. A procedure is developed for estimating schedule reliability. This procedure is applied to a set of 8 routes emanating from a Boston area terminal where it was found that with pooling the fleet size could be reduced by 11% while at the same time improving schedule adherence.

690   Statistical Analysis of Specification Compliance

Charles H. Snow, Member, ASCE, (Regional Construction Engr., Federal Highway Administration, 708 S.W. Third Ave., Portland, Oreg. 97204)

Journal of Transportation Engineering, Vol. 111, No. 3, May, 1985, pp. 280-291

There has been little use of statistical analysis for evaluating the degree to which highway construction materials comply with non-statistical specification requirements. The increased use of statistics offers potential benefits in the areas of project management and improved project quality. Using the standard deviation of a series of individual materials test results, an estimate of the percentage of material within specification limits can be made. Region 10 of the Federal Highway Administration has demonstrated the use of this analysis for evaluating specification compliance and for pinpointing points of change in materials characteristics for further investigation by identifying significant changes in standard deviation and quality level.

691   Development of Color Pavement in Korea

Kang W. Lee, Member, ASCE, (Asst. Prof., Dept. of Civ. Engrg., King Saud Univ., P.O. Box 800, Riyadh, Saudi Arabia), **Ju Won Kim**, (Dir., Han Kook Pavement Construction Co., Ltd., Seoul, Korea) and **Dae Woong Kim**, (Chf. Chemical Engr., Korean Inst. of Chemistry, Seoul, Korea)

Journal of Transportation Engineering, Vol. 111, No. 3, May, 1985, pp. 292-302

A development of color pavement in Korea is presented. It includes an examination of pavement materials used, preparation of specimen, comparison tests between regular and colored mixture, mixing at asphalt plant, and placing colored mixtures in field. A synthetic resin binder with light yellow color was developed, and proved compatible for colored pavement through the several comparison tests: Marshall stability test, wheel tracking test, ravelling test, and accelerated weathering test. For the pigment, several different inorganic products were used: $Fe_2 O_3$ (IOR), $Fe_2 O_3$ (IOY), $Cr_2 O_3$, and ultra-marine blue. The mixing operation and paving method of color mixture were same as regular asphalt concrete mixture, but the quantity of pigment replaced that of mineral filler. After a successful trial placing, the mixture was placed in the plaza of Great Seoul Children Park and the sidewalk of new Banpo Bridge. The utilization of color pavement will provide better roadway environment and traffic safety, and will be accelerated by the beautification plan for 1986 Asian and 1988 World Olympic Games in Seoul.

692   Large Trucks in Urban Areas: A Safety Problem?

James O'Day, (Interim Dir., Univ. of Michigan, Transportation Research Inst., Ann Arbor, Mich. 48109) and **Lidia P. Kostyniuk**, Member, ASCE, (Visiting Assoc. Prof., Dept. of Civ. Engrg., Michigan State Univ., East Lansing, Mich. 48824)

Journal of Transportation Engineering, Vol. 111, No. 3, May, 1985, pp. 303-317

The direct safety effects of increasing the number of large trucks in urban areas are explored. A simple theoretical model of the consequences of mixing trucks with cars is presented. The model, supported by recent detailed data from national in-depth accident investigation programs, indicates that the physical difference of mass between the two types of vehicles necessarily leads to a larger number of fatalities unless there is a concomitant reduction in the probability of such collisions. A comparison of urban and rural truck accident experience shows that the most severe urban accidents occur on urban interstate roads. Therefore, traffic engineers will be challenged by the problems associated with an increased truck population and will need to continue developing ways of reducing the chances of contact between the two types of vehicles in traffic flow.

693   Improvement of Congestion Detection on Expressways

Masata Iwasaki, (Asst. Prof., Dept. of Civ. Engrg., Musachi Inst. of Tech., Sotagaya-ku, Tokyo, Japan), **Masaki Koshi**, (Prof., Inst. of Industrial Sci., Univ. of Tokyo, Minato-ku, Tokyo, Japan) and **Izumi Okura**, (Assoc. Prof., Dept. of Civ. Engrg., Faculty of Engrg., Yokohama National Univ., Yokohama, Japan)

Journal of Transportation Engineering, Vol. 111, No. 4, July, 1985, pp. 327-338

This paper develops a method of improving congestion detection in urban motorway traffic surveillance and control systems in Japan. When developing motorway traffic surveillance and control systems, accurate congestion detection is often difficult or impossible because of the disturbances due to the oscillation of congested traffic flow. To improve the accuracy of congestion detection, it is necessary to distinguish noncongested flow from congested flow. Traffic flow data for the Tokyo Expressway and the Keiyo Expressway were collected. Computer experiments were carried out to develop a test to improve the accuracy of congestion detection on urban expressways. Applying the reference value proposed in this study greatly improved the accuracy of congestion detection.

694   Estimating Fuel Consumption from Engine Size

Tenny N. Lam, Member, ASCE, (Prof., Dept. ofCiv. Engrg., Univ. of California, Davis, Calif. 95616)

Journal of Transportation Engineering, Vol. 111, No. 4, July, 1985, pp. 339-357

Previous studies have demonstrated a useful relationship between fuel consumption and journey speed, vehicle weight, and idling fuel rate under urban driving conditions. The engine displacement has also been found to be strongly related to vehicle weight. The results of a statistical analysis of these known relationships with published data, are presented. The objective is to establish equations for estimating fuel consumption from travel survey data such as journey speed and engine size. Two equations were found to give good estimates of journey fuel consumption. One equation is for urban driving conditions and the other is for steady speed conditions such as in rural and motorway driving. The correlations between fuel consumptions under the two different driving conditions offer a possibility for developing an empirical equation for journey fuel consumption by taking into consideration the general patterns of travel and journeys.

695   Control Valve Flow Coefficients

William Rahmeyer, Member, ASCE, (Asst. Research Prof., Dept. of Civil Engrg., Colorado State Univ., Fort Collins, Colo.) and **Les Driskell**, (Consulting Engr., Pittsburgh, Pa.)

Journal of Transportation Engineering, Vol. 111, No. 4, July, 1985, pp. 358-364

The flow coefficient for a control valve is an important hydraulic parameter that is used to select and design the controls for a closed conduit system. The parameter is used to determine flow capacities, valve positioning, and energy losses of the system. The coefficient is also used for the analysis of the flow system for other phenomena such as cavitation and transients. It is important to understand the definition of the flow coefficient, how it is used, and how it is determined for a control valve. At present there are two sets of testing standards to determine the flow coefficient for a valve. The standards are different in the definition of how the coefficient is determined and can result in two different sets of flow coefficient data for a valve. Confusion has resulted in identifying which set of standards was used to generate the valve coefficient information, and the importance of designing a system using the same definition of the coefficient.

696   Acceptable Walking Distances in Central Areas

Prianka N. Seneviratne, (Asst. Prof., Dept. of Civ. Engrg., Technical Univ. of Nova Scotia, P.O. Box 1000, Halifax, Nova Scotia, Canada B3J 2X4)

Journal of Transportation Engineering, Vol. 111, No. 4, July, 1985, pp. 365-376

Pedestrian walking distances in a central business district are mostly dependent on the arrival mode in that CBD/central business district and the layout of the transportation network on which these modes operate. Changes to the existing pattern of operations, in terms of the location of new transportation facilities or relocation of existing facilities, can have a significant impact on walking distances. However, often the prime concern has been the impact of these changes on the flow of traffic, with little or no emphasis on the pedestrians. As opposed to arbitrarily derived acceptable walking distances suggested by some researchers, an approach based on findings from a series of surveys conducted in Calgary, Alberta, Canada, is proposed. A set of characteristics that influence the distribution of walking distances is identified. Depending on the form of the distribution, the "critical" distance that would be acceptable to a group of people can be derived using simple calculus. The critical distance will be the distance at which the rate of change of the slope of the frequency distribution of the walking distance for that group of people is greatest. Since the walking distance distributions indicate the propensity to walk, they indirectly represent the actual feelings of the people. This approach also facilitates decisions regarding locations, even where information regarding walking habits is scarce or unavailable. The application of Bayesian statistical decision theory enables one to estimate the appropriate distribution and the critical distance associated with it.

697   An Analytical Method of Traffic Flow Using Aerial
Photographs

Yasuji Makigami, (Prof. of Civ. Engrg., Faculty of Engrg. and Sci., Ritsumeikan Univ., Kita-ku, Kyoto, Japan), **Hamao Sakamoto**, (Chief Engr., Section of Traffic Control, Hanshin Expressway Public Corp., Higashi-ku, Osaka, Japan) and **Masachika Hayashi**, (Grad. Student, Civ. Engrg., Ritsumeikan Univ., Kita-ku, Kyoto, Japan)

Journal of Transportation Engineering, Vol. 111, No. 4, July, 1985, pp. 377-394

The outline and results of an aerial traffic survey of an 800-m section of the Hanshin Expressway are described. The objectives of the study were to record congested traffic flow and to determine its causes. Traffic in the study section was photographed by two 35 mm still cameras every 5s for 1h. All vehicles recorded in the southbound traffic flow were numbered and traced in order to project their trajectories in the time and space diagram. Speed and density contour diagrams were based on the theory of three-dimensional representation of traffic flow. The characteristics and causes of traffic congestion were analyzed using these diagrams.

698   Urban Transportation Impacts of Tall Buildings

Jon D. Fricker, Assoc. Member, ASCE, (Asst. Prof. of Transportation Engrg., Purdue Univ., West Lafayette, Ind. 47907) and **Huel-Sheng Tsay**, (Grad. Research Asst., School of Civ. Engrg., Purdue Univ., West Lafayette, Ind. 47907)

Journal of Transportation Engineering, Vol. 111, No. 4, July, 1985, pp. 395-409

Large cities in all parts of the United States are experiencing large development and redevelopment projects in their downtown areas. Regardless of the land use activities involved, the maintenance of basic levels of mobility for travel to, from, and within the central business district is crucial to the development's success. Strangely, a systematic advance evaluation of the ability of the transportation system to serve (or survive) the new activity was not common practice until recently. Given the enormous investments involved in these commercial projects, a coherent procedure to determine how well the transportation system will support the venture seems a prudent step. Given the variety of useful tools with which to build this procedure, investors and public officials alike should be able to analyze the situation within a wide range of precision and cost. This paper cites a number of techniques that are "on the shelf," and illustrates how each component of the downtown transportation system is closely linked with the others. A hypothetical example at an actual location demonstrates several of these points.

699   **Speed (Road) Bumps: Issues and Opinions**

Himmat S. Chadda, (Asst. Chief, Data Analysis Branch, Bureau of Traffic Services, D.C. Dept. of Public Works, Washington, D.C.) and **Seward E. Cross**, (Deputy Bureau Chief, Bureau Chief, Bureau of Traffic Services, D.C. Dept. of Public Works, Washington, D.C.)

Journal of Transportation Engineering, Vol. 111, No. 4, July, 1985, pp. 410-418

Speed (road) bumps are controversial. Despite their effectiveness as a speed deterrent device, the use has not been widespread in the United States. A majority of traffic professionals are strongly opposed to the use of these bumps on public right-of-way. Liability and safety remain the primary concerns. The current trend is to use "humps" instead of "bumps" which are flatter.

700   **Kinematic-Wave Method for Peak Runoff Estimates**

A. Osam Akan, Assoc. Member, ASCE, (Assoc. Prof., Dept. of Civ. Engrg., Old Dominion Univ., Norfolk, Va.)

Journal of Transportation Engineering, Vol. 111, No. 4, July, 1985, pp. 419-425

A set of algebraic equations to determine the design discharge for highway culverts is proposed. The kinematic-wave theory and the concepts underlying the commonly used rational method are adopted to derive these design discharge equations. Various drainage basin configurations, among which are rectangular and converging catchments, cascades of planes and overland flow-channel flow systems are considered. Composite basins with nonuniform slopes, roughness properties and runoff coefficients are included. In order to employ the proposed equations, the rainfall intensity-duration relationships and the physical characteristics of the basin should be specified. Unlike empirical formulas used for the same purpose, the proposed equations are physically based. They can be employed for a variety of drainage situations within the limitations of the kinematic wave theory. A practical application section is included to show the use of the proposed equations.

701   **Multi-Attribute Utility in Pavement Rehabilitation Decisions**

Satish Mohan, Member, ASCE, (Assoc. Prof., Dept. of Civ. Engrg., King Saud Univ., P.O. Box 800, Riyadh, Saudi Arabia) and **Adil Bushnak**, (Asst. Prof., Dept. of Div. Engrg., King Saud Univ., P.O. Box 800, Riyadh, Saudi Arabia)

Journal of Transportation Engineering, Vol. 111, No. 4, July, 1985, pp. 426-440

Priority ranking of pavements for rehabilitation is currently based on benefit-cost analyses and sufficiency ratings in most states. These methods either neglect social factors or assign them dollar values on a linear scale. Such methods rely heavily on subjective judgments, which can vary widely, and lack a mechanism to accommodate road users' preferences. Utility theory provides a decision-making method that optimizes the decision maker's utility or satisfaction, and can convert unquantifiable factors or attributes, such as safety, quality of service, etc., into terms of utility on a zero to one scale. The procedure for using multi-attribute utility in decision making is explained and illustrated in an example. The example shows that utility theory can be used satisfactorily in pavement rehabilitation decisions. Some weaknesses of utility theory are also pointed out.

702   Red Turn Arrow: An Information Theoretic Evaluation

S. Kullback, (Prof. Emeritus, Dept. of Statistics, George Washington Univ., Washington, D.C. 20052) and John C. Keegel, (Assoc. Prof., Dept. of Mathematics, Univ. of the District of Columbia, Washington, D.C. 20008)

Journal of Transportation Engineering, Vol. 111, No. 4, July, 1985, pp. 441-452

An information-theoretic statistical procedure for the analysis of categorical or qualitative variables or count data not necessarily arrayed in a multi-way cross-classification or contigency table is presented. This procedure can be used to provide a uniform approach to both statistical testing and estimation in various kinds of traffic studies. In particular, a statistical procedure is presented for analyzing counts of violations observed at various intersections during cycles of red left turn arrow or red ball signals. Data was collected at fifteen intersections of different types at different times of the day and located in different geographical locations.

703   Transportation Requirements for High Technology Industrial Development

Hani S. Mahmassani, Assoc. Member, ASCE, (Asst. Prof., Dept. of Civ. Engrg., Univ. of Texas at Austin, Austin, Tex.) and Graham S. Toft, Member, ASCE, (Prof., Inst. for Interdisciplinary Engrg. Studies, Purdue Univ., West Lafayette, Ind.)

Journal of Transportation Engineering, Vol. 111, No. 5, September, 1985, pp. 473-484

The belief that high technology industries are transient has led many regional growth strategists and planners to ignore transportation considerations in high tech development. This paper shows that this development has significant transportation implications that are different from those of traditional manufacturing. These requirements are related to an explanatory framework consisting of three developmental stages in high tech industrial activities. High tech professionals are frequent air travelers, thus requiring convenient access to good air service, which may be problematic for smaller nonhub areas. Particular concerns for urban transportation arise because of the potentially rapid rate of high tech development in a given area and its sprawling low-rise development in suburban and exurban settings. Finally, trends in high tech manufacturing and the nature of many high tech products indicate that air freight will be an important logistical component for these industries.

704   Collapse of Thick Wall Pipe in Ultra Deep Water

Edward A. Verner, Member, ASCE, (Vice Pres., TERA, Inc., P.O. Box 740038, Houston, Tex. 77274), Carl G. Langner, (Staff Research Engr., Transportation Research and Engrg. Dept., Shell Development Co., P.O. Box 1380, Houston, Tex. 77001) and Michael D. Reifel, (Pres., TERA, Inc., P.O. Box 740038, Houston, Tex. 77274)

Journal of Transportation Engineering, Vol. 111, No. 5, September, 1985, pp. 485-509

The literature pertaining to the collapse of thick wall pipes in deep water is evaluated. The subject matter is categorized by type of loading: (1) pure external pressure, (2) pressure plus tension, and (3) pressure plus bending. The study focuses on the buckling or yield type of failure rather than fracture; and with well casing as well as conventional high strength line pipe. The failure theories are discussed and summarized; and the data and theories are compared. It is found that elastic stability should be based on mean diameter. Although not sensitive to elastic imperfections, plasticity effects especially prevalent in thick wall pipe are recognized as a cause of substantial imperfection sensitivity with respect to initial ovality. The abundance of data for oil well casing is found to contrast sharply with the lack of data for line pipe. Tension plus pressure appears to be amenable to a tangent modulus approach. Bending plus pressure is found to be the least well documented both in terms of theory and testing.

705 Consistency in Design for Low-Volume Rural Roads

Clarkson H. Oglesby, Honorary Member, ASCE, (Silas Palmer Prof. of Civ. Engrg., Emeritus, Stanford Univ., Stanford, Calif.)

Journal of Transportation Engineering, Vol. 111, No. 5, September, 1985, pp. 510-519

The 2,000,000 miles of low-volume rural roads in the United States are different than the high-volume roads and should be designed differently. Traffic volumes on them are low, averaging about 110 vehicles/day or about one vehicle entering a given mile from both ends every three minutes during peak hours. This contrasts with one vehicle every four seconds at capacity. Geometrics on many of these roads have not changed since they were built in the 1920s and 1930s. Today, road improvements should be based on designs that are consistent and safe, but economical, because needs are great and funds are scarce. Present-day design practices for high volume roads require that each of their features meet a stipulated design speed set by modern surfaces and vehicles. This practice does not fit the low-volume situation since, whenever possible, drivers will exceed any affordable design speed. They must be slowed down when situations warrant it. A consistent approach to design which realizes cheap but safe improvements to low-volume roads is proposed. It involves integrating geometric design and positive guidance approaches. Positive guidance employs striping, signing, and other devices and strategies to mobilize drivers' senses so that they will drive sensibly. Selecting the less costly between geometry and positive guidance techniques will produce safer roads more cheaply.

706 Performance Measures for New York State Intercity Buses

Mark Abkowitz, Member, ASCE, (Asst. Prof., Dept. of Civ. Engrg., Rensselaer Polytechnic Inst., Troy, N.Y. 12180) and **Susan Violette**, Student Member, ASCE, (Research Asst., Dept. of Civ. Engrg., Rensselaer Polytechnic Inst., Troy, N.Y. 12180)

Journal of Transportation Engineering, Vol. 111, No. 5, September, 1985, pp. 521-530

This paper examines the applicability of intracity bus performance evaluation measures to intercity performance evaluation in the state of New York. Intercity performance analysis in the state is receiving greater attention as policymakers examine the fitness of carriers, cost-effectiveness of the existing route structure, and the fairness of the allocation of subsidies to operators on certain routes. An intercity carrier may be more efficient overall if it specializes in either a fixed route or special services, as efficiency tends to be correlated with the type of service in which the carrier gains most of its revenues. The effect of the state operating assistance program was found to sustain the financial health of carriers whose operations would be unable to recover costs otherwise. The New York State (NYS) case study illustrates the difficulties in evaluating intercity services. Only a few accepted measures of intracity bus performance could be applied to fixed-route services operated by intercity carriers due to data limitations. As the need for examining intercity performance becomes more critical, it is essential to collect better information from which to expand on the performance indicators demonstrated in the NYS case study. To achieve that goal, it appears that the reporting requirements and related enforcement must become more stringent in the future.

707   Safe Conversions of Unwarranted Multi-Way Stop Signs

Himmat S. Chadda, Member, ASCE, (Asst. Chief, Traffic Surveys, Data Collection and Analysis Branch, Dept. of Public Works, Washington, D.C.) and **Thomas E. Mulinazzi**, Member, ASCE, (Assoc. Otof., Dept. of Civ. Engrg., Univ. of Kansas, Lawrence, Kans.)

Journal of Transportation Engineering, Vol. 111, No. 5, September, 1985, pp. 531-538

There has been an increasing use of multi-way stop control at urban intersections within the last few decades. Use of multi-way signs where not warranted results in adverse safety, economic, operational, environmental, and social impacts. Many local jurisdictions in the United States have initiated steps for conversions of multi-way stops to less restrictive traffic controls such as two-way stops or, in some cases, yield controls. The local jurisdictions are, however, experiencing difficulties due to the lack of a well-tested and standardized safe conversion procedure. The need for a standardized conversion procedure is critical from the viewpoints of safety and uniformity. The suggested conversion process needs to be implemented in a step-by-step manner. Public relations campaign, information signing, installation of advance intersection warning signs, and enforcement are key elements of a successful conversion program.

708   Ranking Low Guardrail Sites for Remedial Treatment

Bradley T. Hargroves, (Technical Staff, MITRE Corp., 1820 Dolley Madison Blvd., McLean, Va.) and **J. Stuart Tyler**, (Environmental Specialist, Virginia Dept. of Highways and Transportation, Salem, Va.)

Journal of Transportation Engineering, Vol. 111, No. 5, September, 1985, pp. 539-545

A scoring procedure has been developed for prioritizing low guardrail sites for remedial treatment. The relative hazard of individual guardrail sections is estimated based on traffic volume and speed, roadway curvature, soil conditions, and severity of vehicular penetration as well as height, length, type and location of the guardrail. An intermediate step is provided which estimates the expected number of guardrail impacts per year. The procedure draws heavily on the guidelines developed for the installation of permanent and temporary barriers and on empirical results of a variety of highway studies.

709   Overview of Skid Resistance on Ohio Pavements

David C. Colony, Fellow, ASCE, (Prof. of Civ. Engrg., Univ. of Toledo, Toledo, Ohio)

Journal of Transportation Engineering, Vol. 111, No. 5, September, 1985, pp. 546-560

Results of an overview of about 30,000 skid numbers from state highway pavements throughout Ohio are described. Traffic volumes and aggregate types are strongly associated with mean skid numbers within a county. County mean skid numbers vary widely in the state, but are correlated with physiographic types. Skid number variations within each of 12 Ohio Department of Transportation administrative districts can be modeled by suitably selecting the parameters of a beta distribution. Limited data are available as corroboration, but they indicate that the distribution of skid numbers within a district is largely invariant with time, apparently a steady-state condition associated with two complementary stochastic processes: Deterioration of skid numbers under traffic and weather conditions and improvements due to maintenance and reconstruction. The equalization of mean skid numbers among districts is cited as a desirable policy goal; and a program of research aimed at the implementation of that goal and based on the results of the present study is briefly outlined.

710   **Extension of CBR Method to Highway Pavements**

Jacob Uzan, (Sr. Lect., Dept. of Civ. Engrg., Technion, Israel Inst. of Technology, Haifa, Israel 32000)

Journal of Transportation Engineering, Vol. 111, No. 5, September, 1985, pp. 561-569

The CBR design method for flexible airfield pavements was modified in 1971, on the basis of full-scale tests. A new load repetition factor and Equivalent Single Wheel Load computation scheme were introduced and implemented in the modified CBR design method. In the present paper, the method is extended to flexible highway pavements. The effect of loading conditions (which are different in highway and airfield pavements), is verified using AASHO Road Test results. The load repetition factor is then adjusted for the heavy traffic range. Pavement thicknesses obtained with the extended CBR design method are compared with those obtained with the current CBR method for highways, AASHO Road Test, SHELL and the British Roate Note 29. It is found that the extended CBR design method leads to: (1) A substantial reduction of pavement thickness as compared to the current CBR method; (2) A slightly thicker pavement in the light and medium traffic range and a slightly thinner pavement for heavy traffic as compared to other design methods. Design curves are presented for different subgrade CBR values.

711   **Leakage from Ruptured Submarine Oil Pipeline**

C. Kranenburg, (Sr. Scientific Officer, Lab. of Fluid Mechanics, Dept. of Civ. Engrg., Delft Univ. of Technolocy, Delft, Netherlands) and **E. Vegt**, (Grad. Student, Fluid Mechanics and Coastal Engrg. Groups, Delft Univ. of Technology, Delft, Netherlands)

Journal of Transportation Engineering, Vol. 111, No. 5, September, 1985, pp. 570-581

The rupture of a submarine oil pipeline gives rise to various mechanisms leading to an oil spill. Among these mechanisms, the leakage of oil driven by the difference in specific gravities of oil and seawater is difficult to estimate. A two-layer mathematical model and results of laboratory experiments concerning the buoyancy-driven leak rates are presented. The mathematical model is predictive in that no adjustable constants are introduced, and takes account of the effects of (laminar or turbulent) friction, angle of inclination of the pipeline, and inertia of the fluid. Gas or volatile components are assumed to be absent. The experiments were made in a model pipeline at various angles of inclination. The agreement between theoretical and observed leak volumes is satisfactory. Theoretical results for some prototype pipelines are also included.

712   **Generating a Bus Route O-D Matrix From On-Off Data**

Jesse Simon, (Statistical Analyst, Scheduling Dept., Southern Caliornia Rapid Transit District, Los Angeles, Calif.) and **Peter G. Furth**, (Asst. Prof., Civ. Engrg. Dept., Northeastern Univ., Boston, Mass.)

Journal of Transportation Engineering, Vol. 111, No. 6, November, 1985, pp. 583-593

The accuracy of route origin-destination estimates generated from boarding-alighting data was tested against actual origin-destination data. The estimates of trip length distributions and origin-destination matrices did not statistically differ from the actual data in tests of both simple and complex (branching) bus lines. While a note of caution was offered about applicability to extremely complex lines, the procedure is to be recommended as both an inexpensive and accurate method of estimating route O-D matrices for existing lines. Limitations and applications of the method are examined.

713   Comparison of Two Rigid Pavement Design Methods

David L. Guell, Member, ASCE, (Assoc. Prof., Dept. of Civ. Engrg., Univ. of Missouri-Columbia, Columbia, Mo. 65211)

Journal of Transportation Engineering, Vol. 111, No. 6, November, 1985, pp. 607-617

A comparison of the design thickness of rigid pavement slabs as determine by the AASHTO and PCA methods is presented. The comparison is given for a wide range of truck volumes and axle weights to represent the loadings that are likely to occur on facilities ranging from residential streets to major freeways. The effect of foundation strength on slab thickness is also examined for each design method.

714   Dynamics of Falling Weight Deflectometer

Boutros Sebaaly, (Formerly, Grad. Student, Dept. of Civ. Engrg., State Univ. of New York at Buffalo, Buffalo, N.Y. 14260), **Trevor G. Davies**, (Asst. Prof., Dept. of Civ. Engrg., State Univ. of New York at Buffalo, Buffalo, N.Y. 14260) and **Michael S. Mamlouk**, (Assoc. Prof., Dept. of Civ. Engrg., Arizona State Univ., Tempe, Ariz. 85287)

Journal of Transportation Engineering, Vol. 111, No. 6, November, 1985, pp. 618-632

An elastodynamic analysis of pavement response to Falling Weight Deflectometer blows is presented. The analysis is based on a Fourier synthesis of a solution for periodic loading of elastic or viscoelastic horizontally layered strata. The method is applied to selected flexible AAHSO test sections for which high quality experimental data are available in the literature. The results show that inertial effects are important in the prediction of the pavement response. Conventional static analyses yield significantly different results and, therefore, yield erroneous (unconservative) predictions of pavement moduli back-calculated from deflection data. Elastodynamic analyses, based on fundamental material parameters (Young's modulus, mass density) appear to provide a useful vehicle for correlating pavement response between different loading modes (impulse, vibratory, etc.). Since resonance is a less important factor in the displacement response characteristics of pavements subjected to transient loading, deflection data obtained from transient loading devices are in general easier to interpret.

715   A Tale of Two Cities: Light Rail Transit in Canada

Robert Cervero, (Asst. Prof., Dept. of City and Regional Planning, 228 Wurster Hall, Univ. of California, Berkeley, Calif. 94720)

Journal of Transportation Engineering, Vol. 111, No. 6, November, 1985, pp. 633-650

Toronto and Montreal are often cited for having effectively tied together rail transit and land-use planning. This paper probes whether two other Canadian cities, Calgary and Edmonton, have been equally successful with their recent light rail transit systems. Both Calgary and Edmonton have pioneered the development of the modern-day version of turn-of-the-century streetcar technology. Moreover, both have pursued creative approaches to zoning, joint development, cost sharing, and parking policies. An examination of before and after data, however, suggests that LRT's impacts on densities, residential construction, and mixed-use development have been quite modest in both places. Moreover, costs per passenger have risen steadily during the post-LRT period in both places, while total transit patronage levels have stabilized. It is apparent that the recent downturn in both Calgary's and Edmonton's petroleum-sensitive economies have overshadowed any effects of LRT on urban form or travel behavior. Some local observers, however, remain optimistic that the longer term consequences of LRT in both communities will be substantial.

716   Modern Material Ropeway Capabilities and Characteristics

Edward S. Neumann, Member, ASCE, (Prof., Dept. of Civ. Engrg. and Dir., H. O. Staggers National Transportation Center, West Virginia Univ., Morgantown, W. Va.), **Sam Bonasso**, Fellow, ASCE, (Pres., Alpha Associates, Inc. Consulting Engineers, Morgantown, W. Va.) and **Abel D. I. Dede**, Assoc. Member, ASCE, (Grad. Research Asst., Dept. of Civ. Engrg., West Virginia Univ., Morgantown, W. Va.)

Journal of Transportation Engineering, Vol. 111, No. 6, November, 1985, pp. 651-663

Modern material-handling ropeway system characteristics are described. Extremely limited in number in the United States, ropeways offer desirable characteristics in rugged terrain and in environmentally sensitive areas. Five examples from various parts of the world are presented to demonstrate ropeway capabilities. Economic data for recent systems are presented, including equipment costs, installation costs, annual operating costs, and unit transport haul costs.

717   Planning Development with Transit Projects

John H. Page, Assoc. Member, ASCE, (Asst. Prof., Virginia Military Inst., Lexington, Va. 24450) and **Michael J. Demetsky**, Member, ASCE, (Prof., Dept. of Civ. Engrg., Univ. of Virginia, Charlottesville, Va. 22901)

Journal of Transportation Engineering, Vol. 111, No. 6, November, 1985, pp. 665-678

One of the principal objectives of a transit project is to stimulate economic development. It is desirable to have early involvement of private sector developers in the transit planning process. However, the private sector seldom becomes involved before a transit project is nearly completed, and transit planners and public officials seldom incorporate developers' decision processes into their planning in order to increase the range of economic development opportunities. This study presents a site-development model which simulates the long- and short-range decision processes of developers. These decision processes involve four steps: (1) Determination of development demand for route alignment, station location and specific parcels development; (2) analysis of site constraints on development; (3) analysis of various design and marketing options; and (4) the financial analysis of an individual project. Three uses of the model for typical development problems are illustrated using data from the King Street Station in Alexandria, Virginia, a station on the Washington, D.C., Metro System. The three applications include assessment of development potential for various locations surrounding the transit station, consideration of different project designs at the same site, and the development of the marketing strategy for a preliminary design.

718   NO₂ Exposure from Vehicles and Gas Stoves

Michael D. Rowe, (Sci., Biomedical & Environmental Assessment Div., Brookhaven National Lab., Associated Universities, Inc., Upton, N.Y. 11973)

Journal of Transportation Engineering, Vol. 111, No. 6, November, 1985, pp. 679-691

Estimates of health impacts of reduced NO_x emissions from vehicles in the Chicago area show that transportation policy has little power to affect total exposure to NO_2. Even large decreases in outdoor NO_2 concentrations may only produce small decreases in health effects attributable to NO_2 exposure. Daily histories of exposure to NO_2 were constructed, based on assumptions about time spent at home, commuting, shopping, and at work. Outdoor NO_2 concentrations were estimated using EPA's Climatological Dispersion Model (CDM); concentrations in vehicles were estimated from assumptions about relationships among region-average concentrations, concentrations on highways, and inside:outside concentration ratios in vehicles. Indoor concentrations were estimated from assumptions about infiltration rates and emissions from gas cooking stoves in homes. The resulting exposures for different commuting patterns show that exposures in homes having gas stoves dominate total daily exposure.

719   Urban Transit: Equity Aspects

Michael C. Ircha, (Prof., Dept. of Civ. Engrg., Univ. of New Brunswick, Fredericton, New Brunswick E3B 5A3) and Margaret A. Gallagher, (Research Asst., Dept. of Civ. Engrg., Univ. of New Brunswick, Fredericton, New Brunswick E3B 5A3)

Journal of Urban Planning and Development, Vol. 110, No. 1, November, 1985, pp. 1-9

Evidence from the study of urban transit systems in two maritime cities suggests that certain aspects of this municipal service are inequitably distributed. A linear regression analysis of several equity-measuring variables and transit services indicators demonstrates inequities in service provision, particularly for "transit captives." For example, although the needs of the elderly tend to be met in both systems, the same is not true for youthful patrons. Low income families also tend to be discriminated against through the use of flat fares regardless of the length and time of the trip. Equity aspects in urban transit must be considered along with the system's efficiency.

720   Erosion Risk Analysis for a Southwestern Arroyo

Peter F. Lagasse, Member, ASCE, (Sr. Water Resources Engr., Resource Consultants, Inc., Fort Collins, Colo.), James D. Schall, (Sr. Engr., Simons, Li & Assoc., Inc., Fort Collins, Colo.) and Mark Peterson, Member, ASCE, (Engr., Simons, Li & Assoc., Inc., Fort Collins, Colo.)

Journal of Urban Planning and Development, Vol. 110, No. 1, November, 1985, pp. 10-24

A procedure is outlined for establishing a flooding and erosion buffer zone along an arroyo. The approach is applicable in general to ephemeral stream systems of the Southwest, while the hydrologically based definition of erosion and flood risk is applicable specifically in an urban setting. The risk analysis recognizes both the short-term impacts of flooding and erosion and the cumulative impacts of erosion over the long term. The procedure is based on an understanding of the basic physical processes of an arroyo system and integrates simple qualitative concepts with quantitative analysis of the vertical dynamics of the arroyo bed. Trends in vertical instability such as aggradation and degradation are then extended in the horizontal dimension to provide estimates of lateral erosion and channel migration potential. Application of the procedure to Calabacillas Arroyo northwest of Albuquerque, New Mexico, demonstrates reasonable results considering the documented dynamic behavior of ephemeral stream channels in the Southwest. The procedure answers a current need for engineering analysis techniques to support management and control of arroyos in an urban setting.

721   Residential Real Estate Development in Saudi Arabia

Edward Tieh-Yeu Huang, (City Planning Assoc., Community Redevelopment Agency of the City of Los Angeles, Los Angeles, Calif.)

Journal of Urban Planning and Development, Vol. 110, No. 1, November, 1985, pp. 25-33

The decline of world energy consumption in recent years has contributed to dramatic reductions in oil productivity and, thus, the national revenue for Saudi Arabia, which in turn greatly affects housing development in the country. This paper examines the performance of a housing finance agency established by the government of Saudi Arabia to promote housing production in the private sector. A descriptive analysis of the agency's recent activities shows that the agency has made a significant contribution to the provision of housing stock in Saudi Arabia. Yet real estate development has slowed due to the recent economic recession. Suggestions for urban planners and policy makers on modifying the existing finance policies needed to alleviate the difficulties and to maintain a high level of development are presented.

722   Use of Vegetation for Abatement of Highway Traffic Noise

Roswell A. Harris, Assoc. Member, ASCE, (Assoc. Prof., Dept. of Civ. Engrg., Univ. of Louisville, Louisville, Ky. 40292) and **Louis F. Cohn**, Member, ASCE, (Prof. and Chmn., Dept. of Civ. Engrg., Univ. of Louisville, Louisville, Ky. 40292)

Journal of Urban Planning and Development, Vol. 110, No. 1, November, 1985, pp. 34-48

The high cost of conventional highway noise abatement methodology (i.e., free-standing walls) has made mitigation of many impacted sites economically infeasible. A solution that may prove more economically reasonable for those sites is the use of strategically planted evergreen vegetation to form a dense barrier between the highway and impacted area. Field measurements were made on vegetative barriers planted only for visual screening purposes. The results of these measurements indicate that a 2 to 3 dB decrease in noise levels is possible with a narrow [30 ft (9.1 m)] belt of vegetation. These measurements are supported by the literature review, which indicates that an even further reduction may be possible with a barrier planted and maintained in such a way as to encourage maximum density growth. When coupled with the non-quantifiable psychological effects of blocking the highway from view, and the low construction cost, the potential for solving uneconomical abatement problems is clear.

723   Research Needs for Infrastructure Management

Neil S. Grigg, Member, ASCE, (Prof., Dept. of Civ. Engrg., Colorado State Univ., Fort Collins, Colo. 80523)

Journal of Urban Planning and Development, Vol. 110, No. 1, November, 1985, pp. 49-64

The infrastructure problem receiving so much national attention requires research to solve important policy problems at all three levels of government. Research needs for infrastructure are presented for five problem-solving objectives: to improve management processes, utilize new technologies and materials, reduce code and standard constraints, improve financial capacity, and adjust to future living patterns. The research statements are presented in an interdisciplinary format to link research with needed problem solutions. The principal disciplines included are: management science, engineering and physical science, and economics and finance. A bibliography of 80 references is included.

724   Estimating Recreational Travel and Economic Values of State Parks

Yupo Chan, Member, ASCE, (Assoc. Prof., Dept. of Civ. and Environmental Engrg., Washington State Univ., Pullman, Wash. 99164) and **T. Owen Carroll**, (Assoc. Prof., W. Averill Harriman College for Urban and Policy Sciences, State Univ. of New York, Stony Brook, N.Y.)

Journal of Urban Planning and Development, Vol. 110, No. 1, November, 1985, pp. 65-79

Recreational travel—unlike work trips—is highly discretionary; it is also highly sensitive to the location and attractiveness of a recreational facility. Estimating visitations to state parks therefore requires a demand model that explicitly includes both travel propensity to the entire system of parks and choice among individual, competing facilities. These two components—corresponding to the complementary and substitutional effects among recreational sites—are incorporated in a model specification. A decomposition procedure calibrates each of these two components individually and then integrates them in the final model. The procedure is successfully applied toward the waterfront state parks in the New York/Long Island area. The case study illustrates the mathematical properties of the model using a comprehensive set of sensitivity analyses. These analyses also yield estimation on visitation, revenue and direct benefits (as measured by consumers' surplus) corresponding to user-charge increases, park improvements and closings.

725   Design and Construction of Dade County METRO-MOVER

Kyaw Myint, Member, ASCE, (Vice Pres., Parsons Brinckerhoff Quade & Douglas, Inc., Miami, Fla.)

Journal of Urban Planning and Development, Vol. 110, No. 1, November, 1985, pp. 80-92

The background, funding, design features and system characteristics of the Dade County METROMOVER system, the first automated guideway transit system in a U.S. downtown environment, are described. Design features include the route alignment, vehicle, guideway superstructure, guideway substructure and the stations. System characteristics describe the capacity, headways and run time of the system. Some construction-related issues and maintenance of traffic during construction are also addressed.

726   Marsh Enhancement by Freshwater Diversion

George H. Ward, Jr., (Assoc., Espey, Huston & Assoc., Inc., P.O. Box 519, Austin, Tex. 78767)

Journal of Water Resources Planning and Management, Vol. 111, No. 1, January, 1985, pp. 1-23

The delta of the Nueces River, Texas, is a low-lying coastal marsh with dendritic distributaries and flats, infrequently inundated by flood events overbanking from the river. This study examines the feasibility of improving the ecological value of the marsh by creation of diversion works to increase the frequency and duration of marsh inundation. To this end, the morphology and statistical occurrence of flood events in the Nueces were examined using long-period gage records on the river. A canonical parameterization of flood events was used as input to a numerical hydrodynamic model of the deltaic system, which predicts the time evolution of flow and water level throughout the delta, and includes provision for simulating distributary confluence and difluence, inundation and dewatering of floodplains and flats, and activation of transient channels. The response of the marsh to specfied flood events was computed for various alternative physiographies, including existing conditions and candidate combinations of weirs and diversion channels. Coupled with the statistical frequency of occurrence of the specified flood event, the potential ecological value of each diversion alternative was quantitatively established in terms of total inundation period in key segments of the marsh.

727   Pricing and Expansion of a Water Supply System

Graeme C. Dandy, (Sr. Lect., Dept. of Civ. Engrg., Univ. of Adelaide, Australia, and Visiting Assoc. Prof., Univ. of Waterloo, Waterloo, Ontario), **Edward A. McBean**, (Prof., Dept. of Civ. Engrg., Univ. of Waterloo, Waterloo, Ontario) and **Bruce G. Hutchinson**, (Prof., Dept. of Civ. Engrg., Univ. of Waterloo, Waterloo, Ontario)

Journal of Water Resources Planning and Management, Vol. 111, No. 1, January, 1985, pp. 24-42

A general model for constrained optimum water pricing and capacity expansion is applied to the twin cities of Kitchener-Waterloo (KW), Ontario. The model identifies the water price and water supply capacity which maximizes the present value of net economic benefits over a planning horizon. Constraints on the rate of price change and financial cost recovery are included. Results for KW indicate that significant economic benefits can be achieved by jointly optimizing decisions about water pricing and capacity expansion. It is also shown that optimum policies are compatible with the goal of financial cost recovery on the part of the supply authority. The benefits of optimum pricing and capacity expansion are likely to be greatest in water supply systems which exhibit economics of scale or in cities where the rate of population growth is small.

728   Cleaning and Lining Versus Parallel Mains

Thomas M. Walski, Member, ASCE, (Research Civ. Engr., U.S. Army Engr., Waterways Experiment Station, Vicksburg, Miss. 39180)

Journal of Water Resources Planning and Management, Vol. 111, No. 1, January, 1985, pp. 43-53

Cleaning and lining a water main is often an economical alternative to installing a parallel main or spending large amounts of money on energy to overcome large head losses due to friction. This paper first presents methods for selecting the length of pipe to be cleaned and lined, and the diameter and length of the parallel pipe to be installed to meet a target head loss. Then, the costs of each approach are compared. In general, the decision to clean and line a main depends mostly on the head loss in the main at the design flow rate.

729   Research Agenda for Floods to Solve Policy Failure

Stanley A. Changnon, (Chief, Illinois State Water Survey, Champaign, Ill.)

Journal of Water Resources Planning and Management, Vol. 111, No. 1, January, 1985, pp. 54-64

For the first 60 yr of the 20th Century, the U.S. policy regarding floods was aimed at flood control. In the past 15 yr, policy has shifted to a goal of flood hazard mitigation. However, flood losses continue to rise and the Congress and others have raised questions about the causes of policy failure. The answer is complicated because flood policy involves 4 currently changing issues, including the shift from federal to local–state responsibilities; the National Flood Insurance Program; the shift to nonstructural approaches for flood mitigation; and developing programs for emergency assistance. A comprehensive assessment of flood research needs reveals that policy must have a view of efficient use of flood plains, not just loss reduction. The socioeconomic data base is considered inadequate for many policy decisions, and our knowledge of floods is uneven, with much more known in the physical sciences than in the social sciences. Attention to interdisciplinary research involving economists, sociologists, political scientists, and geographers is needed if better policy making and flood hazard mitigation are to be achieved.

730   Significance of Location in Computing Flood Damage

William K. Johnson, Member, ASCE, (Civ. Engr., The Hydrolic Engrg. Center, Corps of Engrs., Davis, Calif. 95616)

Journal of Water Resources Planning and Management, Vol. 111, No. 1, January, 1985, pp. 65-81

Expected annual flood damage for individual residential structures may be estimated using generalized depth-damage and elevation-frequency relationships developed by the Federal Insurance Administration (FIA). Damage computation using these data for different locations of the structure in the flood plain show a significant difference between structures located within the 25 yr flood plain and structures located outside. Within the 25 yr flood plain, especially within the 15 yr flood plain, damage for one and two-story structures with and without basements are exceedingly high: up to ten times greater than those outside. Outside the 25 yr flood plain, expected annual damage decreases gradually with location. These observations have important implications for flood plain management. First, economic feasibility of flood control projects will normally require that a significant number of structures be located within the 25 yr flood plain. Second, removal of structures with high damage potential has the secondary effect of reducing economic feasibility of potential flood control projects. Third, outside the 25 yr flood plain, expected annual damage is relatively insensitive to variations in location, depth-damage function and frequency. Lastly, estimates of expected annual damage using generalized FIA data is useful in preliminary estimates of the damage reduced.

731 Modeling the California State Water Project

Ilwhan Chung, (Systems Analyst, Dept. of Water Resources, Sacramento, Calif.) and **Otto Helweg**, Member, ASCE, (Assoc. Prof., Dept. of Civ. Engrg., Univ. of California, Davis, Calif.)

Journal of Water Resources Planning and Management, Vol. 111, No. 1, January, 1985, pp. 82-97

Two disadvantages of optimization models in water resources planning and management are the simplifications often required to construct the model and user relucatance to rely on optimization models alone. Both of these disadvantages may be overcome by combining a simulation model with an optimization model. In this study, a popular multipurpose, multireservoir system simulation model, HEC-3, developed by the Hydrologic Engineering Center (HEC) of the Corps of Engineers, is combined with a dynamic programming model. One problem in using discrete differential dynamic programming (DDDP) is determining the initial trajectory. In this project, conventional dynamic programming (DP) was used to estimate the initial trajectory, and then a DDDP model was used. Consequently, the optimization model combines conventional dynamic programming and discrete differential dynamic programming to solve the complex problem of operating the California State Water Project's (SWP) Lake Oroville and San Luis Reservoir, given the constraints of the Federal Central Valley Project (CVP) and delta flow requirements. Preliminary results suggest that SWP revenue may be almost doubled by adopting the suggested optimal operation.

732 Optimal Design of Detention and Drainage Channel Systems

Michael S. Bennett, Assoc. Member, ASCE, (Engr., Freese and Nichols, Inc., Fort Worth, Tex.) and **Larry W. Mays**, Member, ASCE, (Assoc. Prof., Dept. of Civ. Engrg., The Univ. of Texas, Austin, Tex. 78712)

Journal of Water Resources Planning and Management, Vol. 111, No. 1, January, 1985, pp. 99-112

An optimization model that determines the minimum cost detention and drainage channel system for a watershed is described. The model determines the location and size of detention basins, the size, type, and number of outlet structures in addition to the design of downstream channel modifications. This new model is based upon dynamic programming for nonserial systems. The model is applied to a watershed for the purpose of illutrating its capabilities.

733 Water Management for Small Urbanizing Watershed

Ronald l. Rossmiller, Member, ASCE, (Assoc. Prof., Dept. of Civ. Engrg., 351 Town Engrg. Bldg., Iowa State Univ., Ames, Iowa 50011)

Journal of Water Resources Planning and Management, Vol. 111, No. 2, April, 1985, pp. 123-136

In 1976 and 1977, the city of Ames, Iowa had to ration water due to a locally severe drought. Some relief was obtained by artificially recharging the depleted aquifer by pumping water from a nearby sand and gravel quarry. Since the drought, city engineering and Iowa State University personnel have conducted studies to make the quarry a permanent part of the city's sources of water supply. These studies have included both water quantity and water quality studies. The results of these studies with emphasis given to the water quantity studies and results are explored. This account of how one city turned the solution of a water supply problem into an opportunity to incorporate water quality, flood control, and recreation considerations into a more complete overall solution can serve as a model for other cities to use as they strive to find solutions to their own unique water-related problems.

734   Irrigation System Study in International Basin

M. C. Chaturvedi, (Prof., Dept. of Applied Mechanics, I.I.T. Delhi, New Delhi, India) and U. C. Chaube, (Reader, Water Resources Development Training Centre, Univ. of Roorkee, Roorkee (U.P.), India 247 667)

Journal of Water Resources Planning and Management, Vol. 111, No. 2, April, 1985, pp. 137-148

Before taking up a comprehensive study of regional irrigation development, it is necessary to identify and understand the technological options, the physical process and the allocation of water resources to pertinent projects. Linear programming models have been used for such a reconnaisance study of the Inoo-Nepal region of the Ganga basin with boundary conditions by Bangladesh. The region is a very large system in terms of resources potential, developmental options and associated constraints. The irrigation model study is carried out at the sub-basins' level (Level I) and at the basin level (Level II). The issues analyzed at level I are: (1)implications of scale of project developments, (2) implications of resource conservation measures such as reducing seepage and evaporation losses and improving irrigation efficiency, (3) implications of conjunctive surface water, natural recharge and artificial recharge use, and 940 implications of crop water demand. At level II the issue is the implication of water demand at Farakka (by India and Bangladesh). The study identifies the operating constraints and relative impact of various issues and options on the irrigation development in physical terms.

735   Instream Flow Protection in Riparian States

W. Douglass Dixon, (Research Asst., Water Resources Management Lab., Univ. of Arkansas, Fayetteville, Ark.) and William E. Cox, (Assoc. Prof. of Civ. Engrg., Virginia Polytechnic Inst. and State Univ., Blacksburg, Va.)

Journal of Water Resources Planning and Management, Vol. 111, No. 2, April, 1985, pp. 149-156

Maintenance of minimum flows necessary for protection of instream water uses has been recognized as a significant water management issue in some of the eastern states. The riparian doctrine, a primary component of the water low of several eastern states, offers some protection to instream uses since it offers protection to riparian land values and requires a sharing of water among landowners along the length of each stream. In addition, legal provisions for recognition of instream flows in federal water management programs apply, and many of the eastern states have adopted applicable legislative measures such as direct controls over water use, programs for preservation of scenic rivers, and minimum release requirements for impoundments. However, such measures do not apply uniformly among the riparian states, and existing legal mechanisms may be inadequate in some cases. Therefore, additional institutional development will likely be necessary to assure continuance of a socially desirable balance between instream and offstream water uses as water demand increases over time.

736   Extended Streamflow Forecasting Using NWSRFS

Gerald N. Day, Assoc. Member, ASCE, (Research Hydro., Hydrologic Research Lab., National Weather Service, Silver Spring, Md. 20910)

Journal of Water Resources Planning and Management, Vol. 111, No. 2, April, 1985, pp. 157-170

Extended forecasting using the National Weather Service River Forecast System (NWSRFS) is done with the NWS' Extended Streamflow Prediction (ESP) program. This paper examines the theory, capabilities, and potential applications of the ESP procedure. ESP uses conceptual hydrologic/hydraulic models to forecast future streamflow using the current snow, soil moisture, river, and reservoir conditions with historical meteorological data. The ESP procedure assumes that meteorological events that occurred in the past are representative of events that may occur in the future. Each year of historical meteorological data is assumed to be a possible representation of the future and is used to simulate a streamflow trace. The simulated streamflow traces can be scanned for maximum flow, minimum flow, volume of flow, reservoir stage, etc., for any period in the future. ESP produces a probabilistic forecast for each streamflow variable and period of interest. The procedure was originally developed for water supply forecasting in snowmelt areas, but it can also be used to produce spring flood outlooks, forecasts for navigation, inflow hydrographs for reservoir operation, and time series needed for risk analysis during droughts.

737   Satisfying Instream Flow Needs under Western Water Rights

Jay M. Bagley, Member, ASCE, (Prof., Utah State Univ., College of Engrg., Utah Water Research Lab., Logan, Utah 84322), **Dean T. Larson**, (Research Sci., Utah Water Research Lab., Utah State Univ., Logan, Utah 84322) and **Lee Kapaloski**, (Attorney at Law, Kapaloski, Kinghorn, and Peters, Salt Lake City, Utah)

Journal of Water Resources Planning and Management, Vol. 111, No. 2, April, 1985, pp. 171-191

The appropriation system of water rights has been criticized for failure to provide adequate protection of instream flow values. The appropriation system is measured against 13 fundamental principles of good state water law. Within the context of this comparison, the implications with respect to accommodating instream flow uses are examined. It is concluded that the appropriation system can equitably incorporate instream flow uses, but is constrained by lack of "litigation proof" methodologies and technologies to project impacts and tradeoffs. The integration of instream flow rights is also retarded by lack of proper recognition of certain hydrologic imperatives that must be observed in order to correctly define the instream flow right. Instream flow rights must not upset the integrity of other rights within the common system.

738   Impact of Conservation on Rates and Operating Costs

Nishith R. Bhatt, (Engr., Dauphin Consolidated Water Supply Co., Harrisburg, Pa.) and **Charles A. Cole**, (Prof., Engrg. Dept., Pennsylvania State Univ., The Capitol Campus, Middletown, Pa. 17057)

Journal of Water Resources Planning and Management, Vol. 111, No. 2, April, 1985, pp. 192-206

The potential impact of 20% reduction in water usage through water conservation is investigated for a small investor owned utility serving the rural residential community of Stewartstown Borough, Pennsylvania. The analysis of short-term impact shows that the average bill would be reduced by 16% if rates had not been changed; however, the company's operating expenses would only have been reduced by 2%. The long-range impact is investigated using present worth analysis. A 20% reduction in water use would have saved $71,280 in present worth (1976) for storage and source facilities and operating costs. The savings in capital and O & M costs would eventually translate into savings in water bills of the customers over the years.

739   **Exporting Colorado Water in Coal Slurry Pipelines**

Dean T. Massey, (General Attorney, Natural Resource Economics Div., Economic Research Service, U.S. Dept. of Agr., Madison, Wis. 53706)

Journal of Water Resources Planning and Management, Vol. 111, No. 2, April, 1985, pp. 207-221

The San Marco Pipeline Company has proposed a 1,000-mile (1,610-km) pipeline system to move coal slurry from southeastern Colorado to several electrical generating plants in the Texas Gulf Coast area. This area of Colorado is experiencing severe water shortage problems due to the large amount of water used for irrigation. Colorado statutes restrict the diversion of both surface and ground water for out-of-state use unless such water is credited to interstate compacts. Water has been found by the courts to be a commodity or article of commerce, therefore, subject to federal constitutional scrutiny under the commerce clause. The constitutionality of Colorado's statutes depend upon the nature of the restrictions imposed, severity of the burden created, and the local purposes served by the statutes. Statutes will be upheld only where they incidentally burden or discriminate against interstate commerce. States may impose severe restrictions on diverting water for out-of-state uses provided the same type of restrictions are imposed on instate use.

740   **Optimal Operation of California Aqueduct**

M. Hossein Sabet, Assoc. Member, ASCE, (Operation Research Specialist, California Dept. of Water Resources, Sacramento, Calif. 95495), **James Q. Coe,** (Sr. Engr., California Dept. of Water Resources, Sacramento, Calif. 95495), **Henry M. Ramirez,** (Assoc. Electrical Utility Engr., California Dept. of Water Resources, Sacramento, Calif. 95495) and **David T. Ford,** Member, ASCE, (Hydrologic Engrg. Consultant, Davis, Calif. 95616)

Journal of Water Resources Planning and Management, Vol. 111, No. 2, April, 1985, pp. 222-237

The goals and objectives of operation of the California State Water Project (SWP) changed significantly on March 31, 1983, when long-term, low-cost contractual agreements for energy purchase for the Project terminated. Now energy required for pumping must be purchased at a substantially increased cost. However, energy generated in the project can be sold with maximum revenues at times of peak electrical demand. To determine the most efficient operation schedules for the SWP, a coordinated set of models was developed for execution in "real-time." This set of models includes network flow programming models, a number of other simulation models, and a large-scale linear programming (LP) model designed to determine optimal flows throughout the SWP aqueduct system, given the energy demands for pumping, energy generation capabilities and water demands. This LP model may be executed with a weekly or daily time step. Solution of the LP problem is accomplished with the nonproprietary XMP package, using the dual-simplex capability to accelerate solution. The model has been used successfully for management of the SWP system operation.

741   Capacity Expansion of Sao Paulo Water Supply

Benedito P. F. Braga, Jr., (Asst. Prof., Departamento de Hidraulica-EPUSP, Cidade Univ., Sao Paulo, Brazil), **Joao G. L. Conejo**, (Dir. Divisao, Dept. of Water and Power, Sao Paulo, Brazil), **Leonard Becker**, (Asst. Research Engr., Civ. Engrg. Dept., Univ. of California, Los Angeles, Calif.) and **William W. -G. Yeh**, Member, ASCE, (Prof., Civ. Engrg. Dept., Univ. of California, Los Angeles, Calif.)

Journal of Water Resources Planning and Management, Vol. 111, No. 2, April, 1985, pp. 238-252

A capacity expansion model has been developed to facilitate the planning and the optimal timing and sizing of the proposed Juquia River system in Sao Paulo, Brazil. The proposed system consists of the construction of a series of reservoirs and pumping stations in the Juquia River basin to develop firm water supplies to meet the projected future water demands for the city of Sao Paulo. A monthly simulation model that utilizes rational operation rules and simulates the optimal operation mode is imbedded in the capacity expansion model so that the optimized results are consistent and hydrologically feasible. Forty-two years of historical streamflows are used in the simulation model. Future water demands up to a time horizon are specified, and a forward dynamic programming algorithm is used to minimize the present worth of total project costs.

742   Water Quality and Regional Water Supply Planning

Alan J. Lauwaert, Member, ASCE, (Civ. Engr., Dept. of the Army, South Pacific Div., Corps of Engrs., 630 Sansome St., Room 718, San Francisco, Calif. 94111)

Journal of Water Resources Planning and Management, Vol. 111, No. 3, July, 1985, pp. 253-267

A method for sizing a flow-way to remove phosphorus from surplus rainfall runoff is presented. The flow-way area would consist of a seasonal wetland community wherein the plants take-up the nutrient. The paper presents the integration of scientific water quality study results with engineering surface water quantity procedures. Benefits of the plan include: (1) Use of surface water reduces agricultural use of groundwter; (2) use of surplus runoff reduces use of existing surface water resources; (3) use of a flow-way maintains regional water quality; and (4) use and management of existing surface water provides economic benefits.

743   Determination of Urban Flood Damages

Stuart J. Appelbaum, Assoc. Member, ASCE, (Civ. Engr., Planning Div., U.S. Army Corps of Engineers, Baltimore Dist., Baltimore, Md. 21203)

Journal of Water Resources Planning and Management, Vol. 111, No. 3, July, 1985, pp. 269-283

A new methodology and computer program, DAPROG 2, have been developed by the Baltimore District, Corps of Engineers to estimate urban flood damages. Stage-damage relationships are computed for individual residential and commercial properties. These individual relationships are then aggregated into composite stage-damage relationships for an entire reach. For residential properties, replacement values for the structure and contents are estimated and applied to depth-percent damage relationships in order to develop stage-damage relationships. Regression equations have been developed to estimate structure replacement cost from physical attributes which are easily obtained in the field. Contents value are also estimated from the data obtained in the field. For commercial properties, standard stage-damage relationships for 67 types of businesses are used. Utility and transportation damages and emergency care costs are also computed. The computer program is interactive and allows the user to correct data errors online. The program has several optional features and reports which are selected by the user. DAPROG 2 has been successfully used by the Baltimore District in several flood control studies. DAPROG 2 has also been used by other Corps Districts and state agencies.

744   State Water Supply Management in New Jersey

William Whipple, Jr., Fellow, ASCE, (Asst. Dir., Div. of Water Resources, Dept. of Environmental Protection, Trenton, N.J.)

Journal of Water Resources Planning and Management, Vol. 111, No. 3, July, 1985, pp. 284-292

In New Jersey, existing water supply problems involve water availability during droughts and other emergencies, contamination, especially of ground water, and institutional problems. A drought occurred in 1980, just as the new Water Supply Master Plan was completed; and a large bond issue was approved, together with statutes greatly strengthening the State's authority for water supply management. As a consequence, the State has a very advanced water supply planning, regulatory and management program, which includes both positive and negative incentives to follow State guidance. To date, the State has built and manages only three water supply projects, and the Federal Government, none. Thus, the greater part of new development and of system rehabilitation, is carried out by the 620 public water supply systems of the State (many of which are privately owned). New Jersey policy emphasizes management and problem-solving. Feasibility planning, backed up by regulatory requirements to provide an adequate standard of service, encourages local initiative. Low interest loans, rather than grants, are used to encourage high priority programs such as remedial work in contaminated well fields.

745   Inventorying Ground Water in Crystalline Rocks of
            Piedmont Region

Charles W. Welby, (Prof. of Geology, Dept. of Marine, Earth and Atmospheric Sci., North Carolina State Univ., Raleigh, N.C. 27695)

Journal of Water Resources Planning and Management, Vol. 111, No. 3, July, 1985, pp. 293-302

A methodology is described for estimating a "Ground Water Working Inventory" for areas of the eastern Piedmont Province drawing ground water from crystalline rocks. The approach can improve efficiency in water resource and land-use planning and use of surface and ground water resources. Calculation of the 7-day, 10-year and 7-day, 1-year flows from stream gage data together with application of geologic data and information calculated from 24-h pumping tests allow the outlining of areas more favorable and less favorable for ground water development. The Ground Water Working Inventory basd on the 7-day low-flow calculations allows estimates of available water to be made and permits land-use planners to make rational decisions about residential densities and other water-related issues. Political bodies and planners can use the methodology to guide development according to how large a role ground water is to play in local water supply planning and management. High density development can be discouraged in those areas where ground water is limited and encouraged where commitments for surface water supplies have been made. Additionally, the inventory provides information which strengthens planning for use of ground water as a supplement to surface water supplies.

746   Ground-Water Reservoir Operation for Drought Man-
            agement

Ram S. Gupta, Member, ASCE, (Area Coordinator, Civ. Engrg., Roger Williams College, Bristol, R.I.) and **Alvin S. Goodman**, Fellow, ASCE, (Prof. of Civ. Engrg., Polytechnic Inst. of New York, Brooklyn, N.Y.)

Journal of Water Resources Planning and Management, Vol. 111, No. 3, July, 1985, pp. 303-320

A hydrodynamic groundwater model is integrated with a multi-level management model to formulate a composite model for investigating groundwater reservoir operation for drought management. The hydrodynamic model solves the three-dimensional groundwater flow equation by the finite difference method. A U.S. Geological Survey Model with extensions is used for this purpose. For given hydrogeological conditions, the management model assists the planner to design a system for augmenting low streamflows according to stipulated water supply requirements, by: (1) Arranging the wells within each unit of a multiaquifer system; and (2) establishing the withdrawal pattern for the system. Sensitivity analyses are made of the parameters pertaining to the conjunctive use of water by applying the model to the Mashipacong Island area in the Delaware Basin. The studies consider the hydrogeological properties of the aquifer and streambed, artificial measures to reduce induced infiltration, and arrangements of wells and pumping schedules. Among the infiltration reduction measures studied, a vetical semi-previous core extending through the entire thickness of the aquifer can reduce losses from over 60% to less than 40%. Such a cutoff core, constructed by the slurry trenching process, may be economically attractive.

747   A Uniform Technique for Flood Frequency Analysis

Wilbert O. Thomas, Jr., (Hydro., U.S. Geological Survey, Reston, Va.)

Journal of Water Resources Planning and Management, Vol. 111, No. 3, July, 1985, pp. 321-337

In 1967 the U.S. Water Resources Council (WRC) published Bulletin 15 recommending that a uniform technique be used by all Federal agencies in estimating floodlow frequencies for gaged watersheds. This uniform technique consisted of fitting the logarithms of annual peak discharges to a Pearson Type III distribution using the method of moments. The objective was to adopt a consistent approach for the estimation of floodflow frequencies that could be used in computing average annual flood losses for project evaluation. In addition, a consistent approach was needed for defining equitable flood-hazard zones as part of the National Flood Insurance Program. In 1976 WRC published Bulletin 17 which extended and updated Bulletin 15 but still recommended the use of the "log-Pearson Type III" method. Since 1976, two updates of Bulletin 17 (17A and 17B) have been published which clarify or improve on this base method, or do both. This paper gives a brief historical review of the development of these bulletins and the motivation and justification for the adoption of this uniform technique. Special emphasis is given to Bulletin 17B, the current guidelines used by Federal agencies. Specific techniques examined are the development of regional skew, weighting of regional and station skew, the basis for the low- and high-outlier tests, and the basis for the adjustment of frequency curves using historical information.

748   Water Resources Engineering Education

William R. Walker, Member, ASCE, (Dir., Virginia Water Resources Research Center, Blacksburg, Va. 24060) and **Phyllis G. Bridgeman,** (Research Asst., Virginia Water Resources Research Center, Blacksburg, Va. 24060)

Journal of Water Resources Planning and Management, Vol. 111, No. 3, July, 1985, pp. 338-345

Water resources engineering education may hinder the development of practitioners who must be not only technically proficient but also capable of managing engineering problems in a social, political, and economic context. Professional water resource engineering requires the ability to communicate in interdisciplinary efforts, to plan public information programs, and to factor legal and social input as well as technical know-how into problem solving. Opportunities to improve the water resources engineering education include: (1) new methods of evaluating students that include assessments of communications skills; (2) broadened course requirements linking technical and social science/humanities programs through joint appointments, faculty exchanges, or social science/humanities courses tailor-made for the engineering student; (3) a shift from traditional graduate training to provide an opportunity for graduate research on open-ended problems typically found in water resource engineering; (4) nonoptional co-op experience in the real world; and (5) evaluation outside of the accreditation process to encourage program experimentation. Professional practitioners can only benefit by such educational improvements and should provide substantial external support for basic research evaluating engineering education curricula, as well as support for, and involvement in curriculum development.

749 Ground-Water Contamination in Silicon Valley

Adam Olivieri, (Sr. Water Resource Engr., California Regional Water Quality Control Board, San Francisco Bay Region, 1111 Jackson St., Rm. 6040, Oakland, Calif. 94607), **Don Eisenberg**, (Sr. Water Resource Control Engr., California Regional Water Quality Control Board, San Francisco Bay Region, 1111 Jackson St., Rm. 6040, Oakland, Calif. 94607), **Martin Kurtovich**, (Assoc. Water Resource Control Engr., California Regional Water Quality Control Board, San Francisco Bay Region, Oakland, Calif. 94607) and **Lori Pettegrew**, (Student Asst., California Regional Water Quality Control Board, San Francisco Bay Region, Oakland, Calif. 94607)

Journal of Water Resources Planning and Management, Vol. 111, No. 3, July, 1985, pp. 346-358

In the southern San Francisco Bay Area, a state regulatory agency carried out a questionnaire survey to determine the locations and characteristics of underground chemical storage and handling facilities at industrial sites in the area. Sampling of soil and ground water were required at sites where underground tanks containing solvents were reported. The survey identified 1692 tanks at 388 sites. The largest fraction were pretreatment sumps for industrial wastewater, followed by fuel tanks, solvent tanks, and tanks containing corrosives. Solvent tanks were reported at 96 sites. Subsurface investigations resulted in discovery of soil or ground water contamination at 75 of the 96 sites with solvent tanks or both. Further investigations and remedial actions are underway at these sites, and new regulations have been adopted at the state and local levels to prevent future release of industrial chemicals into soil and ground water.

750 Problems with Modeling Real-Time Reservoir Operations

Emre K. Can, Assoc. Member, ASCE, (Asst. Prof., School of Engrg., Lakehead Univ., Thunder Bay, Ontario, Canada P7B 5E1) and **Mark H. Houck**, (Assoc. Prof., School of Civ. Engrg., Purdue Univ. West Lafyette, Ind. 47907)

Journal of Water Resources Planning and Management, Vol. 111, No. 4, October, 1985, pp. 367-381

Existence of several problems associated with the optimization of real-time operations of reservoir systems is demonstrated. The first problem presented is due to the use of imperfect streamflow (inflow to the reservoir) forecasts. It is shown that there is a significant relationship between the reliability of the forecast information and the operating or forecast horizon to be used in the model. To illustrate the implications of this relationship, an example is given where extending the forecast horizon beyond three days does not improve the performance of the model for real-time operations. Real-time operations models often require routing models that relate reservoir releases to the flows at downstream points. It is illustrated with examples that use of approximate reach routing models may cause unexpected problems and may mislead the decision makers. A method to partially resolve the problem is also suggested.

751   Planning Detention Storage for Stormwater Management

Vasudevan G. Loganathan, Assoc. Member, ASCE, (Asst. Prof., Dept. of Civ. Engrg., Virginia Polytechnic Inst., and State Univ., Blacksburg, Va. 24061), **Jacques W. Delleur**, Member, ASCE, (Prof., School of Civ. Engrg., Purdue Univ., West Lafayette, Ind. 47907) and **Rafael I. Segarra**, (Grad. Student, Dept. of Civ. Engrg., Virginia Polytechnic Inst. and State Univ., Blacksburg, Va. 24061)

Journal of Water Resources Planning and Management, Vol. 111, No. 4, October, 1985, pp. 382-398

A method for estimating detention storage capacity in stormwater management is presented. A generalized storage-overflow relationship is derived. This relationship defines real available storage (empty space in detention basin) on the positive range and overflow volumes on the negative range. By using exponential probability density functions for the independent hydrologic variables runoff volumes, runoff durations, and interevent times, and the generalized storage relationship, a new probability distribution is derived for the treatment plant overflow volumes. The new distribution provides an easy method for estimating the detentions torage and treatment capacity for a design risk level. The methodology has the advantage that it provides easy to use preliminary planning information for stormwater management without the need for extensive simulation.

752   Multi-Objective Analysis with Subjective Information

Richard N. Palmer, Assoc. Member, ASCE, (Asst. Prof., Dept. of Civ. Engrg., Univ. of Washington, Seattle, Wash. 98195) and **Jay R. Lund**, Assoc. Member, ASCE, (Research Asst., Dept. of Civ. Engrg., Univ. of Washington, Seattle, Wash. 98195)

Journal of Water Resources Planning and Management, Vol. 111, No. 4, October, 1985, pp. 399-416

A method is presented for incorporating subjective information into multi-objective evaluations. The method is based upon an eigenvalue and eigenvector analysis and structures multi-objective evaluations into a series of hierarchies in which pairwise comparisons are made. The method is demonstrated in the design of an aquatic monitoring network. Theoretical aspects of the approach are reviewed, including measures of subjective inconsistency, the sensitivity of inconsistency to pairwise comparisons, subjective scaling factors, and sensitivity of final, multi-objective weights. An interactive computer program for the application of the technique is described.

753   Development of a Flood Management Plan

Duncan W. Wood, Member, ASCE, (Engr., Anderson-Nichols Co., Clinton, Mass.), **Thomas C. Gooch**, Assoc. Member, ASCE, (Engr., Freese and Nichols, Inc., Forth Worth,, Tex.), **Paul M. Pronovost**, (Engr., U.S. Army Corps of Engrs., New England Div., Waltham, Mass.) and **David C. Noonan**, Assoc. Member, ASCE, (Engr., Camp Dresser and McKee, Boston, Mass.)

Journal of Water Resources Planning and Management, Vol. 111, No. 4, October, 1985, pp. 417-433

A comprehensive flood plain management plan was developed for the City of Keene, New Hampshire, reflecting the policies of Section 73 of Public Law 93-251. Community attitudes were investigated through questionnaires, interviews with community leaders, and interaction with a community advisory committee. A full range of available nonstructural and structural flood damage reduction measures was surveyed. An initial set of measures was selected based on physical, economic, and political feasibility in Keene given the existing stream conditions, available stage-damage estimates and community attitudes. These initial measures were evaluated in detail,, and three alternative flood plain management plans were developed and analyzed using benefit-cost and enivornmental assessment procedures. The recommended flood plain management plan for Keene included: 1) A technical assistance program to aid homeowners with floodproofing; 2) small dikes to protect selected structures or groups of structures from flooding; 3) channel improvements at constrictions; 4) enhancement of existing storage by restoration and modification of a damaged dam; 5) development of a flood warning and emergency preparedness system. The experience in Keene represents a prototype study implementing the requirements of Section 73 and can serve as a useful model for future nonstructural flood plain management studies.

754   Lower Mississippi Valley Floods of 1982 and 1983

William E. Read, (Executive Asst. to the Pres., Walk, Haydel and Associates, 720 Woodward Ave., Gulfport, Miss. 39501) and **Michael C. Robinson**, (Div. Historian, Lower Mississippi Valley Div. and Mississippi River Commission, P.O. Box 80, Vicksburg, Miss. 39180)

Journal of Water Resources Planning and Management, Vol. 111, No. 4, October, 1985, pp. 434-453

The 1982-1983 floods on the main stem and tributaries of the Middle and Lower Mississippi River spanned a six-month period from December, 1982–June, 1983. The flood season was initially marked by three tributary-centered events followed by a main stem flood on the Lower Mississippi during April, May, and early June. The prolonged period of high water included flash, general, and backwater flooding that prompted requests for U.S. Army Corps of Engineers assistance from states, communities, levee boards, and other local interests. The floods posed a host of difficult challenges for the Lower Mississippi Valley Division and Mississippi River Commission LMVD/MRC of the U.S. Army Corps of Engineers. The task of managing the floods was compounded by the scope and complexity of the nation's largest flood control system. In addition to responding to requests for floodfight assistance, the LMVD/MRC was heavily involved in maintaining navigation, monitoring weather and hydrological data, insuring the integrity of the levee system, as well as operating reservoirs, pumping plants, the Old River Control Structures, and the Bonnet Carre Floodway. The performance of the Mississippi River and Tributaries Project and elements of the LMVD program justified public investments in a reliable flood control system in the Middle and Lower Mississippi basins.

755   Simulating Cost and Quality in Water Distribution

Robert M. Clark, (Chf., Physical and Chemical Contaminant Removal Branch, DWRD, MERL, Cincinnati, Ohio 45268) and **Richard M. Males**, (Pres., RMM Services, Cincinnati, Ohio)

Journal of Water Resources Planning and Management, Vol. 111, No. 4, October, 1985, pp. 454-466

A spatial approach that disaggregates the water supply system into the components of acquisition-treatment and transmission-distribution, allows these components to be studied in isolation and in combination. Each of these components has different cost and physical functions and the cost and performance trade–offs between these functions can provide important insights into regionalization and cost-related issues. The Water Supply Simulation Model (WSSM) described in this paper is a system of computer programs that allows for an evaluation of the physical and economic characteristics of a water distribution system in a spatial framework. The development of the model and its application to a case study situation is presented.

756   Application of Extreme Value Theory to Flood Damage

Pierre Ouellette, (Lect., Dept. of Economics, Univ. of Montreal, Montreal, Quebec, H3C 3J7 Canada), **Nassir El-Jabir**, Member, ASCE, (Assoc. Prof., Ecole de Genia, Univ. of Moncton New Brunswick E1A 3E9 Canada) and **Jean Rousselle**, (Prof., Civ. Engrg. Dept., Ecole Polytechnique of Montreal, Montreal, Quebec, H3C 3A7 Canada)

Journal of Water Resources Planning and Management, Vol. 111, No. 4, October, 1985, pp. 467-477

Flood plain management requires assessment of the costs and benefits of all projects under consideration. The benefits translate mainly into flood damage reduction. This study presents a methodology for estimating flood damage prior to implementation of flood control structures. In this two-stage methodology, a hydroeconomic model for flood damage estimation is first developed, and a flood damage distribution function is then derived from the theory of extreme values in stochastic processes. The distribution function produces an estimation of actualized damages. The Richelieu River basin was selected for a numerical application because of its combined rural and urban characteristics and the fairly extensive sum of knowledge on the basin supplied by previous studies.

757   Optimization of Water Quality Monitoring Networks

Richard N. Palmer, Assoc. Member, ASCE, (Asst. Prof., Dept. of Civ. Engrg., Univ. of Washington, Seattle, Wash. 98195) and **Mary C. MacKenzie**, (Research Asst., Dept. of Civ. Engrg., Univ. of Washington, Seattle, Wash. 98195)

Journal of Water Resources Planning and Management, Vol. 111, No. 4, October, 1985, pp. 478-493

Water quality and biological monitoring provide an indication of the degree to which the natural environment has been affected by anthropogenic activities. Currently, this monitoring is both expensive and time-consuming. A new cost-effective approach to the design of aquatic monitoring networks is presented. Classical analysis of variance ANOVA techniques are reviewed and a modified ANOVA model with control station pairs is suggested. An interactive optimization procedure is presented that incorporates a modified gradient search algorithm to select designs which maximize the statistical power of a network for a specified budget or minimize the cost of a network for a specified statistical power requirement. The sensitivity of the model results are explored as a function of the cost, the number of sampling stations, replicates, and occasions, the Type I and Type II error, estimates of data variance, and cost components for data describing an aquatic species from a New England power facility. It is shown that for specified power and cost, numerous solutions exist providing the designer with a wide selection of alternatives from which to choose.

758   Wave Interference Effects by Finite Element Method

Min-Chih Huang, (Asst. Prof., Dept. of Naval Architecture and Marine Engrg., National Cheng Kung Univ., Taiwan, 700 ROC), **John W. Leonard**, Member, ASCE, (Prof., Dept. of Civil Engrg. and Ocean Engrg. Program, Oregon State Univ., Corvallis, Oreg. 97331) and **Robert T. Hudspeth**, Member, ASCE, (Prof., Dept. of Civil Engrg. and Ocean Engrg. Program, Oregon State Univ., Corvallis, Oreg. 97331)

Journal of Waterway, Port, Coastal and Ocean Engineering, Vol. 111, No. 1, January, 1985, pp. 1-17

A numerical procedure is presented for computing wave interference effects on multiple, surface-piercing rigid structures in an ocean of finite depth by the finite element method (FEM). Viscous effects are neglected and hydrodynamic pressure forces are assumed to be intertially dominated. Within the limits of linear wave theory, a scattered wave potential is numerically computed using radiation boundary dampers. Numerical results are presented for a single vertical cylinder and for multiple vertical circular and square cylinders under varying incident wave angles. Comparisons between the FEM results and results available from both analytical and integral equation numerical methods are good. Estimates of the computational savings in CPU realized by the FEM compared to the integral equation method are provided. The favorable numerical comparisons realized by the FEM using radiation boundary dampers coupled with the computational savings in CPU suggest the application of the FEM to more complicated systems of multiple structures for both rigid and flexible bodies.

759   Deep-Draft Navigation Project Design

Bruce L. McCartney, Member, ASCE, (Hydr. Engr., Hydraulics and Hydrology Div., U.S. Army Corps of Engineers, Washington, D.C. 20314)

Journal of Waterway, Port, Coastal and Ocean Engineering, Vol. 110, No. 1, February, 1984, pp. 18-28

Navigation channel design for large ships has evolved from a rule-of-thumb approach to comprehensive site specific analysis which include navigation requirements, environmental assessment and economic optimization. The design of navigation channels and harbors requires an understanding of the problem, assembly and evaluation of all pertinent facts, and development of a rational plan. The design engineer should be responsible for developing the rational design and sufficient alternative plans so the economic optimum plan is evident and the recommended plan is substantiated. The elements of this modern design analysis are presented with the deseign philosophy which will result in a safe, efficient, least cost project with due consideration of environmental and social impacts.

760   A Harbor Ray Model of Wave Refraction-Diffraction

Howard N. Southgate, (HIgher Scientific Officer, Hydr. Research Ltd., Wallingford, Oxon, U.K.)

Journal of Waterway, Port, Coastal and Ocean Engineering, Vol. 110, No. 1, February, 1984, pp. 29-44

A harbor ray model has been developed which describes the combined wave effects of diffraction around breakwaters and depth refraction. Ray models are computationally well suited to determining the response of large harbor areas to short period waves. This is a situation for which alternative mathematical models can use prohibitively large amounts of computing time and storage. Diffraction cause by two types of breakwater layout, commonly found at harbor entrances, is shown to be capable of being modeled by a ray method. These breakwater layouts are: (1) A small gap between two straight breakwaters; and (2) a single, straight, semi-infinite breakwater. A new ray system is presented for the semi-infinite breakwater problem which overcomes the difficulty of modelling the region around the geometric shadow boundary. Unlike other techniques developed to resolve this difficulty, this method retains a ray plotting solution technique and therefore hence keeps the computational advantages of such a technique. A comparison is made between the ray model and a finite-element model for the two breakwater layouts on constant-depth and sloping sea-beds.

761   **Drag of Oscillatory Waves on Spheres in a Permeable Bottom**

C. David Ponce-Campos, Assoc. Member, ASCE, (Asst. Prof., Dept. of Civ. and Environ. Engrg., Clarkson College of Technology, Potsdam, N.Y.) and **Ernest F. Brater**, Fellow, ASCE, (Prof. Emeritus of Hydr. Engrg., Dept. of Civ. Engrg., Univ. of Michigan, Ann Arbor, Mich.)

Journal of Waterway, Port, Coastal and Ocean Engineering, Vol. 110, No. 1, February, 1984, pp. 45-61

Experimental values of the coefficient of drag at the threshold of motion were determined for spheres resting on a permeable bottom during oscillatory wave motion. The tests were performed in a wave tank in which a model trench was used to simulate the back fill and cover layers of a buried pipe line. The tests were conducted on model cover layers consisting of spheres of various sizes and specific weights. The ambient velocity and acceleration were computed for the wave conditions and phase angle at the instant of incipient motion. The drag coefficient was determined after assuming that a theoretical nonconvective solution for the coefficient of inertia could be used to estimate the inertial force. The data obtained from the tests may be used to estimate the size of stable stone armoring necessary to protect buried pipelines and marine foundations.

762   **Wave Damping by Soil Motion**

Tokuo Yamamoto, Member, ASCE, (Assoc. Prof., Div. of Applied Marine Physics, Rosenstiel School of Marine and Atmospheric Sci., Univ. of Miami, Miami, Fla.) and **Shigeo Takahashi**, (Grad. Student, Div. of Ocean Engrg., Rosentiel School of Marine and Atmospheric Sci., Univ. of Miami, Miami, Fla.; presently, Port & Harbor Inst., Ministry of Transport, Japan)

Journal of Waterway, Port, Coastal and Ocean Engineering, Vol. 111, No. 1, January, 1985, pp. 62-77

Wave damping by wave-soil interactions are examined quantitatively using the Coulomb-damped poro-elastic theory recently developed. The dispersion relation is obtained explicitly. It is found that the Coulomb friction between soil grains is by far the most important wave damping mechanism in soft soil beds, e.g., clays and silts. This mechanism is highly nonlinear owing to the dynamic softening behavior of soils. Large waves damp much quicker than small waves. Simple formulas and charts are presented for estimation of the wave damping by soil motion for wide ranges of soil wave parameters. Wave lenghts are also modified by the bed motion by up to ± 15%.

763   **Numerical Study of Finite Amplitude Wave Refraction**

Im Sang Oh, (Physical Oceanographer, Hawaii Inst. of Geographics and Dept. of Oceanography, Univ. of Hawaii, Honolulu, Hawaii 96822) and **Chester E. Grosch**, (Prof., Dept. of Oceanography, Old Dominion Univ., Norfolk, Va. 23508)

Journal of Waterway, Port, Coastal and Ocean Engineering, Vol. 111, No. 1, January, 1985, pp. 96-110

Water wave refraction for monochromatic waves is considered in order to examine finite amplitude wave effects using Stokes wave theory with the assumption that a locally flat bottom exists. The ray trajectories and refraction coefficients obtained by using the first and the third order wave theory including an excluding energy dissipation are compared for various topographies. Also, some results from various bottoms of constant slope are summarized at the depth near the wave breaking point for the practical use of this model. Third order theory including energy dissipation seems to produce significant differences on the final results of the refraction calculation.

764 Regime of Oscillatory Flow

Suphat Vongvisessomjai, (Assoc. Prof. and Chmn., Div. of Water Resources Engrg., Asian Inst. of Tech., P.O. Box 2754, Bangkok, Thailand)

Journal of Waterway, Port, Coastal and Ocean Engineering, Vol. 111, No. 1, January, 1985, pp. 78-95

A regime of flow is defined as the transition from laminar to turbulent flow which is an inception of turbulence in the flow field. Knowledge of flow regime provides useful information for quantitative description of flow and its analysis. It also serves as the lower limit of sediment transport. Earlier regimes of flow are very compliated because of their various types of flow and bed roughness. In this study an analysis is made of regime of flow using known published data tested in oscillating beds, wave flumes and water tunnels over loose sediment beds and rigid ripple beds, as well as wave flume data tested by the author. The approach is based on a dimensional consideration and a derivation of theoretical regimes of flow. It is found that the theoretical regimes of flow which are expressed in terms of flow Reynolds number and relative smoothness of the bed can be used for loose sediment beds but cannot be used for ripple beds of larger roughness. A unique regime of flow which is expressed in terms of sediment Froude number and relative bed smoothness is obtained from its proper fitting to the experimental data from all types of flow and bed roughness. The appropriate dimensionless parameter "the relative bed smoothness" reveals automatically the condition of its roughness.

765 Wave-Induced Pressure under Gravity Structure

Philip L.-F. Liu, Member, ASCE, (Prof., School of Civ. and Environmental Engrg., Cornell Univ., Ithaca, N.Y. 14853)

Journal of Waterway, Port, Coastal and Ocean Engineering, Vol. 111, No. 1, January, 1985, pp. 111-120

The wave-induced seepage flow in a porous seabed in the neighborhood of a gravity structure is solved exactly by two-dimensional potential flow theory. The distribution of the hydrodynamic pore-water pressure along the base of the structure and the wave-induced vertical force on the structure are computed from the integral solutions. The effects of the thickness of the porous seabed and the width of the structure are examined. Results obtained from the exact theory are compared with experimental data and there is reasonable agreement.

766 Hydrodynamic Coefficients and Depth Parameter

Subrata K. Chakrabarti, Fellow, ASCE, (Dir., Marine Research and Development, Chicago Bridge and Iron Co., 1501 N. Division St., Plainfield, Ill. 60544)

Journal of Waterway, Port, Coastal and Ocean Engineering, Vol. 111, No. 1, January, 1985, pp. 123-127

The in-line forces measured on a small section of a fixed vertical cylilnder in waves are analyzed to determine the effects of the hydrodynamic coefficients on the water depth parameter or the orbital shape parameter, Ω. The coefficients, C_M, C_D as well as an rms force coefficient, C_F derived from these forces are found to be insensitive to the value of Ω in the range of $\Omega = 0.3$ to 0.9. In an earlier study, it was shown from the test data that the values of C_F at $\Omega = 0.9$ were distinctly different from those at Ω = 0.3 - -0.7. Thus, the present study contradicts the results of the previous study. In the range of KC values of 0 - 40, however, the force coefficients compare well with U-tube experimental data.

767   Wave Runup Formulas for Smooth Slopes

John P. Ahrens, Affiliate Member, ASCE, (Oceanographer, U.S. Army Engr. Waterway Experiment Sta., Vicksburg, Miss.) and **Martin F. Titus,** (Computer Programmer, Automation Support Activity, Washington, D.C.)

Journal of Waterway, Port, Coastal and Ocean Engineering, Vol. 111, No. 1, January, 1985, pp. 128-133

A method of estimating wave runups on plane, smooth slopes is presented. The method uses empirical formulas to predict monochromatic wave runups for a wide range of surf conditions. Intuitive arguments are used to support the form of the equations. The analysis shows that wave nonlinearity has a strong influence on the runup of nonbreaking, monochromatic waves. Other factors being equal, the more nonlinear the wave the greater the runup. Comparisons show that the formulas given make good estimates of the observed wave runup elevations.

768   Distribution of Maximum Wave Height

Miguel A. Corniere, (Asst. Prof., Dept. of Oceanographical and Ports Engrg., Univ. of Santander, Spain), **Miguel A. Losada,** (Prof., Dept. of Oceanographical and Ports Engrg., Univ. of Santander, Spain) and **Luis A. Gimenez-Curto,** (Asst. Prof., Dept. of Oceanographical and Ports Engrg., Univ. of Santander, Spain)

Journal of Waterway, Port, Coastal and Ocean Engineering, Vol. 111, No. 1, January, 1985, pp. 134-139

A new expression for the function describing the distribution of the maximum wave height in a sea state is obtained by fitting numerical simulated data, and checked with some physical data measured off the Norwegian and Danish coasts. This distribution is of double exponential type having two parameters which depend on m_0. The distribution can be applied to narrow as well as broad band spectra. This new expression is in good agreement with real data, particularly in the upper tail of the distribution. The proposed model allows a modification to the traditional selection of sea state duration using a probability level (risk criterion).

769   3-D Model of Bathymetric Response to Structures

Marc Perlin, Member, ASCE, (Asst. Lab. Dir., Coastal and Oceanographic Engrg. Lab., Univ. of Florida, Gainesville, Fla. 32611) and **Robert G. Dean,** Member, ASCE, (Grad. Research Prof., Coastal and Oceanographic Engrg. Dept., Univ. of Florida, Gainesville, Fla. 32611)

Journal of Waterway, Port, Coastal and Ocean Engineering, Vol. 111, No. 2, March, 1985, pp. 153-170

A fully implicit finite-difference, N-line numerical model is developed to predict bathymetric changes in the vicinity of coastal structures. The wave field transformation includes refraction, shoaling, and diffraction. The model simulates the changes in N-contour lines due to both longshore and onshore-offshore sediment transport. A new equation for the distribution of sediment transport across the littoral zone is applied. The model is capable of simulating one or more shore-perpendicular structures, movement of offshore disposal mounds and beach fill evolution. The structure length and location, sediment properties, equilibrium beach profile, etc., are user-specified along with the wave climate. Results are presented for 2 example cases, sediment transport of dredge disposal in the vicinity of Oregon Inlet, and simulation of the Longshore Sand Transport Study at Channel Islands Harbor, California. These examples demonstrate the model's ability to simulate various physical situations and that the model predicts realistic shoreline changes.

770   Qualitative Description of Wave Breaking

David R. Basco, Member, ASCE, (Prof., Civ. and Ocean Engrg., Texas A&M Univ., College Station, Tex. 77843)

Journal of Waterway, Port, Coastal and Ocean Engineering, Vol. 111, No. 2, March, 1985, pp. 171-188

A description of major features and patterns of motion in water waves just after breaking is presented. Previous literature is synthesized and new observations utilized to develop a new qualitative picture of the breaking process. Both classic spilling and plunging-type breakers are found to have similar initial breaking motions, but at vastly different scales. Two primary vortex motions are identified. A plunger vortex is initially created by the overturning jet, which in turn causes a splash-up of trough fluid and subsequent formation of a surface vortex similar to the roller in a hydraulic jump. Introduced for the first time is the hypothesis that the plunger vortex translates laterally to push up a new surface wave with vastly different wave kinematics that continues propagating into the inner surf zone. Of primary interest is the outer or transition region where momentum is being exchanged between mean, periodic and random flow processes along with some energy loss. Evidence is presented from the literature to support the new, second wave hypothesis and all other concepts introduced.

771   Water Particle Velocities in Regular Waves

Geoffrey N. Bullock, (Reader, Plymouth Polytechnic, Plymouth, England) and Ian Short, (Engr., Ferranti Limited, Oldham, Greater Manchester, England)

Journal of Waterway, Port, Coastal and Ocean Engineering, Vol. 111, No. 2, March, 1985, pp. 189-200

The Eulerian water particle velocities under paddle-generated regular waves in a closed channel are not always accurately predicted by a conventional application of Stokes' first, second or fifth order wave theory. Phenomena, which give rise to errors, include mass-transport, the partial clapotis formed by reflection from the spending beach and the free second harmonic wave produced by the sinusoidal motion of the paddle. Measurements taken with a laser doppler anemometer indicate that the amplitudes predicted for the second harmonics of the velocity components can be over 100% in error. Furthermore, there is a mean horizontal velocity which is often greater than the amplitude of the second harmonic and can be 20% of the amplitude of the first harmonic. The observed mean velocities are compared with predictions based on Longuet-Higgins' conduction solution. Because the design of a wave facility influences the particle kinematics, it is concluded that local empirical data will generally be required to achieve accurate predictions.

772   Morison Inertia Coefficients in Orbital Flow

John R. Chaplin, (Sr. Lect. in Civ. Engrg., Univ. of Liverpool, P.O. Box 147, Liverpool, England)

Journal of Waterway, Port, Coastal and Ocean Engineering, Vol. 111, No. 2, March, 1985, pp. 201-215

Inertia coefficients for use in Morison's equation in conditions of orbital flow are significantly lower than those appropriate to planar oscillatory flow. Circulation around the cylinder resulting from asymmetrical shedding of vorticity generates a lift which opposes the conventional inertia force. The magnitude of the lift and the effect of Reynolds number are discussed on theoretical grounds and with reference to force and velocity measurements in orbital flow. Consideration of the lift leads to the formulation of a modified Morison equation, which is appropriate to all circumstances of wave loading on a cylinder when the incident velocity vector is not collinear with the incident acceleration vector.

773   A Fifth-Order Stokes Theory for Steady Waves

John Fenton, (Sr. Lect., School of Mathematics, Univ. of New South Wales, Kensington, N.S.W., Australia 2033)

Journal of Waterway, Port, Coastal and Ocean Engineering, Vol. 111, No. 2, March, 1985, pp. 216-234

An alternative Stokes theory for steady waves in water of constant depth is presented, where the expansion parameter is the wave steepness itself. The first step in application requires the solution of one nonlinear equation, rather than two or three simultaneously as has been previously necessary. In addition to the usually specified design parameters of wave height, period and water depth, it is also necessary to specify the current or mass flux to apply any steady wave theory. The reason being that the waves almost always travel on some finite current and the apparent wave period is actually a Doppler-shifted period. Most previous theories have ignored this, and their application has been indefinite, if not wrong, at first order. A numerical method for testing theoretical results is proposed, which shows that two existing theories are wrong at fifth order, while the present theory and that of Chappelear are correct. Comparisons with experiments and accurate numerical results show that the present theory is accurate for wavelengths shorter than ten times the water depth.

774   Velocity Moments in Nearshore

R. T. Guza, (Assoc. Prof., Shore Processes Lab., Scripps Inst. of Oceanography, Univ. of California, La Jolla, Calif. 92093) and **Edward B. Thornton**, Member, ASCE, (Prof., Naval Postgrad. School, Monterey, Calif. 93940)

Journal of Waterway, Port, Coastal and Ocean Engineering, Vol. 111, No. 2, March, 1985, pp. 235-256

Recent models for nearshore sediment transport suggest the importance of various moments of the fluid velocity field in determining transport rates. Using two days of field data from a low slope beach with moderate wave heights (H $\sim$ 70 cm), some low order, normalized moments are compared to results from simple monochromatic and linear random wave models. Not surprisingly, the random wave model is substantially more accurate than the monochromatic model. However, wave breaking and other nonlinearities introduce effects not explained by either formalism. The observed cross-shore velocity variance is decomposed into wind wave and surf beat components. The surf beat contribution is maximum at the shoreline, while the wind wave component is maximum offshore. The total variance is nearly constant across the surf zone. This observation contradicts assumptions that are fundamental to many models of surf zone dynamics and sediment transport. Analysis of a wider range of wave conditions is needed to assess the generality of these preliminary results. Using field data in the sediment transport model of Bailard (1981) suggests that both bed and suspended load are significant cross-shore transporting mechanisms on this low slope beach with moderate wave energy. Asymmetries in the oscillatory wave field tend to transport sediment shoreward, while the interaction of the offshore mean flow with waves produces an offshore sediment flux.

775   Modification of River's Tide by Its Discharge

Gabriel Godin, (Visiting Prof., Centre de Investigacion Cientifica y di Educacien Superior di Ensenada (CICESE), Ensenada, B.C., Mexico)

Journal of Waterway, Port, Coastal and Ocean Engineering, Vol. 111, No. 2, March, 1985, pp. 257-274

The effect of an increased discharge on the tide progressing into a river is evaluated quantitatively by gaging the signal recorded at upstream stations against a reference station, during intervals of effectively constant discharge; this process is repeated for progressively larger values of the discharge. Upstream, the tidal range is reduced by an increased discharge; the time of arrival of low water is accelerated, while high water is retarded. The changes in range and in time may be represented by simple regression relations. Downstream an increased discharge causes a decrease in the effective friction during flood and an increase in it during ebb. Low water is retarded and high water is accelerated, and some tidal components may actually be amplified over a segment of the river.

776   Interaction of Random Waves and Currents

Terence S. Hedges, (Lect., Dept. of Civ. Engrg., Univ. of Liverpool, Liverpool, United Kingdom L69 3BX), **Kostas Anastasiou**, (Research Assoc., Dept. of Civ. Engrg., Univ. of Liverpool, Liverpool, United Kingdom L69 3BX) and **David Gabriel**, (Research Asst., Dept. of Civ. Engrg., Univ. of Liverpool, Liverpool, United Kingdom L69 3BX)

Journal of Waterway, Port, Coastal and Ocean Engineering, Vol. 111, No. 2, March, 1985, pp. 275-288

The effects of steady, uniform currents on random waves, and the associated water-particle kinematics, are investigated. The basic equations describing the interactions between waves and currents are reviewed, with special reference to the changes in the variance spectra of free-surface displacement and of horizontal water-particle velocity. Theoretical predictions are compared with laboratory measurements of random waves propagating onto an opposing current. The theoretical model is shown to be in reasonable agreement with observations.

777   Vortex-Induced Oscillation of Structures in Water

Wilfred D. Iwan, (Prof. of Applied Mech., California Inst. of Tech., Pasadena, Calif.) and **Dirceu L. R. Botelho**, Assoc. Member, ASCE, (Research Engr., Chevron Oil Field Research Co., La Habra, Calif.)

Journal of Waterway, Port, Coastal and Ocean Engineering, Vol. 111, No. 2, March, 1985, pp. 289-303

An analytical-empirical model for the vortex-induced oscillation of cylindrical structural elements in water is presented. The model is based upon force measurements for forced cylinders in a uniform aqueous flow. The frequency and amplitude of response of flexibly mounted cylindrical elements as well as the stability of steady state oscillation is inferred from the model. The model response is compared with experimental data for both water and air. The results raise some strong doubts about the validity of mixing experimental data obtained in such different media.

778   Floating Breakwater Design

Bruce L. McCartney, Member, ASCE, (Hydr. Engr., Hydr. and Hydrology Div., Office Chief of Engrs., U.S. Army Corps. of Engrs., Washington, D.C. 20314)

Journal of Waterway, Port, Coastal and Ocean Engineering, Vol. 111, No. 2, March, 1985, pp. 304-318

Floating breakwaters are inventoried. The various types are separated into 4 general categories, which are Box, Pontoon, Mat, and Tethered Float. The Tethered Float was identified as a special category but lacked sufficient prototype experience for detailed analysis. Advantages and disadvantages of the Box, Pontoon and Mat are presented. Hydraulic model test results and prototype experiencefor these 3 types are presented. Alternative mooring systems and anchorage methods are summarized. The engineering studies usually needed for a suitable design are outlined. Costs and design data for selected prototype installations are tabulated.

779   Scattering of Solitary Wave at Abrupt Junction

Chiang C. Mei, Member, ASCE, (Prof., Ralph M. Parsons Lab., Dept. of Civ. Engrg., Bldg. 48-413, Massachusetts Inst. of Tech., Cambridge, Mass. 02139)

Journal of Waterway, Port, Coastal and Ocean Engineering, Vol. 111, No. 2, March, 1985, pp. 319-328

The effect of sudden convergence of a rectangular channel on the scattering of a solitary wave and the subsequent evolution is assessed. Separation loss is treated approximately by involving a hydraulic loss formula for steady state flows in a nonuniform conduit and by equivalent linearization. The semi-empirical theory is compared with existing experiments.

780   Influence of Marginal Ice Cover on Storm Surges

Tad S. Murty, (Sr. Research Sci., Inst. of Ocean Sci., Dept. of Fisheries and Oceans, P.O. Box 6000, Sidney, British Columbia, Canada V8L 4B2) and **Greg Hollowya**, (Sr. Research Sci., Inst. of Ocean Sci., Dept. of Fisheries and Oceans, P.O. Box 6000, Sidney, British Columbia, Canada V8L 4B2)

Journal of Waterway, Port, Coastal and Ocean Engineering, Vol. 111, No. 2, March, 1985, pp. 329-336

Observational evidence is provided to show that positive storm surges are damped more than negative surges by ice. Data on the influence of an ice layer on tides and circulation is reviewed as supportive evidence for dissipation of long waves by ice. A mechanism is suggested as to why positive surges are more strongly damped. The crest and trough of the surge wave (which is a long gravity wave) are respectively referred to as the positive and negative surge. The observations consisted of tide gage data from 23 locations in eastern Canadian water bodies for the period 1965-1975. Classical studies, while recognizing the dissipation of long gravity waves by ice, do not account for the asymmetric dissipation of the crests and troughs. Here, we suggest a mechanism involving surface contraction and dilation as responsible for the asymmetric damping. This asymmetric damping occurs during the propagation of the surge in the shallow coastal waters and not during the generation. Open water or leads play an important role in this asymmetric dissipation process.

781 **Response of Variable Cross-Sectional Members to Waves**

Kosuke Nagaya, (Prof., Faculty of Engrg., Gunma Univ., Kiryu, Gunma 376, Japan)

Journal of Waterway, Port, Coastal and Ocean Engineering, Vol. 111, No. 2, March, 1985, pp. 337-353

This paper describes a method for solving transient responses of structural members of variable cross-section subjected to wave forces. The member is divided into small segments and the transfer matrix and the Laplace transform method are applied to the equation of motion of the segments including fluid inertias. Since it is difficult to perform the Laplace transformation when the wave force is a complicated function of the time, the analysis utilizes the Fourier series expansion procedure to obtain the result for general dynamic loads. Then the transformed solution becomes a function of the coordinate of the position. The transfer matrix method derives the solution of the member by the combination of the solution of each segment which is expressed in terms of the Laplace transform parameter. The Laplace transform inversion integral transforms the solution in the image domain into the time domain when the residue theorem is applied. As an example, numerical calculations have been carried out for the response of a circular variable cross-section pipe in a fluid subjected to conoidal wave forces.

782 **Congestion Cost and Pricing of Seaports**

Michihiko Noritake, (Assoc. Prof., Dept. of Civ. Engrg., Faculty of Engrg., Kansai Univ., Suita, Osaka, Japan)

Journal of Waterway, Port, Coastal and Ocean Engineering, Vol. 111, No. 2, March, 1985, pp. 354-370

Wharf facilities which are constructed and administered by public authorities of seaports are regarded as social overhead capital. Since the services that they furnish are not brought to the market, the optimum distribution of services from the viewpoint of social benefits is difficult to quantify. Specifically, ships arriving at a port inevitably bring about external diseconomies resulting from congestion in a port. Evidently, it is important to decrease or eliminate such external diseconomies. In this study, a port with public wharfs which handles general cargoes is treated; and the congestion at a port is analyzed both from economic and engineering points of view. Furthermore, preliminary examinations are made of the method for determining the optimum degree of usage of the wharf facilities, and the method for realizing the optimum degree by means of an adequate pricing system.

783 **Threshold Effects on Sediment Transport by Waves**

Richard J. Seymour, Member, ASCE, (Scripps Inst. of Oceanography, La Jolla, Calif. 92093)

Journal of Waterway, Port, Coastal and Ocean Engineering, Vol. 111, No. 2, March, 1985, pp. 371-387

An analytical technique is developed for correcting the time-averaged net sediment transport rate under oscillatory flows for the effects of the cessation of transport whenever the magnitude of the near-bottom velocity is less than some threshold value. This estimation scheme eliminates the analytical complexity of incorporating the threshold effect directly into the transport model. The threshold speeds are calculated using an empirical formulation. Using a simple model of longshore transport across the surf zone and employing some recent findings on velocity distributions, the magnitudes of threshold effects are explored for both broadband and monochromatic waves. An explanation for the observed difference in longshore transport efficiency for laboratory and field observations is suggested. A simple model for cross-shore transport is used to explore the effects of threshold of motion on this transport mechanism. Graphical presentations are used to show under what combinations of wave height, wave period, and sand size threshold effects are significant for various transport regimes.

784   Automated Remote Recording and Analysis of Coastal Data

Richard J. Seymour, Member, ASCE, (Lect., Scripps Inst. of Oceanography, La Jolla, Calif.), **Meredith H. Sessions**, (Development Engr., Scripps Inst. of Oceanography, La Jolla, Calif.) and **David Castel**, (Developmental Engr., Scripps Inst. of Oceanography, La Jolla, Calif.)

Journal of Waterway, Port, Coastal and Ocean Engineering, Vol. 111, No. 2, March, 1985, pp. 388-400

A system is described for sampling coastal data from a remote central station under computer control. The data gathering network handles wave measurements from offshore buoys and nearshore pressure sensors, and velocity components from current meters and anemometers. Coastal station locations range from Hawaii to North Carolina with system interconnection through ordinary dial up telephone lines. Data are objectively edited automatically, analyzed and are available for remote display within a few minutes of the observation. Measuring instruments, system hardware, operations, and reports are described.

785   Finite Water Depth Effects on Nonlinear Waves

Hang Tuah, (Grad. Student, Ocean Engrg. Program, Dept. of Civ. Engrg., Oregon State Univ., Corvallis, Oreg. 97331) and **Robert T. Hudspeth**, Member, ASCE, (Prof., Ocean Engrg. Program, Dept. of Civ. Engrg., Oregon State Univ., Corvallis, Oreg. 97331)

Journal of Waterway, Port, Coastal and Ocean Engineering, Vol. 111, No. 2, March, 1985, pp. 401-416

A representation for nonlinear random waves is obtained from a perturbation expansion method. The first-order wave solution is assumed to be a zero-mean, Gaussian process. The skewness measure and the skewness kernel for the nonlinear second-order waves are examined numerically and are compared with hurricane-generated, real ocean waves. These skewness measures are shown to always be positive and to increase as the water depth decreases. The effects of the angle of intersection between interacting wave trains are also examined numerically.

786   Finite Element Modelling of Nonlinear Coastal Currents

Chung-Shang Wu, (Adjunct Research Asst., Dept. of Oceanography, Naval Postgraduate School, Monterey, Calif. 93940) and **Philip L.-F. Liu**, Assoc. Member, ASCE, (Prof., School of Civ. and Environmental Engrg., Cornell Univ., Ithaca, N.Y. 14853)

Journal of Waterway, Port, Coastal and Ocean Engineering, Vol. 111, No. 2, March, 1985, pp. 417-432

A numerical model describing wave-induced mean sea level variations and coastal currents in the nearshore region is developed by the finite element method. The model includes nonlinear convective accelerations, lateral mixing and bottom friction. To specify the wave refraction field, a wave model is also developed with a semi-discrete Galerkin method. The numerical accuracy of the model is verified with the analytic solutions for one-dimensional longshore currents and two dimensional rip currents. The numerical model is also applied to predict realistic meandering currents occurring on a periodic rip channel. Due to the nonlinear inertial effect, the unaccelerated longshore current profile is stretched and causes a decrease in the magnitude of maximum velocity. A comparison with the analytic solution of a one-dimensional longshore current velocity distribution, indicates that the linear analytic solution significantly over-estimates the maximum velocity. The numerical results quantitatively demonstrate the relative importance of the nonlinear convective terms in the nearshore current problem.

787   Wave Forces on an Elliptic Cylinder

Anthony N. Williams, (Research Fellow, Dept. of Ocean Engrg., Univ. of Rhode Island, Kingston, R.I.)

Journal of Waterway, Port, Coastal and Ocean Engineering, Vol. 111, No. 2, March, 1985, pp. 433-449

Two approximte methods are presented for the calculation of the wave induced forces and moments on a vertical, surface-piercing cylinder of elliptic cross section. Both methods provide a substantial reduction in computational effort when compared with the exact solution which involves the numerical evaluation of Mathieu functions. One method involves the expansion of the exact expressions for the forces and moments for small values of the elliptic eccentricity parameter. The second method is based on Green's theorem and gives rise to an integral equation for the fluid velocity potential on the cylinder surface. Numerical results are presented for a range of relevant parameters and show excellent agreement with the computed values of the exact solution.

Errata: WW July '85, pp. 780-781.

788   Dynamics of Internal Waves in Cylindrical Tank

Amin H. Helou, Member, ASCE, (Asst. Prof., An-Najah Univ., Nablus, West Bank)

Journal of Waterway, Port, Coastal and Ocean Engineering, Vol. 111, No. 2, March, 1985, pp. 453-457

The internal wave problem in a circular cylindrical tank is solved by the standard method of separation of variables. Numerical results for the surface displacement at the interface, as well as the pressure distribution at the wall, are obtained.

789   Depth-Controlled Wave Height

Charles L. Vincent, Member, ASCE, (Coastal Engrg. Research Center, U.S. Army Engr. Waterway Experiment Sta., Vicksburg, Miss.)

Journal of Waterway, Port, Coastal and Ocean Engineering, Vol. 111, No. 3, May, 1985, pp. 459-475

A method for estimating an upperbound on the total energy of the wind wave spectrum in shallow water is extended to define a depth-controlled zero-moment wave height for irregular waves. The method requires an estimate of the peak frequency of the wave spectrum, knowledge of the Phillips' equilibrium coefficient, α, and water depth, h. A method for estimating α from the peak frequency of the sea spectrum and windspeed is given. Results indicate that the depth-controlled zero-moment wave height is generally less than the depth-limited monchromatic wave height, H_d, and appears to vary with the square root of depth.

790 Dredge-Induced Turbidity Plume Model

Albert Y. Kuo, Member, ASCE, (Prof., School of Marine Sci., Coll. of William and Mary, Gloucester Point, Va.), **Christopher S. Welch**, (Assoc. Prof., School of Marine Sci., Coll. of William and Mary, Gloucester Point, Va.) and **Robert J. Lukens**, (Instr., School of Marine Sci., Coll. of William and Mary, Gloucester Point, Va.)

Journal of Waterway, Port, Coastal and Ocean Engineering, Vol. 111, No. 3, May, 1985, pp. 476-494

A model is developed to describe the turbidity plume induced by dredging a ship channel using a hydraulic dredge. The model predicts the suspended sediment concentration within the plume and the resulting sediment deposition alongside the dredged channel. The model applies to a dredging operation in a water body in which the current is primarily along the channel axis and the channel depth is large enough that no significant suspended sediment reaches water surface. Results of field measurements are presented and compared with model. It is shown that the model describes the qualitative feature of prototype data and that the calibrated model parameters agree with independent observations by other investigators.

791 Bore Height Measurement with Improved Wavestaff

David L. Timpy, (Coastal Engr., Dept. of Public Works, City of Virginia Beach, Virginia Beach, Va.) and **John C. Ludwick**, (Prof., Dept. of Oceanography, Old Dominion Univ., Norfolk, Va.)

Journal of Waterway, Port, Coastal and Ocean Engineering, Vol. 111, No. 3, May, 1985, pp. 495-510

An improved capacitance wavestaff is described and used to measure wave height transformation in a surf zone; and the assumption is tested that wave height decays linearly across a surf zone. Acceptable frequency response in the instrument could be obtained only if sensor wires were smaller than or equal to 0.35 mm in diam. Simultaneous measurements of wave height were taken at points distributed across a surf zone during four field experiments at Virginia Beach, Virginia. Pairs of wavestaffs were repositioned along a line perpendicular to the shore line. Wave height decay across this surf zone was found to be non-linear, and the observed pattern appears to be categorizable into three zones: (1) One in which there is a rapid decay in height due to turbulence induced by the breaking process; (2) one in which there is a nearly constant decay due to a balance between turbulence and wave shoaling affects; and (3) one in which there is a rapid decay of remaining wave energy as a wave propagates up the beach slope against the force of gravity. Energy is dissipated less rapidly by spilling breakers than plunging breakers, and, thus, spilling breakers produce greater runup at the shore line.

792 Wave-Induced Forces on Buried Pipeline

Gerard P. Lennon, Assoc. Member, ASCE, (Asst. Prof., Dept. of Civ. Engrg., Lehigh Univ., Bethlehem, Pa. 18015)

Journal of Waterway, Port, Coastal and Ocean Engineering, Vol. 111, No. 3, May, 1985, pp. 511-524

Wave-induced pressures are an important design consideration for oil and natural gas pipelines buried in the marine environment. Realistic problems are three-dimensional in nature and involve waves approaching the buried pipeline at oblique angles and special pipeline geometries. For fluid flow in a sandy soil where liquefaction does not occur and fluid acceleration terms are negligible, Darcy's law can be used. Assuming that the soil structure and fluid are incompressible, results will show the wave-induced pressure in the domain being governed by the Laplace equation with associated boundary conditions on the domain boundaries. The pressure distribution on the pipeline is obtained using the boundary integral equation method (BIEM). The BIEM is economical because the computations are performed only on the two-dimensional surface boundaries of the solution domain rather than throughout the entire three-dimensional domain. The first problem analyzed is the two-dimensional case where the pipeline is parallel to the wave crests. The numerical solutions show good agreement with finite element and analytical solutions. For waves approaching at oblique angles, the periodic pressure distribution on the pipelines is obtained. Finally, solutions are obtained for soils of different properties. Future work is proposed in order to apply the techniques to more general problems.

793   Turbulent Viscosity in Rough Oscillatory Flow

Trond Aukrust, (Physics Engr., Dept. of Physics, The Univ. of Trondheim, Trondheim, Norway) and **Iver Brevik**, (Lect., Luftkrigsskolen, Trondheim, Norway)

Journal of Waterway, Port, Coastal and Ocean Engineering, Vol. 111, No. 3, May, 1985, pp. 525-541

In an earlier paper a three-layer model of an oscillatory rough turbulent layer was simplified to a two-layer model. The model consisted of an overlap layer in which the turbulent viscosity varied linearly with height, and an outer layer in which the viscosity was constant. This paper improves the theory from the earlier paper: Turbulent viscosity is allowed to decrease linearly above the overlap layer until a certain height is reached, after which the viscosity is held constant. The viscosity everywhere is assumed time independent. The theory is compared to the observations by other researchers. The agreement between theory and experiment is reasonably good. Also, a theoretical derivation of the wave friction factor is given, which for the larger excursion amplitudes agrees well with a previously established semi-empirical friction factor formula.

794   Episodicity in Longshore Sediment Transport

Richard J. Seymour, Member, ASCE, (Head, Ocean Engineering Research Group, Scripps Inst. of Oceanography, La Jolla, Calif. 92093) and **David Castel**, (Development Engr., Scripps Inst. of Oceanography, La Jolla, Calif. 92093)

Journal of Waterway, Port, Coastal and Ocean Engineering, Vol. 111, No. 3, May, 1985, pp. 542-551

Seven West Coast sites are selected having from one to three years of nearshore directional wave measurements several times a day during the period 1979-82. Time series of daily net longshore transport are estimated using the energy flux method, based upon measured S_{xy}. Investigations are made on frequency and cumulative distributions of transport, and from these, a number of statistics characterizing the degree of episodic transport are generated. The transport is found to be very episodic, with almost half of the gross transport occurring during only 10% of the time. The maximum transport occurring in a single day each year produced between ten times, and more than 600 times, the mean daily net transport. Inferences concerning the design requirements for sand bypass systems are drawn from the statistics of episodicity.

795   Riprap Stability Under Wave Action

Nobuhisa Kobayashi, Member, ASCE, (Asst. Prof., Dept. of Civ. Engrg., Univ. of Delaware, Newark, Del. 19716) and **Brian K. Jacobs**, (Grad. Student, Dept. of Civ. Engrg., Univ. of Delaware, Newark, Del. 19716)

Journal of Waterway, Port, Coastal and Ocean Engineering, Vol. 111, No. 3, May, 1985, pp. 552-566

A mathematical model is developed to predict the flow characteristics in the downrush of regular waves and the critical condition for initiation of movement of armor units on the slope of a coastal structure. The analysis of water motion on the slope is based on the standing-wave solution of the finite-amplitude shallow-water equations with an implicit account of the effects of wave breaking. This simple model neglects the effects of permeability, bottom friction and water depth. In addition, use is made of an empirical runup relationship. A stability analysis of armor units is performed including the drag, lift and inertia forces acting on an armor unit which varies along the slope with time over the period of wave downrush. Comparison is made with the large-scale test data on riprap stability from an earlier study. The predicted critical stability number follows the same trend as the observed zero-damage stability number, although the agreement is qualitative.

796 Entrained and Bed-Load Sand Concentrations in Waves

Robert J. Hallermeier, (Visiting Coastal Sci., Cyril Galvin, Coastal Engrg., Box 623, Springfield, Va. 22150)

Journal of Waterway, Port, Coastal and Ocean Engineering, Vol. 111, No. 3, May, 1985, pp. 567-586

Using various types of basic empirical results, equations are developed for average volumetric concentration of sediment moving as bed load in nearshore wave environments. Laboratory and natural situations have appreciably different dependences of bed-load concentration on peak flow velocity, with laboratory sediment concentrations being commonly larger than those in nature. Critical examination of published concentration measurements isolates several subsets of near-bed data for quartz sands which allows the two expressions for bed-load concentration to be evaluated. Field and laboratory results in separate categories exhibit distinct functional trends and some quantitative magnitudes agreeing with calculated bed-load concentrations, so that calculation might give a reference value useful for suspended-load computation. Evidence permits a firm conclusion that laboratory near-bed processes and suspended sediment concentrations in usual conditions do not pertain to important field situations.

797 Gravimetric Statistics of Riprap Quarrystones

Hin-Fatt Cheong, Member, ASCE, (Assoc. Prof., Faculty of Engrg., National Univ. of Singapore, Kent Ridge, Republic of Singapore 0511)

Journal of Waterway, Port, Coastal and Ocean Engineering, Vol. 111, No. 3, May, 1985, pp. 589-593

A statistical study of the gravimetric gradation characteristics of quarrystones used for shoreline protection against small wave attack is described. A nonparametric method based on order statistics is used to determine the minimum sample size required so that the probability is θ that at least β of the population will be between the jth and kth observations. It is also found that riprap quarrystone obtained after secondary crushing exhibited the Log-Pearson Type III distribution.

798 Scaling the Weight of Breakwater Armor Units

James J. Sharp

Journal of Waterway, Port, Coastal and Ocean Engineering, Vol. 111, No. 3, May, 1985, pp. 594-597

The use of the stability parameter for calculating the weights of armor units in models of rubble mound breakwaters is shown to have little justification. An alternative scaling method based on similarity criteria is developed and the results of the two methods are compared. It is shown that the use of the stability number causes little error but that this is fortuitous rather than physically meaningful.

799 Field Data on Seaward Limit of Profile Change

William A. Birkemeier, Assoc. Member, ASCE, (Hydr. Engr., U.S. Army Coastal Engrg. Research Center, Field Research Faciity, Duck, N.C. 27949)

Journal of Waterway, Port, Coastal and Ocean Engineering, Vol. 111, No. 3, May, 1985, pp. 598-602

Many coastal engineering problems require an estimate of the seaward limit of sediment transport, defined as the minimum depth at which no measurable change in water depth occurs. A procedure to estimate this limit depth was evaluated using measurements collected at the Coastal Engineering Research Center's Field Research Facility located on the Atlantic Ocean. The data consisted of measured wave characteristics and accurate repetitive nearshore surveys which extended out to a depth of 30 ft (9 m). Ten unique data points were used in the evaluation with measured limit depths ranging from 18 to 21 (3.9 to 6.4 m). These depths were overpredicted by an average of 4.6 ft (1.4 m). This difference could be reduced to 1.3 ft (0.4 m) by adjusting the coefficients in the equation. A reasonable correlation was also obtained using a simple multiple of wave height.

800 Distortions Associated with Random Sea Simulators

Josep R. Medina, (Asst. Prof., Dept. of Ports and Ocean Engrg., E. T. S. Caminos, Univ. Politécnica de Valencia, Valencia, Spain), José Aguilar, (Asst. Prof., Dept. of Ports and Ocean Engrg., E. T. S. Caminos, Univ. Politécnica de Valencia, Valencia, Spain) and J. Javier Diez, (Prof., Dept. of Ports and Ocean Engrg., E. T. S. Caminos, Univ. Politécnica de Valencia, Valencia, Spain)

Journal of Waterway, Port, Coastal and Ocean Engineering, Vol. 111, No. 4, July, 1985, pp. 603-628

Some numerical techniques for simulating Gaussian ergodic stochastic sea models are described, analyzed, and contrasted. A general method for generating all numerical, linear, one-dimensional simulators by wave superposition permits one to describe or create any of these numerical random sea simulators in five steps. The distortions associated with each numerical simulator by wave superposition are analyzed from a general point of view and the arbitrariness of some numerical simulation techniques commonly used is noted. The time-consumed in these Monte Carlo experiments is an important factor. The numerical algorithms used can change indirectly the level of distortions associated with each numerical simulation technique. A special reference has been made to the use of the fast Fourier transform (FFT) for computing and to the second-order autoregressive behavior of each wave component in order to reduce the time-consumed. Three criteria are proposed for qualifying the numerical simulators in order to adapt the requirements of each numerical experiment considered. To explain the variability of random sea, the deterministic amplitude component simulators are rejected while a nondeterministic spectral amplitude simulator (NSA) using a FFT algorithm can be employed.

801   A Method for Investigation of Steady State Wave Spectra
in Bays

Bryan R. Pearce, Member, ASCE, (Prof., Univ. of Maine at Orono, Dept. of Civ. Engrg., 455 Aubert Hall, Orono, Me. 04469) and Vijay G. Panchang, Student Member, ASCE, (Research Asst., Univ. of Maine at Orono, Dept. of Div. Engrg., 455 Aubert Hall, Orono, Me. 04469)

Journal of Waterway, Port, Coastal and Ocean Engineering, Vol. 111, No. 4, July, 1985, pp. 629-644

A technique is presented for modeling the evolution of wave spectra in bays. In particular, the writers are concerned with those bodies of water where bathymetric effects are important and where conventional methods such as ray tracing may not always work. The technique presented is based upon a modified form of the Helmholtz equation, thus incorporating refraction and diffraction. A splitting technique transforms the elliptic Helmholtz equation into a parabolic form which can be solved conveniently. Energy dissipation can be built into the transformed equation. The spectral evolution is investigated by using a set of spectral components. These components are needed for the inclusion of the source terms, i.e., atmospheric input and nonlinear energy transfer. Toward this end, a solution process appropriate for the proposed technique is presented. Comparisons to laboratory and prototype data are presented.

802   Beach Evolution Caused by Littoral Drift Barrier

Deva K. Borah, Assoc. Member, ASCE, (Sr. Hydr. Engr., Tippetts-Abbett-McCarthy-Stratton, The TAMS Building, 655 Third Ave., New York, N.Y. 10017) and Armando Balloffet, Fellow, ASCE, (Consultant, Tippetts-Abbett-McCarthy-Stratton, The TAMS Building, 655 Third Ave., New York, N.Y. 10017)

Journal of Waterway, Port, Coastal and Ocean Engineering, Vol. 111, No. 4, July, 1985, pp. 645-660

A numerical model is developed to predict the shoreline changes at Nome, Alaska due to the construction of a causeway that will connect the offshore port terminal with the onshore facilities. The model is based on a procedure developed earlier by other researchers. Several aspects of the original procedure are elaborated and modified. The resultant model was verified by simulating the present beach condition near the entrance jetties of Nome Harbor. Using the same parameters, the model was applied to the causeway and shore configurations for a 30-yr period were predicted. These applications show that the modifications made on the original model are promising. The model can be applied to other similar situations provided the parameters are calibrated by simulating existing conditions.

803   FEM Solution of 3-D Wave Interference Problems

Min-Chih Huang, Assoc. Member, ASCE, (Asst.Prof., Dept. of Naval Architecture and Marine Engrg., National Cheng Kung Univ., Taiwan, China 700), Robert T. Hudspeth, Member, ASCE, (Prof., Dept of Civ. Engrg. and Ocean Engrg., Oregon State Univ., Corvallis, Oreg. 97331) and John W. Leonard, Member, ASCE, (Prof., Dept. of Civ. Engrg. and Ocean Engrg., Oregon State Univ., Corvallis, Oreg. 97331)

Journal of Waterway, Port, Coastal and Ocean Engineering, Vol. 111, No. 4, July, 1985, pp. 661-677

A numerical procedure for predicting wave diffraction, wave radiation, and body responses of multiple 3-D bodies or arbitrary shape is described. Viscous effects are neglected, and the hydrodynamic pressure forces are assumed to be inertially dominated. Within the limits of linear wave theory, the boundary value problems are solved numerically by the finite element method (FEM) using a radiation boundary damper approach. Both permeable boundary dampers and a fictitious bottom boundary element are included in the finite element algorithm in order to treat both permeable boundary problems and deep water wave problems. Numerical results are presented for a variety of structures to illustrate the following features: fictitious bottom boundary, multiple-structure wave interference, and permeable boundary.

804   Shallow River Pushboat Preliminary Design

Robert Latorre, (Assoc. Prof., Dept. of Naval Architecture and Marine Engrg., Univ. of New Orleans, P.O. Box 1098, New Orleans, La. 70148)

Journal of Waterway, Port, Coastal and Ocean Engineering, Vol. 111, No. 4, July, 1985, pp. 678-692

The preliminary design of a river pushboat is presented. The design method consists of four parts: (1) Estimation of pushboat "push" EP; (2) estimation of barge tow resistance R_T; (3) selection of pushboat horsepower based on equilibrium speed from equating 1 and 2; and (4) for the horsepower obtained from 3, determine the pushboat dimensions and tunnel stern arrangement. Using published data, the writer introduces the following expression for pushboat EP: $EP = A \; HP \; [1.0 + C(h/T-1.78)] \; [1.0 - E.V^2]$lbs. This expression is shown to be in good agreement over the range of $1,800 \leq HP \leq 5,600$: $1.78 \leq h/T \leq 5.0$; $0 \leq V \leq 12$mph. For the deep water resistance ($h/T = 5.0$), a modified Howe's formula is shown to give a good estimate for the barge tow resistance, R_T. This formula is shown to be unsatisfactory for estimating the shallow water ($h/T = 1.78$) resistance, R_T. Therefore, the method of P. A. Apukhtin is extended using published barge tow resistance data. The shallow water resistance predicted with Apukhtin's extended graph is shown to be in good agreement with available data. The estimates of the tow speed and propulsive efficiency are then completed. It is shown that for the two pushboats considered, 1,800 and 2,400 horsepower, the shallow water conditions reduce the propulsive efficiency by 16-23%. Guidelines for estimating the pushboat length, propeller diameter, and tunnel stern arrangement are given. Criteria for insuring adequate tunnel inflow and outflow are presented.

805   Wave Forces on Vertical Walls

John D. Fenton, (Sr. Lect., School of Mathematics, Univ. of New South Wales, Kensington, N. S. W., Australia 2033)

Journal of Waterway, Port, Coastal and Ocean Engineering, Vol. 111, No. 4, July, 1985, pp. 693-718

Formulas are presented to third order in wave height for the force and moment exerted on a vertical wall by the complete reflection of wave with an arbitrary angle of incidence. These expressions show a number of unusual features, some of which have been found previously for the special case of standing waves. They include the following: The maximum force per unit length is caused by obliquely-incident waves rather than standing waves; the second-order contribution to the load may be larger than that at first order without invalidating the solution; the greatest net force is that directed offshore under the wave troughs; and the greatest onshore force sometimes does not occur under wave crests. The formulas presented make the problem of determining the maximum load for design purposes one of finding the maximum of a given function in a space which includes as its dimensions the wave height, wave length or period, angle of incidence, and the wall length relative to the wavelength.

806 Hydrodynamic Interaction of Flexible Structures

Wen-Gen Liao, Assoc. Member, ASCE, (Lead Sr. Engr., S.S.D., Inc., Berkeley, Calif.)

Journal of Waterway, Port, Coastal and Ocean Engineering, Vol. 111, No. 4, July, 1985, pp. 719-731

In order to investigate the nature of hydrodynamic interaction on multiple, flexible structures subjected to earthquake ground motion, a numerical analysis for determining the hydrodynamic forces and the corresponding structural responses on three-dimensional offshore structures is presented. With the assumption of potential flow, the finite element method is developed to solve the boundary value problem in the fluid domain. If the structures have uniform cross sections in the vertical direction, the problem can be reduced to a two-dimensional one. Moreover, the principle of symmetry and antisymmetry can also be applied to the problem to reduce the fluid domain such that the computation becomes much more efficient. A computer program is developed to analyze the structural behavior according to the afore-mentioned approach. From the results of the examples presented, we arrive at the conclusion that hydrodynamic interaction is relevant if the structures are in close proximity. This phenomenon depends on the characteristics of structures, water depth, distance between structures, excitation frequency, amplitude of ground motion, and the direction of excitation.

807 Dominant Shear Stresses in Arrested Saline Wedges

Vassilios Dermissis, (Asst. Prof., Dept. of Civ. Engrg., Aristoteles Univ. of Thessaloniki, Thessaloniki, Greece) and **Emmanuel Partheniades**, Member, ASCE, (Prof., Dept. of Engrg. Sciences, Univ. of Florida, Gainesville, Fla.)

Journal of Waterway, Port, Coastal and Ocean Engineering, Vol. 111, No. 4, July, 1985, pp. 733-752

The shear stresses in saline wedges have been investigated in a 20m long variable slope flume by the following approaches: (1) Direct measurements of velocities and Reynolds stresses through hot film anemometers; (2) integration of the equations of motion; (3) Schijf-Schoenfeld's one-dimensional model; (4) integration of the equations of motion assuming zero bed stress. It was found that the interfacial and bed friction coefficients, f_i and f_o, respectively, can best be correlated with the number $ReFr^2$, in which Re is the Reynolds number, and Fr. is the nondensimetric Froude number and with the relative density. The results are given as a family of curves with each curve corresponding to a specific $\Delta\rho/\rho$. The scattering of data is small and the agreement results of earlier studies is good. Values of f_i determined by the first two approaches agree closely; as do values obtained by the third and fourth approach. However the latter are substantially higher than the first. This is indicative of a strong effect of the bed resistance on the interfacial friction.

808 Concrete Armor Unit Form Inventory

Austin A. Owen, Member, ASCE, (Hydr. Engr., U.S. Army Engr. Waterways Experiment Sta., Coastal Engrg. Research Center, P.O. Box 631, Vicksburg, Miss. 39180)

Journal of Waterway, Port, Coastal and Ocean Engineering, Vol. 111, No. 4, July, 1985, pp. 755-758

Concrete armor units in many shapes are currently being used to protect coastal structures. A significant percentage of the unit's manufacturing cost is the cost of the form. Forms for manufacturing dolosse, tribars, tetrapods, sta-pods, and sta-bars may be obtained from private industry with an inventory and private companies which build forms. Also, some U.S. Government agencies have forms available for their agency use only. Private industry with available forms are located in the Virgin Islands, the states of Oregon, California, and Hawaii, and in Quebec, Canada. Private companies which build forms for sale are located in the states of New York, Iowa, Texas, and Oregon.

809   Predicted Extreme High Tides for California: 1983-2000

Bernard D. Zetler, (Research Oceanographer, Scripps Inst. of Oceanography, La Jolla, Calif. 92093) and **Reinhard E. Flick,** (Asst. Research Oceanographer, Scripps Inst. of Oceanography, La Jolla, Calif. 92093)

Journal of Waterway, Port, Coastal and Ocean Engineering, Vol. 111, No. 4, July, 1985, pp. 758-765

When a combination of high tides and severe storm induced waves devastated California's coast in the winter of 1982-1983, predictions of much higher tides in the early 1990's appeared in the press. Standard harmonic tide predictions are prepared for San Diego, Los Angeles, San Francisco and Humboldt Bay extending until the year 2000. These show that the range between annual extremes at any station is only 0.4 foot (0.12m) with the highest tides predicted during the period 1986-1990. The predictions include: The semiannual beat of tide constituents which produce peak tides each summer and winter; a distinct 4.4 year beating which peaked during 1982-1983; and a peak enhancement of extreme tides in 1987 from maximum contributions of diurnal constituents due to the 18.61 year cycle in the longitude of the moon's node. Future astronomical components of extreme tides will exceed those of 1982-1983 by at most several tenths of a foot.

810   Wind Wave Growth in Shallow Water

Charles L. Vincent, Member, ASCE, (Offshore & Coastal Technologies, Inc., 10378 Democracy Lane, Fairfax, Va. 22030) and **Steven A. Hughes,** Member, ASCE, (Hydr. Engr., Coastal Oceanography Branch, Coastal Engrg. Research Center, U.S. Army Engr. Waterways Experiment Sta., Vicksburg, Miss. 39180)

Journal of Waterway, Port, Coastal and Ocean Engineering, Vol. 111, No. 4, July, 1985, pp. 765-770

New developments in the area of shallow water wave spectra have aided in the derivation of an equation for the limiting wave height in the case of fully saturated wind seas in shallow water. The result is very similar to the empirically derived expression obtained by Bretschneider in 1958. Starting with an equation for the depth controlled zero moment wave height in terms of depth, Phillip's coefficient, and the wave period associated with the peak spectral frequency, the result is obtained using a recently developed empirical experssion for the Phillip's coefficient and a method for specifying the peak spectral frequency at full development in shallow water. The good comparison to Bretschneider's result is satisfying in that further validity is added to the basic equation used in the derivation and to the newer developments in shallow water spectral representations.

811   Rubble Mounds: Hydraulic Conductivity Equation

A'Alim A. Hannoura, Assoc. Member, ASCE, (Assoc. Prof., Dept. of Civ. Engrg., Univ. of New Orleans, New Orleans, La.) and **John A. McCorquodale,** Member, ASCE, (Prof., Dept. of Civ. Engrg., Univ. of Windsor, Windsor, Ontario, Canada N9B 3P4)

Journal of Waterway, Port, Coastal and Ocean Engineering, Vol. 111, No. 5, September, 1985, pp. 783-799

Experimental investigations were undertaken to develop a generalized hydraulic conductivity equation for rubble mounds. The generalized equation takes into account the inertial force due to the unsteadiness of the flow and the effect of entrained air. Two-phase flow models are used to analyze the air/water flow. A modified Morrison-type equation for the unsteady flow forces is presented. The generalized hydraulic conductivity equation is developed for a wide range of Reynolds numbers. The objective of this study was to provide a generalized equation for modeling wave motion in rubble-mound structures, and an outline of the computations required is given.

812   **Rubble Mounds: Numerical Modeling of Wave Motion**

A'Alim A. Hannoura, Assoc. Member, ASCE, (Assoc. Prof., Dept. of Civ. Engrg., Univ. of New Orleans, New Orleans, La.) and **John A. McCorquodale**, Member, ASCE, (Prof., Dept. of Civ. Engrg., Univ. of Windsor, Windsor, Ontario, Canada N9B 3P4)

Journal of Waterway, Port, Coastal and Ocean Engineering, Vol. 111, No. 5, September, 1985, pp. 800-816

A mixed numerical model is developed to simulate wave motion in rubble-mound structures. The mixed model utilizes a combined finite difference method of characteristics scheme to integrate the unsteady continuity and momentum equations in the x-t plane to obtain the internal water levels while the two-dimensional properties of the flow are found from a finite element solution for the internal flow domain, x-y plane, at any time t. The two-dimensional solution is used to update pressure distribution coefficient and hydraulic conductivity values in the x-t plane. The model is applied to the Sines breakwater in order to check the dynamic stability of the seaward slope under severe wave attack. The predicted values for the internal water surface are found to be in fair agreement with a physical model measurement. Furthermore, the model indicates a lower factor of safety than the traditional analysis. Special provisions are included in the model to account for added mass and to detect and correct for internal wave breaking and the entrainment of air near the interface.

813   **Damage to Armor Units: Model Technique**

K. den Boer, (Prof. Engr., Delft Hydraulics Lab., P.O. Box 152, 8300 AD Emmeloord, The Netherlands)

Journal of Waterway, Port, Coastal and Ocean Engineering, Vol. 111, No. 5, September, 1985, pp. 817-827

After a storm on December 31, 1979 and January 1, 1980, with a maximum H_s of 23.3 ft (7.1 m) a large number of broken Dolosse of 15 metric tonnes (15,000 kg) was observed on the west breakwater of Gioia Tauro, Italy. The present paper describes a study in which the storm was reproduced in a model at a scale of 1 to 45.5 and with which a stable modified design was determined. A model technique using single frame film exposures was applied in order to detect rocking of Dolosse. The main conclusion is that the number of rocking plus the number of displaced Dolosse observed in the model corresponds well to the number of broken Dolosse in nature. This conclusion is only valid for the slope above still-water level.

814   **Significant Wave Height for Shallow Water Design**

Edward F. Thompson, Member, ASCE, (Research Engr., U.S. Army Coastal Engrg. Research Center, Waterways Experiment Station, P.O. Box 631, Vicksburg, Miss. 39180) and **C. L. Vincent**, Member, ASCE, (Research Engr., Offshore and Coastal Technologies, Inc., Fairfax, Va. 22030)

Journal of Waterway, Port, Coastal and Ocean Engineering, Vol. 111, No. 5, September, 1985, pp. 828-842

Wave height parameters used in coastal and ocean engineering are grouped into three classes according to their definition bases: height statistics, energy, and monochromatic. Parameters within each class are easily interrelated for most engineering purposes. However, parameters from different classes are difficult to interrelate, particularly for shallow water applications where waves are near breaking. The often-used parameter "significant wave height" has traditionally been based on height statistics but many modern estimates are based on wave energy. A simple empirical method is developed to relate statistical and energy based significant height estimates. The method is developed with CERC laboratory flume data from a 1:30 plane slope, two samples of field data, and stream function wave theory. Since the two significant height estimates differ by over 40% in some laboratory cases, engineers should clearly recognize the distinction between them.

815   Numerical Prediction of Wave Transformation

Philip L.-F. Liu, Assoc. Member, ASCE, (Prof., Cornell Univ., School of Civ. and Environmental Engrg., Ithaca, N.Y.) and **Ting-Kuei Tsay**, (Asst. Prof., Dept. of Civ. Engrg., Syracuse Univ., Syracuse, N.Y.)

Journal of Waterway, Port, Coastal and Ocean Engineering, Vol. 111, No. 5, September, 1985, pp. 843-855

A numerical model based on the parabolic approximation method is developed to calculate the wave characteristics in the nearshore region. The model is developed for monochromatic linear waves and considers refraction, diffraction, and energy dissipation caused by the bottom turbulent boundary layer. A numerical algorithm is also proposed to treat dignitized bathymetry data. The accuracy of the present model is verified by comparing numerical results with two sets of field measurements collected by the Coastal Engineering Research Center along its research pier at Duck, North Carolina.

816   Force Spectra from Partially Submerged Circular Cylinders in Random Seas

William J. Easson, (R. F., Dept. of Physics, Univ. of Edinburgh, Edinburgh, Scotland), **Clive A. Greated**, (Sr. Lect., Dept. of Physics, Univ. of Edinburgh, Edinburgh, Scotland) and **Tariq S. Duranni**, (Prof., Dept. of Electronic Sci., Univ. of Strathclyde)

Journal of Waterway, Port, Coastal and Ocean Engineering, Vol. 111, No. 5, September, 1985, pp. 856-879

The vertical force on a horizontal cylinder partially submerged in random seas shows large secondary peaks in its power spectrum. An equation is derived which predicts these peaks taking account of buoyancy, inertia and drag forces. It is also shown that at critical depths, the mean force on the cylinder is large and negative and components of the force near the mean frequency of the sea spectrum are very small.

817   Stability of Armor Units on Composite Slopes

Nobuhisa Kobayashi, Member, ASCE, (Asst. Prof., Dept. of Civ. Engrg., Univ. of Delaware, Newark, Del. 19716) and **Brian K. Jacobs**, (Grad. Student, Dept. of Civ. Engrg., Univ. of Delaware, Newark, Del. 19716)

Journal of Waterway, Port, Coastal and Ocean Engineering, Vol. 111, No. 5, September, 1985, pp. 880-894

Riprap and sandbag model tests are conducted in a wave flume to investigate the effects of berm-type slopes on the stability of armor units and wave runup as compared to uniform slopes. Measurements of wave runup, rundown, wave height, breaker type and the response of armor units under regular wave action are made for each test run. The uniform and composite slope test results are analyzed using a modified Saville's method which accounts for the overall effects of the slope configuration on the stability of armor units and wave runup. A simple analysis procedure based on the proposed method is developed for a preliminary design of a berm configuration. An example computation is made for a composite slope protected with riprap. The berm width, the berm slope and the water depth at the shallowest point of the berm are varied so as to determine the optimal berm configuration for increasing the stability of riprap under the assumed wave conditions.

818   Hinged Floating Breakwater

Patrick A. Leach, (Assoc. Engr., ABAM Consulting Engrs., Federal Way, Wash. 98003), **William G. McDougal**, Assoc. Member, ASCE, (Asst. Prof., Ocean Engrg. Program, Dept. of Civ. Engrg., Oregon State Univ., Corvallis, Oreg.) and **Charles K. Sollitt**, Member, ASCE, (Assoc. Prof., Ocean Engrg. Program, Dept. of Civ. Engrg., Oregon State Univ., Corvallis, Oreg. 97331)

Journal of Waterway, Port, Coastal and Ocean Engineering, Vol. 111, No. 5, September, 1985, pp. 895-909

An analytical model is developed to examine the response and efficiency of a rigid, hinged floating breakwater. The theoretical model is verified experimentally and is used to develop design curves which may be employed to estimate necessary physical breakwater characteristics to satisfy specified wave attenuation criteria. The utility of these curves is demonstrated in a design problem. It is shown that a structure of reasonable size is an effective dynamic barrier for high frequency waves and an effective kinematic barrier for low frequency waves.

819   Wave Forces on Inclined Circular Cylinder

Anthony N. Williams, (Asst. Prof., Dept. of Mathematics, Oregon State Univ., Corvallis, Oreg. 97331)

Journal of Waterway, Port, Coastal and Ocean Engineering, Vol. 111, No. 5, September, 1985, pp. 910-920

The integral equation method is utilized to calculate the wave-induced loading on a surface-piercing cylinder of circular cross-section inclined at an arbitrary angle to the sea bed. Numerical values of the various force and moment components are presented for a range of wave and cylinder parameters. The computed estimates of the force coefficients show good agreement in the inertial limit with published experimental results. In addition, the estimates confirm that the inertia force on an inclined cylinder may be severely underestimated by using the vectorial form of the Morison equation with a mass coefficient appropriate to the corresponding vertical cylinder case.

820   Wave Statistical Uncertainties and Design of Breakwater

Bernard Le Mehaute, Member, ASCE, (Prof. of Ocean Engrg., RSMAS, Univ. of Miami, Miami, Fla.) and **Shen Wang**, (Prof. of Ocean Engrg., RSMAS, Univ. of Miami, Miami, Fla.)

Journal of Waterway, Port, Coastal and Ocean Engineering, Vol. 111, No. 5, September, 1985, pp. 921-938

A method is developed in which an optimum design of breakwaters is achieved by taking into account wave climatological uncertainties and the potential maintenance risk as function of these uncertainties. Wave climatological uncertainties result from small sampling and errors of measurement or hindcast calculations. The economic impact which results from these uncertainties on the cost of breakwaters clearly evidenced the need for long-term accurate measurement programs. However, because of the required duration, these programs can only be conceived as an investment for future generations.

821   Refraction-Diffraction Model for Linear Water Waves

Bruce A. Ebersole, (Research Hydr. Engr., Research Div., Coastal Engrg. Research Ctr., Waterways Experiment Sta., P.O. Box 631, Vicksburg, Miss. 39180)

Journal of Waterway, Port, Coastal and Ocean Engineering, Vol. 111, No. 6, November, 1985, pp. 939-953

A numerical model is presented that predicts the transformation of monochromatic waves over complex bathymetry and includes both refractive and diffractive effects. Finite difference approximations are used to solve the governing equations, and the solution is obtained for a finite number of rectilinear grid cells that comprise the domain of interest. Model results are compared with data from two experimental tests, and the capability and utility of the model for real coastal applications are illustrated by application to an ocean inlet system.

822   Time Series Modeling of Coastal Currents

David A. Chin, Assoc. Member, ASCE, (Asst. Prof., Dept. of Math., Middle Georgia College, Cochran, Ga. 31014) and **Philip J. W. Roberts**, Assoc. Member, ASCE, (Asst. Prof., School of Civ. Engrg., Georgia Inst. of Tech., Atlanta, Ga. 30332)

Journal of Waterway, Port, Coastal and Ocean Engineering, Vol. 111, No. 6, November, 1985, pp. 954-972

Time and frequency domain models that relate a vector time series at any location to a measured series at another location are presented and evaluated by application to the prediction of coastal currents. The time domain model belongs to the general class of cross-autoregressive integrated (CARI) models based on classical time series analysis methods. The frequency domain model is a spectral component (SC) model, which relates the Fourier spectrum at different stations. The errors resulting from the use of both model types to predict measured currents off San Francisco, California, were computed. It was found that of the CARI models, a six parameter model was best, although a simpler three parameter model was almost as good. The main drawback to the CARI model was the non-stationarity of the formulation, requiring several years of data to develop a stationary model for a particular season. The SC model was more stationary and had smaller errors than the best CARI model. It is concluded that frequency domain models are the best method of predicting two-dimensional vector time series from other series.

823   Circulation Induced by Coastal Diffuser Discharge

E. Eric Adams, Member, ASCE, (Princ. Research Engr. and Lect., Dept. of Civ. Engrg., Massachusetts Int. of Tech., Cambridge, Mass. 02139) and **John H. Trowbridge**, (Asst. Prof., Dept. of Civ. Engrg., Univ. of Delaware, Newark, Del. 19716)

Journal of Waterway, Port, Coastal and Ocean Engineering, Vol. 111, No. 6, November, 1985, pp. 973-984

Submerged multiport diffusers are often used to discharge waste heat resulting from once-through cooling at coastal power plants. A particularly effective design, known as a staged diffuser, involves nozzles oriented offshore, parallel with the diffuser pipe. Potential flow analyses are used to describe the far field entrainment region for a staged diffuser discharging to quiescent receiving water of either constant depth or linear sloping bottom. Boundary conditions, in the form of a distributed sink, are supplied by a simplified near field model. Whereas near field theory shows that dilution can be increased by increasing discharge velocity, extending diffuser length or siting in deeper water, the latter options are shown to be preferable from the standpoint of minimizing induced velocities and the magnitude of nearshore entrainment.

824   Sea Level Rise Effects on Shoreline Position

Craig H. Everts, Member, ASCE, (Moffatt and Nichol, Engineers, P.O. Box 7707, Long Beach, Calif. 90807)

Journal of Waterway, Port, Coastal and Ocean Engineering, Vol. 111, No. 6, November, 1985, pp. 985-999

Using a conservation of sand approach, the effects of a rising sea surface are quantified and separated from other causes of shore retreat. Site-specific data important in predicting shoreline changes are: (1) Initial shoreface and backbeach profile; (2) subsequent backbeach profile; (3) relative sea level rise; (4) grain size distribution of sediment landward of the shoreface; and (5) net quantity of sand-sized material that enters or leaves a specified coastal reach. A key element of the approach, Bruun's assumption of a shoreface in dynamic equilibrium with the sea surface, was evaluated and found to be reasonably accurate. Field application of the method shows that sea level rise accounts for about 53% of the total shore retreat of 5.5 m/yr measured at Smith Island, Virginia, and for about 88% of the measured 1.7-m/yr retreat of the barrier island south of Oregon Inlet, North Carolina. Net sand losses account for the remainder. Because shoreface adjustments are required to maintain an equilibrium profile, sand replenishment is probably the most realistic method to stabilize a shore against the effects of relative sea level rise. Conversely, a negative sediment budget may also be mitigated by structures which hinder the movement of sand away from a problem beach and enhance its deposition there.

825   Internal Seismic Forces on Submerged Oil Tanks

Medhat A. Haroun, (Assoc. Prof., Civ. Engrg. Dept., Univ. of Calif., Irvine, Calif. 92717) and **Rong Chang**, (Grad. Research Asst., Civ. Engrg. Dept., Univ. of Calif., Irvine, Calif. 92717)

Journal of Waterway, Port, Coastal and Ocean Engineering, Vol. 111, No. 6, November, 1985, pp. 1000-1008

Internal hydrodynamic forces on submerged tanks due to arbitrary seismic excitations are evaluated. The transfer function of the lateral force is evaluated using Laplace transform. A mechanical analog duplicating such hydrodynamic forces is developed and checked for several limiting cases. The study shows that sloshing at liquids' interface can be neglected in the evaluation of such forces.

826   Locks with Devices to Reduce Salt Intrusion

P. van der Kuur, (Proj. Adviser, Delft Hydraulics Lab., Rotterdamse weg 185, P.O. Box 177, 2600 mh, Delft, Netherlands)

Journal of Waterway, Port, Coastal and Ocean Engineering, Vol. 111, No. 6, November, 1985, pp. 1009-1021

Navigation locks can strongly influence the water management of an area (both quantity and quality) especially if they belong to a retaining dike separating saltwater from fresh water. Then these locks consume fresh water for the locking process. An extra quantity is also needed for the reduction of salt intrusion. In the past 30 years, various locks with devices to reduce the salt intrusion have been developed especially in the Netherlands. The state of the art of this development is presented. The performance of the various devices will also be described. The major characteristic of the development is a decreasing consumption of fresh water to reduce the salt intrusion combined with an increasing complexity of the devices.

827   **Forces on Horizontal Cylinder Towed in Waves**

Chung-Chu Teng, (Grad. Research Asst., Ocean Engrg. Program, Dept. of Civ. Engrg., Oregon State Univ., Corvallis, Oreg. 97331) and **John H. Nath**, Fellow, ASCE, (Prof., Ocean Engrg. Program, Dept. of Civ. Engrg., Oregon State Univ., Corvallis, Oreg. 97331)

Journal of Waterway, Port, Coastal and Ocean Engineering, Vol. 111, No. 6, November, 1985, pp. 1022-1040

The hydrodynamic forces on a horizontal cylinder, either in waves or in waves and towing, are determined experimentally for a Reynolds number up to 5×10^5. The differences of the wave force coefficients between a horizontal cylinder in waves and in planar oscillatory flow are shown. The effects of a current on the force coefficients are examined. It is shown that hydrodynamic forces on a cylinder towed in a wave field can be simulated if the linear superposition principle is assumed.

828   **Distribution of Suspended Sediment in Large Waves**

Jorgen Fredsoe, (Assoc. Prof., Inst. of Hydrodynamics and Hydr. Engrg., ISVA, Tech. Univ. of Denmark, Lyngby, Denmark), **Ole H. Andersen**, (Master of Sci., Dept. of Civ. Engrg., Tech. Univ. of Denmark, Lyngby, Denmark) and **Steen Silberg**, (Master of Sci., Dept. of Civ. Engrg., Tech. Univ. of Denmark, Lyngby, Denmark)

Journal of Waterway, Port, Coastal and Ocean Engineering, Vol. 111, No. 6, November, 1985, pp. 1041-1059

The distribution of suspended sediment in the combined wave-current motion is theoretically predicted in the case of a plane bed. This will be the case if the wave-induced motion close to the bed is sufficieny strong. The theory is able to predict the average value of the concentration as well as the instantaneous values at a given distance from the bed. The theory is compared with laboratory and field measurements.

829   **Optimal Berth and Crane Combinations in Containerports**

Paul Schonfeld, Assoc. Member, ASCE, (Prof., Dept. of Civ. Engrg., Univ. of Maryland, College Park, Md.) and **Osama Sharafeldien**, (Doctoral Student, Dept. of Civ. Engrg., Univ. of Maryland, College Park, Md.)

Journal of Waterway, Port, Coastal and Ocean Engineering, Vol. 111, No. 6, November, 1985, pp. 1060-1072

A model developed to optimize the design and operation of containerports is described and demonstrated. The model minimizes total port costs, including the costs of dock labor, facilities and equipment, ships, containers, and cargo. It accounts for queueing delays to ships, mutual interference among cranes, minimum work shifts, and storage yard requirements. Model results are mainly used to determine the optimal combination of berths and cranes under various circumstances and to show that total costs per ship or unit of cargo served can be reduced by increasing the number of cranes per berth and berth utilization above present levels.

830   **Comments on Cross-Flow Principle and Morison's Equation**

C. J. Garrison, Member, ASCE, (Consultant in Marine Hydrodynamics, Pebble Beach, Calif. 93953)

Journal of Waterway, Port, Coastal and Ocean Engineering, Vol. 111, No. 6, November, 1985, pp. 1075-1079

Some of the recent test results on steady and oscillatory flow past inclined circular cylinders are discussed. More specifically, the experimental error in results presented by Sarpkaya, et al, for oscillatory flow is described and the formula for the necessary correction factor is given. Contrary to earlier conclusions based on such erroneous results, the corrected results support the cross-flow principle very well, in general, in the case of the inertia coeeficient and at the larger Reynolds number and Keulegan-Carpenter number in the case of the drag coefficient.

Discussion: **T. Sarpkaya**, (Distinguished Prof. of Mech. Engrg., Naval Postgraduate School, Monterey, Calif. 93943) WW Nov. '85, pp. 1087.

831   **Seawall Constraint in the Shoreline Numerical Model**

Hans Hanson, (Lect., Dept. of Water Resources Engrg., Lund Inst. of Tech., Univ. of Lund, Fack 725, Lund S-220-07, Sweden) and **Nicholas C. Kraus**, (Res. Physical Scientist, Coastal Engrg., Research Center, U.S. Army Waterways Expt. Station, P.O. Box 631, Vicksburg, Miss. 39180)

Journal of Waterway, Port, Coastal and Ocean Engineering, Vol. 111, No. 6, November, 1985, pp. 1079-1083

A method is given for representing the action of a seawall in the 1-line numerical model of shoreline evolution. The method involves restriction of the shoreline position and adjustment of the longshore sand transport rate so as to conserve sand volume and preserve direction of transport. Sample calculations are presented, and the results are in accord with experience.

832   **Saving the Lady**

Rita Robison, (Assoc. Ed., *Civil Engineering--ASCE,* New York, N. Y. 10017)

Civil Engineering--ASCE, Vol. 55, No. 1, January, 1985, pp. 30-39

Restoration of the Statue of Liberty is as much an exercise in cooperative management as it is in technical design and execution. Among those sharing management responsibility are: the French-American Committee for Restoration of the Statue of Liberty, Inc.; the Statue of Liberty-Ellis Island Commission, whose working arm is the Statue of Liberty-Ellis Island Foundation, Inc.; and the National Park Service. All have retained architectural and engineering firms. The construction management firm oversees about 50 contractors. The latest technical decision is to strengthen the existing framework of the upraised arm rather than replace it with a design similar to Gustave Eiffel's original intent. The all-aluminum scaffolding design is described.

Discussion: **Clifford, L. Pelton**, Member, ASCE, (Guaynabo, Puerto Rico) CE June '85, pp. 36.

833   Managing Public Works: How the Best Do It

K. A. Godfrey, Jr., (Sr. Editor, *Civil Engineering--ASCE*, New York, N. Y. 10017) and **Virginia Fairweather**, (Editor, *Civil Engineering-- ASCE*, New York, N. Y. 10017)

Civil Engineering--ASCE, Vol. 55, No. 1, January, 1985, pp. 40-43

Management methods of what are considered the best public works organizations are reviewed. Examples of personnel management in some of these organizations include: Fostering the professional growth of management down to the foreman level; offering bonuses to top performers and giving challenging work assignments; and measuring productivity of operator personnel and using the results to publicize the work of top people and helping low performers improve. To minimize cost, bids are sought on as much work as possible and the work contracted out if the private sector is less costly. A state advisory program for city management and a school for water and wastewater plant operators are described.

834   Wharf Stands on Stone Columns

George Munfakh, (Mgr. of Geotech Dept., Parson, Brinckerhoff, Quade & Douglas, New York, N. Y.)

Civil Engineering--ASCE, Vol. 55, No. 1, January, 1985, pp. 44-47

A wharf was constructed in New Orleans on very weak soils. To prevent lateral and vertical movement of the soil, normally the entire wharf and adjacent structure would be founded on steel piles. Here, $1.25 million was saved by using about half the normal number of piles, and replacing some of them with stone columns, each of which is a 3.6-ft. wide column of crushed stone, extending down to a competent layer of soil. The columns support live loads and work together with each other to minimize lateral movement of the squeezing clay. A pile cap unifying the tops of the stone columns was created by mechanically stabilized earth. Elsewhere, wich drains were used to speed and cut the cost of draining/strengthening the weak soil. Discussion: **Richard D. Barksdale**, Member, ASCE, (Prof. of Civ. Engrg., Georgia Inst. of Tech., Atlanta, GA) CE Sept. '85, pp. 31.

835   Changing Jobs Successfully

James C. O'Donnell, (Sr. Vice Pres., Drake, Beam & Morin, Irvine, Calif.)

Civil Engineering--ASCE, Vol. 55, No. 1, January, 1985, pp. 48-49

Advice from a corporate head hunter offers suggestions on how to deal with internal corporate movement, friendship and professional reputation, external corporate movement, personal contacts, professional movement, managerial movement, entrepreneurial situations, and professional associations. The author considers the job search marketing campaign, resume, personal contact network, the marketing letter, want ads, executive search firms, and professional organizations. The real issue is career objective, looking for opportunity, discovering opportunity, and staying in contact with key individuals.

836   **CPM Harnesses Mammoth Powerplant Job**

J. H. Shukla, Member, ASCE, (Proj. Planning Engr., Gilbert Commonwealth, Jackson, Mich.), **R. L. Kudich**, (Structural Engr., Gilbert Commonwealth, Jackson, Mich.) and **R. W. Staffensen**, (Construction Mgr., Ohio Edison Co., Akron, Ohio)

Civil Engineering--ASCE, Vol. 55, No. 1, January, 1985, pp. 50-53

    One of the largest air quality control retrofit projects in the U. S. is at the Sammis Plant of Ohio Edison Co., along the Ohio River 40 miles west of Pittsburgh. The coal-fired plant's 2,233 MW of capacity makes it one of the utility's largest, and the $450 million cost of the retrofit makes it one of the nation's largest. A very narrow site, between a bluff and the river, with the plant and a major highway between, necessitated placing most of the equipment on a structure 900 feet long and over the highway. Key challenge was to meet air quality deadlines set by courts and environmental agencies. Key to meeting the deadlines, in the project was a set of 200 critical path method logic diagrams, which together encompass the 7,500 tasks required by the project.

837   **Ground Freezing for Construction**

Bernd Braun, Member, ASCE, (Special Projects & Operations Mgr., Deilmann- Haniel GmbH Postfach 130220, Haustenbecke, 4600 Dortmund 13, West Germany) and **William R. Nash**, Member, ASCE, (Chf. Mining Engr., Frontier-Kemper Constructors, Inc., P. O. Box 6548, Evansville, Ind. 47712)

Civil Engineering--ASCE, Vol. 55, No. 1, January, 1985, pp. 54-56

    For almost a century, freezing has been used successfully as an art to stabilize ground and control groundwater. First employed by the mining industry in Europe, it has since been widely adopted throughout the world to stabilize all types of excavations including shafts up to 915 m (3,000 ft.) in depth, control groundwater, and underpin structures. Recent improvement in the sciences of frozen ground engineering and refrigeration technology have opened up many new economical opportunities for freezing in construction and mining industries. Therefore, in the last decade ground freezing has become a viable and competitive construction alternative for providing temporary ground support and groundwater control for excavations in difficult soil conditions. A wide variety of ground freezing projects in the United States have been inspected and accepted by OSHA, MSHA, EPA, and numerous other federal, state and local agencies. Discussion: **Edward Yarmak, Jr.**, (Chf. Engr., Arctic Foundations, Inc., Anchorage, AK) CE July '85, pp. 36.

838   **Design Flexibility Cuts Costs**

Raymond J. Blunk, Member, ASCE, (Waterworks Engr., Engrg. Design Div., Los Angeles Dept. of Water & Power, Los Angeles, Calif.)

Civil Engineering--ASCE, Vol. 55, No. 1, January, 1985, pp. 57-59

    The Los Angeles Water Department saved $5 million on a pipeline project by investigating the capabilities of pipe fabricators to furnish the pipe size required and by being open to alternate pipe sizes. Two reservoir reconstruction projects show that by making bidding more competitive the city obtained a lower project cost, and design changes not only brought savings but solved some difficult design problems. Consisting of reservoir bypass, inlet and outlet lines, the pipeline work at the Fairmont Reservoir 2 and the Lower Franklin Reservoir is part of a $20 million pipeline project to replace hydraulic fill dams in a seismic area. For Fairmont Reservoir 2 a new outlet line was sized for maximum hydraulic slope and power generation. The city engineers based optimal pipe size on pipeline cost plus the cost of lost power related to head loss. They questioned pipe fabricators to determine the availability of 132 in. pipe, and found that only one fabricator could supply what they decided was the optimal size pipe. Another firm felt they could be competitive if they could bid a 144 in. prestressedconcrete cylinder pipe. Still another firm wanted the option of supplying 132 or 144 in. ID welded steel pipe. The city included alternates to enable contractors to bid either 132 or 144 in. ID prestressed concrete cylinder pipe and 132 or 144 in ID or OD welded steel pipe. Even though greater design costs and more complicated bidding resulted, allowing these alternatives guaranteed competition between pipe companies. The city realized similar savings on the Lower Franklin Reservoir project.

839   Micros Go Afield

Rita Robison, (Assoc. Ed., *Civil Engineering—ASCE*, New York, N.Y. 10017)

Civil Engineering--ASCE, Vol. 55, No. 2, February, 1985, pp. 32-35

Taking microcomputers into the field is boosting productivity at construction sites and other field offices. Four examples presented in this article include an owner calculating payments to contractors; a contractor recording labor and equipment costs; a state program to provide microcomputer systems to local traffic engineers; and inspection reports produced automatically from inspectors' verbal notes on cassettes.

840   Learning From Failure

Henry Petroski, (Dir. of Grad. Studies, Duke Univ., Durham, N.C.)

Civil Engineering--ASCE, Vol. 55, No. 2, February, 1985, pp. 36-39

Innovative engineering design involves assumptions about the future use and behavior of a structure. By understanding and learning from the infrequent but spectacular large failures, engineers can develop more reliable designs. Several famous failures illustrate the paradox that we learn from failure and that successful designs can prevent future failures.
Discussion: **J. F. Koenen**, Fellow, ASCE, (Pres., Westerberg/Koenen, Inc., Prospect, IL) CE April '85, pp. 30.

841   Trouble in Our Own Back Yard

Bev Jafek, (Communications Mgr., Woodward-Clyde Consultants, San Francisco, Calif.)

Civil Engineering--ASCE, Vol. 55, No. 2, February, 1985, pp. 40-43

One million tanks will leak toxic and hazardous substances into the U.S. soil within the next four years. Prevention begins with rapid discovery and probability analysis. Various cleanup methods, both for surface and subsurface spills, are described.
Discussion: **E. O. Butts**, Member, ASCE, (Pres., Total Containment, Ottawa, Ontario, Canada) CE July '85, pp. 24. Discussion: **Stephen E. Dee**, Assoc. Member, ASCE, (Capt., USAF, Environmental Planning Div., Air Force Logistics Command, Wright-Patterson AFB, OH) CE July '85, pp. 24. Discussion: **C. T. Sawyer**, (Vice Pres., America Petroleum Inst., Washington, DC) CE July '85, pp. 24. Discussion: **Bev Jafek**, (Communications Mgr., Woodward-Clyde Consultants, San Francisco, CA) CE July '85, pp. 24.

842   Carbon Columns: Success at Last

John C. Seeley, Member, ASCE, (Managing Partner (retired), McNamee, Porter & Seeley, Ann Arbor, Mich.) and **Shin Jon Kang**, Member, ASCE, (Mgr., Advanced Technology Dept., McNamee, Porter & Seeley, Ann Arbor, Mich.)

Civil Engineering--ASCE, Vol. 55, No. 2, February, 1985, pp. 44-46

The first large scale application of carbon columns for municipal wastewater treatment has been operating for two years. The columns were designed as part of a program to upgrade an out of date treatment plant to secondary treatment. Design studies showed that a combined physical-chemical precipitation, sand filtration and adsorption by granular activated carbon offered several advantages over conventional biological treatment. The designers of the North Tonawanda, N.Y. plant avoided the serious mistakes of past systems by using several design innovations. In previous projects, carbon transfer in dry form resulted in clogged equipment and high levels of carbon loss due to abrasion. The spent carbon is transferred in slurry form and then the lines are flushed to cut down abrasion. And careful flushing of the lines after the carbon transfer avoids plugging. The carbon columns in the plant were designed to operate upward to eliminate plugging and provide better contact with the carbon. Water is introduced at the bottom of the columns. As a result of these and other innovations, the effluent has consistently met discharge permit requirements for conventional and organic pollutants as well as phosphorous.

843   **Quality Control: A Neglected Factor**

James J. O'Brien, Member, ASCE, (Chf. Exec. Officer, O'Brien-Kreitzberg & Assoc., Inc., Cherry Hill, N.J.)

Civil Engineering--ASCE, Vol. 55, No. 2, February, 1985, pp. 48-49

Quality in construction is too important to leave to chance. A project quality control plan would begin by incorporating testing and inspecting requirements into the specifications. Engineers must educate owners to insist on a quality control plan and pay for comprehensive inspection. 1985 Essay Contest Award of Merit.

Discussion: **Joseph Goldbloom**, Fellow, ASCE, (Fair Lawn, NJ) CE Oct. '85, pp. 38.

844   **Frame Moments with the Takabeya Method**

H. V. Lamberti, Member, ASCE, (Consulting Engr., Reno, Nev.)

Civil Engineering--ASCE, Vol. 55, No. 2, February, 1985, pp. 50-53

The Takabeya method of frame analysis, based on the same slope-deflection equations as the Hardy-Cross system, has the advantage of requiring much less memory and fewer computations than Hardy-Cross. Thus the Takabeya method of solving frame problems permits engineers to use smaller computers with enough memory to solve reasonably sized structures. A digest of the derivations of the formulae and an outline of an algorithm for a computer program, based on this method are included.

Discussion: **Philip L. Gould**, Fellow, ASCE, (Prof. and Chmn., Dept. of Civ. Engrg., Washington Univ., St. Louis, MO) CE June '85, pp. 31. Discussion: **David J. Frederick**, Member, ASCE, (Escondido, CA) CE June '85, pp. 32. Discussion: **David O. Knuttunen**, Assoc. Member, ASCE, (Engr., Souza & True Inc., Watertown, MA) CE June '85, pp. 32. Discussion: **D. Michael Helmich**, Member, ASCE, (Dir., GHI Engineers & Project Consultants, San Francisco, CA) CE June '85, pp. 33.

845   **Rationalizing Land Records, Mapping, Planning**

K. A. Godfrey, Jr., (Sr. Ed., Civil Engineering—ASCE, New York, N.Y. 10017)

Civil Engineering--ASCE, Vol. 55, No. 2, February, 1985, pp. 54-57

The way land records are kept and used in cities and counties is outmoded. The same is true for base mapping and land use planning. The computer is helping streamline all three, as is illustrated in the two case histories in this article. In Wyandotte County (Kansas City), Kans., computerizing and rationalizing of land records and of base mapping has led to collection of $500,000 in delinquent taxes that otherwise wouldn't have been collected so soon, if at all. This was done by bringing together all five county departments that collect and use land records—County Clerk, County Appraiser, Registrar of Deeds, County Surveyor, and County Treasurer—and having the departments jointly decide now to simplify procedures. New York State's Long Island Regional Planning Board used computer mapping techniques to complete a 1981 update of a 1961 land-use study in little more than one-half the time and with one-tenth the manpower as in 1961. Secret here was a faster and cheaper way of digitizing (feeding to the computer) information already on maps, called videodigitizing.

846 **How Do Engineers Learn to Manage?**

Stanley H. Madsen, Fellow, ASCE, (Eec. Vice Pres., Earth Technology Corp., Long Beach, Calif.)

Civil Engineering--ASCE, Vol. 55, No. 2, February, 1985, pp. 58-59

Most engineers prefer design problems to management problems, but if a firm is to prosper, project managers must learn to manage and devote time to management. Earth Technology Corporation in Long Beach, Calif. set out to improve its project management performance. During a period of rapid growth in the firm, many of the senior staff members were responsible for managing projects, but project management skills did not keep pace with company growth, and the result was poor contracts, delays in completing work and cost overruns. The firm's chief executive officer exposed the problem in an all-day seminar on project management personnel. The firm's management also formed a committee to review problem projects, and concluded that their procedures and policies on project management were not clearly defined. The committee prepared a project managers manual, covering company organization, proposals, contracts, client relations, project planning and implementation, invoicing, project marketing, multidisciplinary projects and interoffice projects. The manual was distributed to all seniors and associates, and a series of project managers met to discuss each section. Preliminary results indicate a decrease in overruns of 35% during a two-year period.
Discussion: **Donald K. Gudgeon**, Fellow, ASCE, (GCP Project Consultants, Tanglin, Singapore) CE July '85, pp. 36.

847 **China Builds Record Bridge**

Chun-Nong Hu, (Prof. and Dir., Research Institute of Technology at the Lanzhou Railway College, Lanzhou, China)

Civil Engineering--ASCE, Vol. 55, No. 2, February, 1985, pp. 60-61

China's Hanjiang Railway Bridge has a 1,000 ft (305 m) long main box girder and is the longest bridge of its type in the world. The steel box girder bridge is built with a slant leg on each side, composed of two branches. Cross beams connect the two branches to form a multistory stand, supporting the girder so that the main girder acts like a multi-spanned continuous girder with relatively even bending moment distributed along it. The Hanjiang Bridge was built on the basis of the design studies of earlier steel box girder bridges in China. Engineers for the Hanjiang Bridge solved two stability problems: stability of the web plate of the box girder and the lateral stability of the narrow box section slant-legged rigid frame. The stability of the web plate under bending moment, shear and local compression and the influence of ribs were investigated. The arrangement of the stiffening ribs and the method of stability calculation of web plate used in the design have been proved by the agreement of the experimental results. Both local stability and the overall stability are obtained provided that the position and stability of the stiffening ribs meets the design requirements. The most crucial point in ensuring lateral stability is to find the most effective means to eliminate the lateral displacement at the top of the slant legs. This is accomplished by spreading each leg into two branches with a slope of 1/6 in the transverse direction. The lateral displacement was proved to be within the given limits.

848   The Pursuit of Quality: QA/QC

Virginia Fairweather, (Editor, *Civil Engineering—ASCE*, New York, N.Y. 10017)

Civil Engineering--ASCE, Vol. 55, No. 2, February, 1985, pp. 62-64

Quality assurance and quality control programs are mandated for many federally-funded construction projects. The private sector is also increasingly requiring such programs. The terms are defined, and practitioners interviewed who express opinions both pro and con about the administration of these programs and their net effect. Most agree that inadequate specifications are a major problem in the construction industry that has led to the institution of QA/QC programs.

Discussion: **Joseph Goldbloom**, Fellow, ASCE, (Fair Lawn, NJ) CE Oct. '85, pp. 38.

849   The Hybrid Arena

Rita Robison, (Assoc. Ed., *Civil Engineering—ASCE*, New York, N.Y.)

Civil Engineering--ASCE, Vol. 55, No. 3, March, 1985, pp. 38-41

The University of North Carolina's new basketball arena at Chapel Hill combines steel and fabric in a dome whose design by David Geiger Associates is based on the principles of skewed symmetry. The tension ring is at the bottom, the compression ring at the top. A standard metal deck roof is supported by four arch trusses and infill panels. The fabric lantern, which architecturally defines the basketball courts below, is carried by tubular arches and tied down by cables. The design is an economical one, as the steel weighs only 15 lb/sq ft.

850   Mixed Use, Mixed Materials

Rita Robison, (Assoc. Ed., *Civil Engineering—ASCE*, New York, N.Y. 10017)

Civil Engineering--ASCE, Vol. 55, No. 3, March, 1985, pp. 42-45

Tabor Center is a $330 million office/commercial/hotel complex in Denver's urban renewal area. The two block site is bisected by a street that had to be kept open during construction. The street is now carried on a bridge built on caissons drilled from the surface, then excavated below. Two office towers were designed as steel/concrete composites, with their loads carried down through subterranean garages to bedrock. One garage is post-tensioned concrete, the other precast. A hotel, a 3-story retail structure and a plaza complete the complex.

851   Mixing Steel and Concrete

Hal Iyengar, (Partner, Skidmore, Owings & Merrill, Chicago, Ill.)

Civil Engineering--ASCE, Vol. 55, No. 3, March, 1985, pp. 46-49

For highrise structures, mixed steel-concrete systems are more efficient and cost less than either material alone. The two design approaches are exterior tube structures and core braced structures. Composite tubular systems combine a reinforced concrete framed tube on the exterior with simple steel framing on the interior. The second type combines a concrete shear wall core with a steel framing for floors and exterior columns. Still in the future are composite superframes in the form of a portal. These exterior frames resist all wind forces, leaving the interior free for atriums. More research is needed for all types of composite design.

852   Oral History: Saving the Past

Civil Engineering--ASCE, Vol. 55, No. 3, March, 1985, pp. 50-53

Excerpts of the recorded interviews of three leading American civil engineers are presented. In the first interview sanitary engineer Samuel A. Greeley recalls how he supervised the construction of an Army camp for 35,000 troops in 1917 and how his supervision of sanitary engineering helped protect the occupants from the great flu epidemic. Solomon Cady Hollister tells how his vision of needed reforms in engineering education redirected the 1955 Grinter Report. in the final interview Abel Wolman tells how he learned to communicate his purposes in sanitary engineering in lay language in order to convince town planners and semi-professionals of the validity of his proposals.

853   Proportional Weirs for Stormwater Pond Outlets

Arne Sandvik, Member, ASCE, (Consulting Engr., EGC-Southwell, Inc., Englewood, Colo.)

Civil Engineering--ASCE, Vol. 55, No. 3, March, 1985, pp. 54-56

Many jurisdictions mandate that new developments must not cause runoff to flow downstream, during and after rainfalls, at rates higher than preconstruction. A common for controlling runoff is the stormwater detention pond. Downstream flooding potential may be further reduced by adopting the more sophisticated outlet, the proportional weir. It reduces peak flows in storms of intensities between those for which conventional, alternative outlets are designed.
Discussion: **Walter Amory**, Fellow, ASCE, (Pres., Amory Engrs., Duxbury, MA) CE Aug. '85, pp. 30. Closure: CE Aug. '85, pp. 30. Discussion: **Walter T. Sittner**, Fellow, ASCE, (Annapolis, MD) CE Aug. '85, pp. 33. Discussion: **James D. Caufield**, Member, ASCE, (J. D. Caufield & Assoc., Inc., Portland, OR) CE Aug. '85, pp. 33.

854   *Skyscraper* **(review)**

Civil Engineering--ASCE, Vol. 55, No. 3, March, 1985, pp. 57-59

The 1984 novel *Skyscraper,* by civil engineer-writer Robert Byrne, tells the suspenseful story of the collapse of a new, 66 story skyscraper in New York City. As well as being compelling fiction (the book was reprinted in condensed form by Readers Digest Condensed Books), the novel is of interest because the author did extensive research into high-rise design and construction and problems of building failures. His hero is a civil-forensic engineer, and the attractiveness of a civil engineering career in that specialty is played up. The building, 850 ft tall and exceptionally narrow for its height, collapses because, in the name of economy, considerable weight is pared off floors and curtainwall. But in the face of the design wind load, this reduced weight leads to an inadequate factor of safety against overturning. Excerpts from the book are presented.

855   Bid-Rigging: An Inside Story

Civil Engineering--ASCE, Vol. 55, No. 3, March, 1985, pp. 60-63

William Carter, engineer and ASCE member, spent five months in prison in 1980 on conviction of conspiracy under U.S. antitrust laws in a case involving a Tennessee highway project for which his construction firm offered an illegal bid. Carter is one of more than 150 contractors who served prison sentences as the result of a 1979 investigation by the U.S. Department of Justice of "bid-rigging" in public highway contract awards. In 1982 Carter's case came before ASCE's Committee on Professional Conduct. He was suspended for one year, then sought and was granted reinstatement in 1983 on the grounds that he had cooperated with the Committee by agreeing to talk about his experience as an example to others. A transcription is presented of a November 1984 roundtable discussion with Carter and two attorneys; Richard Braun, formerly of the Department of Justice, who prosecuted Carter, and Walter McFarlane of the Attorney General's Office of the State of Virginia. ASCE was represented by George Barnes, Past Chairman of the Committee on Professional Conduct, Lawrence Whipple, ASCE staff liaison to the Committee at that time, and Virginia Fairweather, Editor of Civil Engineering magazine.

Discussion: **Richard H. McCuen**, Member, ASCE, (Prof., Dept. of Civ. Engrg., Univ. of Maryland, College Park, MD) CE June '85, pp. 36. Discussion: **Thomas A. Allbaugh**, Member, ASCE, (Proj. Engr., McNamee Porter & Seeley, Ann Arbor, MI) CE June '85, pp. 36. Discussion: **Paul Cella**, Fellow, ASCE, (LaGrange Park, IL) CE Aug. '85, pp. 26.

856   Making Treatment Plants Work

Enos L. Stover, (Prof. of Environmental Engrg., Oklahoma State Univ., Stillwater, Okla.) and **Brent Cowan**, (District Mgr., Scholler, Inc., Philadelphia, Pa.)

Civil Engineering--ASCE, Vol. 55, No. 3, March, 1985, pp. 64-67

For the most part, past design criteria that have been used were based primarily on historical information from plants "that worked" or general "rule of thumb" criteria. With the increasing concern over facilities not being able to meet their effluent limits, although they are only several years old, the design community needs to reassess their methodologies and demonstrate their capability to properly design secondary facilities. Guidelines are needed for future plants that will address specific pollutants criteria as well as the surrogate standards of BOD, COD or TOC. Detailed wastewater characterizations are essential in defining the variability to be encountered and should be carried out through the pilot or laboratory scale evaluations. The resultant data, properly evaluated, will incorporate the specificity and variability into the design of the full scale facility.

857   Shop Drawing Review: Minimizing the Risks

Robert A. Rubin, Fellow, ASCE, (Attorney and Partner, Construction Law, Postner & Rubin, New York, N.Y.) and **Marcy E. Ressler**, (Law Clerk, Postner & Rubin, New York, N.Y.)

Civil Engineering--ASCE, Vol. 55, No. 3, March, 1985, pp. 68-70

Individual designers, professional liability insurers, and professional organizations have tried to define the designers approval responsibilities for shop drawing review in clear, workable terms. In the absence of a workable industry-wide definition of the scope of a designers review, architects and engineers have sought to limit their obligations and excuse them from liability. The problems associated with shop drawing review and the professional responsibility, liability and risks are reviewed and solutions are recommended. The Hyatt Regency walkway collapse is cited as an example of legal ramifications.

Discussion: **Leon Stein**, Fellow, ASCE, (Consulting Struct. Engr., Laguna Niguel, CA) CE Sept. '85, pp. 31.

858   Auditors Eye the Environment

Robert A. Corbitt, M, (Dept. Mgr., Jordan, Jones & Goulding, Inc., Atlanta, Ga.), **David C. Garrett, III.**, (Partner, Powell, Goldstein, Frazer & Murphy, Atlanta, Ga.), **Robert F. Kohm**, (Mgr. of Environmental Planning & Analysis, Aluminum Co. of America, Inc., Pittsburgh, Pa.), **Dale I. Patrick**, (Chf., Program Support Bureau, State of New Mexico Environmental Improvement Div., Santa Fe, N.M.) and **Donald R. Perander**, (Environmental Engr., Armco, Inc., P.O. Box 600, Middletown, Ohio 45043)

Civil Engineering--ASCE, Vol. 55, No. 4, April, 1985, pp. 39-41

1985 ASCE PUBLICATIONS

Environmental audits are being used by an increasing number of companies, and are providing a substantial new market for environmental engineers. Environmental audits are formal self-appraisals that businesses conduct to assess their compliance with environmental regulations, and to reduce environmental problems and liabilities. Some companies are conducting audits through their attorneys to keep them private. Such confidentiality is based on the work products doctrine and attorney client privilege. Several states including New Jersey, Massachusetts, and New Mexico have enacted laws promoting their use.

859   Soil Nailing Supports Excavation

Peter L. Nicholson, Member, ASCE, (Nicholson Construction Co., Bridgeville, Pa.)

Civil Engineering--ASCE, Vol. 55, No. 4, April, 1985, pp. 44-47

Excavation for the PPG Headquarters complex in Pittsburgh, Pa. was critical because the site is adjacent to three historically important buildings. Soil nailing, a method of in situ earth reinforcement, was used at these three areas. One method, which had been tested at the University of California, involves use of shotcrete and wire mesh to support the excavation face with reinforcing members grouted into the soil mass behind the face. This, however, was not satisfactory for the Pittsburgh site, so a modified method was used. Columns were drilled, pressure grouted and reinforced; then the reinforcing nails were drilled and pressure grouted. No movement of the three buildings was detected during excavation and construction of the new buildings.

860   Cay Clay Liners Work?

David E. Daniel, (Asst. Prof. of Civ. Engrg., Univ. of Texas at Austin, Austin, Tex.)

Civil Engineering--ASCE, Vol. 55, No. 4, April, 1985, pp. 48-49

Nearly all liners for landfills and surface impoundments were constructed from compacted clay until about five years ago. But this is no longer so. In the past few years, clay liners used by themselves have fallen out of favor with the Environmental Protection Agency. Recent research examines liner permeability and chemical attack of the liner. Clay liners may prove useful if some caveats are observed. Be aware of the type of waste that will be placed in any one cell. Test the compatibility between liner material and either the actual leachate or waste liquid or a test leachate. Don't expose the liner to concentrated acids, bases or organic chemicals. Use a thick liner to keep effective stress to a minimum. Cover the liner to prevent it from drying out, and avoid construction during freezing weather since this may cause cracking.
Discussion: **Eugene A. Zwenig**, Member, ASCE, (Box 529, Gordon, CA) CE June '85, pp. 26. Discussion: **Andrew K. Phelps**, Assoc. Member, ASCE, (Sr. Engr., Bechtel National, Inc., Oak Ridge, TN) CE June '85, pp. 30.

861   Lift Bridge looks Like Sculpture

Rita Robison, (Assoc. Ed., *Civil Engineering—ASCE*, New York, N.Y. 10017)

Civil Engineering--ASCE, Vol. 55, No. 4, April, 1985, pp. 50-52

A lift bridge built in 1926 over the U.S. Canal at Kaukauna, Wis. was replaced by a sleek new lift bridge whose sculptured shape encloses all the machinery within welded steel box sections. This eliminated the potential for trapped water that can cause corrosion pockets. The lift span consists of single section welded plate girders, and weighs a total of 481,000 lbs. Counterweights are heavy concrete enclosed in steel boxes. The two-lane bridge cost $3,324,000.

862   **Municipal Refuse: Is Burning Best?**

K. A. Godfrey, Jr., (Sr. Editor, *Civil Engineering—ASCE,* New York, N.Y. 10017)

Civil Engineering--ASCE, Vol. 55, No. 4, April, 1985, pp. 53-56

Communities are increasingly worried about polluting groundwater, and it is increasingly difficult to get permits to start new landfills. For these reasons, there is a trend back to municipal incinerators. Municipalities show preference for the 30 year old "mass burn" incinerators, updated largely by addition of better stack-gas cleaning systems, over the innovative designs of the 1970s, many of which fail to work reliably or cost-effectively. In a growing share of the cases, the systems contractor operates the plant, and in some cases owns part or all of it. Dioxin, highly toxic, was detected in the stack emissions of several municipal incinerators, causing great concern four years ago; but recent risk assessment have concluded the dioxin constitutes an insignificant health risk. To reduce the refuse reaching municipal incinerators, New Jersey is taking the national lead in encouraging (perhaps soon mandating) that communities source-separate up to 25% of the tonnage—newspaper, cans and plastic containers, yard waste, etc. There is debate as to whether providing market incentives or mandating source separation is preferable.
Errata: CE July '85, pp. 36. Discussion: **Robert W. Herrmann**, (Assoc. Program Dir., National Ecology, Inc., Timonium, MD) CE Sept. '85, pp. 24.

863   **Hammering Out a New RCRA**

Corinne S. Bernstein, (Asst. News Ed., Civil Engineering-ASCE, New York, NY 10017)

Civil Engineering--ASCE, Vol. 55, No. 4, April, 1985, pp. 57-59

Recent amendments to the Resource Conservation and Recovery Act (RCRA) add thousands of new waste generators and many new wastes to the list of those to be regulated. Some say RCRA is the toughest law passed since the inception of the Environmental Protection Agency (EPA). To date, RCRA is the only major environmental law passed during the Reagan administration. The law, which awaits EPA interpretation, has been the subject of much speculation since its November 1984 passage. Many question whether the EPA, the states, and industry can meet the law's rapidly advancing deadlines and strict standards. The law adds 72 new provisions to the original law and its amendments, and directs EPA to carry out 58 of them in the next two years. Among the legislation's hallmarks are restrictions on land disposal, small quantity generators, burning and blending of wastes, underground storage tanks, interim status facilities, inspections and citizen suits. The act eliminates land disposal for many wastes, keeps it to a minimum for all others and clearly outlines a legislative preference for alternatives, such as treatment and reduction of wastes.
Discussion: **Eugene A. Zwenig**, Member, ASCE, (Box 539, Gordon, CA) CE June '85, pp. 26. Discussion: **D. Allan Firmage**, Fellow, ASCE, (Firmage & Assoc., Provo, UT) CE June '85, pp. 38.

864   **Is There Grit in Your Sludge?**

George E. Wilson, Member, ASCE, (Pres., Eutek Systems, Carmichael, Calif.)

Civil Engineering--ASCE, Vol. 55, No. 4, April, 1985, pp. 61-63

Getting the grit out of municipal sludge can lower the total cost of operating a wastewater plant. Grit escalates maintenance costs and abrasive-laden sludge is costly to dispose of. Because conventional grit control systems remove only abrasives larger than 200-300 microns sand, 90-95% of the abrasives that enter a wastewater plant end up in the sludge. A new system can now eliminate all abrasives, reducing dry solids by 25% and O&M costs for sludge treatment more than 50%. The new unit is based on free vortex accelerated boundary layer grit collection and classification. It imposes an acceleration field within a boundary layer on settleable particles contained in wastewater, separating particles of different densities that have the same settling velocity in water.

865   Hundreds of Bridges—Thousands of Cracks

John W. Fisher, (Prof. of Civ. Engrg., Fritz Engrg. Lab., Lehigh Univ. Bethlehem, Pa.) and Dennis R. Mertz, (Bridge Design Engr., Modjeski & Masters, Harrisburg, Pa.)

Civil Engineering--ASCE, Vol. 55, No. 4, April, 1985, pp. 64-67

In the past 10 years hundreds of welded steel bridges, most of them relatively new, have suffered thousands of cracks due to "secondary stresses," that is, the cracks are caused by deflections in the structure that are caused by loads. No crack has led to collapse of a bridge, and in no case has a bridge had to be closed or load-limited. The cracks were found in steel-girder web gaps between vertical connection plates and the girder flanges where a gap was left between connector and the girder's flange. Causing the cracks were lateral or out-of-plane bending of the web, caused by transverse beams or lateral bracing. Any of three corrective measures has stopped the cracking in existing bridges: (1) A hole is drilled at each end of each cracks; (2) the unstiffened web gap is lengthened, reducing stress concentration, by cutting away a piece of the offending stiffener; or (3) the web gap is closed by bolting (not welding) the connection plate to the web. New bridges are being designed to prevent this type of cracking.

866   A Tale of Six Cities

Dennis E. Palmer, Member, ASCE, (Vice Pres., Barr Engrg. Co., 6800 France Ave. S., Minneapolis, Minn. 55435)

Civil Engineering--ASCE, Vol. 55, No. 4, April, 1985, pp. 68-70

Battle Creek is a four mile urban stream that had an erosion problem that was accelerating with urbanization of its watershed. Erosion created unstable ravine slopes over 60 feet high, destroyed a valuable regional park, endangered public utilities and damaged private property. A watershed district, a form of regional government based on watershed divides, was created to plan the construction and financing of the necessary channel stabilization. Over a seven year period, the District defined the scope of the project, prepared construction plans and implemented construction. After considering the positions of city, county and regional agencies, the district coordinated agency input and established a successful cost allocation formula for the project. Several erosion control techniques were employed including grade control, pipe enclosure, and a combination of grade control and pipe. In the worst areas, an 8 ft diameter pipe was buried in the ravine, which was completely regraded and backfilled to restore stable slopes without removal of remaining vegetation. Above the buried conduit, the base flow of the creek was carried in a channel, which featured six waterfalls for grade control and park aesthetics. A formula was developed to allocate the costs of channel stabilization on the basis of contribution to peak discharge, contribution to runoff volume, location, effects of upland storage and mainstream reservoirs, land use and parcel size. The assessment formula was applied to more than 6,000 parcels. Only 11 of the 6,000 parcels filed appeals to the allocated cost. Although administrative costs were high, the lengthy public participation helped implement a project which deferred up to $6 million in sewer relocation costs and permitted redevelopment of a regional park.

867   Reshaping the Future of Plastic Buildings

Howard Smallowitz, (Asst. News Ed., *Civil Engineering—ASCE,* New York, N.Y.)

Civil Engineering--ASCE, Vol. 55, No. 5, May, 1985, pp. 38-41

Plastic structures have been around for 40 years, but have for the most part been showcases with little use in the mainstream of the building industry. To be used effectively, plastics must be reinforced, and the reinforcements should be placed within each structural section to compensate for the material's weaknesses, and to maximize its advantages. Only then can it become cost effective to use, and still in only very special applications. More research, communication between members of the field and education will be necessary before it becomes a widely used construction material.

Discussion: **Andrew Green**, Member, ASCE, (Pres., Composite Technology, Inc., Outside, TX) CE July '85, pp. 36.

868   New Ways with Concrete

Rita Robison, (Assoc. Ed., *Civil Engineering—ASCE,* New York, N.Y. 10017)

Civil Engineering--ASCE, Vol. 55, No. 5, May, 1985, pp. 42-45

When NASA builds on the moon and in space, the structures will be concrete, say researchers at the Portland Cement Association's Construction Technology Laboratories. They foresee producing concrete on the moon from local materials. On earth, silica fume additives are proving to be more valuable for durability, impermeability and resistance to chemicals than for high strength per se. Silica fume concrete was chosen for these qualities as the exterior wall of Global Marine Development Co.'s CIDS offshore drilling platform, located in the Bering Sea. Roller compacted concrete, suddenly popular for constructing gravity dams, is being used for heavy duty pavement in the U.S. First civilian example is Burlington Northern Railroad's intermodal hub facility in Houston.

869   Timber Bridge Decks

Civil Engineering--ASCE, Vol. 55, No. 5, May, 1985, pp. 47-49

The portion of the nation's 500,000 bridge decks that are made of timber has been declining for years, as the old timber ones on secondary road bridges are rebuilt, in most cases with concrete. But some highway departments are finding that, in some situations, timber is the deck material of choice. In very short spans under 30 or 40 ft, deck-only timber "bridges" on low-traffic-volume roads are often the least costly acceptable alternative. A larger bridge near Pittsburgh needed its old timber deck replaced, and the decision was to re-deck in timber. It cost only half as much and is expected to last far longer. The Canadian province of Ontario has prestressed 15 timber bridge decks, some old and some new, and thus sharply increased their load capacity.

870   High Strength Steel

K. A. Godfrey, Jr., (Sr. Ed., *Civil Engineering—ASCE,* New York, N.Y. 10012)

Civil Engineering--ASCE, Vol. 55, No. 5, May, 1985, pp. 50-53

A *Wall Street Journal* story in January 1984 described a number of bridges, buildings,

1985 ASCE PUBLICATIONS

automobiles, buses and other structures made of high-strength steel that had failed. The reader might conclude that high strength steel's use should be stopped or curtailed. Experts do not agree with that viewpoint but do say high strength steel structures must be specified, designed, built and used more carefully than if the steel had been lower strength. This article includes case histories of some of the structural failures described in the *Journal* article, and reviews the properties of high strength steel and some precautions that should be followed in its use.
Discussion: **R. H. R. Tide**, Member, ASCE, (Consultant, Wiss, Janney, Elstner Assoc., Inc., Northbrook, IL) CE Nov. '85, pp. 30.

871   Checking Off CADD Priorities

Celal N. Kosten, Member, ASCE, (Prof. of Civ. Engrg., Lehigh Univ., Bethlehem, Pa. 18015)

Civil Engineering--ASCE, Vol. 55, No. 5, May, 1985, pp. 54-55

Diving head on into new computer aided design and drafting (CADD) hardware and software without considering their effects on firm or university's operation is as bad as ignoring their usefulness as technical tools. Initiation, installation and integration of CADD and CAE require careful planning. There have been high expectations of the potential contributions of CADD and computer aided engineering to civil engineering practice. Feasibility studies, covering strategic planning, the organization's objectives, a rough time table and funding availability help determine long and short term CADD goals. Completing studies before signing contracts and installing the systems avoids disaster. Some chronological steps help avert disaster.

872   New Pavement from Old Concrete

Gordon, K. Ray, Fellow, ASCE, (Concrete Pavement Consultant, Arlington Hts., Ill. 60004)

Civil Engineering--ASCE, Vol. 55, No. 5, May, 1985, pp. 56-58

Recycling portland cement concrete has gained momentum as an alternative to resurfacing aging highway and airfield pavements. Recycling cuts down rawmaterial requirements, reduces the need to haul concrete across long distances and solves the problem of didposing of large amounts of pavement in an environmentally acceptable manner. Additional savings come from the decreased amount of energy required to haul, mix and place the recycled portland cement concrete. Recycling methods, equipment and applications are presented.

873   Monitoring Saves a Site

Donald R. McMahon, Member, ASCE, (Geotech Engr., Goldberg-Zoino Assoc., Buffalo, N.Y.) and **Donald B. Abrams**, Assoc. Member, ASCE, (Chf. Engrg. Geologist, Goldberg-Zoino Assoc., Buffalo, N.Y.)

Civil Engineering--ASCE, Vol. 55, No. 5, May, 1985, pp. 59-61

A comparison is made of two supported excavations that were similarly designed and constructed. Both excavation support systems consisted of a soldier pile and lagging sheeting wall supported by three levels of struts or tiebacks. The excavation depths were both about 40 feet (12.2 m), one excavation was made in a silty sand, and the support system performed adequately. However, when the same design and construction procedures were used to construct the excavation support system in a sandy gravel, problems ensued causing concern for the safety of workers and surrounding structures. Tiebacks in both excavations were installed by augering through the sheeting wall and pressure grouting as the auger was withdrawn. They were prestressed to 95% of the design capacity to limit ground movements. During the augering for middle level tiebacks in the sandy gravel excavation, large amounts of soil sloughed down from

307

behind the excavation support system, onto the auger flights, leaving large voids. Load losses were observed in the upper level tiebacks as the support system deflected away from the excavation. As the excavation was advanced and lower level tiebacks were installed, it was feared that development of full active earth pressures might overstress the lower level tiebacks and lead to a progressive catastrophic failure. Instead, repairs were made to the support system, the construction techniques were modified, and the project was completed safely.

874   Critical Load Program

Peter Carr, (Engr., Atkins Oil & Gas Engrg., London, England)

Civil Engineering--ASCE, Vol. 55, No. 5, May, 1985, pp. 62-64

     The elastic critical load of a structural frame is a fundamental parameter in assessing its stability. It is an upper bound to the failure load of the frame, in the same way that the Euler load is an upper bound to the failure load of a pin-ended strut. A program for computing the critical load is described. Written in a simple form of Basic, the program will run on most microcomputers and is designed to be short, reliable in convergence and easy to use. Errata: CE Oct. '85, pp. 40.

875   Building on Muck

Jack Fowler, Member, ASCE, (Research Civ. Engr., Waterways Experiment Station, Corps of Engineers, Vicksburg, Miss.)

Civil Engineering--ASCE, Vol. 55, No. 5, May, 1985, pp. 67-69

     The Craney Island Disposal Area at Norfolk, Va. is a 2,500 acre depository for material dredged from ports and channels in the Hampton Roads. The U.S. Army Corps of Engineers extended its useful life by building interior dikes. Because its base is very soft, previous construction methods required displacements 8 to 10 volumes down for one volume above the surface. Floating the interior dikes on geotextile provided the reinforcement necessary to prevent rotational foundation failure or embankment spreading until the soft foundation was sufficiently consolidated. Three test sections were built before construction began on the two main dikes. Discussion: **James R. Schneider**, Member, ASCE, (CH2M Hill, Inc., Portland, OR) and **Lawrence H. Roth**, Member, ASCE, (CH2M Hill, Inc., Portland, OR) CE July '85, pp. 36.

876   A Road to Recovery

Civil Engineering--ASCE, Vol. 55, No. 5, May, 1985, pp. 70-73

     Inspections of the Golden Gate Bridge revealed that the bridge had localized corrosion near its expansion joints, corrosion of its reinforcing steel, and was severely contaminated with corrosion promoting chlorides. Engineers determined that the deck could not be salvaged, and had it replaced. Replacement had to be done at night, so that traffic would not be interfered with. To accomplish this, the deck was replaced with prefabricated orthotropic steel panels. A few were installed each night, and opened to traffic in time for the morning rush hour.

877   Waste Cleanup: Lessons Learned

M. John Cullinane, Jr., (Research Civil Engr., U. S. Army Waterways Experiment Station, Vicksburg, Miss.) and **Richard A. Shafer**, (Research Civil Engineer, U. S. Army Waterways Experiment Station, Vicksburg, Miss.)

Civil Engineering--ASCE, Vol. 55, No. 6, June, 1985, pp. 41-43

Since hazardous waste mitigation is a new science, engineers need to pass on their experiences. The U. S. Army Waterways Experiment Station, seeking to expand the hazardous waste knowledge base, surveyed 150 public and private sector personnel involved in remedial projects. The study, prepared for the Naval Energy and Environmental Support Activity in Port Hueneme, Calif., identified some of the problems in hazardous waste cleanups and the lessons learned. The respondents' early experiences with hazardous waste mitigation included project delays and other rude surprises. The key, they concluded, was project planning. They recommended setting cleanup criteria, defining project scope, hiring experts, defining contractor responsibility, setting realistic schedules and communicating effectively with the public.

878   Staggered Truss Adapted to High Rise

Socrates A. Ionnides, Member, ASCE, (Dir. of Design, Stanley D. Lindsey & Assoc., Nashville, Tenn.) and **Stanley D. Lindsey**, Member, ASCE, (Pres., Stanley D. Lindsey & Assoc., Nashville, Tenn.)

Civil Engineering--ASCE, Vol. 55, No. 6, June, 1985, pp. 44-47

The 34-floor Nashville Convention Center Hotel required a structural system that would overcome the tendency to vibrate in the wind. The solution was a modified staggered frame truss, the first of its kind, that behaves as a combined rigid frame and braced frame. Gravity loads are transferred to the exterior columns in order to resist the high overturning moments induced. Lateral loads are distributed so that they assist in resisting the overturning. Behavior of the three-panel truss is attributed to the rigid connection at the vierendeel panel. In addition, the traditional checkerboard staggering pattern was changed to provide better torsional resistance and eliminate diaphragm shear to the central bay.

Discussion: **Louis F. Sokol**, (U.S. Metric Assoc., Inc., Boulder, CO) CE Sept. '85, pp. 28.

879   Opportunity Knocks from Behind Bars

Howard Smallowitz, (Asst. News Ed., Civil Engineering--ASCE, 345 East 47th St., New York, N.Y. 10017)

Civil Engineering--ASCE, Vol. 55, No. 6, June, 1985, pp. 48-51

As more and more lawbreakers are put behind bars, more and more prisons are needed, so the opportunities for civil engineers are expanding. Three major types of construction are being used to add cells to the nation's penal system: new construction, modular construction, and renovation. Each has its advantages and disadvantages. All, however, are radically different in design and construction management than non-prison structures.

880   **Micro-CAD Systems: A Dream Come True**

Philip V. DiVietro, (Managing Ed., *Civil Engineering--ASCE*, New York, N.Y. 10017)

Civil Engineering--ASCE, Vol. 55, No. 6, June, 1985, pp. 52-55

Computer-aided design and drafting systems that run on inexpensive microcomputers are examined. Several major systems (products) are covered via the users point of view. Many small civil engineering firms were interviewed. Productivity gains, quality of working drawings and other advantages are included. All of the interviewees are using hardware/software packages that are priced below $15,000. Some comparisons are made to a minicomputer system. Article includes three sample quotes that add up to $10,695; $12,889; and $26,653.

881   **Hydraulic Fill Dam Made Earthquake Resistant**

H. J. Billings, Member, ASCE, (Mgr. of Engrg. Services, East Bay Municipal Utility District, Oakland, Calif.)

Civil Engineering--ASCE, Vol. 55, No. 6, June, 1985, pp. 56-59

San Pablo Dam, near the San Andreas fault, constructed in 1919-21 by hydraulic fill methods, was found to be susceptible in places to liquefaction under seismic loading. The dam has been made much more earthquake-resistant by adding a buttress fill covering the dam's upstream face, and drains in the dam and in the new buttress to dissipate more speedily seismically induced increases in soil pore pressures. The investigation, corrective design and construction procedures are described.

882   **Coping with Litigation**

George L. Reed, Fellow, ASCE, (Adm., Shelby County Engrs. Office, Memphis, Tenn. 38103)

Civil Engineering--ASCE, Vol. 55, No. 6, June, 1985, pp. 60-61

Engineers should learn legal philosophy, even though it seems unrelated to traditional technical practice. Monitoring changes in legal theory that may increase engineers' liability and understanding the rationale behind legal processes helps strengthen their defenses against possible suits. Civil engineers are prime targets for lawsuits nowadays because of the deteriorating infrastructure, acid rain, chemical spills, polluted water, deficient buildings and bridges and aging highways. Engineers can take a number of steps to help avoid lawsuits. Perhaps the best step is to close the gap between engineers and attorneys.

883   **Floating Bridge for 100 Year Storm**

Thomas R. Kuesel, (Chmn. of the Board, Parsons, Brinckerhoff, Quade & Douglas, Inc., New York, N. Y.)

Civil Engineering--ASCE, Vol. 55, No. 6, June, 1985, pp. 62-65

In 1979 the west half of the Hood Canal Floating Bridge, built in 1960, in Seattle broke up and sank during a storm that lasted eight hours with winds of 80 mph and gusts over 100 mph. After considering different replacement alternatives, the Washington State Department of Transportation concluded that a floating bridge was the best and least costly coice for the site--a 7500 ft crossing with water depths up to 34 ft, a tide range of 16 ft, and exposure to wind and waves. The configuration of the replacement bridge including draw span design and cable anchorages is described. Methods of financing and constructing the bridge are also presented.

884   Building in Space

Virginia Fairweather, (Editor, ASCE 345 E. 47th St., New York, N. Y. 10017)

Civil Engineering--ASCE, Vol. 55, No. 6, June, 1985, pp. 66-69

Work on the nation's first permanently manned space station has begun. Almost $200 million in contracts for conceptual design for the station were awarded by the National Aeronautics and Space Administration in the spring of 1985. The largest "package" is for work on the truss structure that will form the backbone for the space station. There are two design/construct options at this time: erectable and deployable. Both would be made of graphite epoxy. An erectable truss system would have elements shipped to orbit on the space shuttle where astronauts would assemble the truss in space. A deployable system would be preassembled with compression springs, folded down for shipping to orbit, then released and "exploded" in space. Modules for power, fuel, laboratory activities and living quarters, etc. would then be attached to the truss.

885   Get Involved with Cladding Design

Roger J. Becker, (Vice President, Computerized Structural Design, Milwaukee, Wisc.) and **Rita Robison**, (Associate Ed., *Civil Engineering--ASCE*, New York, N. Y. 10017)

Civil Engineering--ASCE, Vol. 55, No. 6, June, 1985, pp. 70-73

Structural engineers should get involved with cladding design for their buildings. Including disclaimers for responsibility in the design contract will not prevent trouble or lawsuits, and the structural engineer is in the best position to coordinate the design between the architect and the supplier. The structural engineer should carefuly check shop drawings, which should include all of the forces used at connection points of the cladding. A checklist for decisions during planning, design, and shop drawing review is included.
Discussion: **Charles Kilper**, Member, ASCE, (Sr. Assoc., Heitmann and Assoc., Inc., St. Louis, MO) CE Oct. '85, pp. 36. Errata: CE Dec. '85, pp. 33.

886   The Tunnel That Transformed Philadelphia

Rita Robison, (Assoc. Ed., *Civil Engineering--ASCE*, New York, N.Y. 10017)

Civil Engineering--ASCE, Vol. 55, No. 7, July, 1985, pp. 38-41

The 1985 Outstanding Civil Engineering Achievement is the Center City Commuter Rail Connection, a 1.7 mile tunnel connecting two former stub-end railroad stations in Philadelphia. The $330 million project required snaking the four-track tunnel through the densely packed substructure of the city, underpinning 12 buildings along its path, all without disrupting utilities and existing subways. The former Suburban Station was refurbished and a new Market East Station was constructed under the existing Reading Terminal. This serves as the catalyst for redevelopment of the Market St. East business district, where more than $1 billion in new construction is scheduled. Construction management, a joint venture of three civil engineering firms, involved supervision of 46 multi-discipline construction contracts and more than 300 subcontracts.
Errata: CE Sept. '85, pp. 28. Discussion: **Karl B. Weber, III.**, Affiliate Member, ASCE, (Doylestown, PA) CE Dec. '85, pp. 24. Discussion: **George J. Stanley**, Member, ASCE, (Sr. Program Mgr., Southeastern Pennsylvania Transportation Authority, Philadelphia, PA) CE Dec. '85, pp. 24.

887   **Dam Pioneers Concrete Variant**

Civil Engineering--ASCE, Vol. 55, No. 7, July, 1985, pp. 42-45

      The Willow Creek Dam in Oregon is the nation's first dam made entirely of roller compacted concrete. The design and construction of this dam are described and special problems with leakage and seepage control are covered. The work was completed in 124 days at one-third of the cost of a conventional concrete dam or one-half the cost of a conventional fill dam.

888   **A Space Age Test Center**

Civil Engineering--ASCE, Vol. 55, No. 7, July, 1985, pp. 46-48

      The Air Force's Aeropropulsion System Test Facility (ASTF) is the most expensive project the Air Force has ever built. Its massive heaters, coolers and compressors will treat air, then force it over aircraft engines in simulations of actual flights. Engine designers will be able to see how the engines will behave at speeds up to Mach 3.8, altitudes of 100,000 ft, and temperatures from -100° F to 1,020° F. The project has received an OCEA award of merit from ASCE.

889   **Biggest Highway Rebuild**

Civil Engineering--ASCE, Vol. 55, No. 7, July, 1985, pp. 49

      The reconstruction of the Wisconsin Interstate 90 and 94 involved 32 miles of 6-lane pavement. Most of the concrete aggregate for the job was recycled concrete pavement and all pavement reinforcing steel was epoxy coated. It was the largest concrete recycling job ever attempted.

890   **Helms Pumped Storage Project**

Civil Engineering--ASCE, Vol. 55, No. 7, July, 1985, pp. 50

      The powerhouse at California's Helms Pumped Storage Project is 1,200 ft below the surface of the Sierra Nevada mountains. Linking existing reservoirs, the project provides 1,200 Mw of electrical peaking capacity. More than 1 million cu yd of granite was excavated. The 1,800 ft inclined shaft is one of the largest of its type ever excavated and concreted. OCEA nominee.

891   **Tapping Deep Water Resources**

Civil Engineering--ASCE, Vol. 55, No. 7, July, 1985, pp. 51

      Lena is the first commercial guyed tower. Its jacket is the longest ever fabricated, loaded out and launched in one piece. Lena stands in 1,000 ft deep water and has a three-level oil drilling and production deck and two drilling rigs. Twenty steel guy lines extend from the structure to weights on the ocean floor to hold the tower upright. The jacket was barged to the Gulf of Mexico site in one piece. Production on the oil platform began in 1984.

892   **LAX Engineers Beat the Clock**

Civil Engineering--ASCE, Vol. 55, No. 7, July, 1985, pp. 52

Foreign and domestic crowds flying in for the Olympics were no problem for the Los Angeles Airport (LAX), thanks to the expansion and improvements completed a month earlier. LAX more than doubled its terminal space, increased its central terminal capacity by 60% and completed major airfield rehab. The Los Angeles Department of Airports built international and domestic terminals, a roadway to serve them and new cargo handling facilities. The agency expanded its utility plant and parking facilities, and made runway and taxiway improvements. LAX was the first major airport in the U. S. to undertake an expansion of this size.

893   **Mini Hydro at Low Cost**

Civil Engineering--ASCE, Vol. 55, No. 7, July, 1985, pp. 53

The Kingsley Dam is an existing irrigation dam that has been retrofitted with electric generating capacity. The project was completed 6 months ahead of schedule without interrupting irrigation flows during construction and was $14 million under budget. Per longterm kilowatt-hour it is said to provide the least constly electricity in the state.

894   **Multi-Truss Design for Tower**

Civil Engineering--ASCE, Vol. 55, No. 7, July, 1985, pp. 53

A modified structural steel system of outrigger trusses and belt trusses bolster the 54-story Equitable Tower West. Located in New York City, the building's structural system has two levels of outrigger trusses at the 11th and 36th floors. The transfer exterior columns at those floors take windloads at those floors and engage all the columns across the building. When the building deflects, the load comes through the outrigger trusses which bring the exterior columns into play. The design decreases deflection and eliminates many moment connections on the girders framing into the columns in the north/south direction. Belt trusses connect to the outrigger trusses and assist in the distribution of wind shear, moments, and eccentric loading.

895   **Project Mixes Oil and Water**

Civil Engineering--ASCE, Vol. 55, No. 7, July, 1985, pp. 54

The Prudhoe Bay Seawater Treatment Plant was nominated for an OCEA award, for its potential to coax from Alaska's north slope 1 billion barrels of oil which otherwise would have gone untapped. The facility takes water from the Beaufort Sea, filters it, heats it and then pumps it to wells across the oil field. When injected into the wells, the water offsets pressure lost when oil is pumped out, thus making the wells more productive. The facility was built on a barge in Korea, and transported to its site in the arctic, where it was sunk into place. It was designed to accommodate the extreme cold and ice loads typical of the site. An extensive marine life recovery system was incorporated into the plant to minimize its impact on the environment.

896   **Bath County Pumped Storage**

Civil Engineering--ASCE, Vol. 55, No. 7, July, 1985, pp. 55

The Bath County Hydroelectric Pumped Storage Project provides 2,100 Mw of peak power to Virginia Electric & Power Co. and Allegheny Power Co. Called a bargain at $1.6 billion, it includes two earthfill/rockfill dams, two reservoirs, conduits, a 980 ft vertical shaft, and various access, drainage and diversion tunnels. The powerhouse, located in the lower reservoir, is completely submerged during each pumping cycle. OCEA nominee.
Errata: CE Sept. '85, pp. 28.

897   **Shrimp Farm Made Feasible**

Civil Engineering--ASCE, Vol. 55, No. 7, July, 1985, pp. 56-57

Engineers on a limited budget developed a facility on the island of Oahu, Hawaii, where Mexican blue shrimp are being raised commercially. A series of "U" shaped sections were laid parallel to each other on the farm's site. Within the section the shrimp are raised. Between the sections, workers can move to tend to the shrimp. The sections were made of extruded concrete to avoid costly formwork. Inflatable fabric roofs cover the sections—a system that costs only about one-fifth that of a conventional roofing system. For its engineering innovations, the facilty was awarded an OCEA nomination.

898   **Record Bridge Replacement**

Civil Engineering--ASCE, Vol. 55, No. 7, July, 1985, pp. 56-57

Jacks Run Bridge had to be rebuilt in one year instead of the usual two to three. So the Pennsylvania Department of Transportation spearheaded the quickest replacement of a major bridge in the state's history. Building the new structure on the existing alignment eliminated major right of way and environmental impact negotiations. Other hallmarks of the fast track design and construction included speeding design time to 30 days, squeezing bidding for a contractor to 12 days and contracting for winter construction. The three-span plate girder bridge carries four traffic lanes and a 5 ft pedestrian walkway. Two piers, 182 and 142 ft high, carry the bridge above Jacks Run.

899   **Master Plan for Irvine**

Civil Engineering--ASCE, Vol. 55, No. 7, July, 1985, pp. 58

Irvine, Calif. started 24 years ago with 68,000 acres and a master plan that emphasized controlled development and a mix of public and private enterprise. More than 25,000 acres have been developed to date. OCEA nominee.

900   **Martin Marietta Complex**

Civil Engineering--ASCE, Vol. 55, No. 7, July, 1985, pp. 58

Great pains were taken to ensure the environmental integrity and beauty of the area surrounding this light manufacturing and office complex. Located near Orlando, Fla., some 80 percent of the site for this facility was set aside for open space and environmental preservation.

901   **Bridging Twin Ports**

Civil Engineering--ASCE, Vol. 55, No. 7, July, 1985, pp. 59

The Major Richard I. Bong Memorial Bridge, part of U. S. Highway 2, connects harbors at Duluth, Minn. and Superior, Wisc. The new highway link was designed to relieve congestion on nearby roadways and to provide access to sites essential for marine development without disrupting residential and commercial areas. The bridge's S shape enables it to cross a shipping channel at nearly right angles to keep navigation safety high and main span costs low. Difficult soil and foundation conditions, winter weather and sensitive environmental conditions were the major problems overcome.

902   **Hazardous Waste: Closing the Insurance Gap**

James A. Thompson, (Attorney, Wickwire, Gavin & Gibbs, Suite 400, 8230 Boone Blvd., Vienna, Va. 22180), **Julie C. Becker**, (Attroney, Wickwire, Gavin & Gibbs, Suite 400, 8230 Boone Blvd., Vienna, Va. 22180) and **Michael C. Loulakis**, (Attorney, Wickwire, Gavin Gibbs, Suite 400, 8230 Boone Blvd., Vienna, Va. 22180)

Civil Engineering--ASCE, Vol. 55, No. 7, July, 1985, pp. 60-62

Insurers are withdrawing from the hazardous waste field because of large claims and increased risks. Therefore, cleanup contractors are being forced to look for alternative ways to protect themselves. Indemnification, a written agreement whereby one party agrees to be responsible for any judgements entered against a second party, is the best available alternative. But there are limitations to indemnification. Legislation and litigation will be required to clarify many of the issues in the field, and provide guidance to insurers, regulators and contractors.

903   **Who Should Design Bridges?**

Richard Heinen, (Partner, Sanbar Group, New York, Munich, London)

Civil Engineering--ASCE, Vol. 55, No. 7, July, 1985, pp. 63-65

For many years Design-Build-Proposals have been common practice in Europe (except in the UK). Structures in these countries are 20% more economical than similar projects in the U. S., after engineering cost. Recent developments in the U. S. show a trend to alternative designs. Value engineering proposals are becoming accepted in recent years and current practice even allows contractors to make proposals on their own designs. The European experience shows that more economical projects and a larger variety of structures as far aesthetics, safety and impact on the environment is concerned, will result. There are many obstacles in present U. S. procedures to this system, but the possible gains should help promote the acceptance of this system in this country.
Discussion: **David L. Narver, Jr.**, Fellow, ASCE, (San Marino, CA) CE Oct. '85, pp. 28. Discussion: **George M. Lostra**, Member, ASCE, (Chronic & Associates, Boise, ID) CE Dec. '85, pp. 33.

904   **Engineering Education: An Update**

Virginia Fairweather, (Ed., *Civil Engineering—ASCE*, New York, N. Y. 10017)

Civil Engineering--ASCE, Vol. 55, No. 7, July, 1985, pp. 67-69

ASCE held an engineering education conference in April, 1985, entitled, "Educators and Practitioners—Where Are We Going?" Participants agreed that a greater interchange is beneficial between faculty and students at the university and practitioners in the business world.

On other subjects the conference attendees were less than united. Among issues addressed were the role of computers in civil engineering education, a five-year engineering program, the need for better communications skills among engineering graduates, a "divorce" of civil engineering and the construction industry, a dearth of research funds, and the reputation of civil engineering as a "low technology" profession.

Discussion: **Tobert M. Sykes**, Member, ASCE, (Prof. of Civ. Engrg., Ohio State Univ., Columbia, OH) CE July '85, pp. 27.

905   Management Secrets of Top Consultants

K. A. Godfrey, (Sr. Editor, *Civil Engineering—ASCE*, New York, N. Y. 10017)

Civil Engineering--ASCE, Vol. 55, No. 7, July, 1985, pp. 70-73

Leaders of five consulting civil engineering firms, which are considered among the best managed, discuss management strategies. In all cases, carefully selected personnel is considered the primary asset. Teamwork and consensus style decisionmaking are emphasized. Those interviewed stressed the importance of demanding quality work and a commitment to service. Carefully constrolled growth is also a factor.

Discussion: **George M. Lostra**, Member, ASCE, (Chronic & Associates, Boise, ID) CE Dec. '85, pp. 36.

906   Can We Save the Ogallala?

Texas Tech Univ., Lubbock, Tex. 79409 Sweazy,Robert M., (Dir., Water Resources Center)

Civil Engineering--ASCE, Vol. 55, No. 8, August, 1985, pp. 36-39

The useful life of the Ogallala Aquifer, which underlies seven western states, can be extended indefinitely through water conservation in the widest sense of the term. In addition to optimization of crops, soil moisture control and irrigation management now practiced, methods are being developed for recharging stormwater to the aquifer. Research is also underway on secondary recovery techniques that will augment the quantity of stored water available for pumping.

907   Hazardous Waste Cleanup: The Preliminaries

C. Kenna Amos, Jr., (U.S. Environmental Protection Agency, 26 Federal Plaza, New York, N.Y. 10278)

Civil Engineering--ASCE, Vol. 55, No. 8, August, 1985, pp. 40-43

Court-ordered negotiations rather than a trial took five years but led to a flexible plan for cleaning up an inactive hazardous waste disposal site in Niagara Falls, N.Y. Contamination of the city's water treatment plant was traced to the site known as S-Area, owned by Occidental Chemicals Corp. (formerly Hooker Chemicals & Plastics Corp.). The settlement agreement, approved by the U.S. District Court in April 1985, provides for a phased approach in defining, evaluating and correcting the problem. No future technology is ruled out, but the remedy focuses on containment based on a novel hydraulic gradient concept, plus extensive monitoring for at least 35 years.

908   **Danger: Natural System Modeled by Computer**

Martin Kurtovich, (Water Resources Engr., California Water Quality Control Board)

Civil Engineering--ASCE, Vol. 55, No. 8, August, 1985, pp. 44-45

Computer (mathematical) models are written to simulate such natural systems as those leading to acid rain, flooding, or toxic waste migration from disposal sites into groundwater. In some cases (for example, movement of toxic wastes off site) they help predict phenomenon better than any other means. However, computer models can also be overrated. An inadequate understanding of the natural phenomena can prevent reliable predictions, incorrect assumptions about the model can also affect the outcome. It is recommended that standards for model writing and documentation be established and that users should be involved in writing the model. The user should understand the natural system being simulated in order to correctly interpret the results of the computer model.

909   **Hydropower's Newest Generation**

Howard Smallowitz, (Asst. News Editor, Civil Engineering Magazine, 345 East 47th St., New York, N.Y. 10017)

Civil Engineering--ASCE, Vol. 55, No. 8, August, 1985, pp. 46-49

Higher energy prices and strong incentives from the federal government have caused hydropower developers to consider sites that just a few years ago were scoffed at as mere trickles. Intense regulation is still considered a problem to developers, however, a number of projects of all sizes are coming closer to being realities. A project on Alaska's Susitna River should begin producing power by 1997, and will be expanded in stages to its full capacity of 1020 Mw by 2008. It is perhaps the last huge hydro project being planned in the U.S. The 22.5 Mw Swan Lake project already is supplying power to the small, electrically isolated town of Ketchikan. Developers are finishing the Eldred L. Field hydroelectric project, a 15 Mw addition to the historic canals which once powered the industrial revolution in Lowell, Mass. The economics of hydropower have changed so radically, that the owners of an abandoned dam and powerhouse in Bumcombe County, N.C. are bringing the site out of retirement. However, the $7 million renovation of the 2.4 Mw plant has been delayed by state and federal environmental agencies. In Jersey City, N.J. planners hope to tap the energy generated when the city's tap water flows from a reservoir into an aqueduct.

910   **A Space Frame Forecast**

Lev Zetlin, (Pres., Zetlin-Argo Associates, 60 E. 42nd St., New York, N.Y. 10017) and **Virginia Fairweather**, (Editor, Civil Engineering-ASCE, 345 E. 47th St., New York, N.Y. 10017)

Civil Engineering--ASCE, Vol. 55, No. 8, August, 1985, pp. 50-54

New York City's Jacob K. Javits Convention Center has the largest space frame of its kind in the world. There are several unique design features in the space frame, among them joints of hollow nodes into which tensioned rods within members are bolted. An erection system of "stitching" preassembled portions of the space frame is another. The implications of all are addressed. The structure has been extensively tested, partly because of its public nature and partly because of suggestions from the author who serves as peer consultant to the owner, working with both the architect and the structural engineers of record. Testing is also described. Discussion: **Erling Murtha-Smith**, Member, ASCE, (Assoc. Prof., Univ. of Conn., Storrs, CT) CE Nov. '85, pp. 32.

911   Computer Keeps New Orleans: Head Above Water

Michael A. Ports, (Chief Hydraulic Engineer, Daniel, Mann, Johnson & Mendenhall, 512 South Peters St., New Orleans, La. 70130)

Civil Engineering--ASCE, Vol. 55, No. 8, August, 1985, pp. 56-58

The city of New Orleans has been beset with drainage problems since its founding in 1718. A major drainage system was installed in the city in the late 1800s, but it has grown increasingly obsolete as the city has sunk and development has grown more intense. To analyze the best strategy for upgrading the system, Daniel, Mann, Johnson & Mendenhall simulated the current system on a computer. The model was tested with data from actual storms to ensure that it reflected reality. Then the model was modified to reflect various improvement plans to the system. The model simulated how effectively each plan would handle hypothetical storms of varying intensities, allowing planners and the public to decide which plan was the most cost effective.

912   Two Projects Became One

W. Martin Roche, Member, ASCE, (Chf., Div. of Planning, Bureau of Reclamation, Durango, Colo.) and **John W. Davison**, (Civ. Engr., Guam Water Supply Project, Agana, Guam)

Civil Engineering--ASCE, Vol. 55, No. 8, August, 1985, pp. 64-65

Nontraditional funding and a new law were required before two Cortez, Colo. projects could be combined. Building a new canal and enlarging and lining an existing one to move water in essentially the same direction would be costly and doubly disruptive to the environment and communities adjacent to both sites. The Bureau of Reclamation was investigating the McElmo Creek Unit, part of a federally mandated program to lessen the Colorado River's salt content. The agency found that combining this project with the nearby Dolores Project a water resources plan, would be more cost effective and less time consuming than completing the projects separately.

913   Blind Drilling Down Under

Paul Richardson, (Santa Fe Shaft Drilling Co., P.O. Box 1401, Orange, Calif. 92688)

Civil Engineering--ASCE, Vol. 55, No. 8, August, 1985, pp. 66-68

Australia's largest blind drilled shaft is a 14 ft diameter fresh air shaft for the Agnew nickel mine 500 miles northeast of Perth. It was drilled in one pass using air-assisted reverse circulation to remove the drill cuttings. Drilling began on March 22, 1982 and reached a total depth of 2,460 ft on Feb. 12, 1983. Penetration rates were better than projected, but cutter life was considerably shorter than expected, resulting in somewhat excessive cutter costs. The project is notable for its many firsts and for its safety record. During 328 days there was only one serious lost time accident.

914   Cost Engineering for Disputed Work

James J. O'Brien, Fellow, ASCE, (Chf., O'Brien-Kreitzberg & Assoc., Inc., Merchantville, N.J. 08109)

Civil Engineering--ASCE, Vol. 55, No. 8, August, 1985, pp. 69-71

Engineers and architects tend to underestimate the cost of disputed work, especially those connected with delay. The engineer, using industry standards and estimating rules of thumb, can develop a reasonable value of the cost of delay. Entitlement to payment for disputed work is established in essentially the same manner used for change orders.

915   Drainage Tunnels Save Freeway Link

Robert D. Miller, Member, ASCE, (Partner, Howard Needles Tammen & Bergendoff, Phoenix, Ariz.), **Richard E. Schwab**, Member, ASCE, (Chief, Hydraulics Section, Howard Needles Tammen & Bergendoff, Kansas City, Mo.) and **Timothy P. Smirnoff**, Member, ASCE, (Manager, Tunnel and Underground Engineering Dept., Howard Needles Tammen & Bergendoff, Kansas City, Mo.)

Civil Engineering--ASCE, Vol. 55, No. 8, August, 1985, pp. 72-74

A 15 mile section through downtown Phoenix will complete Interstate 10. After years of delay, the project is now close to completion. The project required construction of 6.5 miles of drainage tunnels. The tunnels were selected from some 21 alternate designs to solve the problem of drainage, particularly impacting a 3 mile depressed freeway section. Local rainfall conditions, a high degree of urbanization and the location of the freeway between an extensive portion of watershed and the natural outlet for runoff would create potential for flooding. Three tunnels will convey runoff without disrupting traffic and the downtown economy. The cost of this alternative is half of that of conventional drainage systems. Two of the tunnels with outfalls at the Salt River will act as inverted siphons, thereby overcoming a problem caused by the flat topographic relief of the Phoenix Basin. Runoff is intercepted above the freeway by a conventional near-surface sewer system and conveyed into the tunnels. Runoff falling on the depressed freeway will be pumped to the near-surface conveyance and on into the tunnels. Since the tunnels will be subjected to surge conditions, they are designed as pressure conduits.
Errata: CE Oct. '85, pp. 40.

916   Small Water Users—Planning Crisis

Steven C. Harris, Member, ASCE, (Owner, Harris Water Engineering, 954 Second Ave., Durango, Colo. 81301) and **Andrea Sklarew Maynard**, (Aff., Harris Water Engineering, 954 Second Ave., Durango, Colo. 82301)

Civil Engineering--ASCE, Vol. 55, No. 8, August, 1985, pp. 76-78

Planning to rehabilitate small water systems—both irrigation and municipal water supply—often is ignored. In contrast, planning for new construction and for operations/maintenance work usually is well provided for in budgets and planning. As for small irrigation districts, since their dams were built decades ago, downstream communities have sprung up in their flood plain. Because property could be destroyed and lives lost if the dam fails, the state dam safety agency has applied stricter safety criteria—especially to spillway adequacy—to these dams that have led the state to draw down the reservoirs of some 200 dams. As for town water supplies, inadequate treatment has led residents of some to use bottled water at certain times of year. This article suggests sources of planning money, encourages the Bureau of Reclamation and Corps of Engineers to make their Western area data on hydrology and geology micro-computer accessible, and suggests how a small water user should organize to plan for rehabilitation of its infrastructure.

917   **Restoring the Capitol's West Front**

Virginia K. Dorris, (50 Sterling Pl., Brooklyn, N.Y. 11217)

Civil Engineering--ASCE, Vol. 55, No. 9, September, 1985, pp. 36-38

      After 25 years of debate, preservation (rather than extension) of the West Front of the U.S. Capitol building in Washington, D.C. has begun. A network of stainless steel rods is being placed to shore up the spalling and deteriorated facade in a project that will be completed in 1987. A total of 12,000 ft. of drilling is required, all of it through unknown characteristics of the original sandstone walls. Grouting and stainless steel will also reinforce the foundation. About 25% of the sandstone on the facade will be replaced by limestone, painted to match the original.

918   **Romanesque Reused**

Rita Robison, (*Civil Engineering,* New York, N.Y. 10017, Associate Ed.)

Civil Engineering--ASCE, Vol. 55, No. 9, September, 1985, pp. 41-44

      When Union Station in St. Louis, Mo. was built in the 1890s it was the world's largest railroad station. Having been abandoned by the railroads in the 1970s, it was reopened in August 1985 as a hotel/retail complex developed by the Rouse Company. The iron and steel train shed was strengthened and now, partially open to the sky, covers two new hotel wings, glass-enclosed retail spaces and a landscaped garden. The original Headhouse and adjacent Terminal Hotel also had to be structurally rehabilitated before their interiors could be transformed into a modern hotel. These interiors have had their elaborate Victorian decorations restored.

919   **Solutions in the Pipeline**

B. Jay Schrock, (Pres., JSC International Engineering, 2261 Grove Ave, Sacramento, Calif. 95815)

Civil Engineering--ASCE, Vol. 55, No. 9, September, 1985, pp. 46-49

      Water and sewer agencies across the country are beginning to feel more comfortable about pipeline rehabilitation. Where once replacement was the ony alternative considered for aged, clogged and failing pipes, rehabilitation is now used, sparing the communities that own the pipelines the cost and inconvenience of replacement. Four cases of rehabilitation are examined.

920   **Private Funds, Public Projects**

Corinne S. Bernstein, (Assistant News Edit., Civil Engineering—ASCE, 345 E. 47th St., New York, N.Y. 10017)

Civil Engineering--ASCE, Vol. 55, No. 9, September, 1985, pp. 50-53

      Tapping private sources for public services is not new, though the concept has advanced from traditional contracting for services, which has been around since pre-Revolutionary War days. In 1985, privatization generally refers to a financial partnership between the public and private sectors for competing projects that federal, state and local governments have traditionally funded and municipalities have owned and operated. Privatization is growing slowly and receiving guarded acceptance, as financial, technical and legal consultants study its options. Municipalities have taken their first peeks at the design-

1985 ASCE PUBLICATIONS

build-operate form of privatization in the resource recovery, cogeneration, district heating and hydropower industries. Wastewater and water treatment, rail and prison projects are newcomers to this sort of funding.

921   Pier Review

Dennis V. Padron, (Partner, Han-Padron Associates, 1270 Broadway, New York, N.Y. 10001)

Civil Engineering--ASCE, Vol. 55, No. 9, September, 1985, pp. 54-55

Upgrading piers to accommodate ships larger than those for which the piers were originally designed is a common problem facing civil engineers. Three case histories illustrate different solution concepts that enable piers to resist the larger berthing impact force. At Exxon's Baytown refinery Dock 1 the new fender system is completely independent of the existing dock, imposing no loads on it. Pier 17 at the U.S. Navy's Submarine Base, New London, was upgraded by using modern computer techniques to determine its capacity to resist horizontal forces and providing a fender system that fully utilizes this capacity. At International Marine Terminals' Central Dock on the lower Mississippi, upgrading could be achieved by substituting highly efficient buckling type rubber fenders for the existing tubular rubber fenders.

922   Clearing the Decks

Virginia Fairweather, (Editor, Civil Engineering—ASCE, New York, N.Y. 10017)

Civil Engineering--ASCE, Vol. 55, No. 9, September, 1985, pp. 56-59

Cathodic protection, an electrical-chemical process that prevents corrosion of steel rebars on bridge decks, is rapidly gaining acceptance by highway engineers and state highway departments. The Federal Highway Administration strongly endorses the method and 100% federal funding for installations is avilable. Several methods are currently used, among them the coke-asphalt overlay, the FHWA slot system, the mount system, a variation on the slot, distributed anodes, and zinc metallizing. All are described, along with advantages and disadvantages as seen by the experts and system users interviewed. The first such application of the method was on a bridge deck in California in 1973; there are now about 125 cathodic protection systems installed. In addition, cathodic protection is increasingly being used on bridge substructures and this application is also described.

Discussion: **Joseph A. Lehmann**, Affiliate Member, ASCE, (Pres., Porter Corrosion Control Services, Inc., Houston, TX) CE Nov. '85, pp. 36.

923   New Safety for Old Dams

Rita Robison, (Associate Ed., *Civil Engineering*, —ASCE, New York, N.Y. 10017)

Civil Engineering--ASCE, Vol. 55, No. 9, September, 1985, pp. 60-63

Shortly after the Phase I inspection program conducted by the U.S. Army Corps of Engineers in 1978-81, the Pennsylvania legislature authorized $300 million for repair of dams, ports and other water supply infrastructure. Funded by general obligation bonds, the money is available as loans to both public and private owners. The first loan was made in January 1984, and by mid-1985 about $45 million had been loaned under the Water Facility Loan Program. The Department of Environmnental Resources has the full backing of the Commonwealth Court in negotiating consent agreements with owners and then bringing unsafe dams into compliance.

924   Foam Grout Saves Tunnel

Dan Grimm, (Eng., Micon Services, 32 Fifty First St., Pittsburgh, Pa. 19201) and **W. C. Pete Paris**, (Sr. Geotechnical Eng., Parsons Brinckerhoff Quade & Douglas, 1 Oliver Place, Pittsburgh, PA 15222)

Civil Engineering--ASCE, Vol. 55, No. 9, September, 1985, pp. 64-66

Pittsburgh's 81 year old Mt. Washington tunnel has been made safe for many more decades of light-rail transit use, with the use of foamed chemical grout. The job is the first major U.S. civil engineering application where foamed polyurethane grout was specified from the start. Over much of its length the tunnel was built without grout or backfill in the space between the rock and the brick lining. This made it less than optimally safe over the long term, since a brick lining is not strong in bending. Normally cement grout is used, but before setting it might easily overload the lining. The lower the grout's weight, its other properties being acceptable, the better. This led engineers to foamed polyurethane grout. In the formulation chosen it is very light, 5 pcf, yet it has nearly 3 times the required 35 psi compressive strength.

925   New Cables for Old

Rita Robison, (Assoc. Ed., *Civil Engineering—ASCE,* New York, N.Y. 10017)

Civil Engineering--ASCE, Vol. 55, No. 9, September, 1985, pp. 68-71

In an eingineering first, the main suspension cables will be replaced on New York City's Williamsburg Bridge without closing it to traffic at any time. According to the design by Ammann & Whitney, a single new cable will be placed between each of the existing pairs and anchored in new concrete blocks. The new cables, 24 in. diameter, will be attached to the suspended structure by a pair of ropes at each stiffening truss panel point. As load is added to the cables, the new saddles will be jacked in steps. At the same time, the existing saddles will be jacked in the opposite direction as the existing cables are unloaded. The load will be jacked into new suspender ropes at alternate panel points, then the ropes in between will be loaded and attached. The engineers believe that this design for rehabilitation is preferable to construction of a replacement bridge.

926   Computers Aren't Infallible

Corinne S. Bernstein

Civil Engineering--ASCE, Vol. 55, No. 10, October, 1985, pp. 44-45

Blindly accepting answers just because they are computer-generated does not make for good engineering. Wrong answers may result from software, computer or input problems. Users and programmers are the main players in checking out computer answers. But reviewing and understanding the entire audit trail of a software package is not feasible for the average user. Verifying computer software is difficult at best, consultants agree. But, should we scrap our computers? No, they agree. As with any method of problem solving, verification of results rests with a thorough review of the computer product by experienced professionals.
Discussion: **David L. Narver, Jr.**, Fellow, ASCE, (San Marino, CA) CE Dec. '85, pp. 30.

927   **Micro-CAD: Can It Pay for Itself**

Tracy Lenocker, Member, ASCE, (Vice Pres., The Technical Group, Inc., Anaheim, CA)

Civil Engineering--ASCE, Vol. 55, No. 10, October, 1985, pp. 46-47

      A simple financial analysis is presented that forecasts the payback period on an investment in a microcomputer drafting system. The payback period is calculated in months using productivity usage ratio and the operator's annual salary. Other variables include the fixed costs of hardware and software, overhead costs, profit, and depreciation.

928   **The High-Tech Civils**

Howard N. Smallowitz, (Asst. Editor, Civil Engineering Magazine, 345 E. 47th St., New York, NY 10017)

Civil Engineering--ASCE, Vol. 55, No. 10, October, 1985, pp. 48-49

      Computers are entering virtually every aspect of our lives. Two civil engineers are easing the way computers enter the life of the engineer. Charles Miller, president of CLM/Systems, Orlando, has developed, and continues to develop COGO (Coordinate geometry), the universal computer language of the civil engineer. James Lambert, founder of Artecon Systems, Inc., is bringing CADD into the world of artificial intelligence.
Errata: CE Dec. '85, pp. 33.

929   **Putting CADD to Work**

Rita Robison, (Assoc. Ed., Civil Engineering—ASCE, New York, NY 10017)

Civil Engineering--ASCE, Vol. 55, No. 10, October, 1985, pp. 50-54

      Case histories show that engineering with CADD is no longer experimental. Projects reviewed in this article include an office/shop building for an electrical utility, a flood plain site reclaimed for a residential golf club community, extensive renovations and additions to a manufacturing complex, detention ponds for flood control at an urban park, and preliminary designs for a domed stadium that were made for a developer's promotion efforts.
Errata: CE Dec. '85, pp. 33.

930   **Engineering with Spreadsheets**

David E. Kleiner, Member, ASCE, (Vice Pres., Harza Engineering Co., Chicago, IL)

Civil Engineering--ASCE, Vol. 55, No. 10, October, 1985, pp. 55-57

      With an electronic spreadsheet, the geotechnical engineer has a powerful tool for solving routine engineering problems. The ability of the electronic spreadsheet to use the calculus of finite differences by relaxation or other techniques opens an entirely new application for the personal computer. Some of the uses are memorandum, linear regression analysis, design guides, stability analysis, quantity takeoff and project management. An example is given for seepage analysis by relaxation methods.

931   Project Management on a Micro

Howard Smallowitz, (Asst. Ed., Civil Engineering—ASCE, 345 E. 47th St., New York, NY 10017)

Civil Engineering--ASCE, Vol. 55, No. 10, October, 1985, pp. 58-61

Increasingly, project management programs that once required the power of a mainframe computer are being modified to work on micros. Some of the programs being used at a number of firms are reviewed. The examples cover project scheduling, cost control, purchasing, critical path method scheduling, and computer graphics.

932   Portables Pay Off

Peter J. Tarkoy, Member, ASCE, (Geotechnical Consultant, Sherborn, Mass.)

Civil Engineering--ASCE, Vol. 55, No. 10, October, 1985, pp. 62-65

A portable microcomputer with appropriate software, utilities and copies of job related records can establish a field office for almost any project. Examples of applications for the onsite microcomputer include a revised engineering analysis of site conditions, analysis of equipment performance and downtime, project schedule control, inspection report writing, data collection and transmission to the home office mainframe computer, and electronic mail communications.

933   Rating Pavements by Computer

Donald P. Curphey, Member, ASCE, (Geotechnical and Pavement Engr., Earth Technology Corp., 3777 Long Beach Blvd., Long Beach, CA 90807), **Donald K. Fronek**, (Chmn., Dept. of Electrical Engrg., Louisiana Tech Univ., Ruston, LA 71272) and **John H. Wilson**, (Data Acquisition and Process Control Engrg., Earth Technology Corp., 3777 Long Beach Blvd., Long Beach, CA 90807)

Civil Engineering--ASCE, Vol. 55, No. 10, October, 1985, pp. 66-69

Pavement rating is a necessary and labor-intensive job and one with a high turnover rate of personnel. Also, rating scores are often inconsistent from day to day, or between different raters. A system is described that is able to solve both of these problems associated with pavement rating. The system and its operators fit into a van. Three electronic sensors and associated optical equipment scan a strip of highway across one pavement lane. The system can automatically recognize different types of pavement distress. The concept has been shown to work in the laboratory.

934   Accounting for CADD Center Costs

Francis J. Connell, Fellow, ASCE, (Pres., McFarland-Johnson Engineers, Inc., Binghamton, NY)

Civil Engineering--ASCE, Vol. 55, No. 10, October, 1985, pp. 70-71

The purpose of developing a CADD cost center is to segregate all costs related to the CADD operation so that you may receive fair compensation for CADD usage and to ensure fair charges to the client. Another basic reason is to keep "percent overhead" from exceeding artificial limitations imposed by various clients. When setting up a cost center, establish and maintain an adequate auditable accounting system to develop schedule rates. The accounting system should allow separate accounting of all identifiable CADD costs, provide for allocable costs as they relate to CADD and/or as they properly benefit CADD operations and one that will allow for pooling of costs to develop consistent rates for the various items on the rate schedule. The same system of allocating costs must be applied to all projects and the costs allocated to the cost center must be based on the general cost principles contained in the federal acquisition regulations and the particular clients' cost reimbursement policies.

935   **Data Base in Your City's Future?**

K. A. Godfrey, Jr., Member, ASCE, (Sr. Editor, Civil Engineering—ASCE, 345 E. 47th St., New York, NY 10017)

Civil Engineering--ASCE, Vol. 55, No. 10, October, 1985, pp. 72-73

The city of Bellevue, Wash. computerized its public works maps in 1979 because Bellevue was using the county's maps, and their updating wasn't keeping up with suburban growth. Now this geographic database is being used as a framework that could grow to become the country's first truly integrated municipal infrastructure database. In addition to map data, Bellevue's computers are storing data on the infrastructure of the city's water and sewer systems, stormwater drainage utilities, land use planning regulations and building permits pending, street pavement, and maintenance needs and projects for all public works departments.

936   **Slashing Tunnel Costs**

Reinhard Gnilsen, (Chf. Design Engr., Law/Geoconsult International, Inc., 1140 Hammand Drive NE, Suite 5250, Building E, Atlanta, GA 30328)

Civil Engineering--ASCE, Vol. 55, No. 11, November, 1985, pp. 38-41

The second U.S. application of the New Austrian Tunneling Method at Washington D.C.'s Metro system is described. The innovative method has been used overseas, but American engineers have been conservative about adopting the system here. At the Wheaton Station and tunnel, the contractor proposed the method as a value engineering change and saved several million dollars, mainly due to less excavation and by eliminating the need for a separate steel-reinforced concrete tunnel liner. The method takes both geotechnical and structural aspects of the ground formation into account. The excavation is considered as both a load exerting and load carrying ring. After blasting, immediate support elements control ground movement and take advantage of the ground's self-carrying capacity.

937   **Tunneling a Bridge**

Rita Robison, (Associate Ed., Civil Engineering—ASCE, New York, NY 10017)

Civil Engineering--ASCE, Vol. 55, No. 11, November, 1985, pp. 42-44

Linking an addition to the Georgia World Congress Center in Atlanta required a pedestrian tunnel through a railroad embankment 100 ft wide and 40 ft high, with no disruption of train service during construction. Nannis & Associates, Atlanta consulting engineers, designed the tunnel to be built in two stages and from the top down. In the first phase, half the embankment was excavated to the level where a steel bridge structure and tunnel ceiling could be built. Sheet piling driven between two of the three tracks, tie rods and continuous walters prevented damage to the existing structure. Abutment walls for the bridge are reinforced concrete caissons. After completion of the second phase in the same manner, the tunnel was excavated from below.

938   Advances in Short Span Steel Bridges

Geerhard Haaijer, Member, ASCE, (Vice Pres. & Dir. of Engrg., American Institute of Steel Construction, Chicago, Ill.)

Civil Engineering--ASCE, Vol. 55, No. 11, November, 1985, pp. 45-47

Applying recent research to the design of short span steel bridges can lower their cost and maintenance. Such research involves improved limit state criteria, more uniform lateral load distribution, wider spacing of beams and girders, prefabricated composite units, improved bearings and elimination of joints. The proposal to improve limit state criteria in the load factor design (LFD) method considers three levels of loading: service load, overload and maximum load. A proposed design method, called autostress design (ASD) permits a continuous span bridge to undergo small plastic deformations at a pier that will stabilize after a few cycles, then respond elastically to all subsequent loads not exceeding the overload.

939   Second-Guessing the Engineer

Kris R. Nielson, (Pres., The Nielsen-Wurster Group, Inc., Princeton, NJ) and **Patricia Galloway**, (Vice Pres., The Nielsen-Wurster Group, Inc., Princeton, NJ)

Civil Engineering--ASCE, Vol. 55, No. 11, November, 1985, pp. 48-49

The relationships between utilities and the bodies that regulate them have changed drastically in the past few years. No longer can any costs associated with the construction of new power plants be passed directly on to consumers. Public service commissions increasingly are scrutinizing construction project to make sure that they were done "prudently." Imprudent expenditures are increasingly being kept from utilities' rate bases. Engineers are finding themselves involved, not only as those being second guessed, but occasionally, as the consultants who are doing the second-guessing.

940   Rail Planning—Texas Style

Corinne S. Bernstein, (Asst. News Editor, Civil Engineering—ASCE, New York, NY 10017)

Civil Engineering--ASCE, Vol. 55, No. 11, November, 1985, pp. 50-51

The Dallas Area Rapid Transit (DART) Authority has a 25 year plan to build a 147 mile rail network—all without federal funds. When DART hits its 2010 target date for completing the rail line, Dallas will have the nation's second largest mass transit network. Other cities are eyeing the Dallas transit authority's success to date and its plans to solve an age old Texas size traffic problem. The rail line will be a pre-metro system, or a hybrid between light and heavy rail. Some 318 cars will travel at an average of 35 mph and will be powered by an overhead catenary system.

941   Negotiating Contracts

O. C. Tirella, (Treasurer, CH2M Hill Inc., Denver, CO)

Civil Engineering--ASCE, Vol. 55, No. 11, November, 1985, pp. 52-54

Good contract negotiations require an understanding of psychology and the dynamics of group interactions. Communication skills necessary to resolving key issues include identifying non-verbal messages being sent by the other parties: confidence, readiness, self control, defensiveness, territorial dominance, frustration and acceptance. The first quality good negotiators should have is wide ranging knowledge about the regulations covering contract and subcontract negotiations, business law, accounting and pricing, and data about the client firm. A team composed of a principal negotiator, a technical representative and a contracts/financial person is recommended.

942 Accounting for Every Pipe

David W. Wright, Member, ASCE, (Vice Pres., Presnell Associates, Inc., 717 West Main Street, Suite 300, Louisville, KY 40202-2633) and **G. Stephen Ballard**, Member, ASCE, (Prin. Civ. Engr., Dept. of Utilities, 810 Union Street, Room 603, Norfolk, VA 23501)

Civil Engineering--ASCE, Vol. 55, No. 11, November, 1985, pp. 55-57

The Norfolk Sanitary Sewer Inventory and Appraisal System contains individual records for each of the 20,000 pipe segments in the Norfolk sewer system. The data, assembled from construction drawings, utility maps and field investigations, was transferred to planimetric maps. These maps became the master data record for the sewer system, from which all data was transferred to input forms and entered into the computer database. The inventory system allows sewer information to be analyzed, provides for testing hypotheses, provides for capital improvements programming and budgeting.

943 Predicting Sinkhole Collapse

Byron E. Ruth, Member, ASCE, (Prof., Civ. Engrg. Dept., Univ. of Florida, Gainesville, FL 32611), **Thomas F. Beggs**, Assoc. Member, ASCE, (Engr., Soil and Material Engineers, Inc., 606 S. Military Trail, Deerfield Beach, FL 33442) and **Janet D. Degner**, (Asst. Engr., Remote Sensing Applications Lab, Dept. of Civ. Engrg., Univ. of Florida, Gainesville, FL 32611)

Civil Engineering--ASCE, Vol. 55, No. 11, November, 1985, pp. 58-60

Sinkholes erupt when a cavern or opening in the limestone rock below opens up to the surface. Florida's sinkholes have received most U.S. attention recently, but sinkhole geology is found also in Pennsylvania, Tennessee, Georgia, Alabama, Missouri, Virginia, Michigan and Kentucky. Elsewhere, the Mediterranean countries, South Africa and the Peoples Republic of China also have major sinkhole problems. This article focuses on Florida's sinkhole-prone geology, and conditions promoting their opening up to the surface. It describes how mapping of "lineaments," surface manifestations of bedrock fracture, can help predict where sinkholes may open up. Use of the technique is suggested before siting a major facility. One source of more information on sinkholes is given—the Florida Sinkhole Research Institute, at the Civil Engineering Dept., University of Central Florida, Orlando.

944 What Do Pavements Cost?

Leon Noel, (Chf., Federal Highway Administration Pavement Branch, FHWA, 400 Seventh St. NW, Wash., D.C. 20590)

Civil Engineering--ASCE, Vol. 55, No. 11, November, 1985, pp. 61-63

Life-cycle costing is a rational tool used to forestall problems with highway infrastructure. Currently this tool is still an imprecise one. The Federal Highway Administration has a system that considers an economic analysis as well as an engineering analysis of governing factors. Initial cost, maintenance costs, user costs and rehabilitation costs are among these factors. The principal barrier to accurate life-cycle cost analysis is the lack of knowledge about pavement performance and the varying costs of different maintenance strategies. Discount rates and their impact on such analyses are discussed; in general, low rates favor improvement with high initial cost, and high rates, short-term improvements. The time period for life-cycle analysis means at least thirty years for new pavements. Iowa's experience with life-cycle costing is further described.

945   Jointless Bridge Decks

Clellon L. Loveall, Member, ASCE, (Engrg. Dir., Tennessee Dept. of Highways, James K. Polk Building, Nashville, TN 37219)

Civil Engineering--ASCE, Vol. 55, No. 11, November, 1985, pp. 64-67

Deck joints in a bridge are the source of many problems over its lifetime. In time, chances are good the joint will leak, permitting water and salt to leak through. Often this leads to deterioration of the concrete pier cap and superstructure beneath. The solution of Tennessee DOT is to eliminate nearly all bridge deck expansion joints and superstructure expansion bearings. For bridges up to about 400 ft long with steel superstructures, and 800 ft long in concrete, this means no expansion joints even at the abutments (with some exceptions). For bridges longer than this, expansion joints are used but only at the abutments. Tennessee DOT reports almost no problems have resulted from using this approach. A few structural details are given to illustrate how Tennessee designs "integral abutments," and provides fixed connections of the superstructure to the piers.

946   Unearthing Mt. Baker Tunnel

Edgar B. Johnson, Fellow, ASCE, (Partner in Charge, HNTB, Seattle, WA), **Lee J. Holloway**, Member, ASCE, (Chf. Structural Engrg., HNTB, Seattle, WA) and **Georg Kjerbol**, Member, ASCE, (Senior Project Engineer, HNTB, Seattle, WA)

Civil Engineering--ASCE, Vol. 55, No. 12, December, 1985, pp. 36-39

A new tunnel under Mt. Baker in Seattle is part of major improvements to I-90 between Bellevue and Seattle's business district. Feasibility studies began in 1963, but it was not until 1977 that the concept of a single bored tunnel was accepted by WSDOT. Conventional full face tunneling was not suitable for the site, which had silty clays beneath a residential area. Instead, the stacked drift liner system was used, with 24 bores driven in a circular pattern and filled with concrete before excavation began on the 63 ft diameter tunnel. The liner was designed as a semi-flexible support system capable of adjusting to non-uniform external loading by deforming until equilibrium is reached. Structurally, it acts as a ring in compression. Because they are only 9 ft in diameter, the drifts were excavated with conventional equipment. Access pits 90 ft in diameter and 90 ft deep were converted to vertical retaining walls for the cut-and-cover end portions of the 1,500 ft tunnel.

947   Tampa Does It with Mirrors

K.A. Godfrey, Jr., Member, ASCE, (Sr. Editor, Civil Engineering—ASCE, 345 E. 47th St., New York, NY 10017)

Civil Engineering--ASCE, Vol. 55, No. 12, December, 1985, pp. 40-43

As one of the nation's fastest growing cities, Tampa needed far more public works dollars. But the traditional source, the real estate tax, can no longer be relied on so heavily, nor can federal monies. Tampa is a leader in finding new sources, particularly new user charges for transportation and drainage. Tampa expects to create a storm drainage utility in two or three years, which will charge land owners user fees based on runoff. As for transportation facilities, in 1980 the state legislature authorized its counties to levy gasoline taxes, making it one of few states to do so. Tampa public works people are also making ends meet by cutting costs, for example, by reaming and relining tuberculated water pipe rather than replacing it, and by constraining the rate of growth of the public works budget.

948   Slurry Walls Protect Harvard Square

Eldon L. Abbott, Member, ASCE, (Asst. Vice Pres., Parsons Brinckerhoff Quade & Douglas, Boston, MA 02116), **William H. Hansmire,** Member, ASCE, (Sr. Prof. Assoc., PBQD, Boston, MA 02116) and **Robert P. Rawnsley,** Assoc. Member, ASCE, (Staff Engineer, PBQD, Boston, MA 02116)

Civil Engineering--ASCE, Vol. 55, No. 12, December, 1985, pp. 44-47

Harvard Square Station, Cambridge, Mass., had to be rebuilt when Boston's Red Line Subway was extended further out to the suburbs. For the first time, tied back slurry walls built to protect adjacent historic buildings were incorporated into the permanent subway structure. The station structure was designed for the maximum anticipated total loads. Lateral forces are taken by the reinforced concrete walls with beam action between the roof and invert support points. Vertical loads are transmitted to the foundation soil or rock. Regroutable, multi-strand tiebacks were installed to design capacities ranging from 70 to 140 tons. A monitoring program confirmed design predictions that there would be no significant movement within the slurry walls and adjacent ground.

949   Pipe Laying Comes Out of the Trenches

Tom D. O'Rourke, Member, ASCE, (Assoc. Prof., 265 Hollister Hall, Cornell Univ., Ithaca, NY 14853), **E. W. Flaxman,** (Partner, Binnie & Partners, London, UK) and **Ian Cooper,** (Deputy Dir., Water Research Centre, Swindon, UK)

Civil Engineering--ASCE, Vol. 55, No. 12, December, 1985, pp. 48-51

Recent developments have shown that trenchless construction offers an effective and economically attractive alternative to conventional excavation. These findings are especially important for congested city areas, where open cuts will disturb adjacent facilities, traffic, and local business. In this paper various trenchless techniques are discussed with respect to pipeline diameter, construction length, suitable ground conditions, and steering accuracy. Remotely controlled microtunneling and on-line replacement are discussed in detail. Information is presented on excavation disturbance, comparative costs, and international applications and support of trenchless methods.

950   Engineering with Fabric

Rita Robison, (Assoc. Ed., Civil Engineering—ASCE, New York, NY 10017)

Civil Engineering--ASCE, Vol. 55, No. 12, December, 1985, pp. 52-55

Geotextiles are lending their strength to construction projects where conventional methods won't work. At Washington National Airport, an embankment was built into the Potomac River by first stabilizing the soft silty bottom with a fabric mat 600 × 700 ft in area that was installed in one piece. Less than a year after completion, the embankment, a runway extension, prevented an airplane from skidding into the water. Geotextiles are being used in street rehabilitation to keep water from eroding the roadbase and retard cracking. The city of New Hope, Minn. is rebuilding all streets this way. In new construction, geotextiles permit road alignments over "unbuildable" soils. A peat bog in Wisconsin and red clay in Georgia are examples.

951   Beam Analogy for Vertical Curves

A. Abdul-Shafi, (Assoc. Prof., Dept. of Civ. Engrg., South Dakota State Univ., Brookings, SD 57007)

Civil Engineering--ASCE, Vol. 55, No. 12, December, 1985, pp. 56-57

The vertical curve is analyzed as a moment diagram for a uniformly loaded single beam with a concentrated moment at one end. This contributes some flexibility in approach, and enhances efficienty of solving the curve. It also renders analysis of vertical curves a familiar process to a wider circle of engineers, and makes it less likely to be forgotten, especially by those who encounter these curves only occasionally.

952   San Francisco Outfall: The Champ?

G. J. Murphy, Fellow, ASCE, (Asst. Vice Pres., Parsons Brinckerhoff Quade & Douglas, 1625 Van Ness St., San Francisco, CA 94109) and **Y. Eisenberg**, Member, ASCE, (Professional Assoc., Parsons, Brinckerhoff Quade & Douglas, 1625 Van Ness St., San Francisco, CA 94109)

Civil Engineering--ASCE, Vol. 55, No. 12, December, 1985, pp. 58-61

The Southwest Ocean Outfall, with an overall length of 23,400 ft and capacity of 450 mgd, will be major element of the Clean Water Program of the City and County of San Francisco. Offshore, the outfall crosses one of the world's major active fault zones, the San Andreas. Construction started in 1981 and is scheduled to be completed by mid-1986. The shoreward 3,000 ft was built from a pile-supported trestle; offshore, the conduit sections were placed using a specially built barge. The most interesting and unusual earthquake provisions (including a joint with provision to accept shortening and elongation), other design provisions, and construction highlights (including the 17 month delay after 26 ft waves incapacitated the barge) are discussed, along with the Clean Water Program, one of whose prime goals is to largely restore San Francisco Bay to her earlier condition (today shrimp and commercial fishing have returned).

953   Tunnel Shotcreting—Strength in Fibers

Don Rose, Member, ASCE, (Consulting Engr., San Mateo, CA)

Civil Engineering--ASCE, Vol. 55, No. 12, December, 1985, pp. 62-63

Steel fiber reinforced shotcrete (SFRS) costs less, has a lower rate of rebound and allows modern excavating equipment to advance with fewer interruptions than conventional shotcrete. SFRS has become popular in the U.S. and Canada in the last five to seven years. Conventional concrete design, equipment and laboratory tests have heavily influenced shotcrete mix design. SFRS can be made by either the dry or wet mix methods. Adding microsilica, also known as silica fume, to the reinforced shotcrete mix enhances the SFRS properties.

954   Ethics in the Field

Elizabeth M. Endy, (Inst., Cabrini College, Radnor, PA 19087) and **P. Aarne Vesilind**, (Chmn., Civ. Engrg. Dept., Duke Univ., School of Engrg., Durham, NC 27706)

Civil Engineering--ASCE, Vol. 55, No. 12, December, 1985, pp. 64-66

Just as a child matures through various stages of moral development, the engineer too undergoes professional moral development described by psychiatrists. Development comes in six stages and three levels. These stages and levels of professional moral development are described and illustrated with examples.

955   **Advances in the Art of Testing Soils Under Cyclic Conditions**
Proceedings of a session sponsored by the Geotechnical
Engineering Division in conjunction with the ASCE Convention in Detroit, Mich., Oct. 24, 1985

Vijay Khosla, (editor), Member, ASCE, (Dir. of Engrg., Herron Consultants, Inc., 5555 Canal Rd., Cleveland, Ohio 44125)

New York: ASCE, 1985, 294pp.

The session on "Advances in the Art of Testing Soils under Cyclic Loading Conditions" was organized by the Soil Dynamics committee of the American Society of Civil Engineers in an endeavor to evaluate the state of practice governing testing equipment and methodology, and behavioral analysis of materials. The topics included centrifuge modeling of soils, torsional triaxial and direct shear devices and their use with hollow cylindrical soil specimens, applications of the constant volume cyclic simple shear device, monotonic and cyclic loading behavior for triaxial stress conditions, effects of irregular and multi-directional load application, evaluation of liquefaction flow failure during earthquakes, influence of testing techniques on soil properties in a cyclic triaxial apparatus, evaluation of shear wave velocity and shear modulus using cubic soil samples or bender elements, and mathematical modeling of nonlinear behavior in a resonant column test.

956   **Advances in Underground Pipeline Engineering**
Proceedings of the International Conference sponsored by
the Pipeline Division of the American Society of Civil
Engineers

Jey K. Jeyapalan, (editor), Member, ASCE, (Prof., Univ. of Wisconsin-Madison, Dept. of Civ. and Environmental Engrg., Madison, WI 53706)

New York: ASCE, 1985, 599pp.

The field of underground pipeline engineering has undergone a rapid growth during the last two decades. The analysis, design, construction, operation, and maintenance of underground pipelines have seen several developments, and new materials have entered the underground pipeline industry. The papers included in this book cover the following topics: Metal pipe-soil interaction; installation of clay pipes; concrete pipes for sewers and watermains; new design concepts for metal pipes; design and long term performance of thermoplastic pipe materials; general considerations in pipeline design; design and performance of fiber reinforced plastic pipes; pipeline backfill design and installation; durability of rigid sewer pipes; and design and testing of concrete pipes.

957   **Automated People Movers: Engineering and Management in Major Activity Centers**
Proceedings of a Conference sponsored by the Urban
Transportation Division of the American Society of Civil
Engineers, Miami, Fla., Mar. 25-28, 1985

Edward S. Neumann, (editor), Member, ASCE, (Prof. of Civ. Engrg., West Virginia Univ., Morgantown, WV 26506) and Murthy V. A. Bondada, (editor), Member, ASCE, (Proj. Mgr., Gannett Fleming Corddry & Carpenter, P.O. Box 1963, Harrisburg, PA 17105)

New York: ASCE, 1985, 856pp.

The high levels of interest currently being shown by a number of cities and special land uses indicates that automated people movers (APM) have established a secure position on the spectrum of feasible passenger-carrying transportation technologies. Automated people movers include a wide range of technologies, beginning with simple cable powered systems, through the more complex rubber-tired systems, up to the highly sophisticated magnetically levitated systems. Another example is accelerating walkway systems, which are still in an early evolutionary stage. The papers included in this book collectively cover system concepts, capabilities and requirements, financing, design, construction, operation, and maintenance of APM.

958   **Avoiding Contract Disputes**
Proceedings of a Symposium sponsored by the Construction
Division of the American Society of Civil Engineers in
conjunction with the ASCE Convention in Detroit, Mich.,
Oct. 21-22, 1985

Thomas A. Poulin, (editor), Member, ASCE, (Construction Group Leader, U.S. Forest Service, 1720 Peachtree Rd., NW, Atlanta, GA 30309)

New York: ASCE, 1985, 157pp.

Contract disputes among parties involved in the construction project are very expensive in both time and dollars. Owners, designers, construction managers and contractors all share in the responsibility of improving their efforts to reduce disputes and ultimately claims. Methods successfully employed in reducing the risk of unnecessary claims include constructibility reviews, proper specifying with the "or equal" clause, use of standard specifications, addition of a claims specialist on the design team, effective communication, thorough documentation, improved quality of inspection, and the role of the engineer in the construction phase.

959   **Challenges to Civil Engineering Educators and Practitioners—Where Should We Be Going?**
Proceedings of the Conference

George K. Wadlin, Jr., (editor), Fellow, ASCE, (Administrator, Education Services, American Society of Civil Engineers, 345 East 47th St., New York, N.Y. 10017)

New York: ASCE, 1985, 564pp.

The papers included in this publication present an overview of the current state of civil engineering education and resolutions for future development in the field. Six key areas are covered: (1) Human resources, including both faculty and students, with discussions on women and minorities in the field; (2) financial resources for program development as well as physical facilities; (3) program models for communications, professionalism, technical areas, and special developments; (4) quality control for student recruitment, teaching methods and curricula; (5) application of microcomputers and other high technology in civil engineering education; and (6) education/practice interchange. These topics are addressed in the form of case reports, theoretical discussions, and reviews and summaries of surveys.

960   **Civil Engineering in the Arctic Offshore**
Proceedings of the Conference Arctic '85

F. Lawrence Bennett, (d, Fellow, ASCE, Dept. Head, Dept. of Engrg. Mgmt., Univ. of Alaska, Fairbanks, Alaska 99701) and **Jerry L. Machemehl,** (editor), Member, ASCE, (Sr. Staff Engr., Arco Oil & Gas Co., P.O. Box 2819, Dallas, Tex. 75221)

New York: ASCE, 1985, 1276pp.

1985 ASCE PUBLICATIONS

The 123 papers included in this publication describe the present state of practice, consider emerging concepts and requirements, define research and development needs and critique present educational programs regarding effectiveness in preparing and motivating civil engineers to assume leadership in developing the Arctic offshore. Topics include arctic structures and construction, artificial islands, ice, the arctic environment, exploration and site investigations, surveying, soil properties, and pipelines. These subjects are covered in theoretical studies and case reports. Special design requirements in the arctic environment are also covered.

961   Coastal Zone '85
Proceedings of the Fourth Symposium on Coastal and Ocean Management

Orville T. Magoon, (editor), Member, ASCE, (Chf. Branch, U.S. Army, South Pacific Engrg. Div., San Francisco, Calif. 94126), **Hugh Converse**, (editor), Member, ASCE, (Civ. Engr., Corps of Engrs., San Francisco, Calif. 94126), **Dallas Miner**, (editor), (U.S. National Ocean Service, NOAA), **Delores Clark**, (editor), (U.S. National Ocean Service, NOAA) and **L. Thomas Tobin**, (editor), (Seismic Safety Commission)

New York: ASCE, 1985, 2720pp. (2 vols.)

Coastal Zone '85 was a multidisciplinary conference for professionals, interested citizens, and decision-makers to exchange information and views. Coastal zone management and ocean resources issues were addressed in relation to use, protection, and development. Papers included in this publication discuss improved jurisdiction arrangements, conservation and design considerations, enforcement policies, investigtion and planning methods, data collection, and research efforts. The need for greater public understanding and involvement in ocean and freshwater issues is emphasized. Environmental aspects, natural coastal processes, and coastal and offshore structures and activities (e.g. offshore drilling, ocean dumping) are covered.

962   Composite and Mixed Construction
Proceedings of the U.S./Japan Joint Seminar

Charles W. Roeder, (editor), Member, ASCE, (Assoc. Prof. of Civ. Engrg., Univ. of Washington, Seattle, Wash.)

New York: ASCE, 1985, 345pp.

Steel and reinforced concrete are frequently combined in composite or mixed structural systems. This combination generally results isn greater economy and safety than could be achieved by either material alone. However, the combination of two dissimilar materials results in serious design problems, and it is important that structural designers and researchers keep up with the latest developments in this field. This special publication presents a comprehensive state of the art for composite construction in the United States and abroad. The articles should be of interest to practicing engineers, educators and researchers, since they describe recent research developments and new innovations in the design and construction of composite structures. The papers draw from the professional practice of the United States and numerous other countries.

963   Computer Applications in Water Resources
Proceedings of the Specialty Conference sponsored by the Water Resources Planning and Management Division

Harry C. Torno, (editor), Member, ASCE, (Sr. Staff Engr., EPA, Science Advisory Board, Washington D.C. 20460)

New York: ASCE, 1985, 1447pp.

The use of computers in all engineering disciplines has, in the past decade, increased to the point where virtually all engineers now use these powerful tools. This is especially true in the water resources field. The papers included in this book cover computer applications in planning, analysis, forecasting, modeling, operation, and management. These applications are discussed in relation to water resources, water supply, reservoirs, pollutant transport, groundwater flow, hydraulic networks, water utilities, water distribution systems, river basins, lakes, and flood control. Interactive graphics, computer-aided design, data base management, and expert systems are also covered.

964   **Concrete Face Rockfill Dams—Design, Construction, and Performance**
Proceedings of a Symposium sponsored by the Geotechnical Engrg. Div. of the American Society of Civil Engineers in conjunction with the ASCE Convention in Detroit, Mich., Oct. 21, 1985

J. Barry Cooke, (editor), Fellow, ASCE, (Owner, J. Barry Cooke, Inc., Holiday Plaza, Suite 400, San Rafael, CA 94903) and **James L. Sherard**, (editor), Fellow, ASCE, (Consulting Engr., 3483 Kurtz St., San Diego, CA 92110)

New York: ASCE, 1985, 664pp.

Concrete face rockfill dams are being used with greater frequency and to greater heights in the last decade. The developments leading to this progress are many and are addressed in the papers included in this book. Major features contributing to progress are the development and use of the smooth drum vibratory roller and design improvements in the cut-off to the foundation and concrete face slabs and joints. Papers on the performance of the existing modern dams, rockfill zoning, dam construction, and seismic analysis, and dams currently under design are included. Details are presented on dams constructed of gravel and low compressive strength rock and dams on poor foundations, as well as on dams of high compressive strength rockfill on excellent non-erodible foundations.

965   **Connection Flexibility and Steel Frames**
Proceedings of a session sponsored by the Structural Div. of the American Society of Civil Engineers in conjunction with the ASCE Convention in Detroit, Mich., Oct. 24, 1985

Wai-Fah Chen, (editor), Member, ASCE, (Prof. and Head, Dept. of Structural Engrg., School of Civ. Engrg., Purdue Univ., West Lafayette, IN 47907)

New York: ASCE, 1985, 128pp.

The papers included in this book introduce and present methodologies for the analysis and design of frames which take into account the effect of joint flexibility on the response characteristics of the frames. The behavior of such frames under monotonic and cyclic loadings is also discussed. The word 'joint' used in the present context refers to both the connection and panel zone as it has been demonstrated analytically and experimentally that both connection and panel zone deformations have a significant influence on frame behavior. The papers are divided into five categories: Design considerations; behavior of connections; elastic buckling behavior of flexibly-connected frames; strength and flexibility of flexibly-jointed frames; and joint behavior under cyclic loadings.

966   Construction Cost Control
                    (Manual and Report No. 65)

Task Committee on Revision of Construction Cost Control Manual of the Construction Division of the American Society of Civil Engineers, Gerard J. Carty, chmn.

New York: ASCE, 1985, 114pp. (rev. ed.)

The second edition of *Construction Cost Control*, the first revision in thirty-three years, has been updated to provide engineers and contractors with a system of cost control which represents good current practice. Computer applications are given as well as the manual methods upon which the computer applicatons are based. A secondary purpose of this edition is to serve as a teaching guide. The manual is intended primarily to aid those in construction cost control; to round out their knowledge and increase their contribution to their employers. Many of the methods outlined are in common usage—indeed, they have been selected because they have stood the test of time. Thus it can serve as a review for experienced construction personnel. The principles and procedures given here can be advantageously integrated, in whole or in part, with existing systems. This manual is not a text on accounting, though the system outlined adheres to accepted accounting principles. Explanation of accounting processes is limited to that required for cost control.

967   **Construction QA/QC Systems that Work: Case Studies**
              Proceedings of a session sponsored by the Construction
              Division of the American Society of Civil Engineers in
              Conjunction with the ASCE Convention in Denver, Colo-
                           rado, May 1, 1985

George Stukhart, (editor), Member, ASCE, (Assoc. Prof., Civ. Engrg. Dept., Texas A&M Univ., College Station, Tex. 77843)

New York: ASCE, 1985, 48pp.

Four case studies discuss the issue of responsibility for quality assurance and quality control by describing actual situations where quality work is being done in the field. The case studies include: (1) An evaluation of a quality management program used by the U.S. Army Corps of Engineers; (2) the role of the quality assurance manager on a large commercial project; (3) the construction management organization of the Dallas North Tollway Extension project; and (4) a management training program designed to enhance attitudes about quality in individual members of an organization. A commentary on quality in the constructed project is also given.

968   **Construction Research Applied to Practice**
              Proceedings of a session sponsored by the Construction
              Research Council of the Technical Council on Research of
              the American Society of Civil Engineers in Detroit, Mich.,
                           Oct. 23, 1985

C. William Ibbs, Jr., (editor), Assoc. Member, ASCE, (Asst. Prof., Univ. of Illinois, 2149 Civ. Engrg. Bldg., 208 North Romine St., Urbana, IL 61801)

New York: ASCE, 1985, 95pp.

The five papers compiled in these proceedings of the ASCE technical session "Construction Research Applied to Practice" represent important contributions to the construction industry. Four of the papers deal with various forms of computer simulation, reflecting the current research activity in this profession. The first article reports on research into micro-computer-based simulation of construction site operations, such as equipment movement or labor activity. Videotape data acquisition is discussed in some detail. Another paper describes an implementation of a resource allocation procedure predicated on user assigned criteria. These criteria include user-defined resource profiles and such profiles as traded-off against schedule slippage. Simulation of corporate activities such as financial, organizational, and personnel decisions is the subject of a third paper. This particular simulation exercise has been shown to be valuable training exercise for project management personnel. The fourth article is an application of probabilistic risk analysis and dynamic programming concepts to tunneling. The last presentation focuses on research into lightweight concrete designs that have both considerable structural characteristics and superb insulative properties for building projects.

969   **Current Research and Research Needs in Deep Foundations**
A report of the Subcommittee on Research Needs of the Deep Foundations Committee of the Geotechnical Engineering Division, ASCE

Peter K. Taylor, (Stone & Webster Engrg. Corp., Boston, Mass. 02107)

New York: ASCE, 1985, 45pp.

This report summarizes the current range of research on deep foundations now being performed in the United States. The report is divided into two main sections: (1) Current research or related work in deep foundations; and (2) greatest perceived needs for future research. A substantial amount of research is being conducted in the area of deep foundations. Much of that work has been inspired by the need to better understand the behavior of offshore platforms used in the oil industry. There is economic motivation for this research, i.e., better understanding may result in more economical design with acceptable factors of safety. Consequently, the bulk of this work is being financed by the petroleum industry, rather than by federal agencies. Conversely, research on on-shore deep foundations is being financed in large part by the federal government. The implication is that there is no economic incentive for private industry to finance research in the on-shore aspects of deep foundations.

970   **Damage Mechanics and Continuum Modeling**
Proceedings of 2 sessions sponsored by the Engineering Mechanics Division of the American Society of Civil Engineers in conjunction with the ASCE Convention, Detroit, Mich., Oct. 22, 1985

Norris Stubbs, (editor), (Assoc. Prof., College of Architecture and Dept. of Civ. Engrg., Texas A&M Univ., College Station, TX 77843) and **Dusan Krajcinovic**, (editor), (Prof., CEMM Dept., Univ. of Illinois, Chicago, IL 60680)

New York: ASCE, 1985, 136pp.

Historically, the structural and mechanical engineers are conditioned to model a material in a purely phenomenological sense with little or no regard to underlying physical phenomenon on the microscale. Only after numerous attempts to model the behavior of essentially brittle solids using the plasticity theory proved to be inadequate the attention shifted to a formulation of a theory that will focus on the microcracking as a source of the material nonlinearity. This novel branch of the continuum mechanics, that became known as the continuum damage mechanics, attracted the attention of numerous investigators both in USA and Europe. Four papers on continuum damage theory and effective moduli illustrated some of the developments and achievements in this field. Five more papers on continuum modeling of discrete structures describe continuum modeling of periodic truss structures and discrete structures with geometric nonlinearities, the use of computer graphics in the analysis of large space structures, and damage detection in periodic structures.

971   **Design of Structures to Resist Nuclear Weapons Effects**
(Manual and Report No. 42)

Task Committee on Updating Manual 42 of the Committee on Dynamic Effects of the Structural Division of the American Society of Civil Engineers, M. S. Agbabian, chmn.

New York: ASCE, 1985, 321pp.

This manual provides guidance to engineers engaged in designing facilities intended to resist nuclear weapons effects. Although emphasis is placed on blast-resistant design, other effects are treated in some detail. Specific topics addressed include: airblast, cratering, and ground motion; facility requirements; blast and shock loading on structures; behavior of structural elements; choice of structural systems and design procedures; dynamic analysis; facility interface elements and shock isolation systems; design adequacy under other weapons effects; and slanting, evaluation and upgrading of existing structures for nuclear effects shelter.

972   **Development and Management Aspects of Irrigation and Drainage Systems**
Proceedings of the Specialty Conference sponsored by the Irrigation and Drainage Division of the American Society of Civil Engineers

Conrad G. Keyes, Jr., (editor), Fellow, ASCE, (Prof., New Mexico State Univ., Las Cruces, N.M. 88003) and Tim J. Ward, (editor), Member, ASCE, (Prof., New Mexico State Univ., Las Cruces, N.M. 88003)

New York: ASCE, 1985, 518pp.

The 1985 Irrigation and Drainage Division Specialty Conference held in San Antonio, Texas on July 17-19, 1985 covered major topics on watershed management and water conveyance, drainage system maintenance, current research and research planning, transmission losses in channels, on-farm irrigation management, water rights interaction, institutional questions on surface water management, agricultural runoff effects on water quality in lakes and reservoirs, management of irrigation and drainage projects, weather modification in the south, effects of agricultural drainage on surface water quality, drainage system maintenance, project formulation and developments, ground water management problems and solutions in the southwest, and irrigation water requirements.

973   **Developments in New and Existing Materials**
Proceedings of a session sponsored by the Materials Engineering Division of the American Society of Civil Engineers in conjunction with the ASCE Convention in Detroit, Mich., Oct. 23, 1985

Jack S. Haston, (editor), Member, ASCE, (Branch Mgr., Trinity Engrg. Testing Corp., P.O. Box 223571, Dallas, TX 75222)

New York: ASCE, 1985, 75pp.

Materials engineering has emerged as an identifiable professional practice within civil engineering in the past few years. Although materials engineering is not new, the concept that this practice is enough specialized to merit individual attention as a sub-discipline of civil engineering has only recently been recognized. ASCE's Technical Activities Committee formed the Materials Engineering (ME) Division in July, 1984. The papers contained in these Proceedings are the first to be presented at a session sponsored by this new ME Division. Their topic is "Developments in New and Existing Materials." A wide variety of subjects are addressed including polymer impregnation, fly ash and clay stabilization, tensile fracture and fatigue of cement stabilized soil, and rigid pavement rehabilitation.

974   Dewatering--Avoiding Its Unwanted Side Effects

J. Patrick Powers, (editor), Member, ASCE, (Vice Pres., Moretrench American Corp., Rockaway, N.J.)

New York: ASCE, 1985, 86pp.

The undesirable side effects that sometimes occur when dewatering for construction or mining are presented, including: ground settlement due to dewatering; deterioration of timber piling; depletion of ground water supplies; salt water intrusion; expansion of contaminant plumes; release of polluted ground water into the environment; damage to wetlands; sinkholes. Conditions where such side effects may occur are described. Methods are given for preventing side effects by restricting the influence of dewatering with ground water cutoffs and artificial recharge. Legal and contractual implictions are discussed.

975   **Dredging and Dredged Material Disposal**
Proceedings of The Conference Dredging '84

Raymond L. Montgomery, (editor), Member, ASCE, (Special Asst., Environmental Engrg. Div., USAE Waterways, Experiment Station, Vicksburg, MS 39180) and **Jamie W. Leach**, (editor), (Technical Editor, USAE Waterways Experiment Station, Vicksburg, MS 39180)

New York: ASCE, 1984, 1137pp.

Waterborne commerce is a major factor in the economy of many countries. Because waterborne commerce depends on the ability of nations to provide adequate navigation channels to transport cargo, dredging to develop and maintain waterways, ports, and harbors is an important engineering activity. In the United States, the federal government is responsible for improving and maintaining the navigation channels of the nation's ports, harbors, and inland waterways. This responsibility presently involves approximately 25,000 miles of federal channels and over 1,000 harbors. These channels serve 130 of the nation's 150 largest cities and are used to transport over one-fourth of the nation's ton-miles of domestic cargo. The traffic on U.S. waterways has been increasing at a compound rate of more than 5 percent per year. Some predictions suggest that the volume of this traffic, such as grain, coal, ores, chemicals, fuels, and construction materials, will increase from four to six times in the next 50 years. Approximately 400 million cubic yards of dredged material are dredged from the nation's channels and harbors each year to meet the needs of waterborne commerce. The average total expenditure for dredging over the past three years approaches $550 million. It is clear that dredging is a major engineering activity supporting the economic growth of the United States and many other countries. Specific topics covered in the papers presented in these proceedings include dredging equipment, management, economics, regulation and environmental impacts. Case reports in special dredging projects are presented. Planning, design, monitoring, and special techniques are also considered.

976   **Drilled Piers and Caissons II**

Clyde N. Baker, Jr., (editor), Fellow, ASCE, (Sr. Prin. Engr., STS Consultants, Ltd., 111 Pfingsten Rd., Northbrook, Ill. 60062)

New York: ASCE, 1985, 160pp.

Drilled piers or caissons, also called drilled shafts or bored piles, have long been a very cost effective foundation solution for many soil profile and load situations. A distinct advantage over pile foundations is the ability of a single unit to support an entire column load. Improved techniques for construction under slurry have greatly broadened the range of soil profiles where drilled pier construction is practical. Potential disadvantages can include the impracticality and high cost of load testing due to the much greater loads involved compared to piles and the difficulty of confirming the integrity of concrete placed under water or under slurry. This special technical publication on drilled piers and caissons addresses construction under slurry non-destructive integrity evaluation, load testing and geotechnical behavior under loads.

977   **Energy Forecasting**
Proceedings of a session sponsored by the Energy Division
of the American Society of Civil Engineers in conjunction
with the ASCE Convention in Detroit, Mich., Oct. 24, 1985

Terry H. Morlan, (editor), (Mgr., Demand Forecasting Northwest Power Planning Council, 850 S.W. Broadway, Suite 1100, Portland, OR 97205)

New York: ASCE, 1985, 63pp.

When the traditional stability of growth in demand for electricity was disrupted in the mid-1970's, forecasters and planners were forced to reconsider their existing procedures. Since that time utility forecasting and planning have undergone significant change. Papers in this volume of proceedings examine several aspects of forecasting the demand for electricity. Many alternative approaches are available for forecasting the demand for electricity. The advantages and disadvantages of specific methods are discussed, including the choice of forecasting methods most appropriate for different forecasting problems. The papers also address the changing role of demand forecasts in resource planning. Uncertainty in forecasts have led to changes in planning practices that seek to minimize risk and control demand growth and patterns.

978   **The Engineering Aesthetics of Tall Buildings**
Proceedings of the Session Sponsored by the Structural
Division of the American Society of Civil Engineers in
Conjunction with the ASCE Convention, Denver, Colorado,
May 2, 1985

Joseph G. Burns, (editor), Assoc. Member, ASCE, (Assoc. and Proj. Structural Engr., Skidmore, Owings & Merrill, Chicago, Ill.)

New York: ASCE, 1985, 64pp.

The Proceedings of a session on the Aesthetics and Design of High-Rise Buildings at the ASCE Convention in Denver is included in its entirety. The session was chaired by Lynn Beedle and papers presented on this subject by Joseph Colaco; Myron Goldsmith; Jeppe Larsen and T. Y. Lin; and Charles Thornton. The engineering aesthetics of tall buildings is discussed in relation to various building materials, building form, individual members, joints and connections. The effects of scale were also discussed with respect to the appropriateness and efficiency of various structural systems. High rise buildings' social context is explored, as well as its function in the urban environment. The importance of collaboration between architects and engineers is shown to be an important requirement in the creation of buildings with a strong engineering aesthetic; examples are given of such buildings. The purpose of this session and these proceedings is to encourage the active participation of engineers in the design of large scale buildings, to include the search for new building systems and their appropriate aesthetic expression.

979   **Engineering Surveying Manual**
(Manual and Report No. 64)

Committee on Engineering Surveying of the Surveying Engineering Division of the American Society of Civil Engineers

New York: ASCE, 1985, 262pp.

This manual is directed at the engineering manager who is responsible for planning and executing engineering projects which require the assistance of specialists in engineering surveying. The manual emphasizes the planning, execution, and control of engineering surveying work rather than the details of the techniques involved. There are three essential divisions in the manual. The first part deals with topics requiring attention in the proper direction of most engineering surveys, including project planning; coordinate systems and computer applications; survey error analysis and adjustment; surveying instruments; survey specifications; and safety. The second part deals with engineering surveys for specific types of projects, such as roads and streets; high speed surface transportation facilities; airports; tunnels; dams, levees, and channel rectifictions; high-rise structures; and cross-country utility lines. The third part covers special types of engineering surveys or with special problems relating to engineering surveying, such as hydrographic surveys, mining survey, optical tooling, structural and terrain movements, environmental and resource surveys, and extraterrestrial surveying and mapping.

980   **Environmental Engineering**
Proceedings of the 1985 Specialty Conference sponsored by
the Environmental Engineering Division, ASCE

James C. O'Shaughnessy, (editor), Member, ASCE, (Dept. of Civ. Engrg., Northeastern Univ., Boston, Mass.)

New York: ASCE, 1985, 1153pp.

The purpose of the annual Conference on environmental engineering is to pursue scientific knowledge and promote sound engineering judgement and practice in the solution of environmental engineering problems. The program is arranged so as to include all areas of environmental engineering as defined by the Environmental Engineering Division. The four major technical committees (Air, Noise and Radiation Management; Solid Waste Management; Water Pollution Management; and Water Supply and Resource Management) aided in defining the current technical areas of interest to the environmental engineering profession. A major theme in this year's conference is the impact that toxic and other hazardous materials have on the environment. The topics covered in the *Proceedings* include; domestic and industrial wastewater treatment; water quality modeling; groundwater contamination; transport and treatment; sludge dewatering and management; air quality problems; environmental assessments; environmental auditing; infrastructure management; wastewater collection systems; low-level radioactive waste control; energy recovery systems; landfill controls; water treatment technologies; and small system management.

981   **Federal Policies in Water Resources Planning**

Task Committee on Federal Policies in Water Resources Planning of the Management Resources Planning and Management Division of the American Society of Civil Engineers, Robert L. Smith, chmn.

New York: ASCE, 1985, 65pp.

The adequacy of current federal policies and institutional arrangements for water resources planning is assessed. Conclusions and recommendations are presented in five key areas: (1) Water management responsibilities and the federal system and intergovernmental relations; (2) economic and financial issues; (3) planning and the regulatory responsibility; (4) the Water Resources Council Principles and Guidelines; (5) objectives in water resources management. The study was supplemented by an opinion survey on these issues. The results of the survey are included.

982   **Financing and Charges for Wastewater Systems**
A Special Publication

WPCF Task Force on Financing & Charges, Charles W. Keller, chmn.

Chicago: APWA; New York: ASCE; Wash., D.C.; WPCF, 1984, 139pp. (2nd ed.)

A general overview of the current practices and procedures that should be considered for financing and charges for wastewater collection and treatment systems is presented. The publication also serves as a guide for wastewater utility operating personnel, including municipal officials, engineers, financial staffs, city managers, and analysts. Various ways of allocating costs and developing rates and charges that reasonably and equitably reflect the cost of service are illustrated. The complexity of the interrelated considerations involved in developing wastewater system cost allocation and rates for services is stressed. Chapters include: (1) Institutional alternatives; (2) financial management and accounting; (3) financial operating and capital costs; (4) revenue requirements; (5) allocation of costs of service to cost-causative components; (6) determination and distribution of the costs of service to customer classes based on cost causation; and (7) development and design of a schedule of charges.

983   **Foundations in Permafrost and Seasonal Frost**
Proceedings of a session sponsored by the Technical Council
on Cold Regions Engineering of the American Society of
Civil Engineers in conjunction with the ASCE Convention in
Denver, Colorado, April 29, 1985

Albert Wuori, (editor) and **Francis H. Sayles**, (editor), (Research Civ. Engr., Cold Regions Research & Engrg. Lab., Corps of Engineers, Hanover, N.H. 03755-1290)

New York: ASCE, 1985, 68pp.

The papers in this volume cover the following subject areas: A method that utilizes a decision matrix diagram for selecting the most suitable and economical foundations for construction in the arctic; long-term creep tests on piles in frozen soil which show that primary or attenuating creep is predominant in most cases; the thermal effects of snow accumulations on the design of piling when the snow depth is deeper than normal as the result of drifting caused by adjacent facilities; and the creep settlement of large scale strip footing tests where measured settlements are compared with computed values using different analytical methods and constitution equations.

984   **Fracture Problems in the Transportation Industry**
Proceedings of a session sponsored by the Engineering
Mechanics Div. of the American Society of Civil Engineers
in conjunction with the ASCE Convention in Detroit, Mich.,
Oct. 23, 1985

Pin Tong, (editor), (Chf., Structures and Dynamics Div., U.S. DOT Transportation Systems Ctr., Cambridge, Mass. 02142) and **Oscar Orringer**, (editor), (Mech. Engr., Structures and Dynmics Div., U.S. DOT Transportation Systems Ctr., Cambridge, Mass. 02142)

New York: ASCE, 1985, 146pp.

Six cases of damage tolerance assessment are presented, involving structures or vehicles in the U.S. transportation industry. Damage tolerance means arranging design, maintenance, and inspection practices to assure that growing fatigue cracks or cracks caused by foreign-object damage can be detected and repaired before they cause structural failures. Damage tolerance assessment involves the practical application of fracture mechanics principles to complex structures in complex service environments. Four of the six accounts involve specific assessments of service failures; the remaining two describe continuing efforts to embody the damage tolerance philosophy in design practices. The authors of each case study are principal engineers involved in the work reported.

985   **Freezing and Thawing of Soil-Water Systems**
A State of the Practice Report prepared by the Technical
Council on Cold Regions Engineering of the American
Society of Civil Engineers

Duwayne M. Anderson, (editor), Affiliate Member, ASCE, (Assoc. Provost for Research, Texas A&M Univ., College Station, TX 77843) and **P. J. Williams**, (editor), (Prof. and Dir., Geotechnical Science Laboratories, Carleton Univ., Colonel By Drive, Ottawa, Canada K1S 5B6)

New York: ASCE, 1985, 103pp.

This book presents a state-of-the-practice review on freezing and thawing of soil-water systems through ten interrelated papers on this subject. The papers review the following topics: Thawing of frozen clays; modeling and applications of soil freezing and thawing; partial verification of a thaw settlement model; hydraulic properties of selected soils; a continuum approach to frost heave modeling; a model for dielectric constants of frozen soils; a numerical model of subsea permafrost; frost heave of full-depth asphalt concrete pavements; the origin of aggradational ice in permafrost; and experimental investigation of regelation flow with an ice sandwich permeater.

986   **Hurricane Alicia: One Year Later**
Proceedings of the Specialty Conference sponsored by the
Aerospace Division, Engineering Mechanics Division, and
the Structural Division of the American Society of Civil
Engineers

Ahsan Kareem, (editor), Member, ASCE, (Assoc. Prof. and Dir., Structural Aerodynamics and Ocean System Modeling Laboratory, Dept. of Civ. Engrg., Univ. of Houston, Houston, Tex. 77004)

New York: ASCE, 1985, 341pp.

In the pre-dawn hours of August 28, 2983, Hurricane Alicia ripped its way ashore at the western tip of Galveston Island and moved northward to the Greater Houston area. This storm caused more than a billon dollars in damage, representing possibly the second worst hurricane on record with respect to the dollar-value of damage. Both engineered and non-engineered structures suffered damage. This specialty conference was particularly devoted to the areas of meteorology, structural behavior, window glass and curtain walls, testing for hurricane-resistant design, codes and standards, and designing for the future. In addition, the Conference addressed needed modifications to codes regarding windborne debris, inspection procedures and design methods in coastal areas susceptible to hurricane events. The panel discussion addressed the needs of designing for the future.

987   **Hydraulics and Hydrology in the Small Computer Age**
Proceedings of the Specialty Conference sponsored by the
Hydraulics Division of the American Society of Civil Engineers, 2 vols.

William R. Waldrop, (editor)

New York: ASCE, 1985, 1561pp.

The theme of the 1985 ASCE Hydraulics Division Specialty Conference is °Hydraulics and Hydrology in the Small Computer Age.° The goal of this conference is to demonstrate innovative hydraulic applications of the newest generation of the computer technology, the small computer. This latest development in the computere evolution enables all segments of the engineering community to have access to computers. Approximately 250 papers are included in this Proceedings. These papers are arranged into 54 technical sessions on a variety of topics including computerized data acquisition systems in the lab and field, real time monitoring and forecasting, microcomputer software applications in hydraulics and hydrology, microcomputers in civil engineering education, coastal and wetland processes, ground water hydrology, surface water hydraulics and hydrology, and hydropower development.

988   **Hydropower: Recent Developments**
Proceedings of a session sponsored by the Energy Division
of the American Society of Civil Engineers in conjunction
with the ASCE Convention, Denver, Colorado, May 1, 1985

A. Zagars, (editor)

New York: ASCE, 1985, 99pp.

The Foothills Project has been under consideration by the Denver Board of Water Commissions for many years. It is one of the major steps in the continued efforts of the Board to provide adequate water service to their service zone of the Greater Metropolitan Area. There are three principal features in the project: (1) Foothills treatment plant; (2) Foothills tunnel and access road; (3) Strontia Springs diversion dam. The papers presented in this volume provide a record of the design and construction of the principal features of the dam. The geology and geotechnical aspects, hydraulic design, and concrete quality control and quality assurance of the dam are also covered.

989   **Innovative Powerhouse Designs**
Proceedings of a session sponsored by the Energy Division
of the American Society of Civil Engineers in conjunction
with the ASCE Convention in Detroit, Mich., Oct. 23, 1985

Glenn R. Meloy, (editor), (Hydroelectric Design Center, Dept. of the Army, North Pacific Div., Corps of Engrs., P.O. Box 2870, Portland, Ore. 97208-2870)

New York: ASCE, 1985, 48pp.

Since 1978, interest in small scale hydroelectric projects in the U.S. has jumped to a new high. Both new sites and modification of existing dams have received scrutiny by both private interests and Governmental agencies. New hydro sites and retrofit projects both face interesting challenges in meeting construction problems, environmental impacts, and flood protection problems. The three papers presented in this publication describe innovative solutions to the problems facing modern hydroelectric project designs. Two of the papers outline schemes to add power facilities to existing dams by using barge mounted hydro units. The first focuses on a hydroelectric barge that is floated in and out of a spillway opening on a seasonal basis. The next paper outlines a method using an intake barge, siphon penstock, and turbine-generator barge at an existing low-head dam. The final paper describes a powerhouse located at a new re-regulating dam that is combined with the spillway monoliths.

Discussion: **Geoffrey Spencer**, (Providence, RI) CE July '85, pp. 30.

990   **Innovative Strategies to Improve Urban Transportation Performance**

Arun Chatterjee, Member, ASCE, (Prof., Univ. of Tennessee, Knoxville, Tenn. 37996) and **Chris Hendrickson**, Assoc. Member, ASCE, (Prof., Dept. of Civ. Engrg., Carnegie-Mellon Univ., Pittsburgh, Pa. 15213)

New York: ASCE, 1985, 327pp.

The field of urban transportation is facing some severe challenges. Roadways continue to deteriorate, the financial condition of public transit systems is abysmal, and general productivity improvements in the field are lagging. None of these problems are entirely new. Nevertheless, the severity of these problems and their simultaneous occurrence have created a serious situation. Many studies have been performed, and many ideas have been advanced to cope with the situation, covering a wide variety of disciplines including engineering and management. Underlying this variety of issues and solutions certain common themes prevail: (1) There is an urgent need to improve the effectiveness of urban transportation modes simultaneously with a reduction of their costs. Innovation in this regard is essential; (2) there is no panacea for this difficult task, and a variety of strategies should be employed; and (3) many innovative strategies have been successfully implemented in different areas and under varying conditions. These successful strategies should be more widely adopted. These issues are addressed in the context of engineering strategies, transit performance evaluation, management strategies, and case reports.

991   **Issues in Dam Grouting**
Proceedings of the Session sponsored by the Geotechnical Engineering Division of the American Society of Civil Engineers in conjunction with the ASCE Convention, Denver, Colorado, April 30, 1985

Wallace Hayward Baker, (editor), Member, ASCE, (Pres. and Chf. Engr., GKN Hayward Baker, Inc., Odenton, Md.)

New York: ASCE, 1985, 174pp.

Although perhaps the largest number of individual grouting projects do not involve dams, by far the largest volume of grout used is for grouting of new or existing dams. This volume presents eleven papers related to various issues in dam grouting. Four papers focus on material properties related to grout penetrability and quality, particularly in relation to cement grout consistency. Three papers discuss the use of micro-computers to monitor and evaluate the dam grouting process. Three case histories discuss special dam grouting applications, including the use of asphalt grouts, micro-fine cement grouts, and chemical grouts to develop cut-off curtains below dams. Finally, the concept and preliminary test results of compaction grouting for earth dam foundation embankment densification is illustrated.

992   **Making Project Control Systems Work**
Proceedings of a session sponsored by the Construction Div.
of the American Society of Civil Engineers in conjunction
with the ASCE Convention in Detroit, Mich., Oct. 21, 1985

Paul M. Teicholz, (editor), Member, ASCE, (Mgr., Guy F. Atkinson Co., 10 W. Orange Ave., S. San Francisco, CA 94080)

New York: ASCE, 1985, 48pp.

This publication contains four papers that offer both theoretical and practical advice on how to make control systems more effective on design and construction projects. The use of automated cost and schedule control systems is widespread. Yet, the successful use of these systems is often hampered by inadequate understanding of how best to integrate these systems into project management. These papers emphasize the management requirements for control systems and how the computer can be used to achieve these goals.

993   **Managing Computers**
Proceedings of a symposium sponsored by the Engineering
Management Div. of the American Society of Civil Engi-
neers in conjunction with the ASCE Convention in Detroit,
Mich., Oct. 21-22, 1985

David C. Johnston, (editor), Member, ASCE, (Prin. Engr., Kimley-Hornand and Assocs., Inc., 9330 LBJ Freeway, Suite 790, Dallas, TX 75243)

New York: ASCE, 1985, 51pp.

The papers in this book address the impacts of computers on the organization, the individual, and project management. On the organizational level, topics include managing the introduction of CADD and of decentralized computer use. On the individual level we find forecasts on the impacts on the individual, his professional work, his computer work at home, and his role in guiding change. Software and techniques for computerized project management in the engineering firm are presented.

994   **Measurement and Use of Shear Wave Velocity for**
Evaluating Dynamic Soil Properties
Proceedings of a session sponsored by the Geotechnical
Engineering Division of the American Society of Civil
Engineers in conjunction with the ASCE Convention in
Denver, Colorado, May 1, 1985

Richard D. Woods, (editor), Member, ASCE, (Prof., Univ. of Michigan, Ann Arbor, Mich. 48109)

New York: ASCE, 1985, 84pp.

Geotechnical engineers have recognized in the past decade that shear wave velocity is a basic soil property and have begun to use it to characterize sites for many uses. Most notable has been the realization that the so called "dynamic modulus" is simply the low strain value of elastic modulus and this can be used in many "static" as well as "dynamic" applications. Because of its basic nature, it has also been recognized that shear wave velocity is an excellent diagnostic tool which can be used to evaluate the results of soil modification techniques. Three of the papers included in this publication deal with the measurement of shear wave velocity; one through its relationship with Rayleigh Waves, another by coupling with cone penetration sounding, and the third by combined field and lab measurements. All represent significant advances in the art of shear wave velocity measurements but for different reasons. Other papers take advantage of the long known connection between time and frequency domains and the powerful relationships provided by Fourier Duals. Geotechnical engineers are finally beginning to recognize and take advantage of some of the tools which the geophysicist has been using for several decades. The relationships between well known in-situ exploration techniques, cone penetration tests and down-hole seismic tests are presented. The combining of these two should result in data which is greater than the sum of the two methods separately. An application of shear wave velocity in the identification of the potential for liquefaction and in the evaluation of the stability of an earth embankment subject to earthquake shaking is described. An empirical correlation between shear wave velocity and depth of overburden is presented.

995　　　**Negotiation and Contract Management**
Proceedings of the Symposium Sponsored by the Engineering Management Division of the American Society of Civil Engineers in Conjunction with the ASCE Convention in Denver, Colorado, April 29-30, 1985

David C. Johnston, (editor)

New York: ASCE, 1985, 56pp.

A contract is the result of a negotiation leading to an agreement between two parties. A successful contract usually meets the needs of both parties. Satisfaction of needs is the reason for negotiations. A mutually satisfying contract usually means a better project which is easier to manage, has fewer conflicts, and no litigation. The papers included in this publication present guidelines for negotiating project agreements for people and things, the techniques, art and skills of negotiation, the development of a contract and the advantages of standard contract forms.

996　　　**New Analysis Techniques for Structural Masonry**
Proceedings of a session held in conjunction with Structures Congress '85, Chicago, Ill., Sept. 18, 1985

Subhash C. Anand, (editor), Member, ASCE, (Prof. of Civ. Engrg., Clemson Univ., Clemson, SC)

New York: ASCE, 1985, 136pp.

These seven papers on analysis of structural masonry that constitute this Proceedings were presented at a technical session at the ASCE Structural Engineering Congress held at Hyatt Regency Hotel, Chicago, Illinois, on September 16-18, 1985. The various aspects of masonry analysis that have been covered under these papers include the following: 1) Linear and nonlinear finite element modeling, as well as progressive failure of masonry subjected to in-plane loads where failure may occur only in joints or may involve both bricks and joints; 2) a macroscopic finite element model for masonry (without individual modeling of joints and bricks or blocks) which can handle in-plane and out-of-plane loading including both transverse shear effects and nonlinearity due to cracking; 3) cavity masonry walls subjected to out-of-plane loads; 4) a general failure criterion for the biaxial compressive strength of masonry as well as a design method based on the ultimate strength of masonry walls; 5) reduction in the buckling lengths and corresponding increase in the load carrying capacity of masonry walls due to restraining effects of floor slabs; 6) wall/floor-slab interaction in brick walls which yield load eccentricities and possible capacity reductions; and 7) a 2D-finite element model that can account for shear transfer across wythes in a composite masonry wall.

997 **Nineteenth Coastal Engineering Conference: Proceedings of the International Conference**

Billy L. Edge, (editor), Member, ASCE, (Cubit Engrg. Ltd., Charleston, S. C.)

New York: ASCE, 1985, 3358pp. (3 vols.)

Solving engineering problems associated with coastal and offshore areas is one of the most important environmental challenges we face. The many natural resources in these regions require careful management based on sound scientific and technical information. Over 200 papers are included in this publication providing in depth state-of-the-art reviews, and current research and practices in coastal engineering. The five main divisions of coastal engineering presented here are: (1) Theoretical and observed wave characteristics; (2) coastal processes and sediment transport; (3) coastal structures and related problems; (4) coastal, estuarine and environmental problems; and (5) ship motions.

998 **Noise and Vibration Measurement: Prediction and Mitigation**
Proceedings of a symposium sponsored by the Environmental Engineering Division of the American Society of Civil Engineers in conjunction with the ASCE Convention in Denver, Colorado, May 1-2, 1985

William A. Redl, (editor), Member, ASCE, (Assoc., Edwards & Kelcey, Inc., Livingston, N. J. 07039)

New York: ASCE, 1985, 152pp.

The papers included in this publication represent a comprehensive view of three major sources of noise and vibration in our environment: Construction equipment and operations; transportation system operations; and stationary source operations. The papers document both theoretical applications with respect to measurement, monitoring, prediction and mitigation, and case studies where certain techniques were employed to reduce noise and vibration for sensitive receptors.

999 **Optimization Issues in the Design and Control of Large Space Structures**
Proceedings of the Session Sponsored by the Structural Division of the American Society of Civil Engineers in Conjunction with the ASCE Convention in Denver, Colorado

Manohar P. Kamat, (editor), Member, ASCE, (Prof., College of Engrg., Virginia Polytechnic Inst. and State Univ., Blacksburg, Va. 24061)

New York: ASCE, 1985, 68pp.

The optimization and identification issues which arise in the design and control of large flexible space structures are reviewed. The first paper examines the use of several different types of parameters for parameter estimation based on actual test data. The approximation of multivariable control system performance by the redesign of structures for reduced dynamic response is also studied. A third study shows that by designing a structure based on an optimality criterion approach the dynamic characteristics of a closed-loop system or a large space structure can be improved. The final paper emphasizes the need to account for geometric and material nonlinearities in actively controlling highly flexible large space structures.

1000   Organization and Management of Public Transport Projects
Proceedings of the Specialty Conference

George V. Marks, (editor), Member, ASCE, (Project Mgr., Sverdrup & Parcel and Associates, Inc., 801 N. 11th St., St. Louis, Mo. 63101) and Bhagirath Lall, (editor), Member, ASCE, (Prof. of Civil Engrg., Portland State Univ., Portland, Oreg. 97207)

New York: ASCE, 1985, 332pp.

Transit development projects are undertaken to help satisfy escalating transportation needs in modern cities. The projects address an array of technical, social, and environmental considerations. Most become significant factors in shaping cities for the future. Each project tends to be unique and complex; some are time consuming and expensive. Several cities have recently undertaken major new transit systems. Other cities have completed major improvements to existing systems. Both pose major technical and management challenges for the engineering profession. There are unique management and organizational features to each transit project. Understanding the common and unique characteristics for certain projects can help identify successful management approaches. Representative management issues may involve methods to procure engineering services, construction contract packaging, or communication methods within the design team. The papers included in this publication highlight current and recently completed transit projects. Topics include budget and schedule control, decisions and approvals, internal communication, staffing, quality assurance, and project organization. The papers illustrate the need for sound organizational and management practices throughout all phases of project development. They also illustrate that different management methods can be adapted to the specifics of an individual project.

1001   Probabilistic Basis for Design Criteria in Reinforced Concrete

Reinforced Concrete Research Council, Edward Cohen, chmn.

New York: ASCE, 1985, 144pp.

This collection of papers presents the results of research on the variability of reinforced concrete members carried out at the University of Alberta, Edmonton, Canada, during the period 1977-1981. These studies were supported by the Reinforced Concrete Research Council, the National Research Council of Canada, and the U.S. National Bureau of Standards. Basic data on the variability of concrete, reinforcing bars, pre-stressing strands, and as-constructed dimensions were collected and formed the basis of analyses of the variability of concrete members loaded in flexure, shear, and axial load. The results of this research have been used to propose revisions to the strength reduction (ϕ) factors for a future edition of the American Concrete Institute's "Building Code Requirements for Reinforced Concrete (ACI 318)."

1002   Quality in the Constructed Project
Proceedings of the Workshop sponsored by the American
Society of Civil Engineers

Arthur J. Fox, Jr., (editor), (Engineering News-Record, New York, N.Y.) and Holly A. Cornell, (editor), Fellow, ASCE, (CH2M Hill, P.O. Box 428, Corvallis, Ore. 97339)

New York: ASCE, 1985, 208pp.

The purpose of the workshop summarized in this book was to develop an agenda for ASCE actions to promote quality in the constructed project. The papers, discussions and session reports address this subject in two main sections. The first section, quality in planning and design, covers meeting the owners objectives, achieving teamwork in design, dealing with external factors (codes, standards and regulations) affecting quality, and assuring quality in the design phase. The second section, quality in construction, covers organizing and managing for quality in construction, handling conflicts and disputes, dealing with constraints affecting construction quality, and performing quality construction. A summary report of the conference is included and recommended actions resulting from the workshop are listed.

1003   Rebuilding America: Infrastructure Rehabilitation
Proceedings of the Conference

Walter H. Kraft, (editor), Fellow, ASCE, (Vice Pres., Edwards and Kelcey, Inc., One South New York Ave., Atlantic City, N.J. 08404) and **Mary F. Brown**, (editor), (Edwards and Kelcey, Inc., One South New York Ave., Atlantic City, N.J. 08404)

New York: ASCE, 1985, 288pp.

The future of the nation demands a solution to the problem of our deteriorating infrastructure. Civil engineers are called on to act as guardians of our public works facilities that must be maintained, restored, and increased. The papers and discussions included in this publication define the problem of the nation's infrastructure, identify goals, assess priorities, and set standards. Methods of inventorying and inspecting the system are reviewed. Innovative technology and research related to infrastructure are covered. Case reports are presented that describe handling of special problems such as utilization of a facility during renewal. In addition to infrastructure issues and technology, program management, financing and marketing are regarded as important to the success of meeting the infrastructure challenge. Other essential aspects of handling the infrastructure problem are government policies and legislative and regulatory needs.

1004   Reducing Failures of Engineered Facilities
Proceedings of a Workshop sponsored by the National Science Foundation and the American Society of Civil Engineers, Clearwater Beach, Fla., Jan. 7-9, 1985

Committee on Forensic Engineering of the American Society of Civil Engineers, Narbey Khachaturian, chmn.

New York: ASCE, 1985, 112pp.

One of the most serious problems of modern professional practice is the collapse or malfunction of an engineered facility. The tragic consequences of this problem are all too familiar both to the practitioner and the general public. It is especially tragic when one realizes that in most cases the technical knowledge is available for the satisfactory design and construction of the facility. Civil engineers are very much concerned about this problem, although it is not a matter that can be resolved solely by them. Design of a modern facility involves an interdisciplinary effort. There are several teams with different technical expertise that participate in planning, design, and construction of a project to bring it to reality. Yet it is the responsibility of and a challenge for the civil engineering profession to do everything possible to minimize incidents of failure. This workshop was organized to provide a forum for discussing the problems associated with identification of the causes of failures and dissemination of failure information. The workshop proceedings include nine invited papers, and recommendations for action by the profession. These recommendations are in four categories representing the four working groups: failure investigation, dissemination of information, learning from failures and quality assurance/quality control.

1005   **Rehabilitation, Renovation, and Reconstruction of Buildings**
Proceedings of a workshop sponsored by the National
Science Foundation and the American Society of Civil
Engineers, New York, Feb. 14-15, 1985

Lynn S. Beedle, (editor), Honorary Member, ASCE, (Dir. of Lab., Lehigh Univ., Fritz Engrg. Lab., Bethlehem, PA)

New York: ASCE, 1985, 112pp.

This document constitutes the proceedings for a National Science Foundation funded workshop to discuss research needs pertaining to the buildings side of infrastructure. It includes discussion of research needs associated with assessing building condition, repair and rehabilitation techniques, implementation and project monitoring. Papers cover the owners' perspective, international experiences and specific discussions relating to the technological areas of steel and cast iron, concrete, masonry and cladding. In addition, consensus lists of research needs are discussed in the individual reports (steel/cast iron, concrete, masonry and cladding) which resulted from the working group discussions at the workshop. Two clear conclusions can be reached. They are, that much research is needed and that a great deal of information is already available but not being used.

1006   **Research Needs Related to the Nation's Infrastructure**
Proceedings of the Workshop sponsored by the National
Science Foundatio and the American Society of Civil
Engineers

Edward A. Kippel, (Mgr. of Technical Services, ASCE, New York, NY 10017)

New York: ASCE, 1984, 47pp.

The engineering community, which designs and constructs infrastructure facilities, must look for innovative ways to do a more efficient job. Although some research has focused on problems related to the design and construction of new facilities, the need to repair, retrofit and rehabilitate existing facilities must also be given new attention. This workshop itemizes the research needs related to revitalizing and maintaining the physical facilities basic to the efficient and effective support of our civilized society (infrastructure). Results are divided into three discipline-oriented categories—transportation, water supply and wastewater disposal, and highways, streets and bridges and the following two policy and issue related areas—incentives and institutional mechanisms for effective research; and methods of implementing the research results, and approaches to stimulating the research effort.

1007   **Richart Commemorative Lectures**
Proceedings of a session sponsored by the Geotechnical
Engineering Div. of the American Society of Civil Engineers
in conjunction with the ASCE Convention, Detroit, Mich.,
Oct. 23, 1985

Richard D. Woods, (editor), Member, ASCE, (Prof. of Civ. Engrg., Univ. of Michigan, 2322 G.G. Brown Lab., Ann Arbor, MI 48109)

New York: ASCE, 1985, 164pp.

The geotechnical specialty known as soil dynamics has developed during the past 25 years from a minor role to a mainstream discipline. Professor F. E. Richart has played a major role in the development of this specialty; he has progressively led the way to conquering the basic elements required to build a sound and rational approach to problems of soil and foundation dynamics. The papers in this book were presented in recognition of his accomplishments. Topics include the strength of soils in terms of effective stress, recent developments in resonant column testing, limits on dynamic measurements and instrumentation, and shallow seismic exploration in soil dynamics.

1008   **Rock Masses: Modeling of Underground Openings/Probability of Slope Failure/Fracture of Intact Rock**
Proceedings of the symposium sponsored by the Geotechnical Engineering Division of the American Society of Civil Engineers in Conjunction with the ASCE Convention in Denver, Colorado, April 29-30, 1985

C. H. Dowding, (editor), Member, ASCE, (Assoc. Prof., Dept. of Civ. Engrg., Northwestern Univ., Evanston, Ill. 60201)

New York: ASCE, 1985, 195pp.

While soil mechanics has been the dominant interest of Civil Engineers since Karl Terzaghi's recognition of the principle of effective stress, rock mechanics is becoming increasingly important. One only need look at the largest of constructed facilities for a demonstration of the role of rock mechanics in Civil Engineering. All of the world's tallest buildings and longest span bridges are founded on rock. Nuclear waste will be stored in rock at depths of 600m. The world's largest storm water collection facilities in Chicago and a similar system in Milwaukee are tunneled through limestone. Many of the last segments of the interstate system involve rock excavation as in Glenwood Canyon, Colorado. The world's largest existing or planned dams are (or will be) founded on rock. Protective structures for defensive nuclear weapons such as deep-based MX missiles will be built in rock. Rock mass behavior, which for engineering purposes can be described by its strength deformability and hydraulic conductivity, is controlled by the combined behavior of the continuous (intact) rock and the weaknesses (discontinuities). The discontinuities rupture the continuity of intact rock and are planes of smaller resistance and larger deformability and hydraulic conductivity. Since all rock masses are intersected by discontinuities (both as large as shear zones and as small as crystal-sized fractures) no volume of rock is without them. This volume represents an effort by the Geotechnical Engineering Division to collate the diverse rock mechanics in the division. The volume is divided into three sections: computer modeling of underground openings; probabilistic slope stability; and fracture of intact rock.

1009   **The Role of the Resident Engineer**
Proceedings of the Specialty Conference sponsored by the Committee on Contract Administration of the Construction Division of the American Society of Civil Engineers, Tampa, Fla., Apr. 15-17, 1985

Robert Del Re, (editor), Fellow, ASCE, (Construction Services Mgr., Camp Dresser & McKee, Inc., 2280 U.S. 19 North, Suite 202, Clearwater, Fla. 33575) and Harold V. McKittrick, (editor), Member, ASCE, (Pres., McKittrick & Assocs., Inc., Box 441, Oakton, VA 22125)

New York: ASCE, 1985, 103pp.

These proceedings include the papers presented at the specialty conference on resident engineers. The papers address different perceptions of the definition, qualifications, duties, responsibilities, authority and liability of this important project professional. Aspects of resident engineering include contract administration, project organization, cost control, and quality assurance. Ethics communications, and the resident engineer as a member of the construction team are also reviewed. The role of the resident engineer is discussed for projects in both the public and private sectors.

1010   **Roller Compacted Concrete**
Proceedings of the Symposium sponsored by the Colorado Section and Construction Division of the American Society of Civil Engineers in conjunction with the ASCE Convention in Denver, Colorado, May 1-2, 1985

Kenneth D. Hansen, (editor), Fellow, ASCE, (Engr., Portland Cement Assoc. Denver, Colo. 80224)

New York: ASCE, 1985, 148pp.

Roller compacted concrete has emerged recently as an economical material and rapid method of construction for use in gravity dams, overflow structures and heavy duty pavements. The economics of RCC for dam construction, together with the latest design concepts and criteria for providing functional, safe structures are presented. Methods for providing watertightness or controlling seepage in dams are discussed in depth. A combination of soil mechanics and concrete theory has helped develop simple standard tests using conventional testing laboratory equipment to aid with quality control. The methods used and lessons learned from the construction of two spillway structures, Middle Fork Dam in Colorado, several Japanese dams, plus the plans for Upper Stillwater Dam in Utah may influence heavy construction for many years. Similarly the design and construction experience from nearly a decade of experience with RCC heavy duty pavements in British Columbia has already set the stage for many similar projects in the United States. Ten papers divided into two sections, design of compacted concrete water control structures and pavements and construction and quality control for roller compacted concrete, address these subjects.

1011   **Seepage and Leakage from Dams and Impoundments**
Proceedings of a Symposium sponsored by the Geotechnical Engineering Division in conjunction with the ASCE National Convention, Denver, Colorado, May 5, 1985

Richard L. Volpe, (editor), Member, ASCE, (Geotechnical Consultant, Los Gatos, Calif. 95030) and **William E. Kelly,** (editor), Member, ASCE, (Prof. and Chmn., Dept. of Civ. Engrg., Univ. of Nebraska, Lincoln, Neb. 68588)

New York: ASCE, 1985, 324pp.

All major dams and impoundments are usually analyzed to determine the magnitude and location of seepage that can occur under, around or through the structure. More often than not, seepage protection elements within the dam or impoundment, in the form of geotextiles, natural filters and drains, are included in the design in order to control, collect and safely discharge the collected fluids. Darcy's Law has been successfully used by civil engineers for many decades for problems involving saturated flow, especially for the design of water retention dams. More recently, however, a new class of problems that deal with partially saturated leakage from waste impoundments has evolved. This latter class of problem can pose a serious environmental hazard if the leachate is toxic and it is not effectively collected and treated.

1012   **Seismic Experience Data—Nuclear and Other Plants**
Proceedings of a session sponsored by the Structural Div. of
the American Society of Civil Engineers in conjunction with
the ASCE Convention in Detroit, Mich., Oct. 24, 1985

Yogindra N. Anand, (editor), Member, ASCE, (Civ. Engr., Detroit Edison Co., 2000 2nd Ave.,
Detroit, MI 48226)

New York: ASCE, 1985, 91pp.

The Seismic Qualification Utilities Group (SQUG) was formed to explore the use of
earthquake experience data of equipment performance to resolve Nuclear Regulatory
Commission (NRC) initiated Unresolved Safety Issue (USI) A-46. The NRC staff endorses use of
SQUG data for final resolution of A-46. The parameters used in the facility level testing behave
differently in high level acceleration tests resulting in highly insignificant data. Their
identification and proper data adjustment are important in fragility tests. The failure loads
determined in failure analyses are significant in establishing fragility limits. The response for
multiaxis excitation is highly dependent on the test configuration of the fragility testing. A study
was conducted to correlate the seismic performance of typical power plant and industrial
structures with their design criteria. Some components in a nuclear plant are not required for
continued function. However, their failure may impair the safety features of other plant
components. Such items should be termed as Seismic II/I and should be qualified to maintain
structural integrity only. These topics are addressed in the six papers included in this book.

1013   **A Standard for the Measurement of Oxygen Transfer in
Clean Water**

Oxygen Transfer Standards Committee of The American Society of Civil Engineers

New York: ASCE, 1984, 46pp.

This standard was developed to measure the rate of oxygen transfer from diffused gas
and mechanical oxygenation devices to water. The standard is applicable to laboratory scale
oxygenation devices with water volumes of a few gallons as well as to full scale systems with water
volumes typical of those found in the activated sludge wastewater treatment process. It is
intended that this standard be used in the preparation of specifications for compliance testing and
in the development of performance information. This test method is based upon removal of
dissolved oxygen (DO) from the water volume by sodium sulfite followed by reoxygenation to
near the saturation level. The DO inventory of the water volume is monitored during the
reaeration period by measuring DO concentrations at several points selected so that each point
senses an equal tank volume. The method specifies a minimum number, distribution and range of
DO measurements at each point. The data obtained at each determination point are analyzed by
a simplified mass transfer model to estimate the apparent volumetric mass transfer coefficient
and the saturation concentration. Nonlinear regression is employed to fit the model to the DO
profile measured at each point during reoxygenation. Estimates are adjusted to standard
conditions and the standard oxygen transfer rate is obtained. A procedure based on the clean
water test results is recommended for estimation of oxygen transfer rates under process
conditions. Various components of power consumption are defined and methods for
measurement of gas rate and power consumption by the oxygenation device are given. Energy
efficiency of the oxygenation device is evaluated as the mass rate of oxygen transferred per unit
power consumed.

1014   Standard: Specifications for the Design and Construction
of Composite Slabs and Commentary on Specifications for
the Design and Construction of Composite Slabs

American Society of Civil Engineers

New York: ASCE, 1985, 72pp.

Composite slab construction is defined as a system comprising normal weight or
lightweight structural concrete placed permanently over cold-formed steel decking in which the
steel deck performs the dual role of acting as a form for the concrete during construction and as
positive reinforcement for the slab during service. These specifications present provisions
applicable to composite slabs. The provide design criteria and formulas for the engineer, test
criteria for the steel deck manufacturer, and recommendations for good construction practice.

1015   Stormwater Detention Outlet Control Structures
A Report of the Task Committee on the Design of Outlet
Control Structures of the Committee on Hydraulic Struc-
tures of the Hydraulics Division of the American Society of
Civil Engineers

Task Committee on Design of Outlet Control Structures, Barnabas R. Urbonas, chmn.

New York: ASCE, 1985, 36pp.

The Hydraulics Division's Task Committee for the Design of Stormwater Detention
Outlet Control Structures was formed to review and report on the state-of-the-art of stormwater
detention pond outlet controls. This study addresses the Committee's findings regarding
hydraulic function, water quality, public safety, maintenance, and aesthetic aspects of outlet
controls, and includes results of an extensive literature search and survey of stormwater
management professionals. A checklist of design considerations is given as a design aid to
practitioners in stormwater management. A discussion on the accuracy of stormwater detention
design and the differences between single and multi-frequency control provides the practitioner
with a basis for understanding the state of the art limits in accuracy of design. Also, the
interrelations between the appearance of outlet structures and the considerations related to
economics, public safety and hydraulic function are discussed. Research needed to advance the
state of the art is outlined.

1016   Structural Design, Cementitious Products, and Case
Histories
Proceedings of 3 sessions sponsored by the Structural Div.
and Michigan Section of the American Society of Civil
Engineers in conjunction with the ASCE Convention in
Detroit, Mich., Oct. 22-24, 1985

Yogindra N. Anand, (editor), Member, ASCE, (Civ. Engr., Detroit Edison Co., 2000 2nd Ave.,
Detroit, MI 48226)

New York: ASCE, 1985, 135pp.

The papers in this book represent deverse topics in structural engineering. In design of
beam-columns by the AISC interaction equations, it is demonstrated that an inadequacy exists
when the Stress Equation is more critical than the Stability Equation. Interaction diagrams are
presented for unsymmetrically reinforced concrete columns. The history behind the working

stress and the strength design of reinforced concrete members is analyzed leading to a rational formatting of design procedure. The behavior and use of advanced cementitious products, such as low cement high strength concrete, polymer concrete, fiber reinforced concrete, cement stabilized fly ash, are presented. A procedure is presented to detect an unanticipated source of vibration during a vibration investigation. The foundation design for the elevated guideway of the Detroit Downtown People Mover has large loads and stringent settlement criteria. An automated storage and retrieval system and an automated guided vehicle system are applied to a central maintenance facility.

1017   **Structural Plastics Selection Manual**
(Manual and Report No. 66)

Task Committee on Properties of Selected Plastics Systems of the Strucctural Plastics Research Council of the Technical Council on Research of The American Society of Civil Engineers, F. C. McCormick, chmn.

New York: ASCE, 1985, 598pp.

This manual presents information for the structural engineer in the selection of the proper material or combination of materials that will provide those properties (mechanical, physical, thermal, or whatever) upon which design assumptions and calculations are based. Essential differences between plastics and other structural materials are described along with the associated differences in material selection philosophies. Guidelines for materials selection procedures are given. Factors affecting properties during production and use are examined. A section on reliability and quality control is also included. Test procedures for physical and mechanical properties and environmental effects are described. This manual supplements the "Structural Plastics Design Manual" (ASCE manual and report number 63).

1018   **Structural Safety Studies**
Proceedings of the Symposium sponsored by the Structural
Division of the American Society of Civil Engineers in
Conjunction with the ASCE Convention in Denver, Colo-
rado, May 1-2, 1985

James T. P. Yao, (editor), R. Corotis, C. B. Brown, (editor) and F. Moses, (editor)

New York: ASCE, 1985, 212pp.

Structural engineers are most concerned with the safety aspect of structures. The primary objective of structural engineers is to produce structures which are satisfactory in meeting certain human and societal needs. Moreover, the complexities of many current projects and the increasing attention to existing infrastructures often require reevaluation of structural safety during the lifetime of a particular structure. In so doing, it is necessary to idealize complex natural phenomena. Such idealized models require frequent updating and improvement. The Symposium on Structural Safety Studies is sponsored by the ASCE Structural Division Administrative Committee on Structural Safety and Reliability. The objectives of this Symposium are to introduce relatively new concepts and methods to the structural engineering community, and motivate the audience and reader in knowing more about these subject matters. Topics covered in this Symposium include human errors, random vibration, structural damage, finite-element methods, structural optimization, evaluation of structures, earthquake and wind effects, elastic foundations, design loads, risk assessment, and bridge inspection.

1019   Thermal Design Considerations in Frozen Ground Engi-
neering
A State of the Practice Report prepared by the Technical
Council on Cold Regions Engineering of the American
Society of Civil Engineers

Thomas G. Krzewinski, (editor), Assoc. Member, ASCE, (Sr. Engr., Dames & Moore, 800 Cordova, Anchorage, AK 99501) and **Rupert G. Tart, Jr.**, (editor), Member, ASCE, (Tart Consultants, 6448 Village Parkway, Anchorage, AK 99504)

New York: ASCE, 1985, 285pp.

The design of engineering projects in frozen ground requires thermal design considerations in addition to standard geotechnical design. Factors which influence the thermal characteristics of a site include climatological data, microclimatic characteristics, local hydrology, soil properties and disturbance. This monograph presents ground temperature observations, procedures for temperature monitoring, analytical methods for ground thermal regime calculations and ground thermal properties. Active and passive techniques for ground temperature control and ground thawing methods are also presented followed by case histories of ground temperature effects.

1020   Transition in the Nuclear Industry
Proceedings of the symposium sponsored by the Construc-
tion and Energy Divisions of the American Society of Civil
Engineers in conjunction with the ASCE Convention in
Denver, Colo., April 29-30, 1985

James H. Olyniec, (editor), Member, ASCE, (Chf., Project Management Services, Tennessee Valley Authority, Bellefonte Nuclear Plant, Hollywood, Ala. 35752)

New York: ASCE, 1985, 244pp.

Significant changes have occurred within the utility industry within the past 10 years. Projected growth in electric power demand has not materialized. Nuclear power plants, with long lead times, were left vulnerable. The nuclear power industry is now in the transition between construction of new plants and modifications to existing plants. Additionally, work continues in the planning for radioactive waste disposal and plant decommissioning. These proceedings address these issues and the future outlook for the nuclear power and the civil engineer.

1021   Tunneling Operations and Equipment
Proceedings of the session sponsored by the Construction
Division of the American Society of Civil Engineers in
conjunction with the ASCE Convention in Detroit, Mich.,
Oct. 22, 1985

D. D. Brennan, (editor), Member, ASCE, (Pres., Elgood Mayo Corp., P.O. Box 1413, Lancaster, PA 17604)

New York: ASCE, 1985, 43pp.

The papers included in the Proceedings describe four diverse tunneling prjects of varying diameters, design criteria, ground conditions and construction techniques undertaken at locations throughout the United States. The uniqe method of multiple drift construction of the Mount Baker Ridge Tunnel in Seattle, Washington, presented a challenge to the contractor never before undertaken. The Flint, Michigan, sewer tunnel constructed over a period of three years was bored through soil conditions that varied from rock to running ground and utilized, by necessity, two different types of tunnel boring machines. Large diameter, continuous, soft ground and hard rock tunnel boring machines were used for the construction of the Greenbelt Tunnels in Washington, D.C. and the deep tunnel TARP Project in Chicago, Illinois. The presentation and favorable acceptance of an extensive Value Engineering Change Proposal in Washington, D.C. and the new generation of tunnel boring machines in Chicago present a study in sound construction management and organization. Throughout the Proceedings the authors describe the experience and construction approach taken by several different contractors for the successful completion of these varied and considerably difficult underground projects.

1022   **Uplift Behavior of Anchor Foundations in Soil**
Proceedings of a session sponsored by the Geotechnical
Engrg. Div. of the American Society of Civil Engineers in
conjunction with the ASCE Convention in Detroit, Mich.,
Oct. 24, 1985

Samuel P. Clemence, (editor), Member, ASCE, (Prof. and Dept. Chmn., Dept. of Civ. Engrg., Syracuse Univ., Syracuse, NY 13210)

New York: ASCE, 1985, 132pp.

This publication includes a series of studies which provide a general and consistent framework for the design of anchors in uplift. The papers are presented in three groups: (1) State-of-the-art review; (2) field and laboratory studies; and (3) a case history. The state-of-the-art paper describes the static uplift behavior of anchors used for foundations in soil. Basic anchor types such as spread anchors, helical anchors, and grouted anchors are described and their behavior is evaluated. Criteria for design parameters are also included. The uplift behavior of helical anchors in sand, silt, and clay is described and design procedures recommended. The uplift resistance of shallow offshore foundations is discussed and a recommended design procedure as well as case histories are presented. A case history describing the design and construction of grouted anchors in weathered rock is also included.

1023   **Vibration Problems in Geotechnical Engineering**
Proceedings of a Symposium by the Geotechnical Engi-
neering Div. in conjunction with the ASCE Convention in
Detroit, Mich., Oct. 22, 1985

George Gazetas, (editor), Member, ASCE, (Assoc. Prof. of Civ. Engrg., Rensselaer Polytechnic Institute, Troy, NY 12180) and **Ernest T. Selig**, (editor), Fellow, ASCE, (Prof., Univ. of Massachusetts, Marston Hall 28, Amherst, MA 01003)

New York: ASCE, 1985, 309pp.

This volume constitutes the Proceedings of a two-session Symposium whose main objective is to address geotechnical issues related to vibration problems other than those associated with earthquakes. The first session, entitled "Analysis and Measurement of Machine Foundation Vibrations," contains eight papers covering analytical, experimental and design aspects of machine foundations. Procedures to predict amplitudes of machine-excited vibrations of shallow, embedded and piled foundations are described. Rigorous and simplified methods are presented, while small-scale experiments show the capabilities and limitations of such methods of analysis. Two different designs of compressor foundations are reviewed and the results of field tests aimed at evaluating their performance are presented. A numerical formulation for assessing the effectiveness of active and passive vibration isolation by means of trenches is outlined. The second session, entitled "Detrimental Ground Movement from Man-Made Vibrations," contains seven papers which provide a variety of useful examples of vibration effects. Settlements caused by pile driving vibrations are examined. Observations on ground vibrations from vehicles on projects where vibrations were a problem to people or equipment are presented. Ground vibrations from various sources including dynamic compaction, vibraflotation, pile driving, machine vibrations and vehicle vibrations are also covered. Ground vibrations from dynamic (impact) compaction operations and building response to construction blast vibrations are reviewed. A study of potential detrimental effects on slope stability of vibrations from blasting is presented.

1024   **Water Quality Issues at Fossil Fuel Plants**
Proceedings of a Symposium sponsored by the Energy
Division of the American Society of Civil Engineers in
conjunction with the ASCE Convention in Detroit, Mich.,
Oct. 24, 1985

William G. Dinchak, (editor), Member, ASCE, (Dir., Portland Cement Association, Energy and Water Resources Dept., 5420 Old Orchard, Skokie, IL 60077) and **Michael J. Mathis**, (editor), Member, ASCE, (Engr., Niagara Mohawk Power Co., 300 Erie Blvd. West, Syracuse, NY 13202)

New York: ASCE, 1985, 96pp.

This publication highlights proposed approaches and actual projects for dealing with such water quality issues at fossil power plants as water supply, water treatment, and the protection of surface water and ground water. Described are use of radial wells and a circulating water tunnel (CWT) as water sources for fossil power plants. Unique aspects of the CWT, for example, are a fish deterrent/fish return system, staged diffuser, and intake velocity cap. Solidification of a flue gas desulfurization slurry, a radically different method of handling and disposing of fly ash and bottom ash, a soil-bentonite cutoff wall for coal pile leachate control, conversion of a wet fly and bottom ash sluicing system to a dry fly ash handling system followed by landfilling, and other waste handling systems are included. In addition, development of a coal pile drainage model, data concerning the leachate characteristics of various fly ashes, and a methodology used to estimate waste quality discharge concentrations are presented. The handling, chemical and toxicological nature of wastewaters from combined cycle coal conversion facilities are also described. This volume is a compendium of state-of-the-art information that should be helpful to designers of treatment systems serving fossil power plants.

1025   **Watershed Management in the Eighties**
Proceedings of the Symposium sponsored by the Committee
on Watershed Management of the Irrigation and Drainage
Division of the American Society of Civil Engineers in
conjunction with the ASCE Convention in Denver, Colorado, April 30- May 1, 1985

E. Bruce Jones, (editor), Member, ASCE, (Pres., Resource Cons. Inc., Fort Collins, Colo. 80522) and **Timothy J. Ward**, (editor), Member, ASCE, (Assoc. Prof. of Civ. Engrg., New Mexico State Univ., Las Cruces, N. M.)

New York: ASCE, 1985, 328pp.

The papers included in this publication discuss state-of-the-art methods and current research on management and present case studies on current projects and practices on applied watershed management, watershed modeling and simulation, watershed management and land stability, water quality and watershed hydrology. These topics are applied to forests, rangelands, and urban areas. Other specific subjects considered include land use, rainfall-runoff relationships, erosion, soil water content, and environmental impacts.

1985 ASCE MEMOIRS

1026   **Gregory P. Tschebotarioff, Honorary Member, ASCE**
1899-1985

Gregory P. Tschebotarioff, world-renowned for his many contributions to the field of soil mechanics and foundation engineering, died April 22, 1985 in Holland, Pennsylvania. The son of General Porphyry G. Tschebotarioff and Valentina Doubiagsky Tschebotarioff, he was born February 15, 1899 in Pavlosk, Russia. His father was an officer in the Cossack Guard Battery and his mother was a close friend of the Russian Empress and her two daughters, the Grand Duchesses Olga and Tatiana Nikolayevna. He graduated from the Imperial Law School in Petrograd, Russia in 1916 and went into the Russian Army until 1917. When the revolution began and the Bolsheviks took power, he joined the Grand Army of the Don and fought with the White Russians against the Red Army until 1920.

His engineering studies which began in Russia were continued in Germany, where he earned the "Diplom-Ingeniuer" degree in 1925 and the "Doktor-Ingenieur" degree in 1952. Structural engineering projects in Paris, Berlin, Bremen, Hamlen and Cairo occupied his attention from 1925 to 1929. In 1929 he began his specialization in soil mechanics research and foundation engineering. For seven years thereafter he specialized in foundation work in the Egyptian Government Service. Since 1936 he participated actively in numerous international conferences on soil mechanics and foundation engineering, including the First International Conference at Harvard, in 1936.

From 1937 through 1964, he was a member of the Princeton University Faculty, where he organized a soil mechanics laboratory and courses on soil and foundation engineering. At Princeton, he also organized and was in charge of pioneering research projects on the effects of vibration of soils for the Civil Aeronautics Administration. His work on large-scale model tests on retaining walls, anchored bulkheads and other waterfront structures for the Bureau of Yards and Docks and for the Office of Naval Research attracted world-wide attention and acclaim.

He married Florence D. Bill of Princeton, New Jersey in 1939.

In his years at Princeton, he wrote many professional papers on soil mechanics. He authored *Soil Mechanics, Foundations and Earth Structures,* a technical best-selling textbook, and *Foundations, Retaining & Earth Structures.* He contributed to the McGraw-Hill book *Foundation Engineering.*

He also wrote the autobiography *Russia My Native Land.* Though he fought with the White Army during the Russian Revolution, and was a staunch citizen of the United States, he became the victim of harrassment during the McCarthy era when he denounced what he called "the malevolent distortions of Russian history emanating from sources close to our U.S. propaganda policies."

Throughout his teaching career he did consulting work and in 1955 became an Associate of King and Gavaris, Consulting Engineers, where he was in charge of soils and foundation work through 1970.

Dr. Tschebotarioff retired from his professorship in 1964. In 1974, Dr. Tschebotarioff donated his technical library and personal archives to Purdue University. The collection is housed in the Tschebotarioff Library along with a draft manuscript of his last book, *Civil Engineering on Four Continents.*

He continued his consulting work as an Associate in the regular highway and harbor engineering work of King and Gavaris and conducted investigations for other consulting engineers, construction firms, and federal and state agencies.

In 1959, he was awarded in Belgium, the honorary "Docteur Honoris Causa" degree. A winner of the Karl Terzaghi Award, he was also named to Sigma Xi. According to the citation that proclaimed him an Honorary Member of the American Society of Civil Engineers in 1977, Dr. Tschebotarioff "advanced the art of soil and foundation engineering through basic research and innovative practice, while fostering good will and mutual understanding among all people."

He was active in the National Society of Professional Engineers, the American Historical Association and the American Association for the Advancement of Slavic Society. His many contributions to ASCE publications were climaxed by his presentation of the second Martin Kapp Memorial Lecture of the Metropolitan Section, given January 21, 1976. His talk "Half a Century of Soil Mechanics—Some Thoughts for the Future in Light of the Past" received a standing ovation. 150 engineers attending the presentation.

He is survived by his wife, Florence D. Tschebotarioff and a sister, Valentine Bill.

(Memoir prepared by Gerald A. Leonards, F. ASCE, and Peter T. Gavaris, F. ASCE.)

1027   **Daniel H. Burnett, Member, ASCE**
             1936-1985

Daniel H. Burnett died on Saturday, January 12, 1985 in Riverside, California. He was an ASCE member since 1962.

Mr. Burnett was born in Detroit, Michigan, on February 26, 1936. He graduated from Southwest High School and in 1959, received a Bachelor Degree in Civil Engineering from Wayne State University.

Immediately after graduation, he was employed by the California Department of Water Resources as the engineering work accelerated for the ultimate construction of the great California Water Project.

In 1960 he began a lifelong association with public works at the local government level when he became a design engineer with the Public Works Department of the City of Riverside, California. In 1963 he joined the City of Colton, California, and became Director of Public Works/City Engineer in 1964. He began his affiliation with Neste, Brudin and Stone, Consulting Civil Engineers in 1973 where he continued to specialize in local public works serving as a contract city engineer and district engineer. He became a part owner and Project Manager for NBS, Inc.

Mr. Burnett was very active in professional and community associations. He was past Chapter President of the California Society of Professional Engineers and had been selected as "Engineer of the Year." At his untimely death, he was serving as chairman of the local branch of the American Public Works Association where he established the "Public Works Department of the Year" award. He was also active in the Riverside-San Bernardino Branch ASCE, California Council of Civil Engineers and Land Surveyors, City and County Engineers Association (Past President), and Public Works Officers, League of California Cities.

Mr. Burnett was past president of the Colton Rotary Club, former District Chairman of Boy Scouts of America, and a dedicated supporter of the Virginia Primrose School for the Severely Handicapped. Other interests included YMCA and Community Chest. Survivors are his widow, Linda and sons, Daniel, Jr. and Paul; mother, Helen of Michigan, brother Allan of Martinez, Ca. and sister, Susan Tabalina of Venice, Ca.

(Memoir prepared by Claude Glenn Wilson, F. ASCE.)

1028   **Edward August Farmer, Fellow, ASCE**
1908-1984

Edward August Farmer, the son of Edward Farmer and Anna (Hagen) Farmer was born at Atchison, Kansas on May 9, 1908. He was graduated from the University of Kansas at Lawrence, in 1929 with a bachelor of science degree in civil engineering.

Immediately after graduation, he joined Black & Veatch, Engineers-Architects, where he served in a variety of civil engineering assignments for over fifty years. His work with Black & Veatch included service as a resident engineer, principal engineer, and executive partner.

During World War II, Mr. Farmer represented the firm, as resident engineer, on five large air-bases. Following the war, he was involved in a variety of projects, principally in the fields of water supply and water pollution control. Major engineering design work included projects at Cincinnati, Ohio; Springfield, Ohio; Greenville, South Carolina; Boulder, Colorado; Lincoln, Nebraska; Washington, D.C.; Lawrence, Kansas; Kansas City and St. Louis, Missouri. During much of his professional career, he was involved in design and construction of public water supply improvements for Cincinnati, Ohio. Mr. Farmer was an active participant in his firm's long-standing association with the Cincinnati metropolitan-area water supply organization.

Mr. Farmer was registered as a professional engineer in 10 states and the District of Columbia. He was active in a number of professional organizations, especially the American Water Works Association and the Water Pollution Control Federation. He was a Fellow and Life Member of the American Society of Civil Engineers. He also held memberships in the New England Water Works Association, the Missouri Society of Professional Engineers, the National Society of Professional Engineers, the American Cosulting Engineers Council, and Tau Beta Pi. He was a Diplomate of the American Academy of Environmental Engineers.

He is survived by his widow, Alice Marie (Milburn) Farmer, a son, Edward Michael Farmer, and four grandchildren.

(Memoir prepared by Paul D. Haney, F. ASCE.)

1029   **Donald Loyd Fritts, Member, ASCE**
1930-1984

Donald Loyd Fritts, son of William A. and Emma Blaser Fritts, was born in Amsterdam, Missouri on August 23, 1930. He received his Bachelor of Science degree in Civil Engineering from Kansas State University in 1957.

Mr. Fritts was employed by the Kansas City District of the U.S. Army Corps of Engineers during the entire course of his professionl career. His service with the Corps began immediately after graduation in 1957. Upon completing his internship in June 1958, he was assigned to the Military Construction Branch. In January 1959, he was transferred to the Survey Section where he was named Chief in February, 1960. In August 1963, Mr. Fritts became Chief, Engineering Service Branch, a position which he held for 12 years. He was subsequently reassigned to the position of Chief of the Relocation Branch. In May 1976, he was promoted to Assistant Chief of the Engineering Division, the position he held at the time of his death.

Mr. Fritts was a member of the American Society of Civil Engineers; the National Society of Professional Engineers; the University of Missouri Engineering Advisory Committee; Longview College Engineering Advisory Committee; and the Johnson County Community College Engineering Advisory Committee. He served as president, Western Chapter, Missouri Society of Professional Engineers, 1983-84, and was president-elect, of the Missouri Society of Professional Engineers at the time of his death.

He is survived by his wife, Nadine, two sons, David G. Fritts, Detroit, and Daniel L. Fritts, and two daughters, Miss Diana L. Fritts and Miss Melanie L. Fritts.

(Memoir prepared by William N. Marshall, M. ASCE.)

1030

Robert Kenneth Harrar, Member, ASCE
1925-1981

Robert Kenneth Harrar, son of Archie and Josephine Soderberg Harrar, was born on March 29, 1925 in Kansas City, Jackson County, Missouri. He graduated from Kansas State University with a Bachelor of Science in Civil Engineering in 1950.

After graduation Mr. Harrar was engaged by Burns & McDonnell Engineers and Architects of Kansas City, Missouri where he rose to the position of Manager of the Civil Department of the company's Special Projects Division, the position he held at the time of his death. Mr. Harrar attained prominence in the firm for his proficiency in design and management of military and aviation projects both in the United States and overseas. He was instrumental in the design of airfield facilities at King Khalid Military City, Saudi Arabia, Chiang Kai-Shek International Airport in Taiwan, as well as the Kansas City International Airport, Portland International Airport, LaGuardia Field in New York and Lambert Field in St. Louis. He also coordinated design for the civil engineering aspects of a number of manufacturing and maintenance facilities related to the steel fabrication and airline industries.

Mr. Harrar was a member of the American Society of Civil Engineers, the National Society of Professional Engineers, the American Society of Aeronautics and Astronautics, and the Society of Military Engineers.

On April 25, 1953, Robert Kenneth Harrar married Beverly Linch. He is survived by his wife, his mother and one sister.

(Memoir prepared by William N. Marshall, M. ASCE.)

1031

Julian Pitts Hinton, Fellow, ASCE
1916-1985

Julian Pitts Hinton died February 7, 1985 in Knoxville, Tennessee. The son of Robert Wood and Mamie Robertson Hinton was born January 23, 1916 in Lumberton, Mississippi. He attended secondary schools in Lumberton.

In June 1934 he was employed by the Tennessee Valley Authority as a rodman on surveying work in connection with the construction of Norris Dam and Reservoir, the agency's first project.

He entered the engineering student cooperative program at the University of Tennessee in the fall of 1934 and continued to work part-time as an engineering aide with TVA's testing laboratory, then at U.T. He was graduated from the university with a bachelor's degree in civil engineering in 1938 and subsequently did graduate work there.

Upon his graduation from U.T., Julian went to work as a full-time engineer with TVA's Power Studies Branch, later a part of the Project Planning Branch in TVA's Division of Water Control Planning.

During World War II, he participated in the Yale OCSC program and obtained the rank of second lieutenant. He served as an aircraft maintenance officer with the Army Air Corps. He served in the North American Campaign and was released from active duty with the rank of first lieutenant in 1946.

He returned to TVA's Project Planning Branch and from March 1954 until retirement in December 1971 served as the head of the Power Studies and Hydrology Section. His duties primarily involved leading a group of engineers engaged in preparing general plans of river development and in making detailed power studies needed in planning for new and existing hydroelectric power projects.

On September 6, 1941 he was married to Neta Lay of Knoxville, Tennessee.

He was a fellow of ASCE and past president of the Knoxville Technical Society. He worked with the Fountain City Recreation Center and coached little league baseball for many years. He was a registered engineer in Tennessee and a member of the Washington Pike United Methodist Church. His hobbies were general sports, golf, woodworking, photography, stamp collecting and fishing.

Julian is survived by his wife and two sons Dr. Robert J. Hinton, Dallas, Texas and Richard I. Hinton of Knoxville. There is a granddaughter Elizabeth Guinn Hinton, of Knoxville; a brother R. W. Hinton, Jackson, Mississippi; and a sister Mrs. H. C. Vance, of Biloxi, Mississippi.

(Memoir prepared by J. B. Perry, M. ASCE.)

1032 **Ritchey Hume, Member, ASCE**
1904-1985

Ritchey Hume died February 6, 1985 in Knoxville, Tennessee. The son of Dr. Alfred and Mary Ritchey Hume was born January 19, 1904 in Oxford, Mississippi. His father was a distinguished professor of mathematics at the University of Mississippi. Ritchey received his secondary schooling in Oxford, and later attended the University of Mississippi from which he graduated with a B.S. in Civil Engineering in 1925. While at the University he was a member of the Sigma Chi.

Following graduation, Ritchey was employed by the American Bridge Company in Gary, Indiana, and remained with that firm for six years detailing, designing and estimating structural steel. With the lean years of the early Depression, Ritchey like many of his fellow engineers, worked on several projects. One of these from 1931 to early 1934 was with the Spring Hill, Tennessee water system.

In mid-1934, Ritchey was employed by the Tennessee Valley Authority in the General Engineering and Geology Division in Chattanooga, Tennessee. Later, he moved to Knoxville to work in TVA's Hydraulic Data Branch. There for more than twenty years, his work involved both engineering and administration. He was greatly admired and respected by the employees of the branch. He frequently took their side when problems arose. In 1964, he was promoted to be the Administrative Officer for TVA's Division of Water Control Planning. The work here involved the preparation and control of complex budgets with large responsibilities. He retired from this position in January 1969.

On June 24, 1939, he was married to Evelyn Holt of Memphis.

He was a life member of ASCE, and during his working years was very active in the Tennessee Valley Section. He was a member of the Technical Society of Knoxville and a registered engineer in Tennessee. He worked with the Boy Scouts of America, especially the Cub Scouts program. He also was an enthusiast for little league baseball. Ritchey was a working member of the Second Presbyterian Church of Knoxville.

He is survived by his wife and three sons: Alfred Hume, Raleigh, N.C.; David Hume, Cookeville, Tennessee; and Ritchey Hume, Jr. of Knoxville, Tennessee. There are four grandchildren.

(Memoir prepared by J. B. Perry, M. ASCE.)

1033 **Ross F. Jarvis, Member, ASCE**
1905-1985

Ross F. Jarvis, the son of Sam M. and Lillian Moore Jarvis, was born in Milton, Kansas on March 12, 1905. Jarvis graduated from high school in Wichita, Kansas, having earned most of his living while in school by working at various part time jobs. Not being able financially to enter a college or university, Jarvis apprenticed himself to a Mr. J. F. Burlie, an Architect, for three years. At the same time Jarvis enrolled in a correspondence school, taking courses in Architecture and Engineering. Being possessed of an unusually inquiring mind and a very retentive memory, Jarvis quickly became proficient in both of these fields.

From 1927 until 1931 Jarvis worked as a draftsman and as an estimator for a number of construction firms. At one time during this period Jarvis worked as an engineer designer for Coffman Monoplanes, Inc. His own company, Jarvis Monoplanes, Inc., survived for only one year, succumbing to the depression that began in late 1931. During the Great Depression, from 1932 until late 1940, Jarvis held jobs with a number of marble and granite companies, designing monuments and mausoleums.

When the United States entered the Second World War in December, 1941, Jarvis immediately volunteered. Because of his training in architecture and engineering, Jarvis was assigned as an Engineer Aide in the United States Army Corps of Engineers. Jarvis remained in the Army Corps of Engineers. Jarvis remained in the Army Corps of Engineers following the war. In 1961, seizing upon an opportunity to go to Washington, D.C. and take oral and written examinations to qualify as a professional engineer, Jarvis passed with very high standing.

Jarvis's entire period of service for the United States Government spanned a period of 30 years. He was with the U.S. Corps of Engineers from 1941 until 1962, and with the National Aeronautics and Space Administration from 1962 until his retirement from government service on October 2, 1971.

During his service with the U.S. Corps of Engineers Jarvis was engaged in the design and construction of numerous military and defense facilities, including the design, construction and inspection of missile site silos. His service with NASA was in the design and construction of the Manned Spacecraft Center in Houston, Texas. During his career Jarvis received letters of commendation from his commanding officers, from his commanding general, and from Robert Gilruth, Director of NASA.

Not being content to merely retire to an inactive life, even though in the meantime he had had to undergo bypass heart surgery, Jarvis took a position with the City of Houston to design and inspect an air traffic control center, building and runways for the rapidly growing Hobby Airport Terminal in the south end of the city. He was with the city from 1972 until his eventual complete retirement in 1978.

Ross Jarvis married Ethel Houston of Henderson, Texas in Ardmore, Oklahoma on December 24, 1939. They had no children.

Jarvis was a Life Member of the ASCE, having joined the society in 1957 and attained the grade of Member in 1959. He was also a member of the American Society of Military Engineers and the Houston Civic Club. He served as a volunteer for the Travelers Aid Society of Houston, Texas.

Jarvis died on May 10, 1985, and is survived by his wife, Ethel Houston Jarvis, and a number of nieces and nephews.

(Memoir prepared by William O. Swift, Jr., F. ASCE.)

1034   Alfred Massey Fisher Johnson, Member, ASCE
1912-1985

Alfred Massey Fisher Johnson died February 24, 1985 in Chattanooga, Tennessee. The son of Marie Belliss and Charles Johnson was born January 7, 1912 in Belding, Ionia County, Michigan. In 1935 he was graduated from the Michigan State College of Agriculture and Applied Science (now Michigan State University) with a Bachelor of Science degree in Civil Engineering. While at the University he was a member of the Mortar and Ball (honorary artillery) Sigma Rho Tau (honorary speech) and the ASCE student chapter.

After graduation he was employed by the Tennessee Valley Authority as a general engineer working in geology and aerial mapping. In 1935, he was employed by the U.S. Engineer Department, Mississippi River Commission, at their hydraulics laboratory in Vicksburg, Mississippi. Later from 1938 to 1941 he worked with the U.S. Engineer office in the hydraulic design section in Vicksburg, Mississippi. Holding a reserve commission from the Michigan State College ROTC, he entered the service in 1941 to serve in the anti-aircraft artillery. He served two tours of overseas duty, first in Trinidad and then the Philippines. He received the Bronze Star award in the Philippines Campaign. He left the service as a Lieutenant Colonel in November 1947.

Alfred Johnson was employed by the U.S. Geological Survey from 1947 until retirement in 1977.

On June 8, 1938 he was married to Ruth Marian Arnold at St. Pauls Episcopal Church in Chattanooga.

He was a member of the National Society of Professional Engineers, the National Railway Historical Society, and a Life Member of the American Society of Civil Engineers. He was a member of the Episcopal Church. In 1963 he received the Silver Beaver Award from the Boy Scouts of America. He was a registered engineer in Tennessee.

Alfred Johnson is survived by his wife and two sons, the Reverend Dr. Charles Ernest Arnold of East Ridge in Chattanooga and William Alfred of Knoxville, Tennessee. There are five grandchildren.

(Memoir prepared by J. B. Perry, M. ASCE.)

1035   **Lyle Perry Pederson, Member, ASCE**
             1929-1984

Lyle Perry Pederson, PhD. P.E., the son of Oscar and Mabel Olson Pederson, was born in Kinbrae, Minnesota on April 23, 1929. He graduated from Fulda High School in 1946, and attended Macalaster College for two years before transferring to the University of Minnesota in 1948, where he obtained a B.C.E. degree in 1952.

After a short time with the Minnesota Highway Department, Lyle served two years in the Pacific with the U.S. Navy "Seabees" (1953-54), spending most of this time on Midway Island. After completing his service, he returned to the Minnesota Highway Department.

In 1954 he returned to the University of Minnesota, as a teaching assistant in the Civil Engineering Department. He developed an interest in soils and completed his M.S.C.E. in 1956. He became an Instructor in Civil Engineering in March 1956, and was later promoted to Assistant Professor.

Working with his advisor, Dr. Kersten, Lyle earned his Ph.D. in Civil Engineering in June 1964. His thesis was on the bituminous stabilization of silty soils. The Minnesota Department of Transportation (MnDot) constructed a field project based on the results of his thesis.

He was active in the Soils and Foundations Conference, and taught special programs such as a summer session for university teachers on asphalt technology, special Master's programs for geologists of the Corps of Engineers, and special programs for MnDOT engineers.

After resigning from the University of Minnesota in 1978, Lyle became a Principal Geotechnical Engineer with Subterranean Engineering Corporation (1978-80). He then formed his own geotechnical engineering consulting firm, Lyle Pederson and Associates (1980-81). In 1981 he joined the consulting engineering firm of Mead & Hunt, Inc. as the Minnesota Branch Manager.

Much of his recent engineering work was on dams, dikes, hydroelectric power development, flood control, lake management, and transportation projects. He was the Geotechnical Engineer in charge of subsurface investigations, foundation design and earth support systems for such clients as Corps of Engineers and Northern States Power Company. Some of his projects included Big Stone Dam on Minnesota River, La Farge Dam in Wisconsin, International Falls Airport runway extension, Winona, Minnesota flood control project, Sauk Centre, Minnesota study for hydroelectric power potential, and the Anoka, Minnesota study for hydroelectric power potential at the Coon Rapids Dam.

Lyle was also an instructor in a special Corps of Engineers workshop—"Inspection and Evaluation of Safety of Non-Federal Dams." He was a Registered Professional Civil Engineer in the State of Minnesota.

Lyle contributed extensive time and enthusiasm to the American Society of Civil Engineers and other professional engineering organizations. Lyle joined the Northwestern Section ASCE (now the Minnesota Section) in 1957. He has served on the National ASCE Committee on Embankments, Dams and Slopes. He was on the Minnesota Section ASCE Board of Directors for several years, serving as President in 1983-84. Lyle's other professional activities included Incoming President and Vice President of the Minnesota Federation of Engineering Societies, a founder and the first President (1969-71) of the Minnesota Geotechnical Society, Geotechnical Committee Chairman of the Consulting Engineers Council of Minnesota, member of the International Society of Soils Mechanics and Foundation Engineers, member of the Association of Soil and Foundation Engineers, scholarship committee member of the Minnesota Surveyors and Engineers Society, and member of the Minnesota Society of Professional Engineers.

Lyle died on December 9, 1984 in Shoreview, Minnesota. He was preceded in death by his first wife Mildred Husnik (they were married on September 3, 1955). He is survived by his wife Lolly (whom he married on June 25, 1976); his children: Michael, Sandi, Gayle, Keith, Steve, Scott, Stuart and Sarah, six grandchildren, his mother, and two brothers.

On December 12, 1984 the Minnesota Section ASCE Board of Directors established the "Lyle Pederson Memorial Fund" in honor of our distinguished friend and colleague. All contributions are used for soils laboratory equipment for undergraduate teaching use at the University of Minnesota's Civil & Mineral Engineering Department.

(Abstract of a memoir compiled by Dr. Miles S. Kersten, Hon. M. ASCE, and Robert J. Yourzak, M. ASCE.)

1036　　　　　　　　**Fred Harold, Jr. Rhodes, Fellow, ASCE**
　　　　　　　　　　　　　1901-1984

Fred Harold Rhodes, Jr. was born September 12, 1901 in Colony, Kansas, the son of Fred and Nellie Rhodes. He completed his college preparatory course at Humboldt High School (Kansas) and entered the University of Washington in 1919, where he received BS degrees in Civil Engineering in June and in Mechanical Engineering in December of 1926. He was employed as a Structural Draftsman by the Wallace Bridge Co. until autumn then served as an Instructor in General Engineering at the UW teaching architecture students in the 1927-28 academic year. Following two years as a Structural Engineer with the Tacoma City Engineer, Bridge Department, he returned to the University as Instructor in Civil Engineering in September, 1930.

He taught Mechanics and Structures and advanced through the academic ranks retiring as Professor (emeritus) in June, 1969. He was in the U.S. Army Reserve from 1926 until retirement as Lt. Col. in 1961, with 27 months World War II active duty in anti-aircraft training commands.

He joined the American Society of Civil Engineers in 1926, and attained the grade of Fellow in 1959. During most of his academic career he served as Faculty Advisor to the student chapter. In the Seattle Section he served as Secretary 1936-40, Vice President in 1942, and President in 1945. He was General Chairman of the Spring National Convention 1949, and District 12 Director 1958-61. He had a similar record of service in the Society of American Military Engineers, serving as National Director, 1952-55.

Highway safety was a long-time collateral interest. He served over twenty years as Trustee, Automobile Club of Washington and directed annual short courses for Driver Education instruction and for Fleet Supervisors. He served also as Conference Director, Motor Vehicle Maintenance Conference, an annual meeting that grew to an international attendance of over 900.

He died May 12, 1984. His late wife, Marguerite, preceded him in death by many years, and they had no children. He left a substantial endowment to the University of Washington to enhance undergraduate education in Civil Engineering.

(Memoir prepared by Richard H. Meese, F. ASCE.)

1037

Guy Woodford Sackett, Fellow, ASCE
1885-1984

Guy Woodford Sackett was born at Fort George Island, Duval County, Florida, on October 31, 1885. He died in Jacksonville, Florida on October 3, 1984. His father was John Warren Sackett, M. ASCE., and his mother was Louisa Hamilton Johnson.

Shortly after his birth the family moved to St. Augustine, FL. Guy attended school at the Catholic Convent, graduating in 1902. While in school Guy worked part-time with survey parties for the Corps of Engineers on river and harbor construction projects and for the Florida East Coast Railroad on railway construction. He took courses in mathematics and Civil Engineering subjects from the International Correspondence School of Philadelphia, PA.

His first full time job was as a deck hand on the Corps of Engineers seagoing dredge, the St. Johns. He also moonlighted as a draftsman at the Corps' field office at Mayport, FL.

Guy married Edith Mildred Wilson at the St. Johns Cathedral in Jacksonville on August 14, 1907.

By the time Guy was 30 years old he had earned a United States Coast Guard Merchant Marine License for First Class Pilot, and a Masters License to command ships of unlimited tonnage on unlimited waters. He renewed these licenses until his retirement in 1973, at age 88. Following completion of dredging for the ports of Jacksonville and Tampa Bay, Florida, Guy's expertise in this field led him to the Mississippi and Missouri Rivers, where he was Resident Manager for the Hillsboro Dredging Co. on their contracts for filling in lowlands and building flood control levees for the cities of St. Louis, Cairo, and New Madrid. During World War I he was Supt. of Construction for the Mason & Hanger Co. on their contract for the Charleston, S.C. Remount Depot and Embarkation Center.

Organizations that he worked for in addition to the Corps of Engineers and his own Contracting/Engineering Co. were The Atlantic, Gulf & Pacific Co., The Hillsboro Dredging Co., the Duquense Slag/Pittsburgh Bridge Co., The Standard Dredging Co., Ulen & Co., The Mason & Hanger/Silas Mason Co.

From 1924 through 1927 he was Resident Manager for the Ulen & Company and supervised the construction contract for jetties on the Magdalena River near Baranquilla, Colombia, South America.

Seeking a change in environment, Sackett worked next on the St. Mary's River near Sault Ste. Marie, Michigan. Here he was Project Manager and Chief Engineer for the Standard Dredging Co. on their contract for subaqueous rock excavation, to increase the channel depth between Lakes Superior and Huron. This was to become a link in what is now the busy St. Lawrence Seaway.

Probably the most valuable contribution that Sackett made to his country was in ramrodding the construction of defense plants during World War II. In the year 1940, when Russia and England were about to be overcome, due to lack of munitions, he went to work again for the Mason & Hanger Co. As a Vice President and Project Manager, he was responsible for construction of three large facilities. The Hercules Powder Co., a prime contractor to the U.S. government, awarded a cost-plus-a-fixed fee contract to Mason & Hanger in September 1940 to construct the Radford Ordnance Works in Southwestern Virginia, a plant to manufacture smokeless powder. The first powder was produced seven months later in April of 1941, 90 days ahead of schedule. The other two ammunition plants were the New River Ordnance Works at Dublin, VA, and the Badger Ordnance Works at Baraboo, WI.

Guy moved back to Jacksonville at the end of the war and operated his own contracting and consulting engineering firm until his retirement in 1973.

Guy was a Registered Professional Engineer, licensed by the States of New York, Wisconsin and Florida. He was also a Licensed Land Surveyor in New York State. He taught Nautical Navigation and Land Surveying to Boy Scout Troops. He was a Mason, Rotarian, a Member of the American Society of Military Engineers, a Life Member of the Florida Engineering Society, the American Ordnance Association, and a member of the Episcopal

Church. Guy was elected a Member of the American Society of Civil Engineers in 1919. He became a Fellow, Life Member in 1959, a position he held until his death.

Surviving members of his family are a son and daughter, Guy, Junior, and Mrs. Evelyn Sackett Hogan, three grandchildren and five great-grandchildren.

(Abstract of a memoir prepared by Guy W. Sackett, Jr., F. ASCE.)

1038   **Thomas D. III. Samuel, Fellow, ASCE**
                              1906-1984

Thomas D. Samuel, III graduated from the University of Kansas in Lawrence in 1929 with a civil engineering degree. He was a member of the Beta Theta Pi fraternity. He joined Black & Veatch, Engineers-Architects in 1929 and spent his entire professional career with the firm.

Mr. Samuel served as a draftsman and as an assistant on appraisals during early years with the firm. Appraisals were conducted for the Evansville, Indiana, Water Department; Cities Service Gas Company; Hope Natural Gas Company; and East Ohio Gas Company.

As resident engineer from 1934-1944, he supervised the construction of water and sewerage improvements including a wastewater treatment plant at York, Nebraska; wastewater treatment plant, pumping stations, force mains and gravity sewers at Bartlesville, Oklahoma; high service water supply pumping station, water works office building, and water treatment plant at Pittsburg, Kansas; wastewater treatment plant and gravity interceptor sewers at Pueblo, Colorado; wastewater treatment plant and gravity sewers at Sheridan, Wyoming; water treatment pland and underground storage reservoir at Wichita, Kansas; water treatment plant and distribution system at Jackson, Mississippi; the entire air base and an auxiliary field for Independence Air Base, Independence, Kansas; the water supply and distribution system, wastewater treatment plant and gravity sewers at Greenville, Mississippi Air Base; and a water supply dam, pumping station, water transmission mains, water treatment plant, and electrical distribution facilities at Camp Livingston, Louisiana.

Following his assignments as resident engineer, Mr. Samuel was responsible for various reports on water supply, treatment, and distribution, and on sewers and wastewater treatment plants.

In 1947 Mr. Samuel became in charge of specifications for the Civil-Environmental Division of Black & Veatch. He wrote and supervised others writing specifications for water supply dams, water treatment plants, water supply pumping stations, water distribution systems, wastewater treatment plants, wastewater pumping stations, force mains, and gravity storm and sanitary sewers.

Mr. Samuel was a founding member and the first president of the Kansas City chapter of the Construction Specifications Institute. He was active nationally in the Construction Specifications Institute, participated in the development of the CSI Format for specifications, and became a vice president. He was a frequent speaker at regional and national meetings and was honored by elevation to Fellow.

He was a life member and Fellow of the American Society of Civil Engineers, a member of the American Water Works Association, the National Society of Professional Engineers, a life member of the Missouri Society of Professional Engineers, a member of the National Association of Corrosion Engineers, Specification Writers Association of Canada, American Society of Testing and Materials, American Concrete Institute, and Sigma Tau honorary fraternity. He was a registered professional engineer in Missouri and was an active member and Elder of the Westport Presbyterian Church.

Mr. Samuel retired from Black & Veatch in 1976 and passed away on October 22, 1984 at the age of 78. His wife, Margaret Ann Nagel Samuel, died in 1981. He left a son, Thomas Duncan Samuel IV, Prairie Village, Kansas; a daughter, Mrs. Mary B. Melms, Leawood, Kansas; a brother, Charles F. Samuel, Overland Park, Kansas; and two grandsons.

(Memoir prepared by Gerald A. Neely, M. ASCE.)

1039   Harry Anthony Wiersema, Fellow, ASCE
                              1913-1985

Harry Anthony Wiersema died on May 4, 1985 in Knoxville, Tennessee. The son of Anthony and Reka Verbeek Wiersema was born on July 8, 1892 in Grand Rapids, Michigan. After graduating from the Sterline-Morton High School in Berwyn, Illinois, he entered the University of Illinois. On June 11, 1913 he was graduated with a Bachelor of Science Degree in Architectural Engineering with final and special honors. He had received preliminary honors in 1911. From 1913 to 1917, he was employed by the Harlem Bridge Company in Memphis, Tennessee. During World War I he was superintendent of engineering in the U.S. Navy Yard in Norfolk, Virginia.

In 1919 he received a Bachelor of Science Degree in Civil Engineering from the University of Illinois, then joined the Morgan Engineering Company in Memphis where he worked 14 years as superintendent of engineering.

When Dr. A. E. Morgan was appointed Chairman of the Tennessee Valley Authority in 1933, he employed Mr. Wiersema as office assistant—the first engineer to be employed by TVA. His first assignment was to determine the optimum elevation of Norris Dam

Soon he was advanced to Assistant Director of the Maps and Surveys Division which prepared survey data and topographic maps, all very essential for planning projects for the unified development of the Tennessee Valley.

He became General Office Engineer in 1935, with duties directed to organizing and staffing the new Engineering and Geology Division which was responsible for the planning, design and construction of all multipurpose projects. In 1937, three major divisions were established under the Chief Engineer (Water Control Planning, Engineering Design, and Construction) and Mr. Wiersema was promoted to the position of Assistant to the Chief Engineer.

As the TVA construction program expanded, especially in World War II and the Korean War, he handled many important management functions in personnel, budgets and estimates, labor relations, technical reports and special technical and administrative assignments.

Throughout his long career with TVA, from 1933 until his retirement in 1960, his services were characterized by a high degree of professionalism and integrity. His tremendous loyalty to TVA—its concepts and goals—were shown in many ways during his years in retirement. He was instrumental in organizing the TVA Retirees Association, and volunteering his services on various TVA-TVARA joint projects. He was an avid reader and writer, serving on the Board of the *Tennessee Valley Engineer,* and a regular contributor of articles.

Upon leaving TVA, he was employed by the Development and Resources Corporation of New York as principal planning engineer. In this position he planned the Salto Grande Dam on Uruguay River. He resided in Buenos Aires, Argentina during his work. Later, he worked on dams in Iran, the Soviet Union and other parts of the world. In 1963 he worked for the R. W. Beck Company in Seattle, Washington, and in 1965 for the British Columbia Hydro in Vancouver, British Columbia.

On June 20, 1925 he was married to Ethel Kellogg of Memphis, Tennessee.

He was an active member of the Unitarian Church, a life member of the American Society of Civil Engineers, the National Society of Professional Engineers and the Knoxville Technical Society. He was a registered engineer in Tennessee. He served as manager of the Knoxville Symphony Orchestra in which he played the cello. He founded the East Tennessee Chapter of the American Civil Liberties Union. He was active in the environmental field and served as head of the Clean Environmental Council. In 1970 he received an award from the Knoxville Roundtable of the National Conference of Christians and Jews.

Around 1930 when the steamship Norman exploded and sank in the Mississippi River near Memphis with much loss of life among ASCE members on an excursion, Harry was the last man off the ship and saved several lives due to his fortitude and skill.

In recent years he has served on a legislative committee to secure state regulations for the operation of nursing homes in Tennessee. He also served as a volunteer for the Knox County Runaway Shelter and as a Hospice volunteer at Fort Sanders Hospital working with the terminally ill in the Oncology Center.

Harry is survived by a son, Harry Jr., three daughters, Mrs. Diane Ehlers,, Mrs. Lila Dannhauer and Mrs. Alice Dews, eight grandchildren and five great-grandchildren.

(Memoir prepared by J. B. Perry, M. ASCE.)

Subject Index

New Constitutive Law for Equal Leg Fillet Welds, 580

Pseudodynamic Method for Seismic Performance Testing, 562

Uplift Force-Displacement Response of Buried Pipe, 292

Disposal

Dredging and Dredged Material Disposal, 975

Dissolved oxygen

Biofilm Growths with Sucrose as Substrate, 71

Dissolved Oxygen Model for a Dynamic Reservoir, 91

Flux Use for Calibrating and Validating Models, 66

Nitrification in Water Hyacinth Treatment Systems, 94

Nitrogen Accountability for Fertile Streams, 75

Simultaneous In-Stream Nitrogen and D.O. Balancing, 74

A Standard for the Measurement of Oxygen Transfer in Clean Water, 1013

Stream Dissolved Oxygen Analysis and Control, 70

Distortion

Distortional Buckling of Steel Storage Rack Columns, 644

Diversion structures

Hydropower: Recent Developments, 988

Dolphins, structures

Segmental Post-Tensioned Dolphin at Tagrin Point, 15

Domes, structural

The Hybrid Arena, 849

Putting CADD to Work, 929

Drafting

Shop Drawing Review: Minimizing the Risks, 857

Drag

Comments on Cross-Flow Principle and Morison's Equation, 830

Drag coefficient

Drag of Oscillatory Waves on Spheres in a Permeable Bottom, 761

Draglines

Production Estimating for Draglines, 24

Drainage

Drainage Coefficients for Heavy Land, 421

Drainage Due to Gravity under Nonlinear Law, 419

Kinematic-Wave Method for Peak Runoff Estimates, 700

Drainage systems

Computer Keeps New Orleans: Head Above Water, 911

Development and Management Aspects of Irrigation and Drainage Systems, 972

Drainage Tunnels Save Freeway Link, 915

Optimal Design of Detention and Drainage Channel Systems, 732

Drawdown

Numerical Determination of Aquifer Constants, 386

Volumetric Approach to Type Curves in Leaky Aquifers, 345

Dredging

Deep-Draft Navigation Project Design, 759

Dredge-Induced Turbidity Plume Model, 790

Dredging and Dredged Material Disposal, 975

Drilled piers

Drilled Piers and Caissons II, 976

Drilled shafts

Blind Drilling Down Under, 913

Drilling

Raise Boring in Civil and Mining Applications, 18

Droughts

State Water Supply Management in New Jersey, 744

Ductility

Elastic-Plastic-Softening Analysis of Plane Frames, 525

Interaction of Plastic Local and Lateral Buckling, 607

Dunes

Geometry of Ripples and Dunes, 318

Dynamic analysis

Dynamic Analysis of a Forty-Four Story Building, 567

Dynamic Analysis of Short-Length Gravity Dams, 190

Dynamic Analysis of Structures by the DFT Method, 636

Dynamic Instability Analyses of Axially Impacted Columns, 178

Dynamics of Trusses by Component-Mode Method, 632

Floor Spectra for Nonclassically Damped Structures, 625

Free Vibration Analysis of Continuous Beams, 163

Nonlinear Dynamic Analyses of an Earth Dam, 281

Steady-State Dynamic Analysis of Hysteretic Systems, 222

Vibrational Characteristics of Multi-Cellular Structures, 559

Dynamic characteristics

Ambient Vibration Studies of Golden Gate Bridge II: Pier-Tower Structure, 154

Axisymmetrical Vibrations of Tanks—Analytical, 146

Dynamic loads

Constitutive Model for Loading of Concrete, 506

Effects of Vehicles on Buried, High-Pressure Pipe, 685

Impact Effect on R.C. Slabs: Analytical Approach, 569

Nonlinear Static and Dynamic Analysis of Plates, 136

Silo as a System of Self-Induced Vibration, 481

Dynamic programming

Capacity Expansion of Sao Paulo Water Supply, 741

Stochastic Optimization/Simulation of Centralized Liquid Industrial Waste Treatment, 100

Dynamic response

Cyclic Testing and Modeling of Interfaces, 276

Estimating Elastic Constants and Strength of Discontinuous Rock, 279

Evaluation of Soil Response to EPB Shield Tunneling, 243

Finite Element Analyses of Lock and Dam 26 Cofferdam, 263

Formulation of Drucker-Prager Cap Model, 176

Stress Intensity Factor Using Quarter Point Element, 138

Finite element method

Application of NLFEA to Concrete Structures, 640

Boundary Element Calculations of Diffusion Equation, 144

Dynamic Response of Tall Building to Wind Excitation, 521

FEM Solution of 3-D Wave Interference Problems, 803

Finite Element Modeling of Infinite Reservoirs, 218

Finite Element Modelling of Nonlinear Coastal Currents, 786

Generalized Finite Element Evaluation Procedure, 497

Hydrodynamic Interaction of Flexible Structures, 806

Instability of Thin Walled Bars, 180

Interactive Buckling in Thin-Walled Beam-Columns, 219

Minimum Weight Sizing of Guyed Antenna Towers, 603

New Analysis Techniques for Structural Masonry, 996

On the Reliability of a Simple Hysteretic System, 221

Review of Nonlinear FE Methods with Substructures, 212

Subregion Iteration of Finite Element Method for Large-Field Problems of Ground-Water Solute Transport, 350

Vibrational Characteristics of Multi-Cellular Structures, 559

Wave Interference Effects by Finite Element Method, 758

Finite elements

Static Infinite Element Formulation, 619

Finite strip method

Composite Box Girder Bridge Behavior During Construction, 507

Fisheries

Impact of Lake Acidification on Stratification, 101

Fixed structures

Stiffness Properties of Fixed and Guyed Platforms, 485

Fixed-bed models

Salt River Channelization Project: Model Study, 333

Flanges

Buckling with Enforced Axis of Twist, 224

Eccentrically Loaded High Strength Bolted Connections, 534

Test Evaluation of Composite Beam Design Method, 477

Flexibility

Connection Flexibility and Steel Frames, 965

Model for Flexible Tanks Undergoing Rocking, 134

Tubular Steel Trusses with Cropped Webs, 553

Flexible pavements

Extension of CBR Method to Highway Pavements, 710

Minimum-Cost Design of Flexible Pavements, 683

Flexural strength

Flexural Limit Design of Column Footings, 613

Flexural Strength of Masonry Prisms versus Wall Panels, 597

Flexure

Conjugate Frame for Shear and Flexure, 508

Flexure of Statically Indeterminate Cracked Beams, 223

Probabilistic Basis for Design Criteria in Reinforced Concrete, 1001

Floating breakwaters

Floating Breakwater Design, 778

Hinged Floating Breakwater, 818

Floating ice

Rigid-Plastic Analysis of Floating Plates, 174

Floating structures

Response of Semi-Submersibles in Ice Environment, 667

Flocculants

Modified Approach to Evaluate Column Test Data, 63

Flocculation

A Critique of Camp and Stein's RMS Velocity Gradient, 96

Floes

Response of Semi-Submersibles in Ice Environment, 667

Flood control

Application of Extreme Value Theory to Flood Damage, 756

Determination of Urban Flood Damages, 743

Development of a Flood Management Plan, 753

Flood Storage in Reservoirs, 418

Lower Mississippi Valley Floods of 1982 and 1983, 754

Overbank Flow with Vegetatively Roughened Flood Plains, 395

Research Agenda for Floods to Solve Policy Failure, 729

Significance of Location in Computing Flood Damage, 730

Flood damage

Application of Extreme Value Theory to Flood Damage, 756

Determination of Urban Flood Damages, 743

Research Agenda for Floods to Solve Policy Failure, 729

Significance of Location in Computing Flood Damage, 730

Flood frequency

Large Basin Deterministic Hydrology: A Case

Marketing
The Client Relationship: Effective Marketing Steps, 446

Markov chains
Markov Renewal Model for Maximum Bridge Loading, 192

Markov process
Extreme Winds Simulated from Short-Period Records, 474
Markov Renewal Model for Maximum Bridge Loading, 192

Marshes
Marsh Enhancement by Freshwater Diversion, 726

Masonry
Flexural Strength of Masonry Prisms versus Wall Panels, 597
Limit States Criteria for Masonry Construction, 476
Mechanics of Masonry in Compression, 524
New Analysis Techniques for Structural Masonry, 996

Mass transfer
External Mass-Transfer Rate in Fixed-Bed Adsorption, 106
A Standard for the Measurement of Oxygen Transfer in Clean Water, 1013
Volatilization Rates of Organic Chemicals of Public Health Concern, 97

Mass transport
Field Comparison of Three Mass Transport Models, 315
Water Particle Velocities in Regular Waves, 771

Materials engineering
Developments in New and Existing Materials, 973

Materials, properties
Field Aging of Fixed Sulfur Dioxide Scrubber Waste, 231
High Strength Steel, 870
Inelastic Buckling of Steel Plates, 475
Structural Plastics Selection Manual, 1017
Third-Variant Plasticity Theory for Low-Strength Concrete, 158

Materials tests
Biaxial Stress-Strain Relations for Brick Masonry, 539
Catalytic Modification of Road Asphalt by Polyethylene, 671
Statistical Analysis of Specification Compliance, 690
Structural Plastics Selection Manual, 1017

Mathematical models
Analysis and Simulation of Low Flow Hydraulics, 408
Bed Topography in Bends of Sand-Silt Rivers, 405
Danger: Natural System Modeled by Computer, 908
Dimensionless Formulation of Furrow Irrigation, 444
Distortions Associated with Random Sea Simulators, 800
Flow under Tilt Surface for High-Rate Settling, 67

Hydrodynamic Pressures Acting Upon Hinged-Arc Gates, 352
Leakage from Ruptured Submarine Oil Pipeline, 711
Mathematical Model to Predict 3-D Wind Loading on Building, 141
Mathematical Models within Geodetic Frame, 658
Model of Dispersion in Coastal Waters, 316
Modeling of Unsteady Flows in Alluvial Streams, 332
Overland Flow Hydrographs for SCS Type II Rainfall, 436
Prediction of 2-D Bed Topography in Rivers, 391
Similarity Solution of Overland Flow on Pervious Surface, 381
Uniaxial Cyclic Stress-Strain Behavior of Structural Steel, 193
Unified Analysis of Biofilm Kinetics, 90
Water and Sediment Routing Through Curved Channels, 356

Maximum load
Wave Forces on Vertical Walls, 805

Meandering streams
Formation of Alternate Bars, 406

Measurement
A Standard for the Measurement of Oxygen Transfer in Clean Water, 1013

Measuring instruments
Automated Remote Recording and Analysis of Coastal Data, 784
Bore Height Measurement with Improved Wave-staff, 791
Temperature Effects on Volume Measurements, 242

Mechanical properties
A Method for Finding Engineering Properties of Sealants, 177

Membranes
Stability of Membrane Reinforced Slopes, 304

Membranes, linings
Plastic Lining on Riverton Unit, Wyoming, 437

Memoirs of deceased members
Burnett, Daniel H., 1027
Farmer, Edward August, 1028
Fritts, Donald Loyd, 1029
Harrar, Robert Kenneth, 1030
Hinton, Julian Pitts, 1031
Hume, Ritchey, 1032
Jarvis, Ross F., 1033
Johnson, Alfred Massey Fisher, 1034
Pederson, Lyle Perry, 1035
Rhodes, Fred Harold, Jr., 1036
Sackett, Guy Woodford, 1037
Samuel, Thomas D. III., 1038
Tschebotarioff, Gregory P., 1026
Wiersema, Harry Anthony, 1039

Meteorological data
Extended Streamflow Forecasting Using NWSRFS, 736

Meteorology

407

Structural reliability

Calibration of Bridge Fatigue Design Model, 548

Lateral-Torsional Motion of Tall Buildings, 627

Mechanistic Seismic Damage Model for Reinforced Concrete, 516

Models for Human Error in Structural Reliability, 554

Non-Normal Responses and Fatigue Damage, 206

Reliability of Ductile Systems with Random Strengths, 551

Sensitivity Analysis for Structural Errors, 579

Sensitivity of Reliability-Based Optimum Design, 577

Stochastic Evaluation of Seismic Structural Performance, 545

Structural Optimization Using Reliability Concepts, 614

Structural Safety Studies, 1018

Structural response

Variability in Long-Term Concrete Deformations, 583

Structural safety

Sensitivity Analysis for Structural Errors, 579

Structural Safety Studies, 1018

Structural settlement

Dewatering--Avoiding Its Unwanted Side Effects, 974

Structural steels

Composite and Mixed Construction, 962

Multi-Truss Design for Tower, 894

Uniaxial Cyclic Stress-Strain Behavior of Structural Steel, 193

Struts

Sparsely Connected Built-Up Columns, 509

Stub girders

Direct Model Test of Stub Girder Floor System, 563

Students

Challenges to Civil Engineering Educators and Practitioners—Where Should We Be Going?, 959

Civil Engineering and Engineering Enrollments, 116

Subdivisions

Subdivision Froude Number, 384

Subgrades

Analysis of Plates on a Kerr Foundation Model, 210

Moisture Curve of Compacted Clay: Mercury Intrusion Method, 296

Orthotropic Annular Shells on Elastic Foundations, 203

Use of NDT and Pocket Computers in Pavement Evaluation, 688

Submarine pipelines

Leakage from Ruptured Submarine Oil Pipeline, 711

Wave Load-Submarine Pipeline-Seafloor Interaction, 686

Wave-Induced Forces on Buried Pipeline, 792

Submerged discharge

Circulation Induced by Coastal Diffuser Discharge, 823

Submerged flow

Transport of Suspended Material in Open Submerged Streams, 363

Substructures

Dynamics of Trusses by Component-Mode Method, 632

Review of Nonlinear FE Methods with Substructures, 212

Subsurface drainage

Drainage Required to Manage Salinity, 429

Subsurface flow

Water Division at Low-Level Waste Disposal Sites, 95

Subsystems

Dynamic Characterization of Two-Degree-of-Freedom Equipment-Structure Systems, 127

Seismic Response of Light Subsystems on Inelastic Structures, 495

Subways

Slurry Walls Protect Harvard Square, 948

Supports

Lateral Stability of Beams with Elastic End Restraints, 155

Surf zone

Bore Height Measurement with Improved Wavestaff, 791

Qualitative Description of Wave Breaking, 770

Velocity Moments in Nearshore, 774

Surface defects

Predicting Sinkhole Collapse, 943

Surface drainage

Water Division at Low-Level Waste Disposal Sites, 95

Surface irrigation

Electronic Clocks for Timing Irrigation Advance, 420

Surface jets

Surface Buoyant Jets in Steady and Reversing Crossflows, 364

Surface waters

Hydraulics and Hydrology in the Small Computer Age, 987

Time Series Models for Treatment of Surface Waters, 51

Water Quality and Regional Water Supply Planning, 742

Water Quality Issues at Fossil Fuel Plants, 1024

Surface waves

Shallow Trenches and Propagation of Surface Waves, 142

Surveying

Definition of the Term "Engineering Surveying", 663

Engineering Surveying Manual, 979

The Surveyor and Written Boundary Agreements, 662

Surveys, data collection

Author Index

Plants, 1012

Structural Design, Cementitious Products, and Case Histories, 1016

Anastasiou, Kostas
see Hedges, Terence S., 776

Anders, John C.
see Highter, William H., 269

Andersen, Ole H.
see Fredsoe, Jorgen, 828

Anderson, Duwayne M., ed.
Freezing and Thawing of Soil-Water Systems, 985

Anderson, Jon Baxter
A Method for Finding Engineering Properties of Sealants, 177

Andersson, Helge I.
Spillage over an Inclined Embankment, 397

Ang, Alfred H. -S.
see Park, Young-Ji, 516

Ang, Alfredo H. -S.
see Park, Young- Ji, 517
see Sues, Robert H., 545

Angeles, Honorato L.
Efficient Water Use in Run-of-the-River Irrigation, 425

Appelbaum, Stuart J.
Determination of Urban Flood Damages, 743

Arditi, David
Construction Productivity Improvement, 17
Railway Route Rationalization: A Valuation Model, 675

Arici, Marcello
Analogy for Beam-Foundation Elastic Systems, 576

Arnold, Gregory C.
see Salamonowicz, Paul H., 661

Arnold, J. G.
see Williams, J. R., 375

Arockiasamy, M.
Response of Semi-Submersibles in Ice Environment, 667

Aronberg, Ralph
Motivating and Managing Engineers, 112

Arora, Jasbir S.
see Ryu, Yeon S., 212

Arulanandan, Kendiah
see Arulmoli, Kendiah, 239

Arulmoli, Kendiah
New Method for Evaluating Liquefaction Potential, 239

Arumugam, Kanapathypilly
see Mason, Peter J., 330

Arya, A. S.
see Krishna, P., 596

Asawa, Girdhari, L.
Radial Turbulent Flow Between Parallel Plates, 359

Asbury, Gregory E.
see Hribar, John P., 460

Assadi, Mahyar
Stability of Continuously Restrained Cantilevers, 217

Astaneh-Asl, Abolhassan
Cyclic Out-of-Plane Buckling of Double-Angle Bracing, 542

Atkinson, Gail M.
see Vick, Steven, G., 283

Au, Tung
see Hendrickson, Chris, 458

Auckle, Dev
see Ramamurthy, A. S., 417

Aukrust, Trond
Turbulent Viscosity in Rough Oscillatory Flow, 793

Austin, Mark A.
Design of a Seismic-Resistant Friction-Braced Frame, 643

Avent, R. Richard
Decay, Weathering and Epoxy Repair of Timber, 490

Ayyub, Bilal M.
Decisions in Construction Operations, 42

Azad, Abdul K.
see Baluch, Mohammed H., 223

Azia, Nadim M.
Sediment Transport in Shallow Flows, 399

Babcock, Susan, M.
see Schreyer, Howard L., 158

Baber, Thomas T.
Random Vibration of Degrading, Pinching Systems, 186

Badie, A.
see Wang, M. C., 289

Bagley, Jay M.
Satisfying Instream Flow Needs under Western Water Rights, 737

Bahndorf, Joachim
see Gründig, Lother, 659

Bakan, Lloyd H.
The Client Relationship: Effective Marketing Steps, 446

Baker, Christopher J.
see Elliott, Keith R., 385

Baker, Clyde N., Jr., ed.
Drilled Piers and Caissons II, 976

Baker, Wallace Hayward, ed.
Issues in Dam Grouting, 991

Balendra, Thambirajah
Vibrational Characteristics of Multi-Cellular Structures, 559

Baligh, Mohsen M.
Strain Path Method. 295

Ballard, G. Stephen

Water and Sediment Routing Through Curved Channels, 356

Chang, Luh-Maan
Evaluation of Craftsman Questionnaire, 48

Chang, Peter C.
Analytical Modeling of Tube-In-Tube Structure, 552

Chang, Rong
see Haroun, Medhat A., 825

Chang, Shoou-Yuh
Generating Designs for Wastewater Systems, 92
see Huang, Ju-Chang, 71

Changnon, Stanley A.
Research Agenda for Floods to Solve Policy Failure, 729

Chaplin, John R.
Morison Inertia Coefficients in Orbital Flow, 772

Char, A. N. R.
see Chaudhari, A. P., 307

Charbeneau, Randall J.
see Wirojanagud, Prakob, 342

Chatterjee, Arun
Innovative Strategies to Improve Urban Transportation Performance, 990

Chaturvedi, M. C.
Irrigation System Study in International Basin, 734

Chaube, U. C.
see Chaturvedi, M. C., 734

Chaudhari, A. P.
Flexural Behavior of Reinforced Soil Beams, 307

Chaudhry, M. Hanif
Analysis and Stability of Closed Surge Tanks, 383

Chavez-Morales, Jesus
see Holzapfel, Eduardo A., 439

Chawla, Amrik S.
Average Uplift Computations for Hollow Gravity Dams, 344

Chen, Albert T. F.
Transmitting Boundaries and Seismic Response, 244

Chen, Carl W.
Effect of Ambient Air Quality on Throughfall Acidity, 72

Chen, En-Sheng
Constitutive Model for Concrete in Cyclic Compression, 173

Chen, Jer-Shi
Appropriate Forms in Nonlinear Analysis, 201

Chen, W. F.
see Liu, X. L., 10

Chen, Wai Fah
see Liu, Xila, 535

Chen, Wai-Fah, ed.
Connection Flexibility and Steel Frames, 965

Chen, Wai-Fah

see Sugimoto, H., 593
see Zhou, Shi-Ping, 637

Chen, Yung Hai
Salt River Channelization Project: Model Study, 333

Cheng, Edmond D. H.
Extreme Winds Simulated from Short-Period Records, 474

Cheong, Hin-Fatt
Gravimetric Statistics of Riprap Quarrystones, 797

Chern, Jenn-Chuan
see Bazant, Zdenek P., 130, 149

Chern, Jing C.
see Valid, Voginder P., 301

Chi, Nai-Yuan
see Rounds, Jerald L., 26

Chi, W. K.
see Voyiadjis, G. Z., 195

Chien, Chi-Hui
Effective Length of a Fractured Wire in Wire Rope, 182

Chin, David A.
Model of Dispersion in Coastal Waters, 316
Outfall Dilution: The Role of a Far-Field Model, 80
Time Series Modeling of Coastal Currents, 822

Chiu, Arthur N. L.
see Nakamoto, Reginald T., 616
see Cheng, Edmond D. H., 474

Chiu, Chao-Lin
see Mizumura, Kazumasa, 327

Cho, Kyo Zong
see Barker, Donald B., 162

Cho, Woncheol C.
see Yeh, Gour-Tsyh, 350

Choi, Chang-Koon
Multistory Frames Under Sequential Gravity Loads, 620

Choi, Y. K.
see Mesri, Gholamreza, 259

Chopra, Anil K.
Simplified Earthquake Analysis of Structures with Foundation Uplift, 527
see Fenves, Gregory, 168, 169, 505
see Yim, Solomon C.-S., 641

Chou, Gee David
see Vallabhan, C. V. Girija, 623

Chou, Karen C.
Nonlinear Response to Sustained Load Processes, 478

Chow, Fong-Yen
see Narayanan, Rangachari, 480

Chow, Y. K.
Torsional Response of Piles in Nonhomogeneous Soil, 285

Christensen, G. Lee

Endy, Elizabeth M.
Ethics in the Field, 954

Ermopoulos, John C.
Stability of Frames with Tapered Built-up Members, 594

Esparza, Edward D.
Trench Effects on Blast-Induced Pipeline Stresses, 299

Ettema, Robert
see Raudkivi, Arved J., 360

Everett, Jess
Bound Glass in Shredded Municipal Solid Waste, 233

Everts, Craig H.
Sea Level Rise Effects on Shoreline Position, 824

Faggiano, Paolo
see Zavelani, Adolfo, 624

Fairweather, Virginia
Building in Space, 884
Clearing the Decks, 922
Engineering Education: An Update, 904
The Pursuit of Quality: QA/QC, 848
see Godfrey, K. A., Jr., 833
see Zetlin, Lev, 910

Fanella, D.
Continuum Damage Mechanics of Fiber Reinforced Concrete, 185

Fangmeier, D. D.
see Yitayew, Muluneh, 426

Farber, Kurt
see Bechteler, Wilhelm, 319

Farid, Foad
Fair and Reasonable Markup (FaRM) Pricing Model, 44

Farquhar, G. J.
see Ellis, J. H., 56, 100

Fattah, Qais N.
Dispersion in Anisotropic, Homogeneous, Porous Media, 365

Faust, Charles R.
see Mercer, James W., 16

Feng, Chuan C.
see Qin, Bosheng, 630

Fenton, John
A Fifth-Order Stokes Theory for Steady Waves, 773

Fenton, John D.
Wave Forces on Vertical Walls, 805

Fenves, Gregory
Reservoir Bottom Absorption Effects in Earthquake Response of Concrete Gravity Dams, 505
Simplified Earthquake Analysis of Concrete Gravity Dams: Combined Hydrodynamic and Foundation Interaction Effects, 168
Simplified Earthquke Analysis of Concrete Gravity Dams: Separate Hydrodynamic and Foundation Interaction Effects, 169

Fessehaye, M.
see Liu, H., 374

Figueroa, Ludwig
Inverted Shear Modulus from Wave-Induced Soil Motion, 240

Finno, Richard J.
Evaluation of Soil Response to EPB Shield Tunneling, 243

Fisher, John
see Koehn, Enno, 114

Fisher, John W.
Hundreds of Bridges—Thousands of Cracks, 865

Fiuzat, Abbas A.
see Chen, Yung Hai, 333

Flaxman, E. W.
see O'Rourke, Tom D., 949

Flick, Reinhard E.
see Zetler, Bernard D., 809

Fogler, H. Scott
see Khilar, Kartic C., 278

Ford, David T.
see Sabet, M. Hossein, 740

Forde, Bruce
Steel Construction Evaluation by MLR-Strategies, 29

Foschi, Ricardo O.
Wood Floor Behavior: Experimental Study, 628

Fowler, David W.
see Nelson, Erik L., 512

Fowler, Jack
Building on Muck, 875

Fowler, John E.
see Chaudhry, M. Hanif, 383

Fox, Arthur J., Jr., ed.
Quality in the Constructed Project, 1002

France, John W.
see Poulos, Steve J., 275

Frangopol, Dan M.
Sensitivity of Reliability-Based Optimum Design, 577
Structural Optimization Using Reliability Concepts, 614

Frank, Andrew U.
Distributed Data Bases for Surveying, 656

Frantz, Gregory C.
see Hsiung, Wayne, 492

Fredsoe, Jorgen
Distribution of Suspended Sediment in Large Waves, 828

Freehling, Dan J.
Use of Computerized Databases in Legal Research, 650

Frenette, Marcel
see Julien, Pierre Y., 400

Fricker, Jon D.

see Palmer, Richard N., 752

Luthy, Richard G.
see Banz, Iris, 69

Lyman, Tracy J.
Heavy Oil Mining—An Overview, 666

Lynn, Brian A.
Wind-Induced Fatigue on Low Metal Buildings, 522

Maaskant, R.
The Stability of Cylindrical Air-Supported Structures, 220

Macaitis, Bill
Managing Productivity During an Attrition Program, 451

Macak, Joseph J., III.
see Haas, Charles N., 85

McBean, E. A.
see Ellis, J. H., 56, 100

McBean, Edward A.
see Burn, Donald H., 336
see Dandy, Graeme C., 727

McBean, Robert P.
Wind Load Effects on Flat Plate Solar Collectors, 491

McCall, Deborah J.
see Walski, Thomas M., 61

McCartney, Bruce L.
Deep-Draft Navigation Project Design, 759
Floating Breakwater Design, 778

McConnell, Daniel R.
Earned Value Technique for Performance Measurements, 454

McCorquodale, John A.
see Hannoura, A'Alim A., 811, 812

McCuen, Richard H.
One Plus One Makes Thirty, 118

McCutcheon, William J.
Racking Deformations in Wood Shear Walls, 486

McDonnell, Archie J.
see Warwick, John J., 74, 75

McDonough, James F.
A Common Sense Approach to the Fundamentals Exam, 123
The Microcomputer in C.E. Education: A Survey, 665

McDougal, William G.
see Leach, Patrick A., 818

McFillen, James M.
see Maloney, William F., 21

McGartland, M.
Expert Systems for Construction Project Monitoring, 39

McGuire, William
see Pesquera, Carlos I., 645

Machemehl, Jerry L., ed.

see Bennett, F. Lawrence, d, 960

MacKenzie, Mary C.
see Palmer, Richard N., 757

McKinney, James
see Koehn, Enno, 114

McKittrick, Harold V., ed.
see Del Re, Robert, ed., 1009

McLaughlin, John
Land Information Management: A Canadian Perspective, 657

McMahon, Donald R.
Monitoring Saves a Site, 873

McNally, Michael G.
see Recker, Wilfred W., 678

McNary, W. Scott
Mechanics of Masonry in Compression, 524

McVay, Michael
Cyclic Behavior of Pavement Base Materials, 234

Madsen, Henrik O.
Extreme-Value Statistics for Nonlinear Stress Combination, 194

Madsen, Stanley H.
How Do Engineers Learn to Manage?, 846

Madureira, Claudio J.
see Garga, Vinod K., 288

Magoon, Orville T., ed.
Coastal Zone '85, 961

Mahin, Stephen A.
Pseudodynamic Method for Seismic Performance Testing, 562
see Lin, Jon, 495
see Popov. Egor P., 611

Mahmassani, Hani S.
Transportation Requirements for High Technology Industrial Development, 703

Mahmood, Khalid
see Haque, Muhammad I., 318

Maison, Bruce F.
Dynamic Analysis of a Forty-Four Story Building, 567

Maiti, Surjya
Rammed Earth House Construction, 306

Majcherek, Hanna
Modeling of Flow Velocity Using Weirs, 320
Submerged Weirs, 326

Makigami, Yasuji
An Analytical Method of Traffic Flow Using Aerial Photographs, 697

Males, Richard M.
Algorithm for Mixing Problems in Water Systems, 329
see Clark, Robert M., 755

Malhotra, Sundershan K.
see Thomas, Babu, 532

Maloney, William F.

Moehle, Jack P.
Confinement Effectiveness of Crossties in RC, 602
Lateral Load Response of Flat-Plate Frame, 605

Mohan, Satish
Multi-Attribute Utility in Pavement Rehabilitation Decisions, 701

Molinas, Albert
Generalized Water Surface Profile Computations, 340

Monasa, Frank
see Abdel-Sayed, George, 600, 601

Montgomery, Raymond L., ed.
Dredging and Dredged Material Disposal, 975

Moore, Ian D.
Analytical Theory for Buried Tube Postbuckling, 181

Moore, James E.
Statistical Designation of Traffic Control Subareas, 684

Morgan, J. R.
see Nelson, J. K., 595

Morgan, Joe Miller
see Walski, Thomas M., 61

Morisako, Kiyetaka
see Ishida, Shuze, 151

Morlan, Terry H., ed.
Energy Forecasting, 977

Morris, Glenn A.
Tubular Steel Trusses with Cropped Webs, 553

Morrison, Denby G.
Dynamic Lateral-Load Tests of R/C Column-Slabs, 514

Moses, F., ed.
see Yao, James T. P., ed., 1018

Moses, Fred
see Ghosn, Michel, 192
see Nyman, William E., 548

Mosquera, J. M.
Impact Tests on Frames and Elastic-Plastic Solutions, 213

Muggeridge, D. B.
see Arockiasamy, M., 667

Mulinazzi, Thomas E.
see Chadda, Himmat S., 707

Munfakh, George
Wharf Stands on Stone Columns, 834

Murphy, G. J.
San Francisco Outfall: The Champ?, 952

Murphy, Peter J.
Bed Load or Suspended Load, 321
Equilibrium Boundary Condition for Suspension, 322

Murray, Louis C., Jr.
Phosphogypsum Waste Anion Removal by Soil Minerals, 93

Murthy, B. R. Srinivasa
see Nagaraj, T. S., 284

Murti, Viriyawan
Stress Intensity Factor Using Quarter Point Element, 138

Murty, Tad S.
Influence of Marginal Ice Cover on Storm Surges, 780

Muscolino, Giuseppe
see Di Paola, Mario, 598

Muslu, Yilmaz
see Uyumaz, Ali, 325

Muspratt, Murray A.
Ethics of Professionalism, 125
Issues in Engineering Education, 120

Myint, Kyaw
Design and Construction of Dade County METROMOVER, 725

Naaman, A. E.
see Harajli, M. H., 570

Nadaskay, Anthony J.
Direct Model Test of Stub Girder Floor System, 563

Nagai, Toshihiko
see Figueroa, Ludwig, 240

Nagaraj, T. S.
Compressibility of Partly Saturated Soils, 284

Nagaya, Kosuke
Response of Variable Cross-Sectional Members to Waves, 781

Nair, V. V. D.
see Kumar, A., 485

Nakamoto, Reginald T.
Investigation of Wind Effects on a Tall Guyed Tower, 616

Nakamura, Tsuneyoshi
Optimum Building Design for Forced-Mode Compliance, 197

Nakayama, Tomoaki
Strength of Resin Mortar Tunnel, 225

Nalini, V. N.
see Chakrabarti, A., 216

Nandakumar, K.
see Masliyah, Jacob H., 372

Nandakumar, V.
see Ahuja, Hira N., 41

Narayanan, Rangachari
Design of Slender Webs Having Rectangular Holes, 519
Effect of Support Conditions on Plate Strengths, 480

Narayanan, Rangaswami
see Hassan, N. M. K. Nik, 403

Nash, Andrew B.
see Furth, Peter G., 689

Olivieri, Adam W.
see Eisenberg, Don M., 76

Olkin, Ingram
see Simiu, Emil, 529

Olyniec, James H., ed.
Transition in the Nuclear Industry, 1020

O'Melia, Charles R.
Particles, Pretreatment, and Performance in Water Filtration, 105

O'Neill, Michael W.
see Nogami, Toyoaki, 165

Oppenheim, I. J.
see Allen, R. H., 557

O'Rourke, Michael J.
Drift Snow Loads on Multilevel Roofs, 488

O'Rourke, Thomas D.
see Trautmann, Charles H., 292, 293

O'Rourke, Tom D.
Pipe Laying Comes Out of the Trenches, 949

Orringer, Oscar, ed.
see Tong, Pin, ed., 984

O'Shaughnessy, James C., ed.
Environmental Engineering, 980

Osmolski, Zbig
see Ponce, Victor Miguel, 393

Ossenbruggen, Paul J.
Time Series Models for Treatment of Surface Waters, 51

Oster, J. D.
Chemical Reactions within Root Zone of Arid Zone Soils, 430

Ouellette, Pierre
Application of Extreme Value Theory to Flood Damage, 756

Overcash, Michael R.
see Loehr, Raymond C., 57

Owen, Austin A.
Concrete Armor Unit Form Inventory, 808

Owens, Emmet M.
see Effler, Steven W., 101

Oxygen Transfer Standards Committee of The American Society of Civil Engineers
A Standard for the Measurement of Oxygen Transfer in Clean Water, 1013

Pacal, Rudy M.
see Greenberg, Maurice S., 681

Padron, Dennis V.
Pier Review, 921

Page, Adrian W.
see Dhanasekar, M., 539

Page, John H.
Planning Development with Transit Projects, 717

Palmer, Dennis E.
A Tale of Six Cities, 866

Palmer, Richard N.
Multi-Objective Analysis with Subjective Information, 752
Optimization of Water Quality Monitoring Networks, 757

Panahi, Zahra
Power Production and Air Conditioning by Solar Ponds, 228

Panchang, Vijay G.
see Pearce, Bryan R., 801

Pande, Pramod K.
see Asawa, Girdhari, L., 359

Pandit, Ganpat S.
see Venkappa, Velpula, 483

Pant, Prahlad D.
see Bhalla, Manmohan K., 674

Papadrakakis, Manolis
Inelastic Cyclic Analysis of Imperfect Columns, 546

Papo, Haim B.
Relative Error Analysis of Geodetic Networks, 660

Parchure, Trimbak M.
Erosion of Soft Cohesive Sediment Deposits, 398

Paris, W. C. Pete
see Grimm, Dan, 924

Park, Young- Ji
Seismic Damage Analysis of Reinforced Concrete Buildings, 517
Mechanistic Seismic Damage Model for Reinforced Concrete, 516

Partheniades, Emmanuel
see Dermissis, Vassilios, 807

Pasche, Erik
Overbank Flow with Vegetatively Roughened Flood Plains, 395

Pashanasangi, Sasan
see Harik, Issam E., 564

Patel, H. S.
see Knight, D. W., 317

Patrick, Dale I.
see Corbitt, Robert A., 858

Patton-Mallory, Marcia
Light-Frame Shear Wall Length and Opening Effects, 610

Paudyal, Guna Nidhi
see Gupta, Ashim Das, 438

Paulson, Boyd C., Jr.
Automation and Robotics for Construction, 32

Pavlovic, Milija N.
On the Computation of Slab Effective Widths, 493

Peacock, Robert T.
see Cassidy, John J., 341

Pearce, Bryan R.
A Method for Investigation of Steady State Wave Spectra in Bays, 801

Pecknold, D. A.

see Shukla, J. H., 836

Stahl, Bernhard
see Knapp, Alphia E., 573

Stanbro, W. D.
see Aggour, M. S., 231

Stansfield, Robert C.
see Davis, Edward C., 95

Stathopoulos, Theodore
see Lynn, Brian A., 522

Stefan, H.
see Akiyama, J., 411

Steffler, P. M.
LDA Measurements in Open Channel, 323

Steffler, Peter M.
Water Surface at Change of Channel Curvature, 369

Steger, Joseph A.
Engineers as Managers, 456

Stelzer, David
see Papo, Haim B., 660

Stephenson, Roger V.
Performance of Surface Rotors in an Oxidation Ditch, 54

Stessel, Richard I.
Separation of Solid Waste With Pulsed Airflow, 102

Stessel, Richard Ian
Demoresearch for Resource and Energy Recovery, 111

Stevens, Herbert H.
see Hubbell, David W., 358

Stewart, Rita F.
Production Estimating for Draglines, 24

Stewart, William
Temperature Effects on Volume Measurements, 242

Stiefel, Ulrich
see O'Rourke, Michael J., 488

Stiemer, Siegfried F.
see Forde, Bruce, 29

Stoll, R. D.
Expanded Shale Lightweight Fill: Geotechnical Properties, 290

Stover, Enos L.
Making Treatment Plants Work, 856

Street, R. L.
see Koseff, J. R., 337

Strelkoff, Theodor
Dimensionless Formulation of Furrow Irrigation, 444

Struiksma, Nico
Prediction of 2-D Bed Topography in Rivers, 391

Stubbs, Norris, ed.
Damage Mechanics and Continuum Modeling, 970

Stukhart, George, ed.
Construction QA/QC Systems that Work: Case

Studies, 967

Sturm, Terry W.
Simplified Design of Contractions in Supercritical Flow, 370

Suaris, Wimal
Constitutive Model for Loading of Concrete, 506

The Subcommittee on High-Speed Rail Systems of the Committee on Public Transport of the Urban Transportation Division
High-Speed Rail Systems in the United States, 673

Subcommittee on Vibration Problems Associated with Flexural Members on Transit Systems, Committee on Flexural Members of the Committee on Metals of the Structural Division
Dynamics of Steel Elevated Guideways—An Overview, 588

Sues, Robert H.
Stochastic Evaluation of Seismic Structural Performance, 545

Sugimoto, H.
Inelastic Post-Buckling Behavior of Tubular Members, 593

Sugiura, Kunitomo
Dynamic Instability Analyses of Axially Impacted Columns, 178

Suidan, Makram T.
Unified Analysis of Biofilm Kinetics, 90
see Baskin, Don E., 50
see Wang, Yi-Tin, 79

Summers, R. Scott
see Roberts, Paul V., 106

Sun, Chang-Ning
Testing of Cement-Mortar Lined Carbon Steel Pipes, 669

Swaddiwudhipong, Somsak
Fiber-Reinforced Concrete Deep Beams with Openings, 575

Swamidas, A. S. J.
see Arockiasamy, M., 667

Sweazy, Robert M.
Can We Save the Ogallala?, 906

Syamal, Pradip K.
see Pekau, O. A., 156

Symonds, P.S.
see Mosquera, J. M., 213

Tadanier, Roman
Soil Security Test for Water Retaining Structures, 500

Tadjbakhsh, Iradj G.
see Constantinou, Michalakis C., 515, 642

Tadros, Maher K.
see Ghali, Amin, 586

Taesiri, Yongyuth
see McVay, Michael, 234

Taigbenu, A. E.
Boundary Element Calculations of Diffusion Equation, 144

Takahashi, Shigeo
see Yamamoto, Tokuo, 762

Takatsuka, Toshio
see Nakayama, Tomoaki, 225

Takewaki, Izuru
see Nakamura, Tsuneyoshi, 197

Talbot, James R.
see Sherard, James L., 9

Tallin, Andrew
Wind Induced Lateral-Torsional Motion of Buildings, 608

Talling, Andrew
see Ellingwood, Bruce, 476

Tanji, K. K.
see Oster, J. D., 430

Tarkoy, Peter J.
Portables Pay Off, 932

Tarpy, Thomas S., Jr.
Continuous Timber Diaphragms, 533

Tarquin, A. J.
see Craver, W., Jr., 121

Tart, Rupert G., Jr., ed.
see Krzewinski, Thomas G., ed., 1019

Task Committee on Design of Outlet Control Structures
Stormwater Detention Outlet Control Structures, 1015

Task Committee on Federal Policies in Water Resources Planning of the Management Resources Planning and Management Division of the American Society of Civil Engineers
Federal Policies in Water Resources Planning, 981

Task Committee on Properties of Selected Plastics Systems of the Structural Plastics Research Council of the Technical Council on Research of The American Society of Civil Engineers
Structural Plastics Selection Manual, 1017

Task Committee on Quantifying Land-Use Change Effects of the Watershed Management and Surface-Water Committees of the Irrigation and Drainage Division
Evaluation of Hydrologic Models Used to Quantify Major Land-Use Change Effects, 413

The Task Committee on Redundancy of Flexural Systems of the ASCE-AASHTO Committee on Flexural Members of the Committee on Metals of the Structural Division
State-of-the-Art Report on Redundant Bridge Systems, 629

Task Committee on Revision of Construction Cost Control Manual of the Construction Division of the American Society of Civil Engineers
Construction Cost Control, 966

Task Committee on the Application of Small Computers in Construction
Application of Small Computers in Construction, 31

Task Committee on Updating Manual 42 of the Committee on Dynamic Effects of the Structural

Division of the American Society of Civil Engineers
Design of Structures to Resist Nuclear Weapons Effects, 971

Tassoulas, John L.
see Gazetas, George, 274

Tatom, Frank B.
Errors from Using Conservation of Buoyancy Concept in Plume Computations, 377

Tavakoli, Amir
Productivity Analysis of Construction Operations, 19

Tavoularis, Stavros
see Prinos, Panagiotis, 394

Tayel, Magdy A.
see Haroun, Medhat A., 145, 146

Taylor, Charles M.
see Emanuel, Jack H., 520

Taylor, Michael A.
Direct Biaxial Design of Columns, 479

Taylor, Peter K.
Current Research and Research Needs in Deep Foundations, 969

Tchobanoglous, George
see Weber, A. Scott, 94

Teicholz, Paul M., ed.
Making Project Control Systems Work, 992

Teng, Chung-Chu
Forces on Horizontal Cylinder Towed in Waves, 827

Thomas, Babu
Behavior of Timber Joints with Multiple Nails, 532

Thomas, David J.
see Tarpy, Thomas S., Jr., 533

Thomas, Wilbert O., Jr.
A Uniform Technique for Flood Frequency Analysis, 747

Thompson, Christopher David
Real and Apparent Relaxation of Driven Piles, 248

Thompson, David Elliot
see Thompson, Christopher David, 248

Thompson, Edward F.
Significant Wave Height for Shallow Water Design, 814

Thompson, James A.
Hazardous Waste: Closing the Insurance Gap, 902

Thorne, Colin R.
Estimating Mean Velocity in Mountain Rivers, 354

Thornton, Edward B.
see Guza, R. T., 774

Timpy, David L.
Bore Height Measurement with Improved Wavestaff, 791

Tingsanchali, Tawatchai
Analytical Diffusion Model for Flood Routing, 343